Natural Conflict Resolution

Natural

Conflict

Resolution

Edited by

FILIPPO AURELI

and

FRANS B. M. DE WAAL

UNIVERSITY OF CALIFORNIA PRESS

Berkeley   Los Angeles   London

University of California Press
Berkeley and Los Angeles, California

University of California Press, Ltd.
London, England

Library of Congress Cataloging-in-Publication Data

Natural conflict resolution / edited by Filippo Aureli and
Frans B. M. de Waal.
    p.   cm.
  Includes bibliographical references and index.
  ISBN 0-520-21671-7 (cloth : alk. paper)—
  ISBN 0-520-22346-2 (paper : alk. paper)
  1. Conflict management. I. Aureli, Filippo, 1962–   .
II. Waal, F. B. M. de (Frans B. M.), 1948–   .
  HM1136.N37   2000   99-043046

Printed and bound in Canada
08 07 06 05 04 03 02 01 00
10 9 8 7 6 5 4 3 2 1

Contents

Preface

During the past two decades there has been a sharp increase in interest in cooperation, peace, and conflict resolution in disparate disciplines, such as anthropology, social and developmental psychology, ethology, political sciences, and legal studies. We have closely followed this development in animal behavior and directly participated in it with our work on nonhuman primates. In the past few years, we have had an increasing number of exchanges with colleagues from different disciplines and realized the common bases underlying these heterogeneous research efforts. This volume aims to bring together the various approaches to the study of conflict management and to emphasize the similarities among them.

Many symposia, roundtables, and workshops on conflict resolution have been organized at soci-etal discipline meetings and at interdisciplinary conferences in recent years. At the XXV International Ethological Conference in Vienna in 1997, in addition to a symposium and a roundtable there were an entire paper session and several posters devoted to animal conflict resolution. In view of the success of this conference we judged the time ripe for a volume that would summarize progress across different areas of investigation. In addition to researchers in animal behavior, we reached out for leading experts in other disciplines who have emphasized in their work the natural bases of the phenomenon. Not surprisingly, given how young and dynamic this field is, there was an overwhelmingly positive response to the project. As a result, we combine in one volume 36 original contributions based on the efforts of 52 authors and coauthors.

Each contribution is a review of a particular aspect of the vast topic of conflict management: some contributions summarize years of research, whereas others present recent developments. Each contribution is written to stand on its own, but it is also a part of the whole. We devoted special attention to the integration of the contributions through the use of common terminology, cross-referencing, and introductions to each part of the book. The result is an interdisciplinary volume that provides an overview of progress on many aspects of natural conflict resolution.

Filippo Aureli
Frans B. M. de Waal
Atlanta, Georgia

Acknowledgments

We thank the authors for their enthusiasm, their responses to our requests and suggestions, and their assistance as reviewers of other contributions. We are especially thankful to Marina Cords, Douglas Fry, Peter Verbeek, and Douglas Yarn for their input on the definitions of key terms presented in Appendix B. We are also grateful to Christopher Boehm, Robert Boyd, and Robin Dunbar, who served as external reviewers for some contributions, and to the anonymous reviewers of the original book proposal for valuable suggestions. We are greatly in debt to Jessica Flack for assisting with various editorial tasks. We also appreciate the help of Doris Kretschmer, Danielle Jatlow, Marilyn Schwartz, Nicole Hayward, Sam Rosenthal, and Alexandra Dahne at the University of California Press, and of Princeton Editorial Associates.

Acknowledgments by the individual contributors are listed below:

Chapter 2: Thanks to Thelma Rowell and Marina Cords for constructive comments, and for the support received from the National Institutes of Health (RR-00165) to the Yerkes Regional Primate Research Center.

Chapter 3: Thanks to Eric Snowden for the illustrations.

Chapter 4: Thanks to the dean and faculty of the College of Law at Georgia State University for their support, and to Filippo Aureli and Frans de Waal for introducing me to this intellectually stimulating field.

Box 4.1: Thanks to President and Mrs. Carter and the members of the International Negotiation Network for the rare privilege of working with and learning from them.

Chapter 5: Thanks to Jessica Flack and Richard Wrangham for discussions of and excellent comments on the manuscript.

Box 5.1: Thanks to V. Hayes for language improvement to the manuscript.

Box 5.3: Thanks to Signe Preuschoft for her comments on earlier drafts, to Felix Zaragoza for the photographs, and for the support received from DGICyT, Spain (PB92-0194 and PB95-0377).

Chapter 6: Thanks to Christophe Abegg, Odile Petit, Filippo Aureli, and Frans de Waal for generously sharing data and for fruitful discussions, and to Ana Ducoing and Jessica Flack for language advice.

Box 7.1: Thanks to Jutta Kuester for comments on an initial version.

Chapter 8: Thanks to Karen Strier for a careful critique that improved our manuscript; to Jeff Fite for his photograph; to Guy Boysen, Jesse Hicks, and Brian Paukert for technical assistance in preparing the chapter; to Ann Savage for sharing her insights on the life of wild tamarin groups; and to our collaborators Rebecca Addington and Tessa Smith for sharing their data, engaging in fruitful discussions, and reviewing earlier drafts.

Chapter 9: Thanks to Peter Kappeler and Peter Verbeek for their comments and for discussion of earlier versions, and for the support received from the National Institutes of Health (R01-RR09797).

Box 9.1: Thanks to Marina Cords and Nicola Koyama for their comments on a previous draft.

Box 9.2: Thanks to Marina Cords and Odile Petit for insightful suggestions on a previous version, and for the support received from the National Institutes of Health (RR-00165) to the Yerkes Regional Primate Research Center.

Chapter 10: Thanks to Dee Higley, Dario Maestripieri, Lisa Parr, Mike Potegal, Signe Preuschoft, Cris Price, Kathy Rasmussen, Norbert Sachser, Robert Sapolsky, Colleen Schaffner, Gabriele Schino, Joan Silk, Alfonso Troisi, Peter Verbeek, and Carel van Schaik for valuable comments and discussion; to Michael Seres for his photograph; and for the support received from the National Institutes of Health (R01-RR09797).

Box 10.1: Thanks to the Office of the President, Republic of Botswana, for research permission; to W. J. Hamilton, M. Mokopi, and J. Silk for their help; and for the support received from the National Science Foundation, the National Institutes of Health, the National Geographic Society, and the University of Pennsylvania.

Box 10.2: Thanks to Gale Foland, Mary Gouwens, and Marge Weaver for their contribution to the research.

Chapter 11: Thanks to Michael Pereira and Janelle Smith for sharing their unpublished observations; to Gloria Gianandrea, Heribert Hofer, and Michael Pereira for their critical reading of the manuscript; and to Thelma Rowell for helpful discussions.

Box 11.2: Thanks to Brookfield Zoo's Seven Seas staff for their many contributions to the research and to Joan Silk and Filippo Aureli for discussions about post-conflict behavior and the sampling protocol.

Box 11.3: Thanks for the support received from the Fritz-Thyssen-Stiftung, the Stifterverband der deutschen Wissenschaft, and the Max-Planck-Gesellschaft, and to Wolfgang Wickler for his continued and tremendous support.

Chapter 12: Thanks to all research assistants and colleagues who contributed to our projects; to the teachers, parents, and children of the various schools and day-care centers; and for the support received from RFHR (96-01-00032), the Bank of Sweden Tercentenary Foundation, the May Flower Annual Campaign, the Claes Groschinsky Memory Foundation, the Hierta-Retzius Foundation, the Alice and Larson Foundation, the Harry Frank Guggenheim Foundation, and the Consortium on Negotiation and Conflict Resolution.

Box 12.1: Thanks for the support received from the Harry Frank Guggenheim Foundation, the National Institute for Neurological Disorders and Stroke (F33 NS09638), and the National Institute of Child Health and Human Development (F33 HD08208).

Chapter 13: Thanks to Hans Veenema and Jan van Hooff for discussions, and to Duncan Castles and David Watts for helpful comments.

Box 14.1: Thanks to Naz Awan, Dorothy Cheney, and David Watts for comments on an earlier version.

Chapter 15: Thanks to Signe Preuschoft for discussions and to Michael Pereira and David Watts for constructive comments, and for the support received from the National Institutes of Health (R01-RR09797).

Box 15.1: Thanks for the support received from CNPq, Fundação Biodiversitas, the Liz Claiborne and Art Ortenberg Foundation, the Lincoln Park Zoo Scott Neotropical Fund, and the University of Wisconsin–Madison Graduate School, and to the many individuals who have contributed to the long-term demographic data.

Chapter 16: Thanks to Oxford University Press for permission to use a quote from P. Draper (1978) as the chapter's epigraph. ("The learning environment for aggression and anti-social behavior among the !Kung" by Patricia Draper, from *Learning Non-aggression*, edited by Ashley Montagu. Copyright © 1978 by Ashley Montagu. Used by permission of Oxford University Press, Inc.)

Chapter 17: Thanks to Alicia Ardila-Rey, William Arsenio, George Bregman, Peter H. Kahn Jr., and Daniel Hart for helpful and invaluable feedback on the manuscript.

Box 17.1: Thanks to Younghee Park for her help with the Korean data collection.

Introduction

CHAPTER 1

Why Natural Conflict Resolution?

Filippo Aureli & Frans B. M. de Waal

The reason we customarily speak of the *need* for cooperation and the *potential* for conflict is because the former is desirable whereas the latter is inevitable. Whether the units are people, animals, groups, or nations, as soon as several units together try to accomplish something, there is a need to overcome competition and set aside differences. The problem of a harmonization of goals and reduction of competition for the sake of larger objectives is universal, and the processes that serve to accomplish this may be universal too. These dynamics are present to different degrees among the employees within a corporation, the members of a small band of hunter-gatherers, or the individuals in a lion pride. In all cases, mechanisms for the regulation of conflict should be in place.

It is sufficient to reflect on our everyday life to find examples. We employ various "rituals," such as handshaking or verbal apology, on a regular basis to prevent or mitigate conflicts. We have developed social rules to regulate interactions within a community and legal procedures to solve disputes when the individuals in conflict are not able to find an agreement by themselves. We are so concerned about the disruptive consequences of conflict that we celebrate its resolution at various levels: within our family, community, and nation and at the international level. Conflict resolution, like conflict and cooperation, appears to be a natural phenomenon. We should then find similarities in its expression and procedures across cultures and species.

During the past two decades we have witnessed a change in interests across disciplines from competition, aggression, and war to cooperation, peace, and conflict resolution. For example, there has been recognition that peace is more than the absence of war and that conciliatory reunions between former opponents play a key role in animal societies. Interest in spontaneous forms of conflict resolution has also increased among developmental psychologists, cultural anthropologists, and political scientists, and the legal system has recently emphasized conciliatory alternatives to the more traditional forms of litigation.

This growing interest in the mechanisms of dealing with conflict requires some explanation. We would like to address this issue in two ways. The first focuses on why natural mechanisms for conflict resolution and conflict management exist. The second follows from the first and explains how we decided to put this volume together and to provide views from various disciplines on the convergent trend toward the study of conflict resolution.

Why Natural Conflict Resolution Exists

Natural history teaches us that when individuals live in a group they gain benefits from the presence of others and from active cooperation in locating food, rearing offspring, or detecting predators. These basic functions are of paramount importance for the survival of the members of the group, whether they are ants, birds, or human hunter-gatherers. In modern societies, cooperation may be expressed in more complex ways (e.g., the cooperative fine-tuning of the LINUX operating system by computer experts at different locations on the globe via the Internet), but the underlying functions are still related to improved survival in a given environment.

Group life also entails costs. Living in close proximity to members of the same species implies the simultaneous exploitation of resources; under these conditions competition is likely. These conditions are easily encountered by various species in their natural environments as well as in various settings of modern human societies. More indirect costs result when group members are obliged to coordinate their activities in order to remain together. This may lead to clashes of interests when individuals of different age, sex, dominance rank, and reproductive condition differ in their needs and, accordingly, would like to follow different courses of action. For instance, in macaques the feeding requirements differ between males and females, and the presence of females with young infants tends to slow down group movement (e.g., van Schaik & van Noordwijk 1986). Accordingly, group members have conflicts over when and where to carry out important activities and over travel decisions (Menzel 1993; Boinski in press). Similarly, scheduling conflicts are a daily occurrence among family members in our societies.

In order to maintain the benefits of group living, individuals need to reduce its costs by mitigating competition and solving conflicts of interest. It follows that mechanisms of conflict management are a critical component of the social life of any group-living species. Natural selection should have favored the expression of the mechanisms best fitting the social organization of each species. This does not imply that these mechanisms are strictly genetically based; in fact, there is ample evidence for learnability and flexibility of expression, as reported in several contributions in the present volume. According to evolutionary theory, it is logical to expect conflict management mechanisms as natural phenomena that function in maintaining the integrity of groups and the associated benefits to each group member.

A balance between costs and benefits is a universal feature of stable social relationships: we would argue that major imbalances threaten the stability. Consequently, so long as an individual is interested in maintaining a cooperative bond, he or she should ensure that the cost-benefit balance for the partner does not tip to the negative side. This implies that whenever a conflict of interest between partners arises, both partners have an interest in constraining exploitation of the other, so as to keep their balances positive. When competing for a resource, group members should therefore take into account not only the value of the resource or the risk of injury but the value of their relationships as well (de Waal 1989a).

Aggression is not a negative social force per se (de Waal 1996). Traditionally, psychologists, social scientists, and evolutionary biologists have presented aggression as an antisocial behavior. The new perspective on conflict management views aggression as an instrument of negotiation between partners. To exchange services and favors or to combine their efforts in cooperative actions, partners need to communicate their relative positions and clarify potential conflicts. Overt expression and especially the threat of aggression (e.g., in the form of punishment) are powerful tools during the bargaining process between partners. Considering the mechanisms for its control and the mitigation of negative repercussions, aggression becomes a well-integrated component of social relationships.

The critical role played by the mechanisms for the control of aggressive expression and conflict resolution explains why they have been readily found in animal societies once researchers started to look for them. We know, of course, that the same mechanisms exist in our own species, and in fact the terminology employed in animal studies borrows heavily from concepts traditionally applied to human social relationships. The beginnings of conflict management skills are present at an early stage in human children, and cross-cultural comparisons indicate the universality of these skills. The recent emphasis on alternative techniques for dispute resolution in the legal system is strongly based on the revival of natural forms of dealing with interpersonal conflict. Even the least spontaneous forms of conflict resolution, such as the mediation of international conflict, are based on the natural foundations of the phenomenon, that is, interpersonal relationships. In sum, conflict among monkeys, people, and even entire groups or nations follows a certain dynamic that is universal, and so is its resolution.

Why a Book on Natural Conflict Resolution?

Both of us have been trained as ethologists—that is, biologists studying animal behavior—and we have mainly studied nonhuman primates. The history of this volume necessarily is influenced by our background and by the perspectives on conflict and its regulation that we have developed in the course of the years. If scientists from other disciplines were to put together an analogous volume, the starting point would certainly have been different, but the end product might be rather similar. Our background and experience with animals have made us appreciate the natural bases of the various human expressions of conflict management and consequently stimulated an interest in the common ground across disciplines. The convergence among various disciplines on the theme of conflict resolution is so compelling that we cannot imagine delivering a different final product at this empirical and theoretical stage, even though the species selection is obviously biased.

Some animal societies, such as eusocial insects, are characterized by great genetic similarity among group members and have apparently little conflict. The conflicts of interest are mitigated at the genetic level and are not as individualized as in human societies. In any society, conflicts of interest certainly have evolutionary bases, and much has been theorized, for instance, on the conflicts of interest between parents and offspring (Trivers 1974; Bateson 1994) or between breeding individuals of the

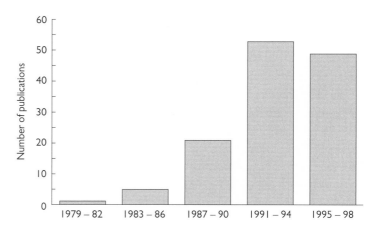

FIGURE I.I. The number of publications on the topic of reconciliation in nonhuman primates in the past 20 years (1979–1998). The figure is based on a literature search using PrimateLit database of the Primate Information Center of the University of Washington, Seattle. Abstracts, media news reports, and dissertations were excluded. No publication was found before 1979. We supplemented the search with four additional publications known by us.

opposite sex (Trivers 1972; Gowaty 1996). Those conflicts often manifest themselves at the behavioral level, and this is the level at which our volume deals with conflict. Our perspective provides a complementary view to the more theoretical approach to conflict (cf. Godfray 1995) by focusing on the behavioral expressions of everyday conflicts and the mechanisms that individuals use to regulate them.

To follow the history behind this volume, we need to go back to the late 1970s when the first ethological study focusing specifically on conflict resolution was conducted on captive chimpanzees (*Pan troglodytes*) by de Waal & van Roosmalen (1979). After this study, many researchers from different backgrounds began to examine various aspects of animal conflict resolution. The behavior that has been most extensively studied is *reconciliation*, that is, a friendly reunion between former opponents soon after an aggressive conflict. Most of this research has focused on nonhuman primates; 27 species have been studied so far, and post-conflict reunions have been found in the vast majority of the studies (Appendix A). Only recently has systematic research started to focus on post-conflict reunions in other animals (see below). Figure 1.1 shows the rapid increase of pub-

lications focusing on these reunions in nonhuman primates in the past two decades. From the virtual absence of systematic studies before 1979, we have witnessed a slow but steady increase of publications in the 1980s and a substantial production in the 1990s.

The increase of interest in conflict resolution has not been limited to primatologists. In fact, primatological studies have stimulated similar research on other taxonomic groups. In 1994, during a plenary lecture of the XXXI Meeting of the Animal Behavior Society, in Seattle, Thelma Rowell concluded that the major contribution of primatology to the general field of animal behavior over the past decades has been the study of reconciliation. The Reconciliation Study Group was formed in 1994 and consists of members from various disciplines and from many countries covering five continents. The group has organized several roundtables and workshops and has its own Internet discussion network and home page. Through discussion sponsored by this group and direct contacts between scientists, the approach and methodology used in studies of nonhuman primates crossed the taxonomic and discipline traditional boundaries. This fruitful exchange has led to research on post-conflict resolution mechanisms in species other than primates and in children (some of the products of this research are included in this volume).

In the past few years, scholars from different disciplines have found one another despite quite different perspectives because of the common interest in problems related to conflict management. Various plenary lectures and symposia at meetings of anthropological and psychological associations and other interdisciplinary initiatives have included speakers on conflict resolution in nonhuman primates. Religious foundations and primatologists have joined forces in the study of the origins of "forgiveness." The Gruter Institute for Law and Behavioral Research has been instrumental in helping biologists interested in conflict resolution and lawyers realize the similarity in recent trends within their disciplines toward more emphasis on cooperation and peaceful resolution.

Interdisciplinary forums have provided, and will continue to provide, opportunities for useful exchanges of methodological and theoretical issues. Ethological observations of naturally occurring behavior typical of animal research on conflict resolution can be applied in other settings, such as children's playgrounds, cross-cultural studies, and even mediation sessions. Noninvasive, carefully designed experiments that had provided pivotal information about the functions and the rules of post-conflict behavior in nonhuman primates (Cords 1994) can be revealing for the study of the basis of human conflict resolution skills. Emphasis on the role of personality and individual differences in psychology can bring useful insight to research on animal conflict regulation. Similarly, concepts such as attachment and separation (Bowlby 1969), so central in developmental and social psychology, appear to share unexplored similarities with aggressive conflict and post-conflict reunions in nonhuman primates. Primatologists and other students of animal behavior would

benefit from learning more about these psychological concepts.

Based on current knowledge, similarities across disciplines are already emerging. One of them is especially interesting. The quality of social relationships plays a key role in the occurrence of conflict resolution not only among nonhuman primates (de Waal & Aureli 1997). As reported by several contributions in this volume, relationship quality also has important theoretical and practical implications in the regulation of conflict among children, in conflict intervention in various cultures, and in mediation processes at various levels including international conflict. Such similarities suggest that contributions from all disciplines (i.e., ethological studies of primates and other animals as well as the various approaches to human conflict resolution in anthropology, psychology, political sciences, and legal studies) would provide evidence for continuity of behavioral mechanisms for conflict regulation across situations and species. This evidence will in turn help to trace the evolution of human skills for conflict management.

From our perspective we have seen the development of a new approach that started from the original study on chimpanzees and expanded first to other nonhuman primates and then to other animals. Human conflict resolution was also being studied, but on a very small scale, so that the curious situation arose that more was known about conflict resolution in nonhuman animals than in our own species. Given current growing interest in the issue, this imbalance will probably soon be corrected. While the number of species studied has increased over time, progress on empirical and theoretical ground has also been made. From the initial focus on post-conflict conciliatory reunions, research interest has grown to explore the mechanisms for controlling aggression with various forms of appeasement or intervention by third parties. Research has also focused on the mechanisms to alleviate distress and reduce tension. In the meantime, we have realized the similarities of research trends in other disciplines and the probable natural bases of these similarities. The idea to put together an interdisciplinary volume on management and resolution of conflict came to us, then, as a *natural* consequence of these developments.

Common Ground and New Perspective

The experience at interdisciplinary meetings and the reading of work by scientists from other disciplines confirmed our intuition regarding the overwhelming similarity in the basic themes underlying conflict regulation. We also realized, however, that the terminology was not always the same across the various disciplines. In order to put together an interdisciplinary volume that would convey the universality and the natural bases of the mechanisms for conflict regulation, we needed to reach a common ground in the terminology. We formed a panel of colleagues from various disciplines with the task of producing definitions of key terms that were compatible across disciplines (Appendix B). We then circulated the list of terms among volume contributors and asked them to conform to the definitions provided while writing their pieces. If authors needed to use a term with a different meaning than that provided in Appendix B, they clearly stated so in their contribution and provided the alternative definition.

Our volume does not exist in isolation. It is part of a recent trend that has emphasized (1) that social conflict should not be equaled with aggression because many conflicts do not have aggressive outcomes and (2) that the view of aggression as an antisocial component of group life is surpassed (de Waal 1989b; Silverberg & Gray 1992; Mason

& Mendoza 1993). There has also been a growing interest in the study of peace viewed as an independent concept and not simply as the absence of war (Gregor 1996). Pioneering work has also been published on the new perspective of social regulation of aggression and competition in primates and other animals (de Waal 1989b; Moynihan 1998). Our volume follows the path outlined by this previous work and integrates detailed knowledge recently accumulated in various disciplines to provide a broad view of the naturally occurring mechanisms of conflict management and the principles underlying them. The emerging picture is a new perspective of social life, a perspective strengthened by the empirical and theoretical convergence from different species and disciplines.

The organization of the volume follows the main research trends to offer a comprehensive overview of the current knowledge. Natural conflict resolution is a rather new field of study, and many of the investigators are relatively junior. In addition, some of the areas of investigation are at an early stage. To include much of the diversity in studies and perspectives, we adopted a volume organization that combines review chapters with shorter contributions (boxes) on specific topics that are embedded within the chapters.

The volume consists of five main sections. It begins with a historical section with broad overviews of the changing emphasis from aggression and authoritative intervention to negotiation and resolution within biology, psychology, and legal studies. The following section deals with current knowledge on the mechanisms for controlling aggression. Many of the contributions in this section report on behaviors that have been previously described in various contexts and are now reinterpreted in view of the new perspective of conflict management. Most of these contributions deal specifically with animal examples, but we believe that most principles underlying the various behaviors have validity for human cases as well.

The section "Repairing the Damage" presents contributions summarizing research that has been the core of the study of conflict resolution, that is, post-conflict conciliatory reunions between opponents. The section consists of chapters and boxes on children, nonhuman primates, and other animals; some contributions review research done over the past twenty years, whereas others present progress on less-studied aspects or species. Building on this knowledge, the section provides both developmental and phylogenetic perspectives and incorporates analyses of the underlying causes and functions of conciliatory reunions.

The following section examines the role of third parties in the process of conflict management. The four contributions are based on nonhuman primates for which more data are available and focus on functional and cognitive issues of third-party involvement during and after conflicts. Viewing third-party intervention as a mechanism for conflict regulation is a rather new development, and these contributions attempt to present a framework for future research. The last section provides the broad scenario needed to appreciate fully what we know about conflict resolution and management in humans and other animals. Socioecological, cultural, and moral issues relevant to conflict regulation are examined in detail. These analyses highlight once more the similarities across species, cultures, and disciplines and strengthen the perspective of conflict resolution as a natural phenomenon.

References

Bateson, P. 1994. The dynamics of parent-offspring relationships in mammals. *Trends in Ecology and Evolution,* 9: 399–403.

Boinski, S. In press. Social manipulation within and between troops mediates primate group movement. In: *On the Move: How and Why Animals Travel in Groups* (S. Boinski & P. Garber, eds.). Chicago: University of Chicago Press.

Bowlby, J. 1969. *Attachment and Loss: Attachment.* New York: Basic Books.

Cords, M. 1994. Experimental approaches to the study of primate conflict resolution. In: *Current Primatology,* Vol. 2: *Social Development, Learning and Behaviour* (J. J. Roeder, B. Thierry, J. R. Anderson, & N. Herrenschmidt, eds.), pp. 127–136. Strasbourg: Presses de l'Université Louis Pasteur.

de Waal, F. B. M. 1989a. Dominance "style" and primate social organization. In: *Comparative Socioecology* (V. Standen & R. Foley, eds.), pp. 243–263. Oxford: Blackwell.

de Waal, F. B. M. 1989b. *Peacemaking among Primates.* Cambridge: Cambridge University Press.

de Waal, F. B. M. 1996. Conflict as negotiation. In: *Great Ape Societies* (W. C. McGrew, L. F. Marchant, & T. Nishida, eds.), pp. 159–172. Cambridge: Cambridge University Press.

de Waal, F. B. M., & Aureli, F. 1997. Conflict resolution and distress alleviation in monkeys and apes. In: *The Integrative Neurobiology of Affiliation* (C. S. Carter, I. I. Lenderhendler, & B. Kirkpatrick, eds.), pp. 317–328. New York: Annals of the New York Academy of Sciences.

de Waal, F. B. M., & van Roosmalen, A. 1979. Reconciliation and consolation among chimpanzees. *Behavioral Ecology and Sociobiology,* 5: 55–66.

Godfray, H. C. J. 1995. Evolutionary theory of parent-offspring conflict. *Nature,* 376: 133–138.

Gowaty, P. A. 1996. Battles of the sexes and origins of monogamy. In: *Partnerships in Birds* (J. M. Black, ed.), pp. 21–52. Oxford: Oxford University Press.

Gregor, T. 1996. *A Natural History of Peace.* Nashville: Vanderbilt University Press.

Mason, W. A., & Mendoza, S. P. 1993. *Primate Social Conflict.* Albany: State University of New York Press.

Menzel, C. R. 1993. Coordination and conflict in Callicebus social groups. In: *Primate Social Conflict* (W. A. Mason & S. P. Mendoza, eds.), pp. 253–290. New York: State University of New York Press.

Moynihan, M. H. 1998. *The Regulation of Competition and Aggression in Animals.* Washington: Smithsonian Institution Press.

Silverberg, J., & Gray, J. P. 1992. *Aggression and Peacefulness in Human and Other Primates.* New York: Oxford University Press.

Trivers, R. L. 1972. Parental investment and sexual selection. In: *Sexual Selection and the Descent of Man, 1871–1971* (B. Campbell, ed.), pp. 136–179. Chicago: Aldine.

Trivers, R. L. 1974. Parent-offspring conflict. *American Zoologist,* 14: 249–264.

van Schaik, C. P., & van Noordwijk, M. A. 1986. The hidden costs of sociality: Intra-group variation in feeding strategies in Sumatran long-tailed macaques (*Macaca fascicularis*). *Behaviour,* 99: 296–315.

PART I

History

Introduction

 The contributions in this section provide the historical background for the rest of the volume. These broad reviews of disparate areas of investigation, such as animal behavior, child psychology, and legal studies, converge on a changing emphasis from aggression and authoritative intervention to negotiation and conflict resolution.

In Chapter 2, while reviewing the earliest animal studies, **de Waal** reports that reconciliation, that is, post-conflict friendly interaction between former opponents, was discovered in the late 1970s in chimpanzees but has since then been demonstrated in a variety of nonhuman primates, both in the field and in captivity (see Appendix A). The term *reconciliation*, with its implication of relationship repair, serves as a heuristic label from which several predictions can be derived, such as that post-conflict interaction should (1) occur preferentially between former opponents, (2) especially between opponents with valuable relationships, and (3) reduce social tension and further aggression. Thus far, these predictions have been borne out by the observational data, and experimental evidence has lent further support. The existence of conciliatory mechanisms has implications for our view of aggressive competition: one of the main constraints on competition among social animals is the value of competitors as partners in cooperative endeavors. Summarized in the Relational Model, this perspective shifts attention away from aggression as the expression of an internal state toward aggression as the product of conflicts of interest. It regards aggressive behavior as the product of social decision making and one of several ways in which conflicts between individuals or groups are negotiated.

Box 2.1 complements de Waal's review by presenting the methodological progress in the study of post-conflict behavior in animals. **Veenema** describes the most-used method of data collection, which is the so-called PC/MC method, in which events during post-conflict observations, or PCs, are compared with those in matched-control observations, or MCs. This method seeks to provide a baseline level of behavior that matches all conditions of the corresponding post-conflict observation except for the previous occurrence of conflict. Veenema discusses the advantages and disadvantages of three methods of comparing PCs and MCs and presents a measure of conciliatory tendency that corrects for baseline level of behavior. These methods can also be used for the demonstration and quantification of other post-conflict management mechanisms and can be employed in other disciplines (see Butovskaya et al., Chapter 12, for an example of such an application in child conflict resolution).

In Chapter 3, **Verbeek, Hartup, and Collins** focus on human children and trace the development of conflict management skills from the toddler years through adolescence in two arenas: the family and the peer group. Within these two arenas conflict can take many forms, each associated with specific management demands. In addition to the development of managing skills for dyadic conflict with parents, siblings, and peers, children need to learn about the role of third parties who often contribute to both the incidence and management of peer conflict. Societal factors also influence children's acquisition of conflict management skills. Across cultures, specific adult patterns of competition and cooperation appear to affect how children manage conflict. Furthermore, socioeconomic factors seem to influence the development of conflict management skills. In a manner similar to de Waal's chapter, this review emphasizes that the development of conflict management skills is best understood within the framework of social relationships.

When from children and adolescents we move to consider conflict management in adults, we realize that, more than any other species, humans have developed elaborate institutions. In Chapter 4, **Yarn** examines the historical and current relationship of law, one of the most elaborate of such institutions, with conciliatory methods of managing conflict. A review of the historical records reveals a long-standing, often competing interaction between the seeking of redress and justice and the seeking of compromise and conciliation. The relative importance of these two contrasting methods changes with shifts in the social and institutional arrangements supporting different conflict management mechanisms. Yarn concludes his chapter by recognizing that a growing interest in conflict resolution across a broad segment of modern society parallels the growing interest in reconciliation and cooperative relationships among biologists.

Social relationships may also play a key role in conflicts between groups. Although the scale of the issues involved in interpersonal or intergroup conflict is quite different, interpersonal and intergroup conflict share certain characteristics. In both cases, for example, when direct negotiations between parties to a conflict may prove impossible, third-party mediation offers an alternative. In Box 4.1, **Neu** reports on third-party mediation in international conflicts. She especially focuses on the mediation efforts of former U.S. president Jimmy Carter in Bosnia and Sudan. In one case (Bosnia) he had no prior relationship with the disputants, whereas in the other (Sudan) he had a long history of interaction with them. Neu points out that well-established relationships and positive interpersonal dynamics are critical for successful conflict mediation; when they do not exist, they must be established to achieve resolution.

CHAPTER 2

The First Kiss

Foundations of Conflict Resolution Research in Animals

Frans B. M. de Waal

Introduction

Over the past half-century, aggression has occupied a pivotal role in debates about human nature. After World War II, aggression became a favorite theme because of Lorenz's (1967) hugely controversial book *On Aggression*. Lorenz's central thesis was that human aggression is an instinct, produced by an innate drive, hard or impossible to control, that seeks an outlet ranging from sports to gang violence. This Freudian message was amplified by popularizers such as Ardrey (1967) and Morris (1967) but countered by psychologists and anthropologists who demonstrated that aggression is learned (e.g., Bandura 1973) and questioned the universality of aggression in human society (e.g., Montagu 1968).

The premises of this debate were fundamentally flawed. It was tacitly assumed that demonstrating either a genetic or a learning component would settle the issue, whereas we now, of course, assume involvement of both influences in almost everything humans do (e.g., Manning 1989). Another serious weakness was the lack of attention to natural checks and balances on aggression: aggression was treated as a separate behavioral category isolated from other aspects of social life. This is all the more surprising as Lorenz himself emphasized the ritualization of aggressive displays in animals and how these displays, and the warning signals they contain, prevent bloodshed. In discussing such regulatory mechanisms, Lorenz exempted our own species perhaps because the world had just witnessed such horrible violence on its part. He speculated that since our ancestors had been peaceful vegetarians, our lineage might not have had the evolutionary time to adjust to the cultural development of deadly weapons: "One can only deplore the fact that man has definitely not got a carnivorous mentality! All this trouble arises from his being a basically harmless, omnivorous creature, lacking in natural weapons with which to kill big prey, and, therefore, also devoid of the built-in safety devices which prevent 'professional' carnivores from abusing their killing power to destroy fellow-members of their own species" (Lorenz 1967, p. 207).

Debate about these ideas was vehement, but the tone changed with increased information. For example, the purported peacefulness of our closest relatives, the apes, was not nearly as complete as both Lorenz and his detractors had assumed. Since the fieldwork by Goodall (1979), lethal intergroup violence is considered a pervasive characteristic of the chimpanzee, a conclusion with possible implications for the origins of human warfare (e.g., Goodall 1986; Wrangham & Peterson 1996). In addition, the chief explanatory framework of Lorenz's days, with its emphasis on species preservation, has since been abandoned. The argument that animals will not kill each other because otherwise their species will die out has been replaced by theories about how individuals and their close kin benefit from constraints on aggression.

Yet one assumption was maintained throughout this entire period and is still very much with us. This assumption is that aggressive behavior is necessarily destructive. Losers of aggressive incidents tend to avoid winners (Scott 1958; Marler 1976), and animals—like birds sitting at regular intervals on a telephone wire—maintain "individual distances" by reacting with hostility to invaders of their space (Hediger 1941). Thus, the one issue *not* under debate was the negative, dispersive impact of aggression.

Many early ethologists worked with territorial species in which the above assumption made sense: territorial aggression is by definition concerned with spatial relations. In group-living animals, however, social cohesion needs to be maintained despite occasional conflict. Whatever the negative effects of aggression, they are usually overcome. In time, my own research, which was greatly inspired by the debates of those days, began to question the traditional dichotomy between socially positive and socially negative behavior. I found that social animals, rather than avoiding each other after a fight, actually seek each other out.

It is worth looking into the discovery behind these insights and how this discovery fit other developments at the time. It is almost as if the time was ripe for a different view, one that considered aggression a social rather than an antisocial behavior, but without, of course, condoning all forms of aggression in human society.

Reconciliation Discovered

Nonhuman reconciliation behavior was first recognized as such in the world's largest captive chimpanzee colony (*Pan troglodytes*) at Burgers Zoo in Arnhem, the Netherlands. As related previously (de Waal 1989a, p. 5), it happened one day when the colony was locked indoors in one of its winter halls. In the course of a charging display, the highest-ranking male fiercely attacked a female. This caused great commotion as other apes came to her defense. After the group had calmed down, an unusual silence followed, as if the apes were waiting for something to happen. This took a couple of minutes. Suddenly the entire colony burst out hooting, and one male produced rhythmic noise on metal drums stacked up in the corner of the hall. In the midst of this pandemonium, two chimpanzees kissed and embraced.

I, the observer, reflected on this sequence for hours before the term *reconciliation* came to mind. This occurred when I realized that the two embracing individuals had been the same male and female of the original fight. The very first documentation of the concept, expressed in the Dutch term *ver-zoening*, is found in an entry of November 19, 1975, in my handwritten Arnhem diaries. In the same month, another entry introduces the Dutch term *troost*, later translated as "consolation," for reassuring contact provided by bystanders to victims of aggression. Once recognized, both interaction patterns were seen on a daily basis. It became indeed hard to imagine that they had gone unnoticed for so long.

In March 1976, several new graduate students came to work at Arnhem. Reconciliation behavior was proposed as one of the research topics. The student assigned to this project was Angeline van Roosmalen, who came to us from the Psychology Department of the University of Amsterdam. This is an interesting detail, as my supervisor, Jan van Hooff, and I were both in the Biology Department of the University of Utrecht. It meant that the student's thesis committee was beyond our control.

The committee members were not altogether enthusiastic about the proposed topic of study. Even though they had never watched chimpanzees and did not come to Arnhem to make up for this deficit, they were a priori convinced that there could be no such thing as reconciliation in nonhuman animals. Their chief alternative hypothesis, developed extemporaneously at one of our meetings, was that social animals go through systematic alternations between positive and negative interactions so as to balance the two: hence every positive encounter is followed by a negative one, and vice versa. This could account for the aggression-affiliation sequences that we (erroneously) interpreted as reconciliation behavior.

The discussion with these psychologists, although frustrating at times, convinced us to redefine van Roosmalen's research topic as a study of post-conflict interaction, and we set out to test whether the very first interaction following a conflict differed behaviorally from second, third, and later post-conflict contacts. If all that was going on was an alternation of different kinds of interactions, the first post-conflict encounter should be affiliative, the second aggressive, and so on. This is, of course, not what we found, because in the life of the chimpanzee there are many more affiliative than aggressive interactions. The question then became whether post-conflict encounters are influenced by the previous conflict. This should be reflected in the timing and frequency of friendly interactions between former opponents: if the tendency to interact would be greater than normal for those two individuals, this would suggest a connection with the previous conflict. In addition, we were interested in whether first post-conflict interactions were of a different nature than subsequent interactions. If first post-conflict interactions served a conciliatory function, they might be unusually intensive.

The study was carried out in the spring and summer of 1976, resulting in a Dutch thesis by van Roosmalen. Reorganized and written up in English by me, the findings were published, in 1979, in a relatively short paper entitled "Reconciliation and Consolation among Chimpanzees" (de Waal & van Roosmalen 1979).

We concluded that certain behavior patterns (e.g., kissing, holding out a hand, submissive vocalization, embracing) are typical of the immediate aftermath of conflict; that is, these behavior patterns occur significantly more often during first than later post-conflict contacts (Figs. 2.1 and 2.2). Other patterns (e.g., play) occurred significantly less often during first post-conflict contacts. Combined with evidence that interindividual proximity following conflict specifically concerns the former opponents, we felt justified in using special labels for post-conflict interaction: *reconciliation* (first interopponent contact) and *consolation* (contact between a recipient of aggression and a third individual). We went on to speculate about

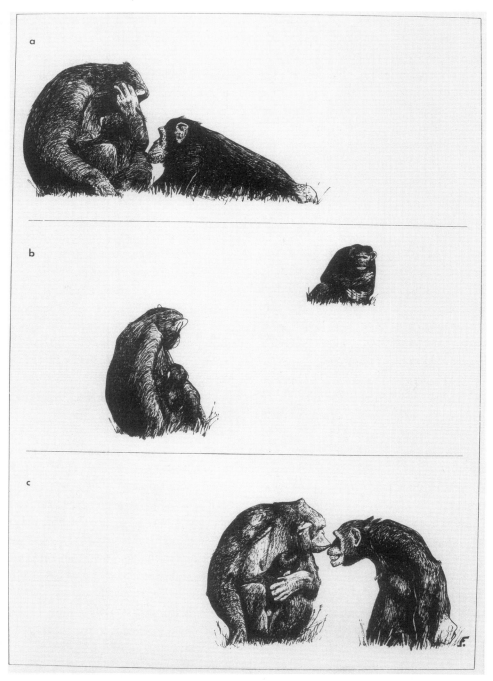

FIGURE 2.1a–c. A reconciliation between adult female Mama (left) and adolescent female Amber (right). (a) Amber is interested in Mama's newborn. (b) Probably Amber came too close to, or even touched, the infant because she was hit by Mama and now screams at a distance. (c) Within one minute, Amber returns to Mama and yelps while receiving a kiss on her nose. After this she is again tolerated close to the infant. Drawing by the author, from de Waal & van Roosmalen (1979).

FIGURE 2.2. Chimpanzees invite reconciliation by means of eye contact and hand gestures. This photograph shows the situation 10 minutes after a protracted, noisy conflict between two adult males at the Arnhem Zoo. The challenged male (left) fled into the tree. He is now being approached by his opponent, who stretches out a hand. Within seconds, the two males have a physical reunion and climb down together to groom each other on the ground. Photograph by the author, from de Waal (1982).

the functional significance of these interactions: "Although 'reconciliation' and 'consolation' are objectively definable interaction types, the terminology is clearly functional rather than descriptive. These terms reflect our impression that such body contacts have a calming effect and serve an important socially homeostatic function" (de Waal & van Roosmalen 1979, p. 65).

Zeitgeist?

The discovery of reconciliation did not occur in a vacuum. If we go back far enough in time, we can already find indications in the literature on human-ape relations. Kellogg & Kellogg (1933) experienced emotional reunions after reprimanding their infant chimpanzee and noted how the infant would express relief in a great sigh during the ensuing embrace. Apes seem ill at ease after conflict with their caretakers and seek to reduce the discomfort by means of affectionate behavior. Although the environment of hand-reared apes is obviously unnatural, current research gives little reason to invoke human influence as an explanation of the following observation by Köhler:

"The little creature, which I had punished for the first time, shrank back, uttered one or two heartbroken wails, as she stared at me horror-struck, while her lips were pouted more than ever. The next moment she had flung her arms round my neck, quite beside herself, and was only comforted by degrees, when I stroked her. This need, here expressed, for forgiveness, is a phenomenon frequently to be observed in the emotional life of chimpanzees" (Köhler 1959 [1925], p. 261).

We had to wait for Mason (1964), van Hooff (1967), and Goodall (1968) for further suggestions in this area, this time including the spontaneous behavior of chimpanzees toward conspecifics. These scientists wrote extensively about appeasement and reassurance signals, emphasizing the arousal-reducing qualities of body contact as well as its importance for social stability.

In 1968, Lindburg presented a paper entitled "Grooming Behavior as a Regulator of Social Interactions in Rhesus Monkeys" at a symposium organized by C. R. Carpenter in Atlanta (Lindburg 1973). While there, Lindburg engaged in lengthy discussions with a colleague, L. Rosenblum, in which both of them agreed that primatologists often ignore the implications of affiliative interaction for the amelioration of disturbed relationships. Lindburg analyzed the role of grooming immediately following aggressive incidents, documenting 37 such sequences in feral rhesus monkeys (*Macaca mulatta*): "In my mind I can still picture some of these events, and intuitively felt that grooming in some way was important in restoring order" (Lindburg, personal communication).

At about the same time as the Arnhem reconciliation studies, Seyfarth (1976) noticed that aggressive conflicts among wild female baboons (*Papio cynocephalus*) are more often followed by approach and affiliative contact between the antagonists if one or both females carry newborn infants. He saw this as an indication that the presence of attractive infants changes the quality of female antagonism. And McKenna (1978) reports his impression of how social grooming helps turn around tense situations among Hanuman langurs (*Presbytis entellus*) at the San Diego Zoo. He concluded: "In one context the recipient of aggression grooms to appease the aggressor and to prevent the aggression from continuing, while in another context the aggressor grooms the victim, thereby assuring the recipient that aggression has ceased and that nonaggressive behaviors can resume between them" (p. 506).

Together with other early suggestions of the tension-reducing qualities of affiliative contact in both primates and non-primates (e.g., Blurton-Jones & Trollope 1968; Ellefson 1968; Poirier 1968; Kummer et al. 1974; Ehrlich & Musicant 1977), it thus appears that in the 1960s and especially in the 1970s there was a gradual move toward the idea of conflict resolution even though the exact function of the observed behavior was not always spelled out.

In the same period, two influential papers pointed out the importance of social relationships. Hinde (1976) clarified that what we observe in

animals are not isolated incidents of behavior. When two individuals have a series of interactions over time, each interaction influences subsequent ones. The two individuals thus build a history of interaction, a relatively stable pattern that we recognize as a *social relationship*.

Kummer (1978) considered the benefits that individual A provides to B as A's value to B. Any individual will try to improve this value: B will select the best available partner, predict this indicidual's behavior, and try to modify its behavior to its own advantage. In other words, B will invest in the relationship with A. Whereas most of B's investments may not lead to quick profits, such as immediately useful actions by A, they may help cultivate patterns of interaction beneficial to both A and B over the long haul.

Thus, by the end of the 1970s, the theoretical groundwork for the study of conflict resolution in nonhuman animals had been laid, containing the following critical elements:

- Indications for an arousal-reducing and stability-promoting function of grooming and affiliative body contact

- Recognition of a connection between aggressive events and subsequent affiliation, including empirical definitions of specific processes such as *reconciliation* and *consolation*

- Distinction between social interactions and long-term social relationships, including recognition of the latter's survival value

Heuristic Terminology

Reconciliation is best regarded as a heuristic concept, that is, a speculative formulation serving as a guide for the solution of a problem. The problem under consideration is how animals manage to maintain group cohesion despite occasional conflict. The reconciliation concept admittedly rests on an anthropomorphic interpretation of animal behavior and as such comes with inevitable human connotations (Asquith 1984; de Waal 1991). These connotations, however, provide a rich theoretical context. Because heuristic concepts are only as good as the evidence in favor of them, much effort over the past twenty years has been invested in testing the validity of the reconciliation concept.

According to my English dictionaries, *reconciliation* refers to the reestablishment of close relationships and the settlement of conflict. In line with this definition, the first aim of reconciliation research has been to compare two contrasting expectations concerning the effect of aggression. The traditional notion that aggression serves a spacing function predicts decreased contact between individuals following aggressive conflict. The reconciliation hypothesis, on the other hand, predicts the exact opposite, namely, that individuals try to undo the damage inflicted by aggression on social relationships. Hence we expect (a) increased friendly interaction following aggression and (b) a highly selective interaction increase, that is, involving not all potential partners indiscriminately but the former opponents specifically. Additional evidence for a connection between conflict and subsequent interaction would come from behavioral differences compared with interactions outside the post-conflict context.

These predictions required the determination of a baseline condition with which to compare post-conflict behavior. The original Arnhem study lacked appropriate controls, but great strides have been made since. One is a matched-control design, introduced by de Waal & Yoshihara (1983), and other advances include more precise controls and improved measurements of the difference between post-conflict (PC) and matched-control (MC) observations (Veenema, Box 2.1).

Controlled studies on a great variety of primate species support the above predictions of increased

Methodological Progress in Post-Conflict Research

Hans C. Veenema

In 1979 de Waal & van Roosmalen showed that following a conflict chimpanzees (*Pan troglodytes*) frequently had affiliative contacts with their former opponents, a behavior they labeled *reconciliation*. However, in order to show that a conflict really influenced the frequency of affiliative contacts, a comparison was needed between the situation directly following a conflict and a "normal" or "control" situation. Such an approach was taken by de Waal & Yoshihara (1983) in a subsequent study on rhesus macaques (*Macaca mulatta*). In this study, a focal sample on one of the two opponents following the conflict (post-conflict observation, or PC) was matched with a control observation (matched-control observation, or MC) on the same animal. This MC was taken on the next observation day at the same time of day to control for seasonal and diurnal patterns in activity.

Besides activity patterns, the likelihood of a contact between two animals may also be influenced by the physical distance between the two former opponents at the start of the observation (Call 1999). Therefore, a number of studies (e.g., York & Rowell 1988) have controlled for distance as well and only started an MC when the distance between the former opponents was similar to the distance at the start of the PC.

This type of data collection, using PCs and corresponding MCs, is now widely used in post-conflict research. The advantage of matching each PC with an MC is that each dyad of opponents is equally represented in the PC and MC data sets. A potential risk of this procedure is a possible dependency between an MC and a PC. One general assumption about the functioning of reconciliation (de Waal, Chapter 2; Cords & Aureli, Chapter 9) is that it influences the relationship between the two former opponents. When an MC is taken on the

BOX 2.1 (continued)

day following a conflict, whether the conflict was reconciled or not may influence the behavior of the two former opponents during the later MC.

Based on this type of data collection, three methods have been developed to investigate the influence of the conflict on the subsequent interactions between the former opponents. (The use of these methods is not limited to the demonstration of reconciliation but can also be used for demonstrating other types of post-conflict interactions, e.g., consolation.)

The first method (*PC-MC method*) compares the timing of the first affiliative interaction between former opponents during one PC with that during the corresponding MC. If this interaction occurs only in the PC or earlier in the PC than in the MC, the PC-MC pair is said to be "attracted"; if the interaction takes place earlier (or only) in the MC, the PC-MC pair is said to be "dispersed." When no interaction takes place in either the PC or the MC, or when the interaction occurs at the same time in both, the PC-MC pair is considered "neutral" (de Waal & Ren 1988). If the conflict does not influence the timing of the first affiliative interaction in the PC, the number of attracted and dispersed pairs should not differ from the 1:1 ratio expected by chance. If the number of attracted pairs is higher than the number of dispersed pairs, we may conclude that the former opponents show higher affiliative tendency in post-conflict situations.

The second method (*rate method*) compares the rate of affiliative interactions between former opponents in the PCs with that during the MCs (Judge 1991) or with that during baseline observations (de Waal 1987). If the rate is higher in the PCs, the former opponents can be said to show higher affiliative tendency in post-conflict situations.

The third method (*time rule method*) compares the frequency of the first affiliative interaction between former opponents as a function of time during the PCs with the equivalent distribution during the MCs (Aureli et al. 1989). This allows one to define a time window after a conflict in which any affiliative interaction between former opponents can be defined operationally as reconciliation (see Das et al. 1997 for a discussion on the statistical issues of this method).

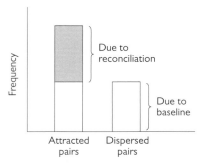

Attracted pairs can be considered to consist of two components. One component reflects the number of attracted pairs that are due to baseline level of affiliative contact and equals the number of dispersed pairs. The second component reflects the tendency of former opponents to contact each other as a result of the previous conflict, that is, reconciliation.

Each of these methods has its own advantages and disadvantages, and the preferred method may depend on the research questions asked. The advantage of the PC-MC method over the other two methods is that in the PC-MC method each PC is directly compared with its matching control observation on the same animal, under very similar conditions, thus controlling for possible differences in the way that different dyads of animals behave toward one another.

The outcome of both the PC-MC method and the rate method is influenced by the duration of the observations (the longer the observation, the higher the chance of contact) and by general contact rates between animals (which can be influenced by variables such as species tendency or cage size). In most studies, the increased attraction between former opponents is limited to the first few minutes after a conflict (reviewed in Kappeler & van Schaik 1992). When data are analyzed from long-lasting observations, both methods may fail to find a significant difference between PCs and MCs. This problem is avoided in the time rule method, in which only affiliative contacts occurring within a certain time window after the conflict are defined as reconciliation. The time rule method is conservative (as contacts outside the time window may still function as reconciliation), but it provides the possibility to investigate the effect of reconciliation on subsequent behavioral patterns, that is, the possible conciliatory function. Using both the PC-MC and the time rule methods is probably the best way to demonstrate the existence of reconciliation.

Besides demonstrating the existence of reconciliation, various studies have focused on differences in the tendency to reconcile between dyads of group members (e.g., kin versus nonkin) or between species. For these comparisons a measure of conciliatory tendency is necessary. Such a measure was first proposed by de Waal & Yoshihara (1983) and was based on the PC-MC method. They proposed to use the number of attracted pairs divided by the total number of PC-MC pairs. However, the number of attracted pairs is influenced by the baseline level of affiliative contact, which may differ between different subsets of individuals (e.g., a higher baseline level of affiliative contact will result in a higher number of attracted and dispersed pairs). Veenema et al. (1994) proposed a measure that is independent of the baseline level of contact. They pointed out that the number of attracted pairs that are due to baseline level of contact is reflected by the number of dispersed pairs. (Note that the number of dispersed pairs cannot be used as an independent measure of baseline affiliation; the number of affiliative contacts in the MCs is a more reliable measure.) By subtracting the number of dispersed pairs from the number of attracted pairs, we obtain an estimate of the amount of attracted pairs that reflects the increase in affiliative interactions due to the preceding conflict. The *conciliatory tendency* can then be calculated as follows:

$$\frac{(\text{Attracted pairs} - \text{dispersed pairs})}{\text{Total PC-MC pairs}}$$

This method is independent of the baseline levels of contacts and of the length of the observations, as longer observations simply mean an increased chance of animals contacting each other. It is therefore an accurate measure for comparing reconciliation frequencies between different dyads, different groups, or different species.

In sum, methodological progress has been made in both operationally defining reconciliation and quantifying conciliatory tendencies. Using these standardized methods, we can now compare reconciliation and other post-conflict interactions across a wide range of species.

interaction between former opponents (Appendix A). Some species further show behavioral specificity; that is, their post-conflict reunions stand out by special gestures and forms of body contact. Dependent on the species, post-conflict reunions may include mouth-to-mouth contact, embracing, sexual intercourse, grooming, hand-holding, and clasping the other's hips (Fig. 2.3). That former opponents engage in such intensive friendly and sexual interactions is contrary to predictions from motivational theory in ethology, which assumes continuity between temporally related events, hence predicts residual antagonism during post-conflict contact (Rowell, Box 11.1).

With regard to the pacifying function implied by the reconciliation concept, several studies have

FIGURE 2.3. Two adult bonobo females engage in genito-genital rubbing. In this species (*Pan paniscus*), reconciliation is typically of a sexual nature. Photograph by the author, from de Waal (1995).

confirmed that the chance of renewed aggression is reduced and tolerance restored following post-conflict reunion. For example, an unpublished study at the Arnhem Zoo found that the probability of revival of a conflict was more than five times lower after reconciled compared with unreconciled conflicts (de Waal 1993). In an experimental study on longtail macaques (*Macaca fascicularis*), Cords (1992) demonstrated that pairs of monkeys that had engaged in a reconciliation following an induced conflict were more tolerant of each other's proximity close to an attractive resource than pairs that had not reconciled, thus suggesting reduced aggression in the dominant and reduced fear in the subordinate. Aureli et al. (1989) further demonstrated that reconciliation among longtail macaques has a calming effect on participants in a previous fight as measured by their rate of self-scratching, a correlate of sympathetic arousal (see Aureli & Smucny, Chapter 10).

Strictly speaking, these findings still do not demonstrate the specific function suggested by the *reconciliation* label, which is to repair damaged relationships. Even if post-conflict interaction reduces aggression and restores tolerance, how do we know that this affects the long-term relationship (Silk, Box 9.1)? First, social relationships are an emergent property of short-term interactions and effects; hence a distinction between the two would seem rather artificial (Hinde 1976; Cords & Aureli 1996). Second, even though primate species vary greatly in conciliatory tendency, in all species reconciliation is most common after fights between partners with close social ties even if we control for the increased level of interaction among these individuals (reviewed by Kappeler & van Schaik 1992; Cords & Aureli, Chapter 9). And perhaps the strongest evidence comes from an experiment by Cords & Thurnheer (1993), who manipulated the degree of cooperation in pairs of monkeys and found that monkeys trained to cooperate dra-

matically increased their tendency to reconcile after a fight. These results are consistent with the idea that reunions after aggression aim at restoring valuable relationships (further see van Schaik & Aureli, Chapter 15).

In short, the reconciliation concept comes with a set of testable predictions most of which have been supported for nonhuman primates (Table 2.1). From the start, however, it has been assumed that these predictions are not necessarily limited to primates: the cognitive requirements for reconciliation are not so demanding as to require enormous brain power (de Waal & Yoshihara 1983). Requirements include an individualized society (i.e., individual recognition and interindividual conflicts of interest), memory of previous fights, and advantages associated with the maintenance of cooperative relationships. A host of social animals meet these requirements, particularly mammals and birds but probably also other taxonomic groups (Schino, Chapter 11).

Evolutionary Explanations

Studies of animal conflict resolution are loaded with theory, albeit not exactly the same theory often encountered in the other animal behavior literature. All too often, this literature assumes all-out competition, whereas conflict resolution assumes that competition is constrained so as to protect cooperative partnerships. Development of this view was inspired by a combination of behavioral observation and the same Darwinian logic that underlies all of theory in modern biology.

Thus far, evolutionary approaches to animal social behavior have been dominated by a false dichotomy between aggression and sociality. Struggle-for-Life language was directly transferred to the social domain resulting in an overemphasis on clashing individual interests. Theorists insisted on cost-benefit analyses, whereas in reality benefit-

TABLE 2.1
Two Contrasting Views of Aggressive Behavior, and Predictions Derived from Them

Predictions	Empirical Definition
1. AGGRESSION AS AN ANTISOCIAL TENDENCY	
Dispersal	Reduced friendly interaction and reduced proximity following aggression
Motivational continuity	Post-conflict contact reflects residual antagonism
Bonding excludes aggression	Aggression is rare among socially close and cooperative individuals
2. RELATIONAL MODEL	
Social repair	Increased friendly interaction following aggression, especially between opponents
Motivational shift	Contact following aggression can be unusually intensive
Conflict as negotiation	Both aggression and post-conflict reunions are common among socially close, cooperative individuals
Calming effect	Post-conflict contact restores tolerance and reduces anxiety

Note: The traditional interpretation of aggression as a socially destructive tendency generates quite different predictions than the Relational Model, which assumes that aggression serves to negotiate the terms of relationships and that post-conflict reunions function as reconciliations. Tested on nonhuman primates, predictions derived from the Relational Model have been overwhelmingly supported.

benefit arrangements seem common. The possibility of *shared* interests was so far from the minds of early evolutionary biologists (except with regard to kin) that when it came to accounting for the rarity of lethal violence, rather than assuming a need for cooperation and stable group life, explanations focused exclusively on the physical risks of combat (Maynard Smith & Price 1973).

This line of argument was taken to the extreme by Popp & DeVore (1979), who wondered why dominant chimpanzees fail to kill subordinates. They suggested as a possible drawback that subordinates might fight for their lives. They added: "By contrast, the only benefit for the potential assassin would be the elimination of just one of many competitors" (p. 329). In this Hobbesian view, then, each individual is surrounded not by friends and family but by competitors. If true, there would be absolutely no need for conflict resolution: victory, defeat, and injury would be the only options worth considering.

Van Rhijn & Vodegel (1980) were the first to demonstrate theoretically that these early models rested on overly simplistic assumptions. They argued that individual recognition is a common capacity in the animal kingdom and that animals remember against which individuals they have won or lost confrontations, resulting in systematic avoidance of stronger individuals. The widespread existence of stable territories and dominance hierarchies attests to the validity of this model.

The models become even more complex if the value of cooperation is factored in, that is, the dependency of social animals on group life in general (van Schaik 1983) and the assistance of certain individuals in particular. Primates commonly engage in alliances, in which two or more parties join each other against a third to reach goals that cannot be reached through individual action (reviewed in Harcourt & de Waal 1992). The existence of alliances has profound implications for competition, not only in the triadic sense (i.e., both contestants need to take into account the presence or absence of potential allies) but also within the dyad itself. Two contestants may have an alliance themselves. If so, they need to consider not only the value of the resource over which they are competing, and the risk of bodily harm, but also the value of their relationship. Specifically, each party needs to weigh the benefits of access to the resource against the possibility that the use of force may reduce the other's tendency (or ability, in case of physical harm) to provide future assistance.

If we follow Kummer (1978) in regarding social relationships as investments, aggressive conflict becomes a potentially deleterious activity that endangers the interest accrued from these investments. Thus, the basic dilemma facing competitors is that they sometimes cannot win a fight without losing a friend and supporter. The same principle underlying all Darwinian theory, that individuals pursue their own reproductive interests, thus automatically leads one to assume that animals that depend on cooperation should either avoid open conflict or evolve ways to control the social damage caused by open conflict (de Waal 1989b).

It is good to realize that this rather obvious-sounding explanation of the evolution of reconciliation behavior was arrived at post hoc, after the phenomenon's discovery. Contrary to the much-envied physics model of theory-based discovery, the field of animal behavior remains very much dependent on actual observation for its advancement. As noted by Mayr (1997), new observations may lead to new concepts, and it is concept formation that ultimately drives theoretical progress. New concepts force us to rethink old assumptions. In evolutionary biology, especially in the game-theory school, assumptions have traditionally been geared toward animals who neither know nor need each other. As a result,

even if the process of reconciliation now appears entirely logical to us, it was never predicted or even remotely considered by modern theoreticians.

The Relational Model

When psychologists coined the term *prosocial* for friendly and cooperative acts, aggression was by implication relegated to the category of antisocial behavior (Krebs & Miller 1985). Aggression is generally presented as a socially negative tendency that poses serious problems to society and hence needs to be contained or eliminated.

Even though aggressive behavior is undeniably social (i.e., aggression is generally defined as harmful behavior directed at members of the same species), psychologists have historically ignored how it arises within and affects relationships. Psychologists implicitly adhere to what may be called an *Individual Model* of aggression—that is, aggressive behavior is viewed as an expression of the individual caused by both internal and external factors, such as experience, genes, hormones, and frustration. Social relationships do not figure prominently in this model. Indeed, research subjects, both human and animal, often have neither a history nor a future of interaction together. Human aggression is typically studied in experiments in which strangers are instructed how to interact (e.g., punish one another), or in which subjects are exposed to visual materials, such as violent movies (reviewed by Berkowitz 1993). Similarly, animals are typically introduced to strangers, sometimes belonging to a different species (e.g., a mouse is placed in a rat's home cage), or electrically shocked so as to study pain-induced aggression (reviewed by Johnson 1972).

Essentially, then, aggressive behavior has been investigated by psychologists in a social vacuum. This neglect of social context is surprising given that most human conflict concerns familiar individuals. However, not all psychologists and social scientists try to divorce aggressive conflict from other aspects of social life or view it as necessarily destructive and antisocial (Lyons 1993). Recently, there are signs of a change in perspective. Gottman's (1994) extensive longitudinal data on human marital relationships, for example, have begun to challenge the conventional notion that conflict is incompatible with lasting marriages.

The conflict resolution perspective breaks with this tradition in that it shifts attention from aggression as the expression of an internal state to aggression as the product of a conflict of interest. It regards aggressive behavior as the product of social decision making: it is one of several ways in which conflicts between individuals or groups can be resolved. This framework will be referred to here as the *Relational Model* (de Waal 1996b; Fig. 2.4), because it concerns the way aggressive behavior functions within social relationships.

The Relational Model views social partners as commodities of variable value. If two individuals compete over, say, a food source or a mate, they need to compare the resource value not only with the risk of injury in a possible fight but also with the damage the fight may cause to the relationship with the opponent and the advantages derived from this relationship. The better armed and stronger the opponent, or the more valuable the relationship between the competitors, the greater the resource value needs to be to make a fight worth the risk. Conversely, if damage to the relationship can easily be reduced through post-conflict interaction—a factor that we may label the *reparability* of the relationship—open conflict becomes more likely. In sum, the Relational Model predicts that the tendency to initiate aggression increases with the number of opportunities for competition, the resource value, and the reparability of the relationship, while it decreases with the risk of injury and the value of the relationship.

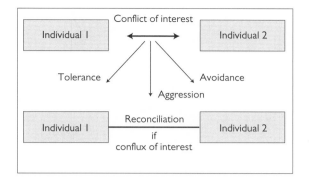

FIGURE 2.4. According to the Relational Model, aggressive behavior is one of several ways in which conflicts of interest are settled. Other possible ways are tolerance (e.g., sharing a resource) or avoidance of confrontation. If aggression does occur, it depends on the nature of the relationship whether repair attempts will be made. If there is a strong mutual interest in maintenance of the relationship, reconciliation is likely. Parties negotiate the terms of their relationships by going through cycles of conflict and reconciliation. From de Waal (1996a).

Within this framework, aggressive conflict is subject to experience-based calculations in which short-term advantages are weighed against both the risk of escalation and long-term social consequences. Paradoxically, the better developed mechanisms of conflict resolution are, the less reluctant individuals will be to engage in open conflict. For example, stumptail macaques (*Macaca arctoides*) are characterized by a high conciliatory tendency, a high degree of social tolerance, an exceptional amount of grooming, but also frequent aggressive conflict (de Waal & Luttrell 1989). The ability to maintain working relationships despite conflict, and to undo damage to relationships, makes room for aggression as an instrument of negotiation. References in the primate literature to negotiation and bargaining can be found in de Waal (1982), Maxim (1982), Hinde (1985), Dunbar (1988), Smuts & Watanabe (1990), Chadwick-Jones (1991), Colmenares (1991), and Noë et al. (1991). De Waal (1996b) explores the connection between negotiation and conflict resolution.

FIGURE 2.5. Weaning compromise in chimpanzees between mother and offspring. The offspring has developed a habit of sucking on a part of the mother's body other than the nipple because of repeated nursing conflicts with the mother. Such a mutually acceptable end situation results from cycles of repeated conflict and reconciliation between two parties with a highly valued relationship. Photograph by the author, from de Waal (1996a).

A typical example is "weaning compromise" in chimpanzees. Some of the older juveniles in our group at the Field Station of the Yerkes Primate Center regularly suckle on their mother's lower lip or ear or nap with the skin right next to her nipple in their mouth. Such substitute nursing is the result of an extended period of discrepant desires between mother and offspring. Access to milk has become increasingly restricted despite vociferous protest by the youngster. Toward the end of the process, youngsters have to content themselves with something that merely resembles nursing (Fig. 2.5).

In this process, the juvenile cajoles its mother with signs of distress, such as pouts and whimpers. If everything else fails, the juvenile may resort to a temper tantrum at the peak of which he almost chokes in his screams or vomits at her feet. The mother has weapons too. Goodall (personal communication) once saw how a female answered her son's tantrums by climbing high up a tree to throw him literally to the ground while at the last instant holding on to an ankle. The young male hung upside down for fifteen seconds or more, screaming his head off, before his mother retrieved him.

Weaning conflict is the first negotiation in a young mammal's social life. It is conducted with a social partner of paramount importance to survival and illustrates all the critical elements of the Relational Model: a combination of conflicting and overlapping interests, and a cycling through positive and negative interactions resulting in some sort of agreement about the "proper" time for nursing (cf. Altmann 1980) and the general terms of the relationship. Mother and offspring share so many interests that the conflict mostly works itself out to their mutual advantage, with increased juvenile independence and the mother's readiness for the next pregnancy.

The question why aggression sometimes escalates to violence, or becomes so frequent as to damage relationships beyond repair, achieves special significance within this model as it automatically translates into questions about the value individuals attach to their relationships and the social skills required to settle disputes in an alternative manner. In the absence of a conflux of interests or when the adversary has been "dehumanized" by political propaganda, there is indeed little basis to contain aggression. Genocide and other atrocities are possible under such circumstances. The Relational Model does not assume, therefore, that aggression is never expressed to its fullest, destructive potential, only that intragroup aggression is, as a rule, an integral part of well-established social relationships. The days of a deterministic view of aggressive behavior separate from other aspects of social life seem behind us (de Waal 1989a; Silverberg & Gray 1992; Mason & Mendoza 1993).

Ramifications of Reconciliation

Once aggressive conflict is viewed as an instrument to negotiate the terms of relationships—an instrument made possible by powerful constraints on the expression of aggression as well as the possibility of social repair—the definition of what is a close or distant relationship changes dramatically. Instead of classifying relationships simply in terms of rates of affiliative and aggressive behavior, the dynamic between the two becomes the critical factor. Relationships marked by high aggression rates may actually be quite close and cooperative.

It has been known since the pioneering fieldwork on Japanese macaques (*Macaca fuscata*) that the strongest ties in macaque society are between female kin (Kawai 1958; Kawamura 1958). Grandmothers, mothers, daughters, and sisters often groom each other, tolerate each other around food and water, and band together against other matrilines. Kurland (1977) was the first to report high rates of aggression among kin in free-ranging Japanese macaques. This was confirmed for other macaque species in outdoor enclosures by Bernstein & Ehardt (1986), de Waal & Luttrell (1989), and Bernstein et al. (1993). Furthermore, kinship ties are not the only ones characterized by frequent aggression: the same applies to nonrelatives with close ties (de Waal & Luttrell 1986).

The number of fights among closely bonded individuals ceases to be exceptional if time spent together is taken into account, that is, the high aggression rate is partly a product of proximity. Yet this applies to other kinds of behavior as well, so that it cannot be used to "devalue" aggression among bonded individuals unless one is prepared to do the same for grooming and other affiliative behavior (Bernstein et al. 1993). In addition, the intensity level of aggression (attack versus threat) does not measurably differ between kin and nonkin. Thus, the closest and most cooperative relation-

ships in a macaque society are characterized by relatively high levels of aggression. The same is true for chimpanzee society, in which adult males engage in frequent coalitions and grooming but also frequent fights among themselves. In both cases, the solution to the apparent paradox is the relatively high conciliatory tendency that characterizes valued relationships (Cords & Aureli, Chapter 9).

The first implication of the Relational Model, then, is that it allows for the full integration of competition and cooperation. This integration is not just an alternation between the two, or an uneasy coexistence; conflict and its resolution may actually contribute to a fine-tuning of expectations between parties, a building of trust despite occasional disagreement, hence a more productive and closer relationship than would be possible if conflict were fully suppressed. The reparability of relationships permits aggression to have a testing quality (de Waal 1996b).

The second implication of the model is that the motivational and emotional underpinnings of conflict are not limited to aggression and fear. The term *agonistic* has been widely employed by ethologists to describe the behavior during aggressive confrontations, covering both aggressive and fearful responses. Conflict may also involve affiliative motivations, however. This is most evident during conflict between mother and unweaned offspring, in which one can actually see the parties drawn between punitive or fearful responses on the one hand and attachment responses on the other (e.g., Weaver & de Waal 1997). But also in adult relationships, fear is often mixed with attraction, as reflected in submissive "greetings" among chimpanzees, in which subordinates actually approach dominants with pant-grunts and bobbing. Along the same lines, Schenkel (1967, p. 319) concluded about canids: "Submission is the effort of the inferior to attain friendly or harmonic social integration."

The issue of how dominance relationships and affiliative patterns are integrated goes back to the writings of Maslow (1940) and is also reflected in Mason's (1964) observation: "One of the most difficult problems is to reconcile the finding that dominance is based on fear-aggression with the observation that the dominant animal is often sought or followed by the other members of the group" (p. 292).

It is only now that we are beginning to acquire the conceptual tools to tackle this paradox. One of the greatest challenges in the study of cooperative animals is to integrate fully the vertical (hierarchical) and horizontal (affiliative networking) dimensions of social organization. Success of this endeavor requires a full analysis of the complex relation between conflict resolution and social dominance and continued study of the dilemma of cooperation among individuals with partially conflicting interests (de Waal 1986).

This dilemma can be investigated through the window of animal emotions. Both the perceived risk of injury and potential damage to the relationship may induce stress and anxiety in combatants. Indeed, Aureli's (1997) research on behavioral indicators of post-conflict anxiety suggests that not only losers but also winners of aggressive conflicts have something to "worry" about. As predicted, signs of anxiety correlate with the quality of the disharmonized relationship (see also Aureli & Smucny, Chapter 10).

Conclusion

The evolutionary debates of the 1960s revolved around the instinctive nature of aggression, whereas the sociobiological revolution of the 1970s emphasized harsh, selfish competition. Both approaches overlooked the complexity introduced into the realm of competition by the need for cooperation and a cohesive group life. This need exists in all group-living animals that survive through mutual assistance despite divergent individual interests. By the end of the 1970s, a small number of students of animal behavior, mainly primatologists, were developing an alternative view that took into account the need of nonhuman primates for calming contact when stressed, their tendency to foster beneficial long-term relationships, and the actual phenomenon of *reconciliation* in which former opponents reunite after a conflict, often in remarkably intense fashion. Out of these developments grew an increasingly detailed theoretical model of conflict resolution that integrated a variety of insights into primate behavior, from aggressive behavior and coalition formation to tension regulation and social dominance. Without denying the competitive nature of primates, the model assumes powerful constraints on conflict for the sake of cooperation. There is no reason why this model should not also apply to other taxonomic groups and to human behavior.

References

Altmann, J. 1980. *Baboon Mothers and Infants.* Cambridge: Harvard University Press.

Ardrey, R. A. 1967. *The Territorial Imperative.* London: Collins.

Asquith, P. J. 1984. The inevitability and utility of anthropomorphism in description of primate behaviour. In: *The Meaning of Primate Signals* (R. Harré & V. Reynolds, eds.), pp. 138–176. Cambridge: Cambridge University Press.

Aureli, F. 1997. Post-conflict anxiety in nonhuman primates: The mediating role of emotion in conflict resolution. *Aggressive Behavior,* 23: 315–328.

Aureli, F., van Schaik, C. P., & van Hooff, J. A. R. A. M. 1989. Functional aspects of reconciliation among captive long-tailed macaques (*Macaca fascicularis*). *American Journal of Primatology*, 19: 39–51.

Bandura, A. 1973. *Aggression: A Social Learning Analysis*. Englewood Cliffs, N.J.: Prentice-Hall.

Berkowitz, L. 1993. *Aggression: Its Causes, Consequences, and Control*. New York: McGraw-Hill.

Bernstein, I. S., & Ehardt, C. 1986. The influence of kinship and socialization on aggressive behavior in rhesus monkeys (*Macaca mulatta*). *Animal Behaviour*, 34: 739–747.

Bernstein, I. S., Judge, P. G., & Ruehlmann, T. E. 1993. Kinship, association, and social relationships in rhesus monkeys. *American Journal of Primatology*, 31: 41–54.

Blurton-Jones, N. G., & Trollope, J. 1968. Social behaviour of stump-tailed macaques in captivity. *Primates*, 9: 365–394.

Call, J. 1999. The effect of inter-opponent distance on the assessment of reconciliation. *Primates*, 40: 515–523.

Chadwick-Jones, J. K. 1991. The social contingency model and olive baboons. *International Journal of Primatology*, 12: 145–161.

Colmenares, F. 1991. Greeting behaviour between male baboons: Oestrous females, rivalry, and negotiation. *Animal Behaviour*, 41: 49–60.

Cords, M. 1992. Post-conflict reunions and reconciliation in long-tailed macaques. *Animal Behaviour*, 44: 57–61.

Cords, M., & Aureli, F. 1996. Reasons for reconciling. *Evolutionary Anthropology*, 5: 42–45.

Cords, M., & Thurnheer, S. 1993. Reconciliation with valuable partners by long-tailed macaques. *Ethology*, 93: 315–325.

Das, M., Penke, Z., & van Hooff, J. A. R. A. M. 1997. Affiliation between aggressors and third parties following conflicts in long-tailed macaques (*Macaca fascicularis*). *International Journal of Primatology*, 18: 157–179.

de Waal, F. B. M. 1982. *Chimpanzee Politics*. London: Jonathan Cape.

de Waal, F. B. M. 1986. Integration of dominance and social bonding in primates. *Quarterly Review of Biology*, 61: 459–479.

de Waal, F. B. M. 1987. Tension regulation and nonreproductive functions of sex in captive bonobos (*Pan paniscus*). *National Geographic Research*, 3: 318–335.

de Waal, F. B. M. 1989a. *Peacemaking among Primates*. Cambridge: Harvard University Press.

de Waal, F. B. M. 1989b. Dominance "style" and primate social organization. In: *Comparative Socioecology: The Behavioural Ecology of Humans and Other Mammals* (V. Standen & R. Foley, eds.), pp. 243–264. Oxford: Blackwell.

de Waal, F. B. M. 1991. Complementary methods and convergent evidence in the study of primate social cognition. *Behaviour*, 118: 297–320.

de Waal, F. B. M. 1993. Reconciliation among primates: A review of empirical evidence and unresolved issues. In: *Primate Social Conflict* (W. A. Mason & S. P. Mendoza, eds.), pp. 111–144. Albany: State University of New York Press.

de Waal, F. B. M. 1995. Sex as an alternative to aggression in the bonobo. In: *Sexual Nature, Sexual Culture* (P. Abramson & S. Pinkerton, eds.), pp. 37–56. Chicago: University of Chicago Press.

de Waal, F. B. M. 1996a. *Good Natured: The Origins of Right and Wrong in Humans and Other Animals*. Cambridge: Harvard University Press.

de Waal, F. B. M. 1996b. Conflict as negotiation. In: *Great Ape Societies* (W. C. McGrew, L. F. Marchant, & T. Nishida, eds.), pp. 159–172. Cambridge: Cambridge University Press.

de Waal, F. B. M., & Luttrell, L. M. 1986. The similarity principle underlying social bonding among female rhesus monkeys. *Folia primatologica*, 46: 215–234.

de Waal, F. B. M., & Luttrell, L. M. 1989. Toward a comparative socioecology of the genus Macaca: Different dominance styles in rhesus and stumptail macaques. *American Journal of Primatology*, 19: 83–109.

de Waal, F. B. M., & Ren, R. 1988. Comparison of the reconciliation behavior of stumptail and rhesus macaques. *Ethology*, 78: 129–142.

de Waal, F. B. M., & van Roosmalen, A. 1979. Reconciliation and consolation among chimpanzees. *Behavioral Ecology and Sociobiology*, 5: 55–66.

de Waal, F. B. M., & Yoshihara, D. 1983. Reconciliation and redirected affection in rhesus monkeys. *Behaviour*, 85: 224–241.

Dunbar, R. I. M. 1988. *Primate Social Systems.* London: Croom Helm.

Ehrlich, A., & Musicant, A. 1977. Social and individual behaviors in captive slow lorises. *Behaviour*, 60: 195–220.

Ellefson, J. 1968. Territorial behavior in the common white-handed gibbon, *Hylobatus lar.* In: *Primates: Studies in Adaptation and Variability* (P. Jay, ed.), pp. 180–199. New York: Holt.

Goodall, J. van Lawick-. 1968. The behaviour of free-living chimpanzees in the Gombe Stream Reserve. *Animal Behaviour Monographs*, 1: 161–311.

Goodall, J. 1979. Life and death at Gombe. *National Geographic*, 155: 592–621.

Goodall, J. 1986. *The Chimpanzees of Gombe: Patterns of Behavior.* Cambridge: Harvard University Press, Belknap Press.

Gottman, J. 1994. *Why Marriages Succeed or Fail.* New York: Simon and Schuster.

Harcourt, A. H., & de Waal, F. B. M. 1992. *Coalitions and Alliances in Humans and Other Animals.* Oxford: Oxford University Press.

Hediger, H. 1941. Biologische Gesetzmäßigkeiten im Verhalten von Wirbeltieren. *Mitteilungen Naturforschungs Gesellschaft Bern 1940*, pp. 37–55.

Hinde, R. A. 1976. Interactions, relationships and social structure. *Man*, 11: 1–17.

Hinde, R. A. 1985. Expression and negotiation. In: *The Development of Expressive Behavior: Biology-Environment Interactions* (G. Zivin, ed.), pp. 103–116. Orlando, Fla.: Academic Press.

Johnson, R. N. 1972. *Aggression in Man and Animals.* Philadelphia: Saunders.

Judge, P. G. 1991. Dyadic and triadic reconciliation in pigtail macaques (*Macaca nemestrina*). *American Journal of Primatology*, 23: 225–237.

Kappeler, P. M., & van Schaik, C. P. 1992. Methodological and evolutionary aspects of reconciliation among primates. *Ethology*, 92: 51–69.

Kawai, M. 1958. On the system of social ranks in a natural troop of Japanese monkey. I: Basic rank and dependent rank. II: Ranking order as observed among the monkeys on and near the test box. *Primates*, 1: 111–148 (in Japanese).

Kawamura, S. 1958. Matriarchal social ranks in the Minoo-B troop: A study of the rank system of Japanese monkeys. *Primates*, 1: 148–156 (in Japanese).

Kellogg, W., & Kellogg, L. 1933. *The Ape and the Child.* New York: McGraw-Hill.

Köhler, W. 1959 [1925]. *Mentality of Apes.* 2nd edn. New York: Vintage.

Krebs, D. L., & Miller, D. T. 1985. Altruism and aggression. In: *Handbook of Social Psychology*, Vol. 2 (G. Lindzey & E. Aronson, eds.). New York: Random House.

Kummer, H. 1978. On the value of social relationships to nonhuman primates: A heuristic scheme. *Social Science Information*, 17: 687–705.

Kummer, H., Götz, W., & Angst, W. 1974. Triadic differentiation: An inhibitory process protecting pair bonds in baboons. *Behaviour*, 49: 62–87.

Kurland, J. A. 1977. *Kin Selection in the Japanese Monkey.* Contributions to Primatology, Vol. 12. Basel: Karger.

Lindburg, D. 1973. Grooming behavior as a regulator of social interactions in rhesus monkeys. In: *Behavioral Regulators of Behavior in Primates* (C. Carpenter, ed.), pp. 85–105. Lewisburg, Pa.: Bucknell University.

Lorenz, K. 1967. *On Aggression.* London: Methuen.

Lyons, D. M. 1993. Conflict as a constructive force in social life. In: *Primate Social Conflict* (W. A. Mason & S. P. Mendoza, eds.), pp. 387–408. Albany: State University of New York Press.

Manning, A. 1989. The genetic basis of aggression. In: *Aggression and War: Their Biological and Social Bases* (J. Groebel & R. A. Hinde, eds.), pp. 48–57. Cambridge: Cambridge University Press.

Marler, P. 1976. On animal aggression. *American Psychologist*, 31: 239–246.

Maslow, A. H. 1940. Relationship-quality and social behavior in infrahuman primates. *Journal of Social Psychology*, 11: 313–324.

Mason, W. A. 1964. Sociability and social organization in monkeys and apes. In: *Advances in Experimental Psychology* (L. Berkowitz, ed.), pp. 277–305. New York: Academic Press.

Mason, W. A., & Mendoza, S. P. 1993. *Primate Social Conflict.* Albany: State University of New York Press.

Maxim, P. E. 1982. Contexts and messages in macaque social communication. *American Journal of Primatology,* 2: 63–85.

Maynard Smith, J., & Price, G. R. 1973. The logic of animal conflict. *Nature,* 246: 15–18.

Mayr, E. 1997. *This Is Biology: The Science of the Living World.* Cambridge: Harvard University Press, Belknap Press.

McKenna, J. 1978. Biosocial function of grooming behavior among the common langur monkey (*Presbytis entellus*). *American Journal of Physical Anthropology,* 48: 503–510.

Montagu, M. F. A. 1968. The new litany of "innate depravity" or original sin revisited. In: *Man and Aggression* (M. F. A. Montagu, ed.). London: Oxford University Press.

Morris, D. 1967. *The Naked Ape.* New York: Bantam.

Noë, R., van Schaik, C. P., & van Hooff, J. A. R. A. M. 1991. The market effect: An explanation for pay-off asymmetries among collaborating animals. *Ethology,* 87: 97–118.

Poirier, F. 1968. Dominance structure of the Nigiri langur (*Presbytis johnii*) of South India. *Folia primatologica,* 12: 161–186.

Popp, J. L., & DeVore, I. 1979. Aggressive competition and social dominance theory: Synopsis. In: *The Great Apes* (D. Hamburg & E. McCown, eds.), pp. 317–338. Menlo Park, Calif.: Benjamin/Cummings.

Schenkel, R. 1967. Submission: Its features and function in the wolf and dog. *American Zoologist,* 7: 319–329.

Scott, J. P. 1958. *Animal Behavior.* Chicago: University of Chicago Press.

Seyfarth, R. 1976. Social relationships among adult female baboons. *Animal Behaviour,* 24: 917–938.

Silverberg, J., & Gray, J. P. 1992. *Aggression and Peacefulness in Humans and Other Primates.* New York: Oxford University Press.

Smuts, B. B., & Watanabe, J. M. 1990. Social relationships and ritualized greetings in adult male baboons (*Papio cynocephalus anubis*). *International Journal of Primatology,* 11: 147–172.

van Hooff, J. A. R. A. M. 1967. The facial displays of the Catarrhine monkeys and apes. In: *Primate Ethology* (D. Morris, ed.), pp. 7–68. London: Weidenfeld.

van Rhijn, J., & Vodegel, R. 1980. Being honest about one's intentions: An evolutionary stable strategy for animal conflicts. *Journal of Theoretical Biology,* 85: 623–641.

van Schaik, C. P. 1983. Why are diurnal primates living in groups? *Behaviour,* 87: 120–122.

Veenema, H. C., Das, M., & Aureli, F. 1994. Methodological improvements for the study of reconciliation. *Behavioural Processes,* 31: 29–38.

Weaver, A. Ch., & de Waal, F. B. M. 1997. The development of reconciliation in tufted capuchins, *Cebus apella. American Journal of Primatolology,* 42: 153.

Wrangham, R. W., & Peterson, D. 1996. *Demonic Males.* Boston: Houghton Mifflin.

York, A. D., & Rowell, T. E. 1988. Reconciliation following aggression in patas monkeys, *Erythrocebus patas. Animal Behaviour,* 36: 502–509.

CHAPTER 3

Conflict Management in Children and Adolescents

Peter Verbeek, Willard W. Hartup, & W. Andrew Collins

Introduction

Successful social development reconciles individuation with social integration and requires the
acquisition of conflict management skills that
afford both. The interplay between individual
and social motives is already apparent in conflict
between toddlers. For instance, Hay & Ross (1982)
found that 21-month-old winners of toys would
often abandon the toy they had just taken from a
peer in order to engage in a new dispute over
another toy held by the former opponent. Such
tendencies were common even when the toy held
by the other was an exact copy of the toy originally won. Earlier, Eckerman et al. (1979) showed
that to one-year-old children the attractiveness of
a toy increased after another person touched it.
Thus toddler conflict may serve to test the "social
waters" and may be instrumental in the acquisi-

tion of knowledge about social relations as well as
ownership.

In this chapter we trace the development of
conflict management skills from toddlerhood
through adolescence, beginning with a brief overview of contemporary conceptualizations and pro-
ceeding with a discussion of what we currently
know about conflict management in two principal
arenas: the family and the peer group. We then
consider how socioeconomic and cultural conditions affect the development of conflict management skills. We present our discussion within
the framework of relationship theory and practical implications regarding adult mediation of
children's conflicts.

Throughout development interpersonal conflict can take many forms, ranging from isolated

TABLE 3.1
Examples of Unilateral and Bilateral Conflict Management and Post-Conflict Events

Conflict Phase	Unilateral	Bilateral
TERMINATION	standing firm	justification
	subordination	negotiation
	power assertion: verbal/physical	cooperation
	coercion	granting appeals
	temper tantrum	disengagement
	avoidance/withdrawal	compromise
		conciliation
IMMEDIATE OUTCOME	distributive	variable
	standoff	integrative
POST-CONFLICT EVENT	separation	continued interaction
		peaceful reunion

incidents to interlocking series of oppositional events. Developmental research commonly focuses on distinct episodes that are separated in time and that can be broken down conceptually into three sequential phases: (1) instigation; (2) termination; and (3) immediate outcome (or *resolution*). The emphasis in this chapter is on conflict termination and immediate outcome, and for the purpose of our review we combine these under the heading of *conflict management*. Conflict management can be unilateral or bilateral. Unilateral conflict management is characterized by opportunism and lack of consideration for the opponent's perspectives and wishes, as well as by subordination. Conversely, bilateral conflict management is characterized by mutual perspective taking and often by dovetailing of opposing goals and expectations (see Table 3.1 for an overview of basic terms).

Instigation is commonly classified according to two main causal domains: moral and social (cf. Killen & de Waal, Chapter 17). For instance, among

preschoolers a conflict may be about ownership of a toy (moral domain), or it may be a dispute about the rules of a game (social domain; Fig. 3.1).

Termination refers to children's strategies aimed at ending conflict. Termination strategies can be unilateral (e.g., power assertion, coercion) or bilateral, for instance, when children negotiate or use conciliation to end their dispute.

Immediate outcome is commonly classified in two main categories: *distributive* and *variable* conflict outcomes. A distributive outcome includes situations during which one child's gain is the other child's loss. A variable outcome refers to situations in which both win or benefit from the resolution (cf. Littlefield et al. 1993). A variant of this latter category is the *integrative* outcome (cf. Anderson 1937). In this situation a shared interest in social interaction provides the basis for a mutually beneficial resolution. For instance, two children may have a similar stake in a particular interaction and may agree to overcome their differences in order

FIGURE 3.1. An example of a conflict in the social domain. Two preschool girls use physical assertion to terminate a dispute over taking turns in pushing a boy on a swing. Illustration by Eric P. Snowden from video observations by Peter Verbeek.

to resume the mutually rewarding interaction. Conversely, two children may share a long-term investment in an established relationship and may reconcile their differences in order to preserve the integrity of their relationship (see de Waal, Chapter 2; Cords & Aureli, Chapter 9; and van Schaik & Aureli, Chapter 15, for similar concepts in nonhuman animals).

Opponent interaction that extends beyond immediate conflict outcome has largely been ignored in the developmental literature. Not long ago we knew little about whether or how children make up with some delay following a conflict-induced separation. In this chapter we review recent evidence of peaceful post-conflict reunions that serve to reconcile former opponents' relations (Table 3.1; see Butovskaya et al., Chapter 12; cf. de Waal, Chapter 2).

Parent-Child Conflict Management

Conflict among family members manifests two basic principles: differentiation and developmen-tal change. Conflicts between children and their parents differ from those between children and their peers in terms of instigation, termination, and ending or resolution (Laursen & Collins 1994). Conflict instigation and management vary as a function of both developmental change and changes in normative expectancies regarding appropriate behavior.

Childhood

Research on parent-child conflict during the first decade of life most often has focused on emotional outbursts, such as temper tantrums (cf. Potegal, Box 12.1), and coercive behavior of children toward other family members as evidence of conflict. The frequency of such behavior begins to decline during early childhood and continues to do so during middle childhood (Goodenough 1931; Newson & Newson 1968, 1976; Patterson 1982). The frequency of episodes during which parents discipline their children also decreases between the ages of three and nine (Clifford 1959). Consequently, research on conflict management in this period has focused on the relative effectiveness of various parental strategies for gaining compliance and managing negative behaviors. As a result there is little descriptive information about the characteristics of conflict between parents and children and the role of each in conflict management.

With young children, parents typically employ distraction and physical assertion for preventing harm and gaining compliance. In middle childhood, however, parents report less frequent physical punishment and increasing use of techniques such as deprivation of privileges, appeals to children's self-esteem or sense of humor, arousal of children's sense of guilt, and reminders that children are responsible for what happens to them (Clifford 1959; Newson & Newson 1976; Roberts et al. 1984). These techniques may reflect changes in parents' attributions about the degree to which children should be expected to manage their behavior and also a greater tendency to regard misbehavior as deliberate intent and, thus, as warranting both parental anger and punishment (Dix et al. 1986).

Several emotional characteristics of six- to twelve-year-olds complicate parent-child conflict management. Compared with preschool children, six- to twelve-year-olds are more likely to sulk, become depressed, avoid parents, or engage in passive noncooperation with their parents (Clifford 1959). Furthermore, children are increasingly likely to attribute conflict with parents to the inadequacy of parental helping behaviors and disappointment in the frequency of parent-child interactions. In preadolescence, children also increasingly blame conflicts on parents' failures to fulfill role expectations and on lack of consensus on familial and societal values (Fisher & Johnson 1990). These attributional patterns may affect children's expectations about what type of conflict management is required.

Adolescence

Conflict management studies that focus on this age group are divided between those focusing on management of actual conflicts and those seeking preferred responses. These two methods often yield different estimates. For instance, Sternberg & Dobson (1987) found clear preference for compromise and mutual discussion in hypothetical disagreements, with disengagement (e.g., wait and see, avoid) being infrequently endorsed. The same subjects, however, reported that recent, real-life conflicts are terminated with disengagement as often as with negotiation (see also Youniss & Smollar 1985).

Naturalistic observations, experimental analogues, and self-reports with both peers (Goodwin 1982; Youniss & Smollar 1985) and family members (Montemayor & Hanson 1985; Schoenleber 1988; Raffaelli 1990; Smetana 1991) show clearly that adolescents' conflicts are most commonly terminated through power assertion and disengagement, rather than through negotiation. Later in adolescence preference for power-assertive techniques declines, thus making more complex bilateral techniques, such as negotiation, relatively more common.

Between adolescents and parents, integrative and distributive resolutions are about equally likely, and conflict rarely has a detrimental impact on future relations with parents (Laursen 1989). Nevertheless, family members are less likely than individuals without familial ties to report positive feelings once a conflict has ended (Laursen 1989; Raffaelli 1990). The importance of satisfying resolutions, however, is indicated by repeated findings that bilateral engagement in terminating conflict, rather than the occurrence of conflict, is a marker of adaptive, well-functioning relationships (e.g., Kelley et al. 1983; Grotevant & Cooper 1986).

Explanations for Age-Related Changes in Conflict Management

Explanations for shifts toward more adultlike, less power-assertive terminations have ranged from hormonal causes to psychological ones. For example, pubertal levels of certain hormones appear to be correlated with individual differences in the intensity of conflicts (Inoff-Germain et al. 1988). Individual variations in pubertal timing may also be tied to the incidence of adolescent conflicts with parents. Conflicts over family rules in early adolescence, for example, have been found to be especially likely in families with early-maturing offspring, both male and female (Hill & Holmbeck 1987; Hill 1988).

Various investigators have predicted that knowledge of appropriate skills and strategies for negotiated terminations should increase from middle childhood to late adolescence, on the assumption that logical competence, including understanding of interpersonal relationships, should be more advanced among older individuals than younger ones. Such changes potentially alter the degree to which the reasoning of children matches the complexity of parental reasoning about conflict. For example, most studies indicate relative congruity in the perceptions of 10- to 11-year-olds and their parents concerning matters in which parents' authority is legitimate (Smetana 1989), whereas incongruity becomes more likely during adolescence. Children's concepts of the basis for parental authority also change with age. Whereas preschoolers view parental authority as resting on the power to punish or reward, children in early middle childhood increasingly believe parental authority derives from all the things that parents do for them. After about age eight, parents' expert knowledge and skill are also seen as reasons to submit to their authority (Braine et al. 1991). Consequently, Maccoby (1984) speculates, parental appeals based on fairness, the return of favors, or reminders of the parents' greater knowledge and experience may increase in effectiveness during middle childhood, with parents less often feeling compelled to resort to promises of reward or threats of punishment. Central to these practices is an emphasis on the implications of children's actions for others (induction), rather than on use of parents' superior power to coerce compliance (Hoffman & Saltzstein 1967; Hoffman 1988).

Summary

During the toddler years parent-child conflict management is primarily a unilateral affair. Coercion, often manifested through temper tantrums, is a common termination strategy used by children at this young age, and it is usually up to parents to use appropriate measures to maintain positive relations with their child (but see Potegal, Box 12.1).

During the preschool years conflict management commonly consists of children's voluntary submission to parental authority. From middle childhood through preadolescence, children come to appreciate parental knowledge and expertise as reasons to submit to parental demands in conflict situations. Parent and child roles are further specified, and role expectations often become points of contention. At this age children are also more prone to sulk during the aftermath of parent-child conflict, and negative affect may imperil the parent-child relationship.

During much of adolescence, power assertion and disengagement become the termination strategies of choice for youth in conflicts with parents. Satisfaction with their relationships becomes especially important to both parents and adolescents. Expectations often differ, however, and parents and adolescents may blame each other for violating expectations. Synchronized expectations depend on synchronized cognitions, and cognitive growth enables older adolescents to negotiate rather than coerce resolutions to conflicts with parents.

Sibling Conflict Management

Conflict is universally embedded in sibling relationships. Since most children have siblings, this means that sibling conflict is widely experienced. These conflicts, however, are both unilateral and bilateral in terms of the management tactics used. The use of unilateral tactics such as coercion has been found to be negatively correlated with cooperation between siblings. Unilateral tactics also negatively correlate with helping, sharing, and sympathy expressed by older siblings toward younger ones (Dunn & Munn 1986). Bilateral tactics, however, have different correlates: attempts to reconcile positions and justify them are positively correlated with sharing, helping, and the comforting of older siblings by younger ones, and conciliation and justification are both positively correlated with cooperation by older children with their younger siblings. Patterns such as these suggest that conflict management is embedded in more general "types" of sibling relationships according to whether the efforts are generally bilateral or unilateral.

Home observations conducted with school-age children suggest generalization of conflict management involving siblings. Patterson (1982), for example, found that the use of coercion between siblings was correlated, first, with the use of coercion in parent-child relations and, second, with antisocial behavior in the children's dealings outside the home. Hetherington (1988) also found that children with coercive, aggressive sibling relationships evinced poorer peer relationships, behavior problems, and trouble at school. Causal interpretations are difficult, however, especially as to whether pathways from relationship to relationship are direct or indirect, but the generality of the child's modes of conflict management across relationships is striking.

Sibling relations differ according to parents' behavior toward their children—separately and together. For example, conflict and hostility in parent-child relations involving one sibling are correlated with conflict between the parents and the other sibling (Radke-Yarrow et al. 1988). When mothers or fathers favor one sibling (or children perceive them that way), greater conflict and more coercive relations ensue between the children. Moreover, psychosocial adjustment is poorer among children who perceive themselves to be less positively treated by their parents than their siblings (Boer & Dunn 1992). Overall, cross-

relational continuities in conflict management suggest that families constitute social systems rather than separate dyads.

Peer Conflict Management

In most societies children and adolescents spend a considerable amount of time interacting with age-mates in peer groups (e.g., Barker & Wright 1955; Whiting & Whiting 1975). Initially, such groups are commonly same-sex groups (e.g., Harkness & Super 1985); later, throughout adolescence, mixed-sex groups become more prominent (e.g., Hartup 1983; Berndt & Savin-Williams 1993).

Much of the literature focuses on conflict taking place in dyads of close peers. In the next section we discuss how concerns about the future of a close friendship can mitigate peer conflict. However, even when not concerned with the integrity of an existing close relationship, children, especially boys, are concerned with maintaining their status in the group, which can affect their choice of conflict management strategy (Maccoby 1996; cf. Strayer & Noel 1986). In fact, when children grow older—and this is true for boys as well as for girls—maintaining peer relationships becomes an increasingly complex task that involves concerns of closeness and friendship, as well as concerns of social status, and peer group integration and functioning (e.g., Coie & Cillessen 1993; Laursen 1993).

Conflict management is closely tied to friendship relations in childhood and adolescence. Conflicts are instrumental in initiating and maintaining friendships, and to a lesser extent in terminating them. Children manage conflicts and adopt peacemaking strategies in order to meet the expectations of themselves and their friends, further their understanding of the obligations and responsibilities of friendship, and maintain these relationships through time. Friendship experiences vary from child to child, and these variations have considerable developmental significance.

Age Trends

Recently, Laursen et al. (1998b) used a meta-analytic approach to take a general look at developmental differences in peer conflict management. The analysis was based on 25 studies dealing with dyadic conflict among peers classified as acquaintances (e.g., classmates, dormitory roommates), friends, romantic partners, and siblings.

The meta-analysis showed that, overall, peers managed conflict more often with negotiation than with coercion or disengagement. Significant developmental contrasts emerged, however. Children (age 2–10) commonly employed coercion, whereas adolescents (age 11–18) frequently employed negotiation as well as coercion. Conversely, young adults (age 19–25) more often resorted to either negotiation or disengagement.

The meta-analysis of Laursen et al. did not consider the aftermath of peer conflict. Recent cross-cultural findings showed that young children transformed a significant percentage of distributive outcomes into integrative resolutions after a "cooling-off" period of a few minutes (Verbeek 1997; Butovskaya et al., Chapter 12; and see below). When such post-conflict reconciliation (cf. de Waal, Chapter 2) is considered, young children appear considerably more constructive in their approach to peer conflict than one would infer from the aggregated findings of Laursen et al.

Conflict Management and Peer Status

Much of the research on peer social status has focused on measures of popularity (*sociometric* status, e.g., Coie et al. 1982) and on measures of social rank (*dominance* status, e.g., Strayer 1976). Sociometric status is measured through peer nominations and dominance status through naturalistic observations. Sociometric status, such as being "popular" or "rejected" among peers, is often seen as a reflection of individual traits (e.g., Coie & Dodge 1983; Dodge et al. 1990). There has been relatively little overlap between these two perspectives on peer status; researchers commonly focus on either sociometric or dominance status (but see La Freniere & Charlesworth 1983; Boulton & Smith 1990; Santos & Strayer 1997).

In contrast to sociometric status, which is based on peer perceptions, dominance status is a direct consequence of interactions with peers. As with sociometric status, there is some evidence for stability of dominance status over time (La Freniere & Charlesworth 1983; Strayer 1992), but there is little or no evidence for stability across settings. Interestingly, young children tend to have inflated views of the extent to which they are accepted by their peers (Patterson et al. 1990), and they commonly overestimate their own rank—and the rank of liked peers—in the dominance hierarchy (Boulton & Smith 1990).

Several studies have established a link between conflict management and sociometric status. For instance, Bryant (1992) found that popular children were viewed by their peers as more conciliatory compared with rejected peers. Rejected peers, in turn, were seen as more coercive in their approach to managing peer conflict. Studies of children's responses to hypothetical conflict scenarios and to limited-resource conditions provide convergent evidence: conciliatory strategies were associated with popularity and coercive strategies with rejection by peers (French & Waas 1987; Putallaz & Sheppard 1990; Chung & Asher 1996).

Limited-resource paradigms have also shed light on the link between dominance status and conflict management. Charlesworth (1996; La Freniere & Charlesworth 1987) devised a situation wherein four same-aged children (age range: four to eight) were given access to an attractive cartoon picture viewer. Conflict was inherent in this situation because in order for a child to see the movie in the viewer, the assistance of two other children was required, thereby relegating the fourth child to a bystander position. Children who ranked high in their classroom hierarchies ("alpha" children) spent considerably more time viewing the cartoon than lower-ranking children. Their success in obtaining the resource (in essence, a distributive resolution) was not strictly based on coercion, however, but rather involved a combination of coercive and cooperative strategies. More recently, Charlesworth observed Indian, Malayan, and South African children in the same situation and found that "alpha" children in each country behaved in a fashion similar to that of the U.S. children of the original study (Charlesworth 1996).

Commonly, when groups of children first meet (e.g., early in the new school year), conflicts, and assertive interactions not resulting in conflict, occur relatively frequently and contribute to the eventual establishment of a dominance hierarchy (Strayer 1992). Once dominance relations are established, rates of conflict and aggression decrease, and triadic support—opportunistic at first—mainly becomes an issue of friends supporting one another (Strayer & Noel 1986). Considering Charlesworth's (1983, 1996) findings, future research should investigate whether the use of a mixed strategy of coercion and cooperation by high-ranking children is indicative of established relations, or whether this "carrot and stick" approach is already apparent during the time that peer relations are first established.

Such an approach could also help to shed light on why it is that first impressions seem to matter in peer groups. Ladd et al. (1988), for instance, found that preschoolers who frequently argued with their peers early in the year were likely to be rejected throughout the entire year. In fact, children who argued early in the year but changed their ways during the year were still rejected later in the year. In a similar vein Denham & Holt (1993) found that peer reputation established early in the year was a significantly stronger predictor of being liked later in the year than actual social behavior.

Taken together, this evidence suggests that peer status both affects and is affected by the use of specific conflict management strategies. A synthesis of sociometric and observational methods is needed to unravel the relative strength and directionality of these relationships.

Conflict Management and Friendship Formation

Once children find themselves in proximity with one another (e.g., on playgrounds, in schools), friendships emerge on the basis of shared interests and attitudes as well as the shared understanding that continued interaction between them is in their mutual interest. Observational studies show that, first, agreements must occur over time within a context of shared interests in order for acquaintances to become friends and, second, certain conflicts and certain modes of conflict management actually facilitate friendship formation. For example, the use of "soft" modes of conflict management (e.g., "weak demands" followed by agreement) are associated with "hitting it off" (Gottman 1983). Classroom studies confirm that children frequently initiate conflicts with their friends (usually through weak demands) that are more likely to lead to integrative than distributive resolutions (Rizzo 1989).

Overall, then, friendship formation among young children seems to be marked by a dialectic between agreement and disagreement, a condition that also characterizes friendship formation among older children (Renshaw & Asher 1983). An agreements/disagreements dialectic may be necessary in friendship formation because comparisons are intrinsic to social attraction. Agreements carry limited information about exchange possibilities expect in contrast to disagreements (Hartup 1992; see de Waal 1986 for a discussion of a similar binding mechanism in nonhuman primates).

Conflict Management and Established Relationships

Three- and four-year-old children manage conflicts with friends differently than conflicts with nonfriends. Four observational studies conducted during free play concur in demonstrating that conflicts between friends are managed in qualitatively different ways than conflicts with nonfriends. Although friends' and nonfriends' conflicts do not differ in frequency, length, or the situations that instigate them, conflicts between friends are less intense than those occurring between nonfriends, and, most important, conflict management differs (Hartup et al. 1988). Conflicts between friends are more likely to involve disengagement and negotiation, whereas conflicts between nonfriends are more likely to consist of standing firm/subordination. Resolutions also differ: integrative resolutions are more common between friends than between nonfriends. Finally, with their friends, children are more likely to remain in proximity and engage in social interaction when the conflict is over. Vespo & Caplan (1993) confirmed these results using a slightly different approach, namely, by comparing friend and nonfriend dyads rather than by comparing the conflicts of friends and nonfriends.

Verbeek & Creveling (unpublished data) extended these findings and found a direct relation between the use of conciliatory termination strategies and subsequent interaction between the children. First, conciliation (invitations to cooperate, apologies, offers to share, and friendly physical contact) occurred more frequently during conflicts between friends than between nonfriends, mainly among girls. Most important, though, the results show a direct connection

between the use of conciliatory termination strategies and continued interaction following the conflict. Friends remain together, it seems, as a direct result of their use of conciliation, negotiation, and disengagement.

In a subsequent study, Verbeek (1997) demonstrated that remaining together following conflict is more common among friends (and acquaintances) than among nonfriends—partly because friends are more likely to be interacting prior to the conflict and partly because they use conciliatory strategies more readily. A second conflict outcome was examined, *post-conflict reunions*, that is, coming together in a peaceful manner in a post-conflict observation (Fig. 3.2). These reunions, which generally occurred within the first four minutes after a conflict-induced separation, did not differ according to the friendship relations existing between the children (cf. Butovskaya et al., Chapter 12). Friendship, then, seems to promote management strategies among young

FIGURE 3.2. An example of a peaceful post-conflict reunion between two former opponents after a brief conflict-induced separation. The boy is an interested bystander and did not mediate the reunion between the two preschool girls. Illustration by Eric P. Snowden from video observations by Peter Verbeek.

children that keep them together through a conflict episode but do not necessarily facilitate reunion once a conflict-induced separation has occurred.

Most observational studies of conflict among school-age children have been conducted in so-called *closed-field* rather than *open-field* situations, that is, children were observed in laboratory tasks for a fixed time with a designated partner rather than in free play or unstructured settings. Consequently, comparisons between studies with older and younger children are not easy to make. Nevertheless, when sharing tasks (Fonzi et al. 1997), distributive games (Matsumoto et al. 1986), or persuasion tasks (Jones 1985) are used, friends spend more time negotiating than nonfriends, make more proposals to one another, compromise more readily, and grant one another's appeals proportionally more often. Conflicts are sometimes more numerous and intense between friends than nonfriends in closed-field situations and in discussion/consensus tasks (cf. Nelson & Aboud 1985), but negotiated and conciliatory terminations are nevertheless favored more by friends than nonfriends.

When the laboratory task is competitive, friends compete more vigorously with one another than nonfriends (Fonzi et al. 1997), especially if they are boys (Berndt 1981); friends also experience more frequent and long-lasting conflicts (Hartup et al. 1993). Nevertheless, in managing conflicts in these situations, friends pay closer attention to the rules than nonfriends (Fonzi et al. 1997) and provide one another with more frequent assertions.

Friends and nonfriends have also been examined in cognitive tasks. In one instance (Fraysee 1995), six- and seven-year-old children working on conservation tasks with friends produced more self-justifications and criticisms than nonfriends did. Nevertheless, conflicts between friends involved more positive reactions than conflicts between

nonfriends, as well as more frequent searches for equitable solutions.

In another investigation (Azmitia & Montgomery 1993), 10-year-old children working on inductive reasoning problems with their friends were found to solve especially difficult problems more readily than nonfriends. The behaviors most closely linked to this outcome were monitoring of outcomes during the task and *transactive conflicts*, that is, dyadic exchanges in which each member of the dyad refers to the partner's reasoning or significantly clarifies one's own strategies. From these studies, as well as the social games described earlier, it becomes clear that friends are more active in their search for solutions, are more task-oriented, and make more active use of conflict to obtain solutions than nonfriends.

Overall, some two dozen published investigations contain data comparing friends and nonfriends in terms of conflict management. Questionnaire studies have been conducted as well as observational studies described here. Meta-analyses based on the entire literature with children ranging from preschool age through preadolescence (Newcomb & Bagwell 1995) confirm the pattern we describe: conflict frequency does not generally differ between friends and nonfriends, but modes of conflict management do.

Among the many gains that the greater use of conciliatory termination achieves for friends is the maintenance of their relationships with one another (Hartup et al. 1988; Laursen 1993; Verbeek 1997; Laursen et al. 1998b; see de Waal, Chapter 2, and Cords & Aureli, Chapter 9, for a discussion of similar findings in nonhuman primates). Having learned that common ground (mutual interests and attitudes) maximizes social rewards and having invested considerable energy in finding and initiating relationships with one's friends, children learn early on to select conciliatory behavior, negotiation, disengagement, and other "soft" modes of conflict management to protect and preserve these relationships.

Although considerable attention has been given to the hypothesis that friends favor conciliatory strategies in order to maintain their interaction and relationships with their friends, other functions may be involved. By being affirmative and relatively noncoercive with one another in their social exchanges, friends become significant sources of consensual validation (Sullivan 1953) and a sense of well-being (Hartup & Stevens 1997). Friends may render especially insightful assistance in cognitive functioning and performance (Hartup 1996). Consequently, investigators should think of the dialectic between friendship and conflict management as related to broad adaptational consequences, not merely the maintenance of the affiliation for its own sake.

Summary

Children gain their first substantial experience with peers during the preschool years. At this age children manifest different conflict management strategies with parents and with peers. Individual differences in conflict management style become apparent and affect peer acceptance. Conciliatory children are generally liked and coercive children disliked among their peers, and first impressions tend to persist. Conflict management also affects the formation of close dyadic relationships: friendships arise out of frequent play in which successive episodes of conflict and bilateral resolutions are embedded. Once established, friendships are maintained in much the same way.

During middle childhood through early adolescence, peer conflict management reflects both established patterns and new developments. Successive agreement/disagreement episodes are instrumental in friendship formation among peers. Friends often compete more vigorously then nonfriends, but they also work things out propor-

tionally more often than nonfriends. In fact, at this age conflict management among friends positively affects social as well as cognitive growth. On the flip side, antisocial children often have antisocial friends, and destructive patterns of conflict management may become the norm across relationships.

Cognitive growth affects conflict management in the context of the increasingly complex peer relations that characterize late adolescence. These developments set the stage for early adulthood, during which variable conflict resolutions tend to become the norm.

Triadic Events

Triadic Interactions within the Family

In families, third parties often contribute to both the incidence and resolution of conflict. Some of these effects are indirect, in that the two principal parties to the conflict behave differently in the presence of the third party than they would under other conditions. For example, mother-son dyads have been found to manifest greater engagement, security, and consistency when the father was present than when mother and son were alone (Gjerde 1986). These differences suggest that in intact families fathers' presence may indirectly facilitate integrative resolutions to conflict.

In other instances, third parties directly affect conflict resolution. Vuchinich et al. (1988) studied conflict during familial dinner-table conversations. The authors observed that all family members frequently intervened in conflict between two other members. Third parties were about equally likely to attempt to end or to continue the conflict. Intervention strategies varied with gender and familial roles. Females were more likely to intervene than males. Fathers tended to use authority; mothers, mediational tactics; and children, distraction. In about half of the interventions the third party attempted to form an alliance with one of the primary parties to the conflict. Third-party interventions usually prolonged conflicts but often led to resolution, perhaps because third parties tended to reciprocate nonconflictual behaviors, rather than conflictual ones. (See also Das, Chapter 13; Petit & Thierry, Box 13.1; and Watts et al., Chapter 14, for examples of triadic interactions in nonhuman primates.)

These findings leave open questions of the conditions under which third-party interventions have more or less positive implications for conflict resolution. A recent study by Perlman & Ross (1997) sheds some light on this. In this study parents of preschool-age children were observed to intervene in the most intense sibling fights, and parental intervention in these instances commonly resulted in integrative resolutions.

Peer Intervention and Mediation

As early as the preschool years peers support one another in conflicts, and such triadic support has been linked to group hierarchy and social status, as well as to friendship (Strayer & Noel 1986; Grammer 1992). When asked, preschool-age children are not always able to identify coalitions and alliances within their peer group (Strayer et al. 1980); however, once they reach middle childhood, children are quite capable of telling who supports whom in peer conflicts (Loots 1985).

We still know relatively little about the effects of peer intervention on resolution. Most of the available evidence associates peer intervention with distributive resolutions (e.g., Strayer & Noel 1986). A recent observational study of preschool conflict (Verbeek 1997) showed, however, that although the initiator and recipient of triadic conflict more often than not failed to reconcile their differences, friendly post-conflict reunions between conflict recipients and supporters were

not uncommon. In fact, such post-conflict peacemaking was especially likely when the supporter and the recipient interacted prior to the conflict. The strategies of the supporters in this study are intriguing as they combine coercion (in support of the initiator) with post-conflict peacemaking (with the recipient). Further research should focus on the cognitive and social-motivational correlates of these triadic strategies in young children and expand this line of research to older children and adolescents.

Little is known about peer mediation, that is, impartial intervention aimed at integrative resolution. Questions such as whether the rate of occurrence changes with age, and whether peer mediation increases the likelihood of integrative resolution, remain largely unanswered. There is some evidence to suggest that peer mediation may be associated with cultural values and expectations. Butovskaya et al. (Chapter 12) report that among Kalmyk and Russian children mediation of peer conflict is perceived as both desirable and honorable. Moreover, naturalistic observations of peer conflict showed that peer mediation occurred significantly more often among Kalmyk and Russian children than among Swedish and U.S. children in two comparable studies (see also Petit & Thierry, Box 13.1, for a discussion of impartial intervention among nonhuman primates).

Teacher Mediation of Peer Conflict

Debates about whether or not teachers should mediate peer conflict of children, and what form such mediation should take, continue to divide teachers and parents alike. The long-standing public interest in teacher mediation of children's conflict has not been adequately matched by research; well-designed studies on this topic remain in short supply.

That teacher mediation not always results in the desired outcome is illustrated by a recent study on conflict management in preschoolers in which it was found that only 8 percent of former opponents continued to associate after being told by teachers "to make peace," compared with 35 percent after an independently resolved dispute (Verbeek & Creveling unpublished data). Adult mediation thus appeared more of a hindrance than a facilitation for peaceful post-conflict interaction.

The limited body of systematic research on teacher mediation of children's peer conflict suggests that in order to be effective teacher mediation must match the causal domain (e.g., moral vs. social) and must be age- and skill-appropriate. Studies by Nucci (1984) and Killen et al. (1994) on teacher mediation of peer conflict have shown that children as young as preschool age are already quite capable of perceiving the relative merit of adult mediation in that they prefer domain-appropriate over domain-inappropriate teacher explanations during mediation.

Bayer et al. (1995) showed that teacher mediation is likely to be most effective when it is targeted within the "zone of proximal development" (Vygotsky 1978), that is, within the limits constituted by children's age-dependent ability to negotiate resolution independently and the outcome they may achieve with the expert assistance of an adult. Malloy & McMurray's (1996) recent study on peer conflict in an integrated preschool classroom suggests that teachers need to utilize different mediation strategies for typically developing children and children with disabilities according to their levels of aggression, communication, perceptual motor ability, and cognitive understanding. Finally, and notwithstanding Malloy & McMurray's findings, Lieber (1994) presents evidence for the fact that even children with a variety of disabilities are often able to manage their conflicts without adult intervention and reiterates that adult mediation strategies should be aimed at helping children manage their conflicts

rather than taking over conflict management from them (cf. Perlman & Ross 1997).

Socioeconomic and Cultural Influences

Socioeconomic Influences

On any given night three-quarters of a million people in the United States, and an estimated three million in the European Union, are homeless, and many of the documented homeless are children, including the very young (UNICEF 1998). Whereas poverty and homelessness among youth were once primarily associated with developing nations, recent statistics show that both are on the rise in Western societies. We still know little about the effects of poverty and homelessness on peer conflict management, but a handful of studies suggest that homeless youth, in particular, may be at a developmental disadvantage compared with their middle-class counterparts.

Several U.S. studies show that homeless children's lack of status and material possessions mediates their rejection by classmates (Gewirtzman & Fodor 1987; Horowitz et al. 1988). Peer conflict may be more frequent among homeless children than in poor, inner-city children with a home (Molnar et al. 1991). Homeless adolescents cited peer conflict as the worst problem facing them on a daily basis (Horowitz et al. 1994). The homeless adolescents in this study frequently appealed to their mothers and teachers for assistance in managing conflict with peers. This is in stark contrast to adolescents from intact, middle-class homes who commonly manage peer conflict without adult intervention.

In summary, homeless adolescents' rejection by their peers, anxiety about frequent peer conflict, and insecurity about how to manage it may have a compounded detrimental effect on their social development and may put them at risk for future social maladjustment and psychopathology (cf. Coie et al. 1992; Coie & Cillessen 1993).

Cultural Influences

Progress toward a better understanding of peer conflict management is hampered by the scant supply of comparative evidence (but see Butovskaya et al., Chapter 12). The relative shortage of both within-culture and between-cultures comparisons limits our ability to make inferences with respect to the universality of developmental patterns observed among primarily Western middle-class youth. Moreover, attempts to isolate the influences of cultural values and expectations on peer conflict management are often limited by an inherent confound between culture and socio-economic status (see above).

Some of the available evidence suggests that the cultural expectations and values associated with different modes of subsistence may affect how children learn to manage conflict with peers. For instance, a within-culture comparison of Mexican children from three distinct ecocultural niches—a commerce-oriented small town, an industrialized urban city, and a small agricultural rural town—produced significant differences in peer conflict management preferences (Kagan et al. 1981). Children from the small commercial town were more competitive in their approach to peer conflict than children from the other two settings. This was illustrated by their clear preference for continued conflict (67 percent of total responses) over disengagement (13 percent). Interestingly, the children from the industrialized urban and the agricultural rural town did not differ in their preferences: both groups showed only a slight preference for continued conflict over disengagement. This latter pattern seems more in line with the conflict management strategies of urban Western middle-class children we mentioned earlier. Sim-

ilar findings come from studies conducted in southern Mexico (Fry 1988). Peer aggression was significantly more prominent among children from a competitive and violent Zapotec community than among children from a neighboring Zapotec community described as generally peaceful. Taken together, these findings suggest that there is a link between niche-specific adult patterns of interpersonal conflict and children's conflict management with peers. These findings also suggest that a simple urban-rural dichotomy may be insufficient to explain the effects of town ecology on peer aggression and conflict management.

Contemporary Western society is characterized by increasing ethnic diversity. Ogbu (1993) proposed that there are "primary" and "secondary" cultural differences among the various ethnic groups in Western societies. Primary cultural differences arise from the fact that members of two populations had their own ways of behaving and thinking before they came in continuous contact with each other. Secondary cultural differences arise in part as coping mechanisms used by minorities to deal with the problems they face in their relationships with members of the dominant group and the societal institutions controlled by the latter.

Studies conducted in the early 1970s suggest that children of a Mexican American cultural background tend to favor integrative resolutions more than Anglo-American children their age (Madsen & Shapira 1970; Kagan & Madsen 1971; McClintock 1974). More comparisons of peer conflict management among Mexican American and Anglo-American children living in the same suburb of a large southern California city confirm these earlier results (Khoram 1994). That is, Anglo-American children reported using power assertion as a conflict termination strategy significantly more often than their Mexican American counterparts. Moreover, the Anglo-American children perceived peer conflict to end in distributive outcomes more than the Mexican American children.

When we adopt Ogbu's (1993) theoretical framework, we are left with the question of whether these fairly robust differences between Mexican American and Anglo-American children are best explained as being "primary" or "secondary" in origin. Considering the evidence of within-culture differences in peer conflict management in the Mexican American children's ancestral country, the case for primacy appears weak. Alternatively, constructive conflict management may be part and parcel of the way Mexican American children cope with their minority status. Future research should focus on the relational significance of the conflict management style of Mexican American children, with regard to the maintenance of peer relations within as well as outside their own ethnic groups.

In summary, within-culture and between-cultures comparisons suggest that peer conflict management is not independent from niche-specific adult patterns of competition and cooperation. Future research in this general area should focus on the link between modes of subsistence (including distribution of resources) and adult patterns of interpersonal conflict management. In addition, we need to learn more about the mechanisms of cross-generational transmission of niche-specific patterns of conflict management. For instance, we need to determine whether transmission takes place through observational learning,

concerted efforts by adults (i.e., instructional learning), or both.

Conclusion

Two kinds of relationships figure prominently in children's lives: "vertical" attachments to adults and "horizontal" relationships with peers. Conflict is inherent in these relationships, and managing conflict with familiar partners provides important socializing experiences.

Certain kinds of conflict management strategies support and maintain children's relationships in and outside the home. Moreover, relationship experience, in turn, contributes to the development of skill in effectively managing conflicts, and effective conflict management is correlated with a variety of developmental indicators.

Adult mediation is most effective when it is explicit, as well as age- and domain-appropriate. Moreover, adult mediators who consider the specific nature of the opponents' relationship can be expected to be most effective.

Socioeconomic and cultural factors affect conflict management style and as such affect children's relationships and development. Especially telling are findings that suggest that homelessness may deprive youth of the opportunity to establish normal relationships and acquire the proper skills to manage interpersonal conflict. Clearly, societies should take steps to ensure that all children have equal opportunity to go through the formative experience of bilateral conflict management.

References

Anderson, H. H. 1937. An experimental study of dominative and integrative behavior in children of preschool age. *Journal of Social Psychology*, 8: 335–345.

Azmitia, M., & Montgomery, R. 1993. Friendship, transactive dialogues, and the development of scientific reasoning. *Social Development*, 2: 202–221.

Barker, R. G., & Wright, H. F. 1955. *Midwest and Its Children*. New York: Harper.

Bayer, C. L., Whaley, K. L., & May, S. E. 1995. Strategic assistance in toddler disputes: II. Sequences and patterns of teachers' message strategies. *Early Education and Development*, 6: 405–432.

Berndt, T. J. 1981. Effects of friendship on prosocial intentions and behavior. *Child Development*, 52: 636–643.

Berndt, T. J., & Savin-Williams, R. C. 1993. Peer relations and friendships. In: *Handbook of Clinical Research and Practice with Adolescents* (P. H. Tolan & B. J. Cohler, eds.), pp. 209–219. New York: Wiley.

Boer, F., & Dunn, J. 1992. *Children's Sibling Relationships*. Hillsdale, N.J.: Lawrence Erlbaum Associates.

Boulton, M. J., & Smith, P. K. 1990. Affective bias in children's perceptions of dominance relationships. *Child Development*, 61: 221–229.

Braine, L. G., Pomerantz, E., Lorber, D., & Krantz, D. H. 1991. Conflicts with authority: Children's feelings, actions, and justifications. *Developmental Psychology*, 27: 829–840.

Bryant, B. K. 1992. Conflict resolution strategies in relation to children's peer relations. *Journal of Applied Developmental Psychology*, 13: 35–50.

Charlesworth, W. R. 1996. Co-operation and competition: Contributions to an evolutionary and developmental model. *International Journal of Behavioral Development*, 19: 25–39.

Charlesworth, W. R., & La Freniere, P. 1983. Dominance, friendship, and resource utilization in preschool children's groups. *Ethology and Sociobiology*, 4: 175–186.

Chung, T., & Asher, S. R. 1996. Children's goals and strategies in peer conflict situations. *Merrill-Palmer Quarterly*, 42: 125–147.

Clifford, E. 1959. Discipline in the home: A controlled observational study of parental practices. *Journal of Genetic Psychology*, 95: 45–82.

Coie, J. D., & Cillessen, A. H. N. 1993. Peer rejection: Origins and effects on children's development. *Current Directions in Psychological Science*, 2: 89–92.

Coie, J. D., & Dodge, K. A. 1983. Continuities and changes in children's social status. A 5-year longitudinal study. *Merrill-Palmer Quarterly*, 29: 261–282.

Coie, J. D., Dodge, K. A., & Coppotelli, H. 1982. Dimensions and types of social status: A cross-age perspective. *Developmental Psychology*, 18: 557–570.

Coie, J. D., Lochman, J. E., Terry, R., & Hyman, C. 1992. Predicting early adolescent disorder from childhood aggression and peer rejection. *Journal of Consulting and Clinical Psychology*, 60: 783–792.

Denham, S. A., & Holt, R. W. 1993. Preschoolers' likability as cause or consequence of their social behavior. *Developmental Psychology*, 29: 271–275.

de Waal, F. B. M. 1986. Integration of dominance and social bonding in primates. *Quarterly Review of Biology*, 61: 459–479.

Dix, T., Ruble, D., Grusec, J., & Nixon, S. 1986. Social cognition in parents: Inferential and affective reactions to children of three age levels. *Child Development*, 57: 879–894.

Dodge, K. A., Coie, J. D., Pettit, G. S., & Price, J. M. 1990. Peer status and aggression in boys' groups: Developmental and contextual analysis. *Child Development*, 61: 1289–1309.

Dunn, J., & Munn, P. 1986. Siblings and the development of prosocial behaviour. *International Journal of Behavioral Development*, 9: 265–284.

Eckerman, C. O., Whately, J. L., & McGehee, L. J. 1979. Approaching and contacting the object another manipulates: A social skill of the 1-year-old. *Developmental Psychology*, 15: 585–593.

Fisher, C. B., & Johnson, B. L. 1990. Getting mad at mom and dad: Children's changing views of family conflict. *International Journal of Behavioral Development*, 13: 31–48.

Fonzi, A., Schneider, B. H., Tani, F., & Tomada, G. 1997. Predicting children's friendship status from their dyadic interaction in structured situations of potential conflict. *Child Development*, 68: 496–506.

Fraysee, J. C. 1995. Combined effects of friendship and stage of cognitive development on interactive dynamics. *Journal of Genetic Psychology*, 155: 161–177.

French, D. C., & Waas, G. A. 1987. Social-cognitive and behavioral characteristics of peer-rejected boys. *Professional School Psychology*, 2: 103–112.

Fry, D. P. 1988. Intercommunity differences in aggression among Zapotec children. *Child Development*, 59: 1008–1019.

Gewirtzman, R., & Fodor, I. 1987. The homeless child at school: From welfare hotel to classroom. *Child Welfare*, 66: 237–245.

Gjerde, P. 1986. The interpersonal structure of family interaction settings: Parent-adolescent relations in dyads and triads. *Developmental Psychology*, 22: 297–304.

Goodenough, F. L. 1931. *Anger in Young Children*. Minneapolis: University of Minnesota Press.

Goodwin, M. H. 1982. Processes of dispute management among urban black children. *American Ethnologist*, 9: 76–96.

Gottman, J. M. 1983. How children become friends. *Monographs of the Society for Research in Child Development*, 48 (2, Serial No. 201).

Grammer, K. 1992. Intervention in conflicts among children: Contexts and consequences. In: *Coalitions and Alliances in Humans and Other Animals* (A. H. Harcourt & F. B. M. de Waal, eds.), pp. 258–283. Oxford: Oxford University Press.

Grotevant, H., & Cooper, C. 1986. Individuation in family relationships. *Human Development*, 29: 82–100.

Harkness, S., & Super, C. M. 1985. The cultural context of gender segregation in children's peer groups. *Child Development*, 56: 219–224.

Hartup, W. W. 1983. Peer relations. In: *Socialization, Personality, and Social Development* (P. H. Mussen, gen. ed.), *Handbook of Child Psychology*, Vol. 4 (E. M. Hetherington, ed.), pp. 103–196. New York: Wiley.

Hartup, W. W. 1992. Conflict and friendship relations. In: *Conflict in Child and Adolescent Development* (C. U. Shantz & W. W. Hartup, eds.), pp. 186–215. Cambridge: Cambridge University Press.

Hartup, W. W. 1996. The company they keep: Friendships and their developmental significance. *Child Development*, 67: 1–13.

Hartup, W. W., & Stevens, N. 1997. Friendships and adaptation in the life course. *Psychological Bulletin*, 121: 355–370.

Hartup, W. W., Laursen, B., Stewart, M. I., & Eastenson, A. 1988. Conflict and the friendship relations of young children. *Child Development*, 59: 1590–1600.

Hartup, W. W., French, D. C., Laursen, B., Johnston, K. T., & Ogawa, J. R. 1993. Conflict and friendship relations in middle childhood: Behavior in a closed-field situation. *Child Development*, 64: 445–454.

Hay, D. F. & Ross, H. S. 1982. The social nature of early conflict. *Child Development*, 53: 105–113.

Hetherington, E. M. 1988. Parents, children, siblings: Six years after divorce. In: *Relationships within Families: Mutual Influences* (R. A. Hinde & J. Stevenson-Hinde, eds.), pp. 311–331. New York: Oxford University Press.

Hill, J. P. 1988. Adapting to menarche: Familial control and conflict. In: *Development during the Transition to Adolescence: Minnesota Symposia on Child Psychology*, Vol. 21 (M. R. Gunnar & W. A. Collins, eds.), pp. 43–77. Hillsdale, N.J.: Lawrence Erlbaum Associates.

Hill, J. P., & Holmbeck, G. N. 1987. Disagreements about rules in families with seventh-grade girls and boys. *Journal of Youth and Adolescence*, 16: 221–246.

Hoffman, M. L. 1988. Moral development. In: *Developmental Psychology: An Advanced Textbook*, 2nd edn. (M. H. Bornstein & M. E. Lamb, eds.), pp. 497–548. Hillsdale, N.J.: Lawrence Erlbaum Associates.

Hoffman, M. L., & Saltzstein, H. D. 1967. Parent discipline and the child's moral development. *Journal of Personality and Social Psychology*, 5: 45–47.

Horowitz, S. V., Springer, C. M., & Kose, G. 1988. Stress in hotel children: The effects of homelessness on attitudes toward school. *Children's Environments Quarterly*, 5: 34–36.

Horowitz, S. V., Boardman, S. K., & Redlener, I. 1994. Constructive conflict management and coping in homeless children and adolescents. *Journal of Social Issues*, 50: 85–98.

Inoff-Germain, G., Arnold, G. S., Nottleman, E. D., Susman, E. J., Cutler, G. B., Jr., & Chrousos, G. P. 1988. Relations between hormone levels and observational measures of aggressive behavior of young adolescents in family interactions. *Developmental Psychology*, 24: 129–139.

Jones, D. C. 1985. Persuasive appeals and responses to appeals among friends and acquaintances. *Child Development*, 56: 757–763.

Kagan, S., & Madsen, M. 1971. Cooperation and competition of Mexican, Mexican-American and Anglo-American children of two ages under four instructional sets. *Developmental Psychology*, 5: 32–39.

Kagan, S., Knight, G. P., Martinez, S., & Santana, P. E. 1981. Conflict resolution style among Mexican children. *Journal of Cross-Cultural Psychology*, 12: 222–232.

Kelley, H. H., Berscheid, E., Christensen, A., Harvey, J. H., Huston, T. L., Levinger, G., McClintock, E., Peplau, L. A., & Peterson, D. R., eds. 1983. *Close Relationships*. New York: Freeman.

Khoram, A. 1994. A comparative study of conflict resolution among Mexican-American and Anglo-American children (doctoral dissertation, California School of Professional Psychology). *Dissertation Abstracts International*, 55: 596B.

Killen, M., Breton, S., Ferguson, H., & Handler, K. 1994. Preschoolers' evaluations of teacher methods of intervention in social trangressions. *Merrill-Palmer Quarterly*, 40: 399–415.

La Freniere, P. L., & Charlesworth, W. R. 1983. Dominance, attention, and affiliation in a preschool group: A nine-month longitudinal study. *Ethology and Sociobiology*, 4: 55–67.

La Freniere, P. L., & Charlesworth, W. R. 1987. Effects of friendship and dominance status on preschoolers' resource utilization in a cooperative/competitive situation. *International Journal of Behavioral Development*, 10: 345–358.

Ladd, G. W., Price, J. M., & Hart, G. H. 1988. Predicting preschoolers' peer status from their playground behaviors. *Child Development*, 59: 986–992.

Laursen, B. 1989. Relationships and conflict during adolescence. Doctoral dissertation, University of Minnesota.

Laursen, B. 1993. Conflict management among close peers. In: *Close Friendships in Adolescence* (B. Laursen, ed.), pp. 39–54. San Francisco: Jossey-Bass.

Laursen, B., & Collins, W. A. 1994. Interpersonal conflict during adolescence. *Psychological Bulletin*, 115: 197–209.

Laursen, B., Coy, K. C., & Collins, W. A. 1998a. Reconsidering changes in parent-child conflict across adolescence: A meta-analysis. *Child Development*, 69: 817–832.

Laursen, B., Betts, N. T., & Finkelstein, B. D. 1998b. The resolution of conflict with peers: A developmental meta-analysis. Manuscript, Florida Atlantic University.

Lieber, J. 1994. Conflict and its resolution in preschoolers with and without disabilities. *Early Education and Development*, 5: 5–17.

Littlefield, L., Love, A., Peck, C., & Wertheim, E. H. 1993. A model for resolving conflict: Some theoretical, empirical and practical implications. *Australian Psychologist*, 28: 80–85.

Loots, G. M. P. 1985. Social relationships in groups of children: An observational study. Doctoral dissertation, Free University, Amsterdam.

Maccoby, E. E. 1984. Middle childhood in the context of the family. In: *Development during Middle Childhood: The Years from Six to Twelve* (W. A. Collins, ed.), pp. 184–239. Washington, D.C.: National Academy of Sciences Press.

Maccoby, E. E. 1996. Peer conflict and intrafamily conflict: Are there conceptual bridges? *Merrill-Palmer Quarterly*, 42: 165–176.

Maccoby, E. E., & Martin, J. A. 1983. Socialization in the context of the family: Parent-child interaction. In: *Handbook of Child Psychology*, Vol. 4 (P. H. Mussen, ed.), pp. 1–101. New York: Wiley.

Madsen, M., & Shapira, A. 1970. Cooperative and competitive behavior of urban Afro-American, Anglo-American, Mexican-American and Mexican village children. *Developmental Psychology*, 3: 16–20.

Malloy, H. L., & McMurray, P. 1996. Conflict strategies and resolutions: Peer conflict in an integrated early childhood classroom. *Early Childhood Research Quarterly*, 11: 185–206.

Matsumoto, D., Haan, N., Yabrove, G., Theodorou, P., & Carney, C. C. 1986. Preschoolers' moral actions and emotions in prisoners' dilemma. *Developmental Psychology*, 22: 663–670.

McClintock, C. G. 1974. Development of social motives in Anglo-American and Mexican-American children. *Journal of Personality and Social Psychology*, 29: 348–354.

Molnar, J., Rath, W. R., Klein, T. P., Lowe, C., & Hartmann, A. H. 1991. *Ill Fares the Land: The Consequences of Homelessness and Chronic Poverty for Children and Families in New York City*. New York: Bank Street College of Education.

Montemayor, R., & Hanson, E. 1985. A naturalistic view of conflict between adolescents and their parents and siblings. *Journal of Early Adolescence, 5:* 23–30.

Nelson, J., & Aboud, F. E. 1985. The resolution of social conflict between friends. *Child Development, 56:* 1009–1017.

Newcomb, A. F., & Bagwell, C. 1995. Children's friendship relations: A meta-analytic review. *Psychological Bulletin, 117:* 306–347.

Newson, J., & Newson, E. 1968. *Four Years Old in an Urban Community.* Chicago: Aldine.

Newson, J., & Newson, E. 1976. *Seven Years Old in the Home Environment.* New York: Wiley.

Nucci, L. P. 1984. Evaluating teachers as social agents: Students' ratings of domain appropriate and domain inappropriate teacher responses to transgressions. *American Educational Research Journal, 21:* 367–378.

Ogbu, J. U. 1993. Differences in cultural frame of reference. *International Journal of Behavioral Development, 16:* 483–506.

Patterson, C. J., Kupersmidt, J. B., & Griesler, P. C. 1990. Children's perceptions of self and of relationships with others as a function of sociometric status. *Child Development, 61:* 1335–1349.

Patterson, G. R. 1982. *Coercive Family Process.* Eugene, Oreg.: Castalia Publishing Co.

Perlman, M., & Ross, H. S. 1997. The benefits of parent intervention in children's disputes: An examination of concurrent changes in children's fighting styles. *Child Development, 64:* 690–700.

Putallaz, M., & Sheppard, B. H. 1990. Children's social status and orientations to limited resources. *Child Development, 61:* 2022–2027.

Radke-Yarrow, M., Richters, J., & Wilson, W. E. 1988. Child development in a network of relationships. In: *Relationships within Families: Mutual Influences* (R. A. Hinde & J. Stevenson-Hinde, eds.), pp. 48–67. New York: Oxford University Press.

Raffaelli, M. 1990. Sibling conflict in early adolescence. Doctoral dissertation, University of Chicago.

Renshaw, P., & Asher, S. R. 1983. Children's goals and strategies for social interaction. *Merrill-Palmer Quarterly, 29:* 353–374.

Rizzo, T. A. 1989. *Friendship Development among Children in School.* Norwood, N.J.: Ablex.

Roberts, G. C., Block, J. H., & Block, J. 1984. Continuity and change in parents' child-rearing. *Child Development, 55:* 586–597.

Santos, A. J., & Strayer, F. F. 1997. A socio-structural analysis of preschool children's affiliative behavior. Paper presented at the XXV International Ethological Conference, Vienna, Austria, August.

Schoenleber, K. L. 1988. Parental perceptions and expectations and their relationship to parent-child conflict and parental satisfaction during the transition to adolescence. Doctoral dissertation, University of Minnesota.

Selman, R. L., & Schultz, L. H. 1989. Children's strategies for interpersonal negotiation with peers: An interpretive/empirical approach to the study of social development. In: *Peer Relationships in Child Development* (T. J. Berndt & G. W. Ladd, eds.), pp. 371–406. New York: Wiley.

Smetana, J. G. 1989. Adolescents' and parents' reasoning about actual family conflict. *Child Development, 60:* 1052–1067.

Smetana, J. G. 1991. Adolescents' and mothers' evaluations of justifications for conflicts. In: *Shared Views in the Family during Adolescence: New Directions for Child Development* (No. 51) (R. L. Paikoff, ed.), pp. 71–86. San Francisco: Jossey-Bass.

Sternberg, R. J., & Dobson, D. M. 1987. Resolving interpersonal conflicts: An analysis of stylistic consistency. *Journal of Personality and Social Psychology, 52:* 794–812.

Strayer, F. F. 1992. The development of agonistic and affiliative structures in preschool play groups. In: *Aggression and Peacefulness in Humans and Other Primates* (J. Silverberg & P. Gray, eds.), pp. 150–171. Oxford: Oxford University Press.

Strayer, F. F., & Noel, J. M. 1986. The prosocial and antisocial functions of preschool aggression: An ethological study of triadic conflict among young children. In: *Altruism and Aggression* (C. Zahn-Waxler, M. Cummings, & R. Iannotti, eds.), pp. 107–134. Cambridge: Cambridge University Press.

Strayer, F. F., Strayer, J., & Chapeskie, T. R. 1980. The perception of social power relations among preschool children. In: *Dominance Relations* (D. R. Omark, F. F. Strayer, & D. Freedman, eds.), pp. 191–203. New York: Garland.

Strayer, J. 1976. An ethological analysis of social agonism and dominance relations among preschool children. *Child Development*, 47: 980–989.

Sullivan, H. S. 1953. *The Interpersonal Theory of Psychiatry*. New York: Norton.

UNICEF. 1998. *The Progress of Nations*. New York: Author.

Verbeek, P. 1997. Peacemaking of young children (doctoral dissertation, Emory University, 1996). *Dissertation Abstracts International*: Section B: the Sciences & Engineering, Vol. 57, 7253.

Vespo, J. E., & Caplan, M. Z. 1993. Preschoolers' differential conflict behavior with friends and acquaintances. *Early Education and Development*, 4: 45–53.

Vuchinich, S., Emery, R. E., & Cassidy, J. 1988. Family members as third parties in dyadic family conflict: Strategies, alliances, and outcomes. *Child Development*, 59: 1293–1302.

Vygotsky, L. S. 1978. *Mind in Society*. Cambridge: Harvard University Press.

Whiting, B., & Whiting, J. 1975. *Children of Six Cultures: A Psycho-cultural Analysis*. Cambridge: Harvard University Press.

Youniss, J., & Smollar, J. 1985. *Adolescent Relations with Mothers, Fathers, and Friends*. Chicago: University of Chicago Press.

CHAPTER 4

Law, Love, and Reconciliation

Searching for Natural Conflict Resolution in Homo Sapiens

Douglas H. Yarn

A Medieval Tale of Reconciliation

The time is the late thirteenth century; the place, England. Geoffrey and Edward, (very) minor nobility with adjoining plots of land, were disputing. How it all began is unclear, but it did involve some wandering grazing cattle, the building of a stone wall, and the subsequent demolition of the same. The escalation of events caused great consternation to Geoffrey, who considered Edward to have always been a reliable and supportive neighbor. They had assisted each other in clearing fields, shared tools, sat at each other's tables, and shared food. They had known each other all their lives, as had their fathers and their fathers' fathers. Their wives were distantly related. To complicate matters further, Edward was a big man known to have a violent and quick temper. Despite his fears and misgivings, Geoffrey confronted Edward in the village one day. Angry words were exchanged, and they would have come to blows had the villagers not intervened. The dispute became the talk of the village, and the community itself split as members sided with the antagonists. Friends and family discouraged them from violence and entreated them not to refer the matter to law. Eventually, the local priest called for a "loveday," and Geoffrey and Edward agreed. Each selected a friend, and together these two friends retired to discuss the matter and devise a mutually acceptable solution. After several failed attempts, a compromise was reached. The former disputants embraced and were reconciled. An audible sigh of relief arose from the village.

A Modern Tale of Litigation

The time is the late twentieth century, somewhere in the United States. Jeff and Ed are neighbors in a new subdivision in a suburb of a large metropolis. Both are midlevel executives at their companies; they have moved to the neighborhood from other places and probably will move again in a few years. In the brief period that they have lived next to each other, they have rarely seen each other, much less exchanged pleasantries. Their wives have met briefly but are too busy with their own jobs and car pooling to interact. The exact nature of the argument regarding their adjoining properties is unimportant, but suffice it to say, the dispute involved roving dogs, brown patches on expensive sod, the erection of a fence, and sediment from drainage. Jeff worried about his property value for resale and was a little uncomfortable about confronting his neighbor, who seemed rather surly, but he registered his complaint on Ed's answering machine, asking him to correct the situation. The next day he received an E-mail response, "So sue me," and he did. Their neighbors neither knew nor cared about the dispute. Jeff contacted a lawyer who drafted a complaint and filed suit against Ed in the local county court. The court docket was so backlogged that the matter would languish for several years while lawyers for Jeff and Ed exchanged interrogatories, took depositions, made motions, and billed their respective clients. Eventually, the case was put on the trial calendar, a jury empaneled, and the evidence presented over a period of several days. Finally, the jury retired and returned delivering a verdict in favor of Ed. Furious at the verdict and at his lawyer, whose final bill tipped the scales, Jeff appealed. While waiting for the appellate court to hear the matter, Ed was transferred to another city and moved.

There is another possible scenario for our tale of modern litigation. It begins in much the same way, with Jeff seeking the assistance of an attorney and filing a suit in court. However, the court orders Jeff and Ed to participate in mediation in an attempt to resolve the matter. During the medi-

ation, Jeff discovers that Ed is not such a bad guy after all, and for the first time Ed understands Jeff's complaint and believes it would not be unreasonable to rectify the situation in return for Jeff's cooperation on another matter. In addition, they discover a common interest in golf. The mediation concludes with an agreement, the suit is dropped, and the neighbors shake hands and agree on a tee time.

What's in These Tales (and in This Chapter)?

These tales, in both their similarities and differences, illustrate some of the different historical and current methods by which we, *Homo sapiens sapiens*, manage and resolve interpersonal conflict. Like other social species, we employ myriad mechanisms. With respect to our own conflicts, we fight, threaten, cajole, entreat, appease, submit, avoid, or withdraw, just to name a few. With respect to the conflicts of others, we ignore, console, join sides, or intervene in various partisan or nonpartisan ways. More than any other species, however, we have developed elaborate social institutions for the expression, management, and resolution of conflict. Using the general organization of this book as a guide, this chapter explores one of the most elaborate of these institutions, law, and its historical and current relationship to conciliatory methods of handling conflict. After defining and contrasting law and these other methods, we return to the historical record, which reveals shifts in the social and institutional arrangements supporting different conflict handling mechanisms and which indicates that each has its own unique ethos.

Law: Controlling Aggression

Law is used to control aggression through the threat or actual use of force. In the context of handling interpersonal conflict, law is an articulated set of norms (laws) with the mechanisms to enforce those norms or to apply them in a given situation (Malinowski 1932 [1926]). In Western societies, a criminal trial or civil litigation is the dominant method of applying laws to a given situation. These are *adjudicative* processes characterized by the use of a third party with the power to impose a solution or binding decision. In the common-law tradition of Anglo-American jurisprudence, the proceedings are formal and public, with an adversarial presentation usually by lawyers hired by the disputants to advocate on their behalf. The disputants have no say in the final decision, which will usually produce a clear winner and loser. The power of the state stands by with the threat of force to ensure compliance with the decision.

What is the relationship of biology to this definition of law? Biological concepts expressed through the notion of "natural law" have had a recurring presence in jurisprudence, and there is a deeply ingrained evolutionary tradition in American legal theory (Elliot 1985). Most of this tradition uses evolutionary concepts to explain how law and legal institutions "adapt" and change over time. An interesting, but misguided, example of this is Sir Henry Maine's (1861) attempt to show that the common-law system of the late nineteenth century was the apex of a natural progression emerging from familial patriarchy. In the past two decades, scholars have begun exploring the biological roots of legal institutions themselves (e.g., Gruter 1991). One line of speculation starts with the proposition that humans, like any other organism, seek individual reproductive success. This drive requires egoistic behavior; however, like many other social species, humans use groups to optimize their reproductive success. Since egoistic expressions such as aggression and dominance produce intragroup conflict, social groups were more successful when these behaviors were

balanced or moderated by altruistic and reconciliatory behaviors perhaps emanating from inclusive fitness and reciprocity principles. Moralistic aggression directed toward free riders, nonreciprocators, and individuals whose egoistic behavior went "too far" promoted this balance and served as a kind of protomorality, the roots of a "sense of justice" (Trivers 1971; Alexander 1987; de Waal 1991; Gruter & Masters 1992; McGuire 1994; de Waal 1996; Killen & de Waal, Chapter 17). This led to articulated norms and moral systems, to which one need only add a method of enforcement, an analogue of which can be found in the conflict intervention of high-ranking animals among some of our primate cousins with hierarchal societies (Goodall 1983; Boehm 1994; Petit & Thierry, Box 13.1), to get "law" as defined above. At its most basic level, this intervention paradigm of law relies on dominance/submission patterns of behavior.

Law is extremely limited when it comes to reconciliation. In fact, litigation does not really function to resolve conflict. Apply a simplified version of this concept of litigation to a simple conflict scenario used in much of the current conflict resolution literature today (see, e.g., Fisher & Ury 1981): Two sisters want the single orange in the household and begin arguing over who should get it. Each wants the orange (conflict of interest), but neither will willingly yield it to the other (incompatible behavior). If one were significantly stronger than the other, she might be able simply to take it by force, or the weaker might decide that satisfying her interests is not worth the effort in the face of potential injury and defeat. Presuming they are equally matched or that neither is willing to risk the physical or social costs of fighting or withdrawing, they have two choices—work it out or call for Mom. So one of the sisters calls out for Mother, who steps in and, after hearing their respective claims for the fruit, determines its fate based on a particular norm, for example, eldest first, finders-keepers, or equitable division. Mom leaves with the admonition, "Don't let me catch you fighting again."

To the extent that it relies on the articulation of a common set of enforceable norms, law may prevent aggression by alerting parties to the consequences of their future actions ("Mom will give it to my little sister and spank me if I take it from her, so why bother"). By setting expectations about normative relationships, law may encourage disputants to adjust their claims and settle; however, law does not prevent conflicts of interest from arising. While the existence of legally enforceable norms and the mechanism of litigation may encourage resolution, law does not address the conflict of interests directly. Rather, it imposes outcomes based on applicable norms to control one or more parties from acting on their interests or engaging in incompatible behavior. In what might be labeled a classic example of intervention by the dominant animal and submission by the lesser, Mom (the adjudicator) has controlled potentially disruptive aggression and imposed an outcome. Although she forced a peace, she did not resolve the underlying conflict of interests. Likewise, law does not seek to resolve the underlying conflict of interests but merely to end the disputing through the imposition of an outcome. In this sense, law is not designed to end conflict and is best categorized as conflict management rather than conflict resolution.

What about reconciliation between the two sisters? Although it is likely to occur at some point because their relationship is probably important to them (see de Waal, Chapter 2), the legal process has done little to promote reconciliation. In fact, one could argue that law stifles reconciliation by not focusing on resolution and by producing a loser—either one sister didn't get any orange, or both got less than they desired—which further

strains the relationship. One of the accepted truisms in law today is that the litigation process promotes adversarialism, increases hostility and animosity, and destroys any semblance of a relationship between the litigants. Other social, political, and economic forces may promote reconciliation, but law has made it increasingly more difficult.

ADR: Repairing the Damage

In sharp contrast to law, humans employ other conflict-handling mechanisms that more directly encourage resolution and thereby improve conditions for interpersonal reconciliation. Currently, lawyers refer to these processes as *alternative dispute resolution*, or *ADR*. ADR is a catchall phrase that includes any peaceful alternatives to litigation. ADR includes dyadic processes, primarily negotiation, and triadic processes, primarily mediation and some forms of arbitration (Yarn 1999). These *conciliatory* processes are characterized by voluntary, consensual participation; private, usually informal, proceedings under mutual control of the disputants; and the application of norms selected by the parties or the creation of relevant norms for the situation. Conciliatory processes achieve the settlement of disputes by adjustment or compromise between the claims, interests, and demands of the parties. *Compromise* implies participation and choice in the solution, which by implication is a "win/win" solution providing something for all the disputants.

Using the conciliatory approach, the sisters might agree to split the orange or trade off in some fashion. They may find that they do not have incompatible interests after all—for example, one wants the juice and the other the rind. Their mother may help them find a solution to their conflict of interests or provide consolation. Thus, a conflict of interests is more amenable to resolution

TABLE 4.1
Comparison of Characteristics of Law and ADR

	Law	ADR
CONFLICT-RESOLUTION APPROACH	dissensus: focus on legal rights and principles	consensus: focus on needs and interests
PRIMARY TRIADIC PROCESS	adjudication: third party determines the outcome	mediation: third party helps parties determine the outcome
SOCIAL OBJECTIVE	control egoistic aggression	preserve social relationships
DEPENDENT UPON	societal norms and power to enforce	social interdependence
BEHAVIORAL PARADIGM	dominance-submission/interference	mutualism, cooperation, altruism

by a conciliatory process. Once the conflict of interests is resolved and both disputants have satisfied their needs, the sisters are more likely to restore their relationship.

What are the biological roots of these mechanisms? Are they the same as for law? At this point, these questions are unanswered. Most sociobiological theory focuses on the roots of moral and ethical behavior (e.g., de Waal 1996) and has yet to explore explicitly the possibility of distinct roots. While they may be two sides of the same coin serving the same goal of intragroup cohesion, the social expressions of law and ADR reflect some fundamental differences. Whereas law responds to egoistic behavior by relying on a pattern of dominance-submission/interference, ADR relies on patterns of mutualism, cooperation, and altruistic behavior. Law sanctions the offender and threatens sanctions to deter potential offenders, but ADR depends on the willingness of the disputants to resolve their own problem. It mobilizes their (dare I say it) innate desire to cooperate and perhaps reconcile (Table 4.1).

Although ADR's appeal is its potential for conflict resolution and reconciliation, it may be ineffective for handling conflict in certain settings. A slight modification to the problem of the orange neatly illustrates the weakness. Instead of two sisters, two unrelated people at a grocery store in a large urban area need a bag of oranges to make a fruit salad for guests that night. They begin arguing over the last available bag. The grocer begs them to be reasonable and to compromise; however, the bag has equal utility for each, neither appears motivated by some nonreciprocal altruism to forfeit their claim, and they have no relationship over which to find trade-offs or to encourage compromise (cf. de Waal, Chapter 2; Cords & Aureli, Chapter 9; van Schaik & Aureli, Chapter 15). Although one may succeed in acquiring the bag through physical or economic aggression (i.e., pay more for it), the conflict of interest cannot be easily resolved, and the dispute will continue unless there is authoritative intervention. Even between two people who are familiar with or related to each other, the animosity could be so great and the interests so in opposition that someone else must impose an outcome to prevent violence.

Triadic Affairs

In both law and ADR, third parties play vital roles (Table 4.1). In law, third parties include the adjudicator (judge and jury) and advocates (lawyers). The adjudicator is empowered by the broader society to decide the dispute. The hiring of an attorney is rather like mustering a coalition to increase one's dominance in the situation. In contrast, the third party in ADR (conciliator, mediator, or arbitrator) is empowered only to the extent that the disputants are willing to relinquish control over the process and outcome. The role of the third party, if any, is not to decide the dispute and thereby impose a settlement but rather to assist the parties in identifying a mutually agreeable solution. There are other third parties, of course, such as friends and family of the disputants who might have considerable effect on the perspectives and attitudes of the principals; however, those persons are more external to the legal and ADR processes and not relevant to this particular examination (see Fry, Chapter 16, for examples in other cultures).

The weakness in law's ability to reconcile might be traced to the nature of triadic interference in handling conflict, and the continuing legitimacy of the intervening adjudicator depends on perceived neutrality (see de Waal 1984 for this principle in chimpanzees). The use of a normative framework allows for neutral decision making and frees the adjudicator from having to stay "middle of the road" or from allying with one of the parties; however, it prevents him or her from handling a pure conflict of interests (Aubert 1963; Shapiro 1981). A third party in an adjudicative process will look backward over time to reconstruct events and classify them in terms of applicable rules or standards, that is, norms. This requires the parties to formulate the dispute as a dissensus, a dyadic conflict of rights and principles framed within the applicable normative structure, before the adjudicator can make a determination.

In contrast, the third party in a conciliatory process will seek to influence the parties to frame the dispute as a conflict of their respective interests in order to reach a mutually acceptable agreement through consensus (Aubert 1963; Eckhoff 1966). Lacking the power to impose an outcome, conciliators must rely on the willingness of the disputants to cooperate. Conciliators encourage cooperation by reminding the parties of their interdependence, by helping to reduce their emotional hostilities, and by improving understanding of each other's point of view (Yarn 1997). This type of triadic involvement is likely to promote reconciliation as it pursues conflict resolution (Table 4.1).

One Historical Socioecological Context

Law or ADR may be more efficient in handling disputes in different settings. An exploration of the historical relationship between law and conciliatory processes reveals social institutions that are sometimes complementary but always functionally distinct.

Law in the Time of Love

Although law does not function directly to resolve conflict or promote reconciliation, it has recognized the value of processes that lead toward reconciliation. This is illustrated by the medieval tale of Geoffrey and Edward's conflict. Several centuries before they had and resolved their dispute, the Anglo-Saxon king Ethelred (ca. A.D. 978–1008; *English Historical Documents* 1955, p. 402) decreed that "a thane has two choices, love or law, and he that chooses love is as much bound by that as he would be by a legal judgment." The Anglo-Saxon word for love, *lufu*, means "amicable settlement" in this context (Toller 1882). No doubt amicable settlement was preferable to some of the prevailing contemporary adversarial alternatives—violent self-help, trial by ordeal, or formal battle. Besides avoiding litigation, there was a social logic to this use of extrajudicial, conciliatory dispute resolution processes. Early English courts lacked independence from the sovereign, and the Crown itself often lacked the resources to enforce a system of justice that was primarily punitive and coercive. The survival of individuals, communities, and the state depended on the cooperation of local society in the administration of justice; therefore, dispute processing had to be sensitive to community cohesion. Reflecting this sensitivity, Henry I (ca. A.D. 1100–1135) decreed that disputants are either "brought together by love or separated by judgment" (Downer 1972, p. 100). In highly interdependent communities, legal judgment has the effect of separating elements of the community without resolving the underlying conflict between the disputants. Law courts, then and now, can make judgments, but they cannot make peace between the parties. Reconciliation was a more desirable outcome; therefore, conciliatory processes were preferable in the maintenance of community.

The early courts recognized the social value of processes that led toward reconciliation and would often refer cases to mediation or arbitration. The Kentish laws of Aethelberht (ca. A.D. 602–603) provided for the voluntary referral to arbitration of disputes over the proper compensation for leg injuries under the *wergeld*, an early form of workers' compensation (Attenborough 1963 [1922]). A clearer statutory support for arbitration is found in the Laws of Hlothhere and Eadric, kings of Kent (ca. 673–685), in which surety is used in conjunc-

tion with the referral and submission of the dispute to arbitrators (*English Historical Documents* 1955). In the stateless conditions of the early Middle Ages, not unlike today, trial was an expensive and risky affair. To make it possible for his subjects to avoid the "incerta penitus alea placitorum" [the utterly uncertain dice of pleas], Henry I declared that amicable settlement was superior to legal judgment ("Pactum legem vincit et amor iudicium"; *English Historical Documents* 1955, p. 164). This principle was echoed later by Glanvill in one of the earliest treatises on the common law: "generaliter verum est quod conventio legum vincit" [it is generally true that agreement prevails over law] (Hall 1965, p. 129). Henry I gave his judges the power to proceed as either conciliators or adjudicators: "all disputes which are brought to the notice of the shire court shall be settled thereat, either by amicable arrangement or by the rigour of judgment" (*English Historical Documents* 1955, p. 459). In describing judicial conduct in the eleventh and twelfth centuries, one legal historian notes: "What was usually expected of a law court was not a clear cut decision, of right or wrong, . . . but much more something in the nature of an effort to bring about settlement of the litigation by an acceptable, honourable compromise. This might be brought about by the mediation of the court . . . or jurors from the neighborhood or arbiters, accepted or even elected by the parties. . . . The feeling seems to have been that a real court decision of right or wrong, excluding anything that looked like a compromise, was a harsh and extreme measure" (Van Caenegem 1959, pp. 41–42). In addition, there was a widespread judicial acceptance of charters or *placitum* to create legal records of extralegal settlements (Wormald 1986). Thus in practice, love dominated law, and courts recognized and promoted compromise more often than granting judgment. "Lovedays" (*jours d'amour*) were an established institution for reconciliation. At the request of the disputants, courts would suspend their proceedings while members of the community facilitated settlement (Maitland & Baildon 1891).

The concept of Christian love and the efforts of the church to secure peace among fellow Christians influenced amicable settlement. Biblical references are numerous, including "love thy neighbor," "love thy enemies," "agree with thine adversary quickly" or he may deliver you to the judge and jail, and the advice of Paul to avoid lawsuits because they must be heard before non-Christian courts (Matt. 22:39 and Mark 12:31; Matt. 5:44; Matt. 5:25; 1 Cor. 6). Clerics were instructed to encourage and not to impede peaceful extrajudicial dispute resolution (Bennett 1958). The gloss on one thirteenth-century English canon states that it is the "duty of prelates to coerce both clerics and laymen to peace (pacem) rather than to Judgment (Iudicium)" (Lyndwood 1679, p. 72). Under early canon law, an arbitrator was expected to function in a conciliatory capacity and reestablish harmony between the litigants. As in the amicable agreement, the arbitrator relied on notions of equity to estimate how much one person might owe another and to suggest solutions to the dispute, but he could not give sentence. The outcome was an *arbitratus*, or transaction, as opposed to an *arbitrium*, or binding judgment given by an arbiter who followed formalities of court proceedings (Fowler 1976; David 1985). Some examples illustrate the emphasis on reconciliation. In a fifteenth-century example, Richard Asser, a barber, accused Nicholas Bradmor, "leech," of malpractice. Asser injured his thumb, which Bradmor undertook to heal. Asser accused Bradmor of negligence after the thumb festered and had to be removed. In the related arbitration, each chose one arbitrator and agreed to abide by their award. The award directed that the disputants "were to kiss each other" and for Bradmor to give Asser a gallon of wine, which they drank together (Sayles 1971, pp. 162–163).

In an earlier medical malpractice dispute, the abbot of Bourne complained that Robert Loke "undertook to cure and remove a sore called a wen which had arisen and grown on the face of the abbot." Apparently, Loke's treatment failed, and "their friends having intervened," the parties chose one arbitrator each who, "having heard the complaints and defenses of the parties . . . with diligent examination," awarded the abbot "ten marks, and that they should amicably embrace without thereafter having any action on either side" (Arnold 1987, pp. 53–54).

In the later part of the fifteenth century, two rival ancient guilds, the Skinners and the Taylors, after fighting over seniority during the lord mayor's procession on the Thames, agreed to abide by the judgment of the mayor, Robert Billesdon. The award provided that the guilds alternate precedence each year and, as their barges approached Westminster, lash them together and drink a toast (Hunter 1985).

Lawyers and the Decline of Conciliatory Processes

How did lawyers fit into this scheme? For the most part, they did not. The legal profession evolved separately from these conciliatory traditions. Historically, lawyers have been nurtured in a tradition that has a very limited perspective on dispute processing and the role of reconciliation. In England, there were no lawyers in the early days of judicial formation. By the time of Henry II (1154–1189), it was possible for a litigant to appoint someone, a *responsalis*, to do his technical pleading. Anyone could act in that capacity; however, by the thirteenth century, attorneys constituted a recognized but poorly organized and trained profession. There were no schools of common law, and the universities considered law too vulgar a subject for scholarly investigation. In 1292,

Edward I issued an edict directing the Common Pleas to choose qualified, trained persons to appear exclusively before the courts (in effect starting the legal monopoly that the profession enjoys today). The bar is a product of the courts that were burdened with the responsibility for its training. It emerged in response to the needs of the courts and of the litigants before those courts. The effect of putting the education of lawyers in the hands of the court was to isolate lawyers from a tradition of reconciliation as the preferred form of dispute resolution. Though lawyers might participate in conciliatory processes as respected members of the community, it was not an exclusive role. The author of one of the earliest treatises on arbitration observed that "most men either have been or may be Arbitrators, or at least have done, or may submit themselves to the Arbitration of others" (March 1647, title page). Since anyone could arbitrate or mediate and lawyers, though sometimes used, were not essential to conduct of the processes, the bar remained fixed in the orbit of the increasingly powerful common-law courts.

Consequently, the bar's history reflects a dearth of understanding as to the value of these processes and a reluctance to tolerate them. During the earliest days of the profession, lawyers were advised to abuse lovedays to gain tactical advantages in the litigation (e.g., de Bracton 1878–1883 [ca. 1268]), and by the seventeenth century arbitration submissions and awards were being vehemently contested in the courts. By the mid-seventeenth century, the courts of common law had greatly expanded their jurisdiction, lawyers were in increasing demand, and conciliatory processes were eroding under social and economic change. England was engulfed in a litigation explosion wherein persons of all classes initiated "frivolous suites, of trifling trialls, which a Common Yeoman were Iudge fitte enough to end in his chaire at home" (Garey 1623, p. 55).

Legal treatises on arbitrament became more common as lawyers began to subvert the process. The concept of reconciliation evaporated from the scene, and it became more common for lawyers to appear as advocates in arbitrations and for arbitrations to imitate litigation. By the end of the seventeenth century, the anonymous author of *Arbitrium Redivivum* refers almost longingly to the time when "in the Saxon or Old English [arbitrament] was called a Love Day, because of the Quiet and Tranquility that should follow the ending of the controversie" (1694, p. 2).

Others have looked back longingly to those halcyon days of love, but for most of its history, the bar has shown little interest in recognizing and much less in practicing conciliatory processes. In fact, whenever an effort was made to revive conciliatory processes, the bar was there in opposition. In "England's Balme," Cromwell's lord chancellor, William Sheppard, suggested that common-law courts provide reconciliation modeled after the ancient loveday as a mandatory prelitigation procedure. Fortunately for the bar that opposed such reforms, the king was invited back before anything came of it (Matthews 1984).

Much later, in 1828, reform-minded Lord Chancellor Brougham proposed the establishment of "Courts of Conciliation" based on contemporary models in Europe (Arthurs 1985). The bar was vehemently opposed, and Brougham tried a more diluted version, reminiscent of Henry I's reforms, authorizing judges in local courts to proceed by conciliation or arbitration as well as by law. The bar argued that a judge could not be everything at once and that the law would be "cheapened" at the expense of making the courts more flexible in their use of process. Such "cheapness," one commentator from the bar opined, would make litigation in England as common as it was in the United States. The commentator went on to note that "if a taste for litigation be encouraged by the cheapness of law, as a taste for dram drinking by the cheapness of gin, . . . [such courts] would be an eternal source of discord, a curse instead of a blessing to the poor" (*Lawyers Magazine* 1831).

As law became the dominant mechanism for handling conflict in England, many settlements in the Americas were founded on utopian religious and social principles in which conciliatory dispute resolution played a major role. They avoided problems with the bar by barring the bar itself. In many colonies attempting to reinstitutionalize conciliatory processes, lawyers were prohibited from practicing (Auerbach 1983). The Puritans integrated notions of Christian brotherly love into their models of dispute processing. In Dedham, the community covenant provided that disputes be resolved by "three understanding men" or by "two judicious men" selected by the disputants or the townspeople. Arbitrators urged the disputants to "live together in a way of neighborly love and do each other as they would have the other do themselves" (Lockridge 1970, pp. 5–6). A decision of a Boston town meeting in 1635 ordered that disputes were to be settled "amicably by arbitration . . . without recourse to law and courts" and that no inhabitants "shall sue one another at lawe" until an arbitration panel had heard the dispute (Rutman 1965, pp. 154–155).

In Sudbury, a Puritan community patterned after a feudal English open-field, or communal, village, not one case was recorded as reaching litigation. The sole surviving case on record indicates that even that case had gone to arbitration (Powell 1963). The Quakers brought their two-stage dispute resolution process—involving a combination of mediation and arbitration—to the colonies. Refusal to abide by the award was grounds for expulsion, and only after disownment could a Quaker pursue a remedy at law. Quakers dominated the governments of Pennsylvania and

western New Jersey until the mid-seventeenth century. The courts in those regions reflected the Quaker philosophy and encouraged arbitration by reference or by court order. George Washington's will has an arbitration clause.

By the twentieth century, however, the institutions that supported ADR had eroded, and litigation had become the dominant, formal mode of dispute resolution in the United States. Despite Abraham Lincoln's famous advice to young lawyers to convince disputants to compromise because "as a peacemaker the lawyer has a superior opportunity to be a good man" (Lincoln 1920 [1850], p. 140), the legal profession had little to do with ADR outside of lawyer-to-lawyer negotiations until recently.

Law and the Resurgence of ADR

In the first scenario of the introductory modern tale of Jeff and Ed, the two neighbors followed the stereotypical pattern of litigation in our modern but highly litigious society. During the last two decades of the twentieth century, however, conciliatory processes have enjoyed a resurgence in the United States and elsewhere. Much of the impetus for this resurgence came from a growing dissatisfaction with the courts. By the late 1970s, the public began to perceive problems with the state of litigation. On a macro level, litigation was identified as one of the primary ills in our society, and the legal profession was considered largely to blame. Commentators pointed to the "shocking" numbers of lawsuits and lawyers, particularly compared with other countries, as the cause of American social and economic decline. Of course, this was and continues to be an oversimplification, but there is little doubt that ours is a litigious society. Whatever the reasons, and there are many, by the mid-1980s, there appeared to be more lawsuits than the American civil justice system could adjudicate economically and efficiently. On a micro level, the resulting backlog was creating and continues to create years of harmful delay in the resolution of a dispute. The litigation process has become very expensive and undeniably stressful for all the participants. The outcomes are rarely satisfactory, even to those who "win," and relationships, personal and commercial, rarely survive intact.

Out of the many recommendations for civil justice reform dealing with the real and perceived ills of litigation, one of the least controversial and most broadly supported has been to provide greater access to ADR processes. A small but vocal group of judges, lawyers, and others concerned with the state of the legal system began to offer their ideas about how the courts could utilize ADR to increase litigant satisfaction, decrease the courts' caseload, and improve the delivery of justice. A pivotal moment for the dissemination of these ideas came with the 1976 American Bar Association Pound Conference (National Conference on the Causes of Popular Dissatisfaction with the Administration of Justice), which drew attention to the possibilities for court-connected ADR and the importance of ADR education within the legal establishment. During the conference, Professor Frank Sander proposed his model of the court as not "simply a courthouse but a Dispute Resolution Center." From this proposal came the idea of the *multidoor* courthouse, in which litigants could be channeled through to the appropriate process, whether that be mediation, the court, arbitration, or any other ADR process. The Pound Conference and the publication of Professor Sander's paper is often credited as the beginning of the ADR movement among the broader bar (Sander 1976). By the mid-1970s, mediation was widely touted as a better way to work through the difficult relationship problems in divorce. By the late 1970s, community or "neighborhood" justice centers, many subsidized by the Department of Justice and local courts, were cropping up to provide ADR to improve community cohesion. By

the mid- and late 1980s, many private companies and individuals were moving their disputing into private ADR processes through ADR agreements and by establishing company policies to use ADR. In the public sphere, state legislatures and courts were creating court-connected ADR systems to divert the caseload. Several professional associations sprang up, and innumerable ADR service providers have opened their doors.

Today, leadership of both the bench and the bar support ADR as a fair and efficient way to reduce case backlog, alleviate the pressures on our courts, and settle cases earlier. During the 1980s, law schools began offering courses in ADR theory and skills, and lawyers began developing themselves as their firms' "ADR specialists." By 1995, the proliferation of "ADR specialists" led to the publication of a directory of mediators and arbitrators. Today, there is an Alternative Dispute Resolution Section of the American Bar Association and for many state bars. So now we have the second scenario to our modern tale of litigation. Today, a court is likely to redirect Jeff and Ed into conciliatory processes before it will allow them to litigate. If all goes well, they will resolve their conflict and perhaps reconcile.

Institutional Dominance, Conflicting Ethos

This brief history of the relationship between law and conciliatory processes began with law respecting the social utility of reconciliation. Law supported and even adapted conciliatory techniques in order to maintain its own social utility as a conflict-handling institution. As social and legal theory begin to emphasize individual rights over community cohesion and interests, law competed with and eventually lost interest in reconciliation and conciliatory processes. Undoubtedly, people continued to use conciliatory processes to resolve their interpersonal and intragroup conflicts, but as a social institution, law has become the dominant, most recognized and highly institutionalized mechanism for handling conflict. The social institutions that provided a framework for triadic, conciliatory processes were eclipsed and disappeared or became virtually invisible. Only recently has the legal system revived an interest in ADR, an interest driven primarily by the actual and perceived inability of the law to handle conflict effectively in today's society, and legal institutions have played a major role in the revitalization of ADR. Today, most ADR occurs deep within the shadow of law and lawyers. One of the more interesting tensions in the modern ADR movement results from the heavy involvement of courts and the bar. Psychologists, social workers, and other non-lawyers, many of whom have been participating in and promoting ADR well before the courts and bar showed any interest, resist the trend toward lawyer domination of the movement and believe that lawyers ignore the value of ADR in promoting reconciliation. The relationship between law, lawyers, and conflict resolution has not always been a happy one, and the current trend of legal domination may prove detrimental to the social utility of ADR as an approach to handling conflict that emphasizes resolution and reconciliation.

Is Conflict Resolution Becoming the Prevailing Zeitgeist?

It is curious to note that lawyers are not the only ones who have recently become more interested in conciliatory processes. In the broader ADR movement, psychologists and social workers are promoting these processes in family and neighborhood disputes. School counselors and other educators are introducing them to reduce school violence, and managers are attempting to use them to reduce conflict and violence in the workplace. Even at the level of international conflict, there is greater recognition of the potential of these processes (see Neu, Box 4.1). This is an immense shift in modern

Interpersonal Dynamics in International Conflict Mediation

Joyce Neu

On the eve of the twenty-first century, there are about 30 major wars taking place (Jongman & Schmid 1998; Sollenberg & Wallensteen 1998). These large-group armed conflicts, most of which are civil wars within the borders of nation-states, have precipitated the deaths of millions and displaced millions more. Because of the sheer size and scope of such conflicts, we tend to ignore the importance of interpersonal dynamics in their resolution. As with conflicts among individuals, the personal interactions and relationships between leaders of the warring factions and between them and intervening mediators can be essential to the cessation of hostilities and the eventual development of peace.

Violent conflict between large groups, however, differs from conflict between individuals in significant ways. First, there has to be a decision taken by the leadership of the group to engage in armed conflict. An individual does not need to lobby support or engage in prolonged decision making to initiate conflict. Second, waging war requires organization, for example, recruiting or mobilizing troops, assembling weapons, and planning strategies. Individuals, on the other hand, may strike out without any plan of action. Third, during the fighting, the progress of the conflict is evaluated to ensure that the constituents of the group support the actions the leadership is taking. Individuals in conflict rarely act in collaboration with large numbers of constituents. Finally, in large-group conflicts, making peace is dangerous. Once a large group is mobilized to support a cause such as war, like a battleship, it takes time to turn opinion around toward peace. As we saw with the assassinations of Anwar Sadat and Yitzak Rabin decades apart, they paid for engaging in peace processes with their lives (each was assassinated by a member of his own group who

BOX 4.1 (continued)

viewed each as a "traitor" for talking peace). Although it may be difficult for individuals to make peace with one another, it rarely puts one's survival into jeopardy.

In large-group conflicts in which the protagonists know each other well and have shared a history together, even perhaps having waged war together against a common enemy, it may be difficult to step back from the conflict and envision solutions (e.g., Eritrea's Isaias Afeworki and Ethiopia's Meles Zenawi have been engaged in a border conflict since May 1998 but once fought together against the former Ethiopian regime). In these cases, as in cases of interpersonal conflict, each party knows the other's strengths and weaknesses and can use this knowledge against the other. A mediator's services may be needed to help restore the relationship (Pruitt et al. 1993). Mediation has been documented as part of human behavior as long ago as the fifth century B.C. (e.g., Mo-tzu, a Chinese philosopher in the fifth century B.C.; *Theseus,* by Plutarch, written A.D. 75; and more recently, the 1899 Hague Convention for the Pacific Settlement of International Disputes) and is a behavior found in other species as well (de Waal 1996).

When a third party is involved as mediator, whether in individual or large-group conflicts, another relationship is added to the equation: the one between mediator and the opponents (Richmond 1998). The third party must be perceived by all sides to the conflict as neutral and must be informed about both the issues at stake and the risks of undertaking the mediation (Waite 1993). In the case of international conflict, the third party must also be viewed as someone with adequate authority and credibility to gain both internal and external support for whatever solution is found to the conflict and must be committed to seeing a solution implemented.

The experience of former U.S. president Jimmy Carter as a third-party mediator provides useful examples of the importance of interpersonal dynamics in this context. During his presidency, Carter spent 13 days at Camp David negotiating a peace accord between Egyptian president Anwar Sadat and Israeli prime minister Menachem Begin. Carter's relationship with Sadat was warm and personal; his relationship with Begin, more distant. On the 13th day of talks, Begin was refusing to sign any peace agreement. Earlier, Begin had requested that Carter

autograph photographs of himself, Carter, and Sadat for his grandchildren. Carter did so; then, realizing the trouble they were in with the Israeli side, Carter's secretary found the name of each of Begin's grandchildren, and Carter personalized each photograph. He presented the photos to Begin, who was visibly moved at seeing each grandchild's name on the photos. A short time later, Begin agreed to remove the last obstacle to the peace accord (Carter 1982). This personal touch may have tipped the balance, causing Begin to rethink his position.

Since Carter's presidency, he has intervened in a number of international conflicts, bringing to bear his authority as a former U.S. president, his access to world leaders, and his reputation for fairness and honesty (Babbitt 1994; Bourne 1997). Recognizing the importance of interpersonal dynamics in conflict resolution, Carter has noted: "I don't think we would have had any breakthrough in North Korea unless we had gone out on a boat for five or six hours with Kim Il Sung and his wife. While public opinion is still against Rosalynn and my being fairly close to Zaire President Mobutu and his wife and his daughter, these kinds of personal connections are very important" (*State of the World Conflict Report, 1995–1996* 1997, p. 86).

As a mediator, Carter helped facilitate cease-fires in Sudan (1995) and Bosnia (1994–1995). In the case of Sudan, where the Carter Center had been involved in health projects for years, a prior, trusted relationship between Carter and the warring factions existed. In the case of Bosnia, there was no prior relationship, and confidence-building measures were taken to create trust.

In March 1995, Carter traveled to Sudan to meet with the government in the north and the rebels in the south to negotiate a two-month cease-fire that allowed health workers entry into formerly inaccessible areas. Carter's seriousness of purpose in maintaining good interpersonal relationships with the heads of the warring factions and his commitment to peace in Sudan were evidenced by his sending his son, Chip Carter, to Khartoum and Nairobi after he returned to the United States and the opening of Carter Center/Global 2000 offices in both cities within two weeks of the cease-fire agreement. Following Chip Carter's trip, President Car-

BOX 4.1 (continued)

During a break in the cease-fire negotiations, Carter receives a traditional Serb musical instrument as a gift from Karadzic. Photograph by Joyce Neu.

ter continued to send representatives to meet with the Sudanese president and the rebel leaders. Carter's willingness to send family and close associates to Sudan as "guarantors" of the cease-fire demonstrated his sincerity and seriousness of purpose. Although the government of Sudan was initially unwilling to consider an extension of the cease-fire, President al-Bashir announced a two-month extension in late May (Carter 1995).

The war in Bosnia had been going on for three years when in December 1994 Carter received an invitation from Bosnian Serb leader Radovan Karadzic to intervene. Carter did not know either Karadzic or the other major actors in the region. To test Karadzic's sincerity and to determine if he, and not the military commander, held the reins of power, Carter secured a pledge from Karadzic by telephone to take several actions within 48 hours. When these steps had been taken, Carter traveled to Bosnia, where he secured a four-month cease-fire and an agreement by the Bosnian Serbs to reengage in negotiations under the auspices of the Contact Group (Neu & Shewfelt in press). Carter used a "single text" approach, an approach he first used during the Camp David talks between Israel and Egypt. In a single text approach, the individual leaders work on the same document, taken back and forth to each by the mediator, for as long as it takes to reach accord. This iterative process allows the actions of each party to be transparent and builds an atmosphere of trust among the participants. This approach permitted Carter, with no previous relationships with the protagonists, to secure the most widely honored cease-fire until the Dayton accords that ended the war in 1995.

Mediating violent conflicts requires the trust of the parties, which is gained preferably through prior personal relationships or through steps taken to establish such relationships. Understanding the significance of developing these interpersonal relationships is still rare in the diplomatic world, but examples of mediators such as Jimmy Carter demonstrate how pivotal interpersonal dynamics can be to building peace.

society's preferences for conflict handling—from reliance on dictation and direction emanating from third parties with the power to impose outcomes to reliance on the disputants themselves to find and comply with an appropriate solution.

Biologists have been making a similar shift in emphasis—from competition and aggression in nature to symbiosis, mutualism, and cooperation. Indeed, there is increased recognition that conciliatory behavior and peacemaking is a necessary concomitant to aggression and fighting in social species (de Waal, Chapter 2). It puts in question the often accepted nineteenth-century picture of nature as red in tooth and claw. Thus, while it may be true that when the going gets tough, the tough survive, it may be equally true that when the going gets tough, the tough cooperate. At least in some species, individual fitness includes the ability not only to compete but also to cooperate, and this promotes survival for the individual, family, group, and species. This theoretical trend, which might be dubbed the "natural conflict resolution" movement, parallels the ADR movement. Both seem to have emerged independently. Biologists and lawyers rarely interact professionally and usually do not spend a great deal of time reading each other's works; however, with this conflux of interests, some professional convergence over conflict resolution would seem productive. Lawyers and others in the ADR movement would certainly benefit from a better understanding of the role of relationships in reconciliation, for example. Ethologists should spend more time observing human behavior in trials and negotiations and study the effectiveness of different triadic techniques in promoting or inhibiting reconciliation behavior. Lawyers and ethologists should discuss their observations on how context affects conflict-handling preferences. Legal theory, which relies primarily on the social science model emphasizing the relationship of mind and environment and individual experience, would benefit from a biological model that recognizes a more complex interaction of genes, mind, and environment and rejects the nature-nurture dichotomy.

Why have the disciplines of law and biology/ethology suddenly developed an interest in peacemaking and reconciliation? Is it becoming a prevailing zeitgeist, and if so, why? Prevailing social theory has a profound influence on both scientific theory and legal theory, and a society's prevailing social theory is profoundly influenced by the society's experience and the environmental conditions in which it finds itself. Has the dispersal of power among diverse groups and individuals undermined recognition of dominance so much that the aggression/control model for managing human affairs and imposing outcomes to conflict has failed? We may no longer live in the medieval agrarian village of Geoffrey and Edward, but as we enter the twenty-first century, we find ourselves living in a complex "global village" in which interdependence and cooperation have taken on a larger significance and the skills of compromise and reconciliation have become more highly valued in a greater number of human relations and endeavors. In a world in which war and violence are increasingly unacceptable alternatives, the exercise of power requires consensus among diverse groups with potentially competing interests, and the legitimacy of every individual's right to the minimal requirements of survival is morally undeniable, there is a need for compromise at every level. It is comforting to think that the necessary basic tendencies and skills are already within us in the form of natural conflict resolution.

References

Alexander, R. D. 1987. *The Biology of Moral Systems.* Hawthorne, N.Y.: Water de Gruyter.

Arbitrium Redivivum: or the Law of Arbitration; Collected from the Law Books Both Ancient and Modern, and Deduced to These Times: Wherein the Whole Learning of Awards or Arbitrements Is Methodically Treated. 1694. London.

Arnold, M. S., ed. 1987. *Select Cases of Trespass from the King's Courts, 1307–1399.* Vol. 2, no. 103, Seldon Society Series. Buffalo, N.Y.: William S. Hein & Co.

Arthurs, H. W. 1985. *"Without the Law": Administrative Justice and Legal Pluralism in Nineteenth-Century England.* Toronto: University of Toronto Press.

Attenborough, F. 1963 [1922]. *The Laws of the Earliest English Kings.* New York: Russell & Russell.

Aubert, V. 1963. Competition and dissensus: Two types of conflict and conflict resolution. *Conflict Resolution,* 7: 26–42.

Auerbach, J. 1983. *Justice without Law?.* Oxford: Oxford University Press.

Babbitt, E. 1994. Jimmy Carter: The power of moral suasion in international mediation. In: *When Talk Works* (D. Kolb & Associates, eds.), pp. 375–393. San Francisco: Jossey-Bass.

Bennett, J. W. 1958. The medieval loveday. *Speculum,* 33: 351–370.

Boehm, C. 1994. Pacifying interventions at Arnhem Zoo and Gombe. In: *Chimpanzee Cultures* (R. W. Wrangham, W. C. McGrew, F. B. M. de Waal, P. Heltne, eds.), pp. 211–226. Cambridge: Harvard University Press.

Bourne, P. G. 1997. *Jimmy Carter: A Comprehensive Biography from Plains to Postpresidency.* New York: Scribner.

Burger, W. 1982. Isn't there a better way? Annual report on the state of the judiciary. January 24.

Carter, J. 1982. *Keeping Faith.* Toronto: Bantam Books.

Carter, J. 1995. Press statement on Sudanese cease-fire. The Carter Center, Atlanta. www.emory.edu/carter_center/RLS95/sudancf.htm

David, R. 1985. *Arbitration in International Trade.* Boston: Kluwer Law.

de Bracton, H. 1878–1883 [ca. 1268]. *De Legibus et Consuetudinibus Angliae.* London: Longman & Co.

de Waal, F. B. M. 1984. Sex-differences in the formation of coalitions among chimpanzees. *Ethology and Sociobiology,* 5: 239–255.

de Waal, F. B. M. 1991. The chimpanzee's sense of social regularity and its relation to the human sense of justice. *American Behavioral Scientist,* 34: 335–349.

de Waal, F. B. M. 1996. *Good Natured: The Origins of Right and Wrong in Humans and Other Animals.* Cambridge: Harvard University Press.

Downer, L. J. 1972. *Legis Henrici Primi.* Oxford: Clarendon Press.

Eckhoff, T. 1966. The mediator, the judge, and the administrator in conflict resolution. In: *Contributions to the Sociology of Law* (B. M. Blegvad, ed.), pp. 148–172. Copenhagen: Scandinavian University Books.

Elliot, E. D. 1985. The evolutionary tradition in jurisprudence. *Columbia Law Review,* 85: 38–94.

English Historical Documents. 1955. New York: Oxford University Press.

Fisher, R., & Ury, W. 1981. *Getting to Yes: Negotiating Agreement without Giving In.* Boston: Houghton Mifflin.

Fowler. 1976. Forms of arbitration. In: *Proceedings of the Fourth International Congress of Medieval Canon Law* (S. Kuttner, ed.), ser. C, 133. Toronto: International Congress of Medieval Canon Law.

Garey, S. 1623. *A Manuall for Magistrates: Or a Lanterne for Lawyers: A Sermon Preached before Iudges and Iustices at Norwich Assizes 1619.* London.

Goodall, J. 1983. Order without law. In: *Law, Biology and Culture* (M. Gruter & P. Bohannan, eds.), pp. 50–62. Santa Barbara, Calif.: Ross-Erikson.

Gruter, M. 1991. *Law and the Mind.* Newbury Park, Calif.: Sage.

Gruter, M., & Masters, R. D. 1992. *The Sense of Justice: Biological Foundations of Law.* Newbury Park, Calif.: Sage.

Hall, G. D. G. 1965. *The Treatise on the Laws and Customs of the Realm of England Commonly Called Glanvill.* Oxford: Oxford University Press.

Hunter, A. 1985. Arbitration procedure in England: Past, present and future. *Arbitration International,* 1: 82–95.

Jongman, A. J., & Schmid, A. 1998. Contemporary armed conflicts: Trends and events in 1997. In: *Prevention and Management of Violent Conflicts: An International Directory* (PIOOM & Berghof Research Institute for Constructive Conflict Management, ed.), pp. 41–45. Utrecht: European Platform for Conflict Prevention and Transformation.

Lawyers Magazine. 1831. Anonymous letter, 5: 33–35.

Lincoln, A. 1920 [1850]. Notes for a law lecture (July 1, 1850). In: *Complete Works of Abraham Lincoln,* vol. 2 (J. Nicolay & J. Hay, eds.), p. 140. New York: Century Press.

Lockridge, K. A. 1970. *A New England Town: The First Hundred Years.* New York: Norton.

Lyndwood, W. 1679. *Provinciale.* Oxford.

Maine, H. S. 1861. *Ancient Law.* London: J. Murray.

Maitland, F., & Baildon, W. 1891. *The Court Baron.* No. 4, Seldon Society Series. Buffalo, N.Y.: William S. Hein & Co.

Malinowski, B. 1932 [1926]. *Crime and Custom in Savage Society*. New York: Harcourt, Brace & Co.

March, J. O. 1647. *Actions for Slander: to which is added Awards and Arbitrements*. London.

Matthews, N. 1984. *William Sheppard, Cromwell's Law Reformer*. New York: Cambridge University Press.

McGuire, A. M. 1994. Helping behaviors in the natural environment: Dimensions and correlates of helping. *Personality and Social Psychology Bulletin*, 20: 45–56.

Miller, A. 1984. The adversary system: Dinosaur or phoenix. *Minnesota Law Review*, 69: 1–37.

Neu, J., & Shewfelt, S. In press. *Eminent Third Party Mediation: The Carter Center Intervention in Bosnia*. Atlanta: Carter Center.

Powell, S. C. 1963. *Puritan Village: The Formation of a New England Town*. Middletown, Conn.: Wesleyan University Press.

Pruitt, D., Peirce, R. S., Zubek, J. M., McGillicuddy, N. B., & Welton, G. L. 1993. Determinants of short-term and long-term success in mediation. In: *Conflict between People and Groups* (S. Worchel & J. A. Simpson, eds.), pp. 60–75. Chicago: Nelson-Hall Publishers.

Richmond, O. 1998. Devious objectives and the disputants' view of international mediation: A theoretical framework. *Journal of Peace Research*, 35: 707–722.

Rutman, D. B. 1965. *Winthrop's Boston: Portrait of a Puritan Town*. Chapel Hill, N.C.: University of North Carolina Press.

Sander, F. 1976. Varieties of dispute processing. 70 F.R.D. 111. St. Paul, Minn.: West.

Sayles, G. O. 1971. *Select Cases in the Court of the King's Bench*. Vol. 7, no. 88, Seldon Society Series. Buffalo, N.Y.: William S. Hein & Co.

Shapiro, M. 1981. *Courts: A Comparative and Political Analysis*. Chicago: University of Chicago Press.

Sollenberg, M., & Wallensteen, P. 1998. Major armed conflicts, 1997. In: *SIRPRI Yearbook 1998: Armaments, Disarmament, and International Security* (Stockholm International Peace Research Institute, ed.), pp. 17–30. New York: Oxford University Press.

State of the World Conflict Report, 1995–1996. 1997. An interview with Jimmy and Rosalynn Carter, pp. 84–86. Atlanta: Carter Center.

Toller, T. N. 1882. *Anglo-Saxon Dictionary*. Oxford: Clarendon Press.

Trivers, R. L. 1971. The evolution of reciprocal altruism. *Quarterly Review of Biology*, 46: 35–57.

Van Caenegem, R. 1959. *Royal Writs in England from the Conquest to Glanvill*. No. 77, Seldon Society Series. Buffalo, N.Y.: William S. Hein & Co.

Waite, T. 1993. *Taken on Trust: An Autobiography*. New York: Quill.

Wormald, P. 1986. Charters, law and the settlement of disputes in Anglo-Saxon England. In: *The Settlement of Disputes in Early Medieval Europe* (W. Davis & P. Fouracre, eds.), Cambridge: Cambridge University Press.

Yarn, D. H. 1997. *Alternative Dispute Resolution: Practice and Procedure in Georgia*. 2nd edn. Norcross, Ga.: Harrison.

Yarn, D. H. 1999. *Dictionary of Conflict Resolution*. San Francisco: Jossey-Bass.

PART II

Controlling Aggression

Introduction

This section highlights mechanisms for the control of aggression, which are needed for any species relying on cooperation for survival. The low rate of aggressive confrontations despite the widespread opportunities for conflict between group members indicates the high efficacy of these mechanisms. Because these mechanisms aim to prevent the occurrence of an event (i.e., aggressive escalation), they are more difficult to study than the ones following an event (such as reconciliation). Consequently, there has been less progress in this area than in the study of post-conflict resolution. The contributions of this section clearly indicate, however, the importance of studying these mechanisms because reducing the probability of escalation is certainly a more efficient way to deal with conflict than repairing the damage afterward.

In Chapter 5 **Preuschoft and van Schaik** examine how communication of power asymmetries and established dominance relationships may serve as conflict management devices by regulating aggressive escalation. They argue that escalation is delayed, even during conflict between strangers, by stepwise assessment of the relative fighting abilities. This assessment is often facilitated by active communication of such abilities and motivation for the disputed resource, or by arbitrary conventions. Specific signals of dominance and submission are especially exchanged by animals with individualized relationships such as territory neighbors and group members. In these contexts, dyadic confrontations may also be influenced by the involvement or mere presence of third parties. Alliances can therefore play a role in maintaining the stability of power relationships and therefore in minimizing the likelihood of escalated fights.

The costs of conflict are not only those associated with injuries due to escalated fights or time investments that could be used otherwise. The new perspective of conflict within the framework of social relationships emphasizes the importance of social costs associated with the disturbance of interindividual relationships. In Box 5.1 **Matsumura and Okamoto** suggest integrating this notion into game theory models of conflict. Individuals can impose social costs on another group member by modifying the social relationships between them. Depending on the socioecological conditions, this possibility may provide power leverage to a weaker individual that can withdraw favors and services from more powerful group members if they escalate the conflict. Such disputes may actually result in tolerance around a resource instead of an escalated fight.

Elaborated interactions may be used to dissipate tension and reduce the likelihood of aggressive confrontation. Two examples are presented in Boxes 5.2 and 5.3. In the first box, **Kuester and Paul** report on the use of infants by adult males of some primate species as a buffer against aggression. These males regularly carry, huddle, and groom certain infants. They also approach other males while carrying an infant; then the two males sit in close contact and show friendly behavior toward the infant. These interactions appear to ease tension between the males. In the other box, **Colmenares, Hofer, and East** focus on greeting ceremonies in conflict-provoking situations when partners are uncertain about what to expect from one another. Even though such greetings are also a clear part of the human behavioral repertoire, the authors examine their nature and function in baboons and hyenas. Their form, the contexts in which they occur, and the nature of the social relationship between the greeting individuals sug-

gest that many greeting ceremonies are driven by motivational ambivalence and that they serve functions such as reassurance of subordinates and appeasement of potential aggressors.

There is great variation in strategies for dealing with conflict in the animal kingdom. This is not surprising because living organisms vary so much from one another. Variation is, however, present also between closely related species and even between members of the same species or group. In Chapter 6, **Thierry** presents the variation of conflict management patterns across closely related macaque species. He examines these patterns within the framework of interspecific differences in dominance style and discusses their covariation with other characters of social organizations. Individual variation is investigated in Box 6.1. Here, **Sapolsky** points out that there are multiple physiological profiles of dominant and subordinate individuals within the same group. These profiles and the consequent susceptibility to stress-related diseases depend on the individual style of anticipating and responding to conflict.

Aggressive escalation can be prevented by increasing tolerance between group members while facing conflict-enhancing situations. In Chapter 7, **Judge** focuses on the various ways in which human and nonhuman primates and other animals deal with crowding. Crowding is usually interpreted as a tense situation in which competition for space and the potential for aggression are increased. Judge reviews the literature to examine whether individuals do indeed increase aggression under high population density and whether they increase conflict management patterns to prevent conflict escalation. He concludes that there is no simple relationship between density and aggression, probably because individuals of various species employ coping strategies to prevent the negative effects of crowding. Another potentially conflict-provoking situation is feeding. In Box 7.1, **Koyama** reviews several studies supporting the view that a selective increase in tension-reduction interactions occurs before a predictable period of conflict, such as scheduled feeding time. These findings suggest that some nonhuman primates, at least, possess the ability to anticipate future events enabling them to implement powerful pre-conflict strategies to dissipate tension and reduce the likelihood of conflict escalation.

Special lifestyles can also prevent aggressive escalation of within-group conflict. In cooperatively breeding species, help of nonbreeding group members is essential for the successful raising of offspring. Cooperation is not limited to the care of infants but is highly important for predator detection and foraging efficiency. In Chapter 8, **Schaffner and Caine** suggest that in these species reproductive inhibition of all group members but the breeding pair is an important proximate mechanism in fostering and maintaining peaceful interactions. As a consequence, the limited aggressive interactions are mild and appear not to disrupt social relationships among group members.

CHAPTER 5

Dominance and Communication

Conflict Management in Various Social Settings

Signe Preuschoft & Carel P. van Schaik

To win without fighting is the best.

Sun Tzu, The Art of War

Introduction

Is dominance a conflict management device? To what extent can dominance be seen as an adaptation to reduce the frequency of aggression and the probability of escalating violence? Much of the theory developed for animal conflict refers to contests between animals unfamiliar to one another. This chapter, however, will be largely concerned with the nature of dominance in groups, where virtually all conflicts occur within social long-term relationships—the very context in which reconciliation and peaceful conflict resolution are functionally most relevant (e.g., de Waal, Chapter 2). To appreciate fully what is special about dominance, it is necessary first to see how conflicts are handled between strangers. After showing that even among strangers escalated fighting is the last resort, we will examine how familiarity, spatial association, and interindividual bonding affect the way in which conflicts are managed. Our point of departure is the familiar framework of decision making in animals: the economics of escalation and assessment of fighting abilities, and the socioecology of resource acquisition.

Conflict of Interest and Modes of Competition

Conflicts of interest can arise between two animals when their goals are incompatible, either because both individuals seek different things but one can suppress the other in realizing its goal (as during weaning), or because two animals seek to have the same thing but only one can have it (as in competition for food; Hand 1986).

Conflicts are not always expressed in behavioral interactions. The strategy chosen to resolve a conflict is best conceived in terms of a cost-benefit analysis carried out by each competitor.

Two basic modes of competition can be distinguished, scramble and contest (Nicholson 1954). Individuals who use scramble competition attempt to maximize efficiency in locating and exploiting the resource so that they may appropriate as much of it as possible. Scramble competition can be viewed as a fallback strategy employed when resources are not economically defendable, for instance, because they are widely dispersed or because they appear in very large patches. Thus, we will see no fighting or other agonistic behavior (i.e., aggressive and submissive patterns) between two animals engaged in scramble competition, even though they are competing.

Individuals who use contest competition, by contrast, attempt to monopolize a resource. Contest competition may involve any tactic ranging from manipulation, to deception, to physical coercion (Huntingford & Turner 1987). Contest competition is recognized when (1) A enhances its own interests at some cost to B, (2) A's behavior is goal-directed (in that it is instrumental to achieve a goal, or in that it is truly intentional; cf. Byrne 1995), and (3) B is forced by A's behavior to incur the cost. The first two conditions together distinguish interference from accidental damaging interactions; the third condition is necessary to exclude altruistic behavior on the part of B. Only contest competition can give rise to dominance relationships.

Individual Attributes and Properties of Relationships

Dominance is one of the most hotly debated concepts in ethology. Thus, before we explore how dominance functions to prevent escalating aggression, we must review some of the issues that have historically plagued discussions of dominance. We suspect that some of the disagreement is due to a failure to distinguish between competitive encounters involving strangers and those involving familiar individuals.

One question that is debated is whether dominance results directly from individual attributes or whether it reflects a relational property between two animals (e.g., Bernstein 1981 and commentaries; Chase 1986; Slater 1986). Those who assume dominance is an individual attribute argue that the capacity to dominate conspecifics is determined by individual attributes such as absolute age, size, weight, or condition (Slater 1986). In any given population, these individual attributes will follow a normal distribution. Hence, individuals who dominate many conspecifics are more likely to be situated in one of the tails of the distribution. By contrast, advocates of the relational property concept point out that dominance is the result of a comparison, a relationship between two individuals (Hinde 1976): even if both individuals are big, the important question is which one is bigger (e.g., Chase 1986; van Hooff & Wensing 1987).

Table 5.1 shows that the determinants of dominance vary with social context. We envisage a continuum from unfamiliar animals, that is, strangers, to animals with individualized social relationships (Lee 1994). Especially when animals frequently encounter strangers, dominance—or rather power asymmetry—depends primarily on individual attributes and can be associated with morphological signals that signify membership in a demographic or reproductive class. By contrast, in species that live in permanent groups with stable composition, dominance is relationship-dependent and largely independent of class membership.

We define dominance-subordination relationships as long-term dyadic relationships that are characterized by an asymmetric distribution of power. Power asymmetry can be measured in the degree of unidirectionality, or directional consistency in the exchange of behaviors—particularly certain signals—in a dyad (de Waal 1986; Preuschoft 1999; see below). A capacity to form dominance-subordination relationships can evolve if it endows its bearers with a fitness advantage. This fitness advantage could result if animals engaged in dominance-subordination relationships are better able to manage social conflicts. Therefore, we will explore how the recognition of an asymmetry in power can help animals to manage conflicts in a way that minimizes both physical costs and damage to the relationship (van Schaik & Aureli, Chapter 15).

When Strangers Meet: Aggression and the Economy of Escalation

Let us now turn to conflicts of interest that arise during encounters of unfamiliar individuals. Because fighting is risky, escalation is not necessarily the best option. Rather, the optimum course of action depends on the behavior of one's opponent. Such dyadic confrontations are therefore amenable to analysis by game theory (Maynard Smith 1982b; Matsumura & Okamoto, Box 5.1).

In general, whether a conflict over a resource develops into overt fighting depends on (1) each competitor's motivation, that is, on the value the resource has to each of them, (2) each competitor's own estimate of how likely it is to win, and (3) the cost each competitor is prepared to incur when escalating the conflict into a fight (cf. Dunbar 1988). Thus, other things being equal, highly motivated (deprived) animals should tend to risk more, and rivals armed with canines, spurs, tusks, and the like should tend to be more cautious. Nonetheless, the benefits of winning a fight can be severely diminished by the cost of even small injuries, so overt fighting is usually only the last resort to settle conflicts. In many cases, therefore, animals should resolve conflicts without fighting (Maynard Smith 1982b).

BOX 5.1

Conflict, Social Costs, and Game Theory

Shuichi Matsumura & Kyoko Okamoto

Each animal in a social group has both common and conflicting interests with other members. Since "optimal" behavior for each individual depends on the behavior of other group members, game theory can be a useful tool for evolutionary analyses of social interactions within a group. Game theory is a mathematical theory that examines the outcome of interactions between two or more players who pursue their own benefits. Game theory originated in the field of human economic behavior (von Neumann & Morgenstern 1944). More recently, however, the theory has been developed in the field of evolutionary biology (Maynard Smith 1982b; Dugatkin & Reeve 1998). It has been used to examine the outcome of contest and cooperation between two or more individuals who pursue their own benefits using various behavioral strategies.

One of these games, the hawk-dove game (Maynard Smith & Price 1973; Maynard Smith 1982b), has clearly shown that it is meaningless to consider the fighting strategies of an individual without also considering the behavior of others. This is a simple game to examine the costs and benefits when opponents can use only two strategies: one strategy is to compete for a resource in an escalated manner; the other is always to avoid escalation. Following this model, various game models for animal conflict have been proposed. Such models are suitable for encounters between animals that do not live in individualized groups, as research on insects, spiders, and fishes has shown (Riechert 1998). They appear, however, not to be sufficient to account for the behavior of group-living species.

A new view of conflict between group members has recently emerged. Group-living animals use various behavioral tactics to manage conflict with other group members. The Relational Model (de Waal 1996; de Waal & Aureli 1997; de Waal, Chapter 2)

BOX 5.1 (continued)

regards aggression as the product of social decision making. In the model, aggression, as well as tolerance and avoidance, are viewed as possible outcomes of conflicts of interest between individuals. Aggression, however, threatens cooperative relationships. This threat is mitigated by post-conflict friendly reunion, or *reconciliation,* which is expected to repair disturbed relationships. This new perspective of animal conflict carries direct implications for the game theory approach, and the need for an integration of behavioral research with game theory has been gradually recognized (de Waal 1996; Pusey & Packer 1997; de Waal, Chapter 2).

This new perspective implies that escalated fights involve an additional type of cost. Most game models have considered only physical costs, that is, costs of injury or energy expenditure during the contest. An individual can, however, impose social costs on another group member by modifying the social relationship between them. Escalated fights may be inhibited not merely by high physical costs but also by the potential cost of losing future cooperation from a partner (de Waal, Chapter 2).

There have been a few attempts to apply game theory to conflicts within social groups. Van Rhijn & Vodegel (1980) examined contest strategies under the condition that every animal identified other members individually. Their simulations showed that players that settled conflicts according to dominance can attain general success. Hand's (1986) contest model included the possibility of *leverage* by subordinate individuals; such leverage could inflict costs for winning, that is, a loss of fitness incurred by winning inappropriately. To demonstrate this point, she used the example of a male who is constrained from driving his mate away from all contestable food items by the cost in reduced fitness that he will incur if her nutritional conditions get worse. In general, dominant individuals are forced to tolerate subordinate individuals around resources if, at other times, participation by subordinates in cooperative actions (e.g., against predators or other groups) positively impacts the fitness of dominants (Vehrencamp 1983). Thus, in this way, subordinate individuals have some implicit power over dominants.

Although these attempts illustrated factors influencing contests within social groups, the models are still insufficient to examine how social costs influence behavioral strategies. Social costs differ from physical costs in two main aspects. First, social costs depend on how the contestants interact after the contest. The subordinate may refuse cooperation with the dominant, whereas the dominant may decrease tolerance toward the subordinate. Second, the social costs are determined by the benefits provided by the social relationship, or *the value of the relationship* (cf. Cords & Aureli, Chapter 9). Asymmetry in the value each partner attaches to the other results in asymmetry in the social costs each can impose on the other.

The following are necessary conditions for conflict models incorporating social relationships: (1) individual recognition among players, (2) repeated interactions within a dyad, and (3) the option to force social costs onto the opponent. In addition, differences in fighting ability, physical conditions, or dominance rank between the opponents should be considered because these differences remain likely to influence the outcome of conflicts among group members. Behavioral strategies in the models must capture the way opponents interact after a contest, as well as the way they behave during the contest.

Although game models with repeated interactions are much more difficult, a promising approach could be to use a framework similar to that for the evolution of negative reciprocity, that is, "the punishment game" (Clutton-Brock & Parker 1995). This is a single-step game that combines a series of interactions. Punishment is a spiteful behavior that involves a small cost to the punisher and a large cost to the recipient. The model assumes punishment to be an aggressive attack. We can envisage replacing the attack with a social form of punishment that has a significant social cost to one partner and a smaller social cost to the other. For example, when many males compete over an estrous female and the female forms a consort relationship with one of them, the female is a valuable partner. A temporal break of the consort appears to have a great social cost to the male and a small social cost to the female because she can easily find another mate. If the female can punish the male socially by breaking the consort, she might win a conflict over a resource against the male.

BOX 5.1 (continued)

Several problems must be solved before constructing a comprehensive model on the basis of the above framework. In particular, answers to questions such as the following are needed: How is the value of social costs determined? Does dominance constrain the choice of strategies? Are counterstrategies possible? How do third parties influence the game?

It would certainly be useful to learn from evolutionary studies of cooperation that use games with repeated interactions, such as the iterated prisoner's dilemma (Axelrod & Hamilton 1981). Evolutionary simulations in which various behavioral strategies interacted with one another are especially important (e.g., Nowak & Sigmund 1993; Axelrod 1997; Boerlijst et al. 1997). In such simulations, players with various behav-

ioral strategies compete with one another, and successful strategies increase in number with each generation. For our purposes, players who impose social cost to the partner could be introduced into such simulations. The "tit-for-tat strategy" may then be interpreted as a strategy that imposes social costs via retaliative defection. Furthermore, iterated games incorporating dominance asymmetry and other ecological parameters can examine the "optimal" amount of social cost in a given condition, the effect of dominance, and the possibility of counterstrategies (e.g., Okamoto 1998). We are still at an early stage, but we believe that such simulation studies will play an important role in the integration of behavioral research with game theory.

Assessment of Relative Fighting Ability

To predict their chances of winning, individuals somehow need to gather information on the strength and motivation of their opponent relative to their own. How can animals decide whether to escalate when they meet for the first time? When strangers meet, stepwise escalation is a means to acquire information about each other's fighting potential at the lowest possible cost (Archer & Huntingford 1994). Therefore, assessment is a predominant activity in animal contest, and information can be assumed to accumulate during confrontations in a way similar to statistical sampling (Enquist & Leimar 1983; Table 5.1: decision rule "a"). If at any stage of assessment one individual senses it is weaker, less motivated, or prepared to incur fewer costs than its opponent, it should seek to withdraw. Stepwise assessment occurs also among familiar animals and has been described for species as diverse as red deer (*Cervus elaphus:* Clutton-Brock et al. 1982) and frogs (*Acris crepitans blanchardi:* Wagner 1992). Contests

last longer when the opponents are more evenly matched, because more probing is needed to settle a contest when the difference between opponents is small. Escalated fighting is most likely to occur when both contestants are of similar power or when they have imperfect information about their power asymmetry (Maynard Smith & Parker 1976).

Gathering information about asymmetries by prolonged probing is time consuming. Both partners lose time for other activities and risk missing out on other, easier opportunities. Particularly when encounter rates are high and the benefits of winning are fairly modest, thorough assessment of individual opponents may become disproportionately expensive, and animals may fare better by using simple rules.

One way to assess one's own fighting ability is to extrapolate on the basis of one's previous fighting success (Table 5.1: decision rule "b1"). Indeed, repeated experience of defeat in contests against various opponents generates individuals who

TABLE 5.1

Decision Rules and Types of Signals Involved in Conflict Management as a Result of Familiarity
and Encounter Rate of Competitors Characteristic of Various Grouping Patterns

Grouping Pattern	Familiarity	Encounter Rate	Decision Rule	Signals
SOLITARY	strangers	rare	(a) individual assessment	badge of demo-graphic class
AGGREGATION	strangers	frequent	assessment by (b1) population average, or (b2) class average*	badge of fighting ability
TERRITORIAL			(c) "uncorrelated asymmetry"	
(i) simultaneous settlement	mostly strangers	frequent	and (a) or (b)	badge
(ii) resident system	mostly familiar	rare	and (d)	coverable badge of status
STABLE ASSOCIATION	familiar animals	frequent	(d) prior experience with individual opponent	transient, behavioral signals

*Assessment is based on extrapolation from total prior experience in fighting success (b1) or from prior experience with opponents of a specific class (b2).

essentially are "trained losers" (Scott 1958). These individuals tend to defer to all rivals subsequently encountered. The opposite, "winner experience," has also been found in several species (Chase et al. 1994). Provided that animals encounter competitors at random, such a decision rule ensures that animals act as if they would compare their own capabilities with the population average (Huntingford & Turner 1987).

When opponents come in well-defined classes, individuals may use information from previous conflicts with similarly classed opponents to predict their likelihood of winning a present conflict. Assessment may be conditional on physical or behavioral attributes, for example, coat color or singing (Table 5.1: decision rule "b2"). In this case, generalization extends to all individuals, familiar or not, with a certain feature. In essence, animals are probably selected for susceptibility to defeats

and victories in previous conflicts as reinforcers of future agonistic behavior (Table 5.1: decision rules "b1" and "b2").

Animals may also settle conflicts on the basis of asymmetries that exist prior to combat but are not correlated with their fighting abilities (e.g., Hammerstein 1981). This settling of conflicts by "conventions" can be based on prior ownership or residence, on relative location, or on signals (Table 5.1: decision rule "c"). "Conventions" can be expected when fighting is dangerous, the prospect of actually winning the resource is limited, or thorough assessment is costly.

These examples show that strangers finding themselves in a competitive situation are loath to escalate to damaging fights and use lengthy and stepwise assessment procedures or rely on simple decision rules when assessment is too costly in relation to the expectable gains.

Negotiation by Communication

Many interactions observed in fighting behavior are only understandable as cooperation in mutual assessment of fighting ability. In the cichlid fish, *Nannacara anomala*, for instance, tail beating requires one fish to act as the beater and the other to act as the recipient. The recipient must stay still so that the beater can direct its behavior (Enquist et al. 1990). In fact, despite their conflict, rivals are often engaged in a mutualistic interaction insofar as prevention of escalation serves both individuals' interests.

Signaling fighting ability is advantageous for both contestants because it enables them to avoid fights with predictable outcomes. Indeed, assessment of fighting ability and motivation is often facilitated by information contained in communicatory displays: ritualized behaviors or morphological features that substitute for behavioral acts. Communicatory displays, or signals, are cheap devices designed by evolution to elicit certain responses from the receiver (Dawkins & Krebs 1978). But what prevents senders from emitting signals that misinform about fighting ability? Whether competing animals do or do not use honest signals has been discussed extensively (e.g., Dawkins & Krebs 1978; Caryl 1981, 1982; Hinde 1985a, 1985b; Grafen 1990). The general finding is that dishonest boasting by bluff signals is surprisingly rare, because of either high costs in signal production (energetic or hormonal) or high costs ensuing from signal emission, for example, in terms of predator/parasite pressure or intraspecific aggression.

Assessment signals are reliably connected with fighting potential (Maynard Smith & Harper 1988). For instance, the development of permanent visual markers, such as plumage patterns, skin coloration, hair styles, or fatty deposits, is usually triggered by hormones. Not surprisingly, then, the majority of these signals code for sex and age (Table 5.1: "b2"). They function to resolve conflicts between the demographic classes without escalation. Consequently, in a species in which males dominate females, a female might theoretically raise her dominance status by exhibiting a male-type signal. This advantage, however, may be achieved only at the cost of masculinization in other respects as well. Therefore, there are rather tight constraints on "faking" such signals.

Since it benefits senders to elicit responses at the cheapest possible means, receivers are expected to evolve some kind of "sales resistance" to signals that elicit responses not beneficial to them, and receivers are expected to evolve high sensitivity to information that is beneficial. For instance, male collared flycatchers, *Ficedula albicollis*, exhibit a conspicuous white forehead patch, which is larger in healthy males. During disputes over territories, males accentuate this permanent patch by fluffing the feathers on the forehead, whereupon the male with the smaller white patch defers (Pärt & Qvarnström 1997). These territory disputes occur primarily during the phase of simultaneous settlement (see Table 5.1) upon return to their breeding sites in Europe. Because signals like the flycatcher's white patch confer a competitive advantage, they are called *badges of status* (Rohwer 1982).

True badges of individual dominance status within demographic classes exist mainly in animals that form flocks with relatively unstable membership (Table 5.1; Whitfield 1987). In stable groups, by contrast, visual markers may simply aid individual recognition and contribute to dominance establishment only in association with individual assessment and memory of past encounters. In many territorial species badges signal territory ownership. In these species, a *permanent badge* may turn out to be profoundly unpractical if floaters prospecting for vacant territories cannot but signal claim of ownership: spotted by a territory holder, they elicit intense attack, since they are perceived as challenging. Several species have resolved this problem by using coverable badges (Hansen & Rohwer 1986). For instance, male red-winged blackbirds, *Agelaius phoeniceus,* bear bright red shoulder patches, "epaulets," that are coverable. Uncovering their epaulets helps owners to defend their territory. Owners are less aggressive to dummies with small or black epaulets. Clipped owners who can no longer cover their red badges are thus unable to stop signaling claim or aggressiveness. These clipped owners become confined to their own territories and receive more aggression when trespassing on their neighbors' territories. Even within their own territories, clipped males suffer more intrusions from neighbors and spend more time chasing intruding neighbors than males not permanently signaling combativeness (Metz & Weatherhead 1992). One might argue that these males were treated as if they had broken the contract of neighborliness by signaling too much combativeness and were therefore no longer accepted as partners for peaceful sharing of boundary zones.

The flexibility of signal emission is even greater for ritualized behavioral displays, the performance of which is totally context-dependent. These transient displays constitute *conventional signals* that bear a purely arbitrary relation to fighting ability. The ritualized agonistic displays performed during contests include signals of offensive and defensive aggression and of submission. In a given interaction, threat signals represent aggressive assertion, whereas submissive displays represent yielding. Often, these displays shade into one another and are graded in intensity, thus allowing a great deal of "mind reading" by the receiver (Krebs & Dawkins 1984). Traditionally, ritualized signals are held to express emotions and tendencies to act (e.g., Hinde 1985b). Threat signals, however, are not highly predictive of attack or winning (Caryl 1979). Maynard Smith (1982a) conceded that

assertive signals may actually mean "I want this resource," and Hinde (1981, 1985a) argued that signals express conditional tendencies to act, for instance, "If you try to get this resource, I will attack." Clearly, then, if the receiver accepts the condition, we will not observe any attacks. Thus, the finding that attack is not a frequent sequel of threat might simply indicate that after a threat escalation becomes obsolete (van Rhijn 1980; Preuschoft & Preuschoft 1994).

Signals thus constitute a means to negotiate the conditions under which a low-cost compromise serving either competitor's interests can be achieved. A threat might function as an opening bid, and the recipient might respond with a submissive display, instead of withdrawing, thus offering a compromise by signaling a readiness to defer but also interest in staying nearby. In this way animals can also improve accuracy of information transmission and can test whether "they read each other's minds" correctly.

Conversely, when there is no room for negotiation, because a contestant is unlikely to stand a trial, it should opt for a sneaky strategy—for example, a surprise attack—rather than a carefully selected sequence of aggravation. When perceiving an intruder while sitting alone on the nest, little blue penguin females (*Eudyptula minor*) tend to remain silent. If, however, they choose to escalate, they use the highest intensity threat without any prior warning (Waas 1991). By contrast, lone males and pairs honestly announce their readiness to escalate: individuals that vocalize are more likely to attack than those that remain silent. It appears that a lone female's low-intensity threat not only helps an intruder to verify that there is, in fact, a suitable nest site nearby but that the threat is also not sufficiently intimidating to deter the intruder. As a result, females should fare better if they remain silent and only make a show of determination once detected.

Although negotiation is expected even among opponents who are strangers (because of the mutual interest to minimize costs of competition), it plays a vastly greater role among animals that compete within the bounds of cooperation.

Contests among Familiar Animals: Dominance

In many species, individuals repeatedly meet the same conspecifics. When this happens, knowledge about fighting ability from previous contests with the same individual should inform decisions about assertion or yielding, and individualized dominance ensues (Table 5.1: decision rule "d"). Such dyadic dominance is not restricted to group-living animals. Among Indian pythons (*Python molurus*), for instance, when two unfamiliar males meet, they typically exchange agonistic acts until one has clearly overpowered the other, which retreats and displays submission. Subsequent encounters between the two are won by the same individual, but fights become progressively shorter until a mere approach by the previous winner induces the other snake to retreat (Barker et al. 1979). This example illustrates that after repeated competitive interactions with the same individual animals tend to settle for conventional low-cost behaviors: a dominance-subordination relationship is established. As in the sequential assessment (Enquist & Leimar 1983) during first encounters, each subsequent encounter adds information, thus improving the assessment of power asymmetries (van Rhijn & Vodegel 1980). Not surprisingly, then, escalated fighting is even rarer among familiar individuals than among strangers (e.g., Karavanich & Atema 1998). Assessments can be updated by intermittent subtle probing (van Rhijn 1980) and by observing interactions between other familiar individuals (Hogue et al. 1996).

Animals are expected to interact preferentially with familiar individuals, because this saves time

and risks involved in assessment and escalation. These benefits may be a major determinant of the stability of associations among many group-living or territorial animals. Rather than fleeing never to return again, competitively inferior animals may gain the superiors' tolerance by signaling submission (e.g., Preuschoft 1995). Even intention movements of withdrawal may end or prevent a confrontation. While threat displays accentuate size and weapons and elicit yielding on part of the recipient, displays of submission reduce apparent size, conceal weapons, and correlate with yielding on the part of the sender. As a rule, such advertised harmlessness effectively appeases the recipient by inhibiting aggression (e.g., Preuschoft 1992). In many group-living species submissive signals give rise to subsequent affiliative interactions (Preuschoft 1995). In essence, then, submission can be viewed as a mechanism of animals in stable groups or neighborhoods to compensate for the dispersive effects of competition.

Familiar animals with competitive-cooperative relationships are found in two major social contexts: those in which animals occupy neighboring territories and those in which they live in the same group. Territoriality and dyadic dominance are conventions, rules that tell an animal when to persist or even escalate and when to defer. In this sense, conventions are devices to manage conflicts, because they provide the rationale that governs the economy of an animal's competitive efforts by maximizing benefits while keeping the costs at a minimum. We now examine these two contexts more closely.

Dominance in Space: Territories

If an animal consistently wins conflicts with a given opponent in one place but loses in another, its dominance is site-dependent. This is called territoriality. The classic territory is an area defended by its "owner," who expels other conspecifics.

Territory owners are attached to a specific geographic area, and their behavior changes from offensive aggression and assertiveness in the center of their territory to submission and retreat beyond the territory boundary. Still, territory owners do not necessarily lead solitary lives. In many cases owners tolerate and even invite members of the other sex in their territory. Indeed, many animals become territorial only once members of the opposite sex are present (Zucker 1994b). Sometimes, owners are physically or behaviorally different from non-owners. For instance, male tree lizards (*Urosaurus ornatus*) with more blue in their dewlaps dominate those with less (Thompson & Moore 1991). When encounter rates are high, that is, at high population densities, dewlap coloration is overridden by a badge that is modifiable within minutes—darkening of the dorsal skin. This badge characterizes dominant territory holders, who also perch high and are sexually active but tolerate residency of several subordinates that do not darken dorsally (Zucker 1994a).

Not all conspecifics are equally likely to be attacked by a territory holder. Strangers of the same sex are often fiercely attacked. On the other hand, familiar, neighboring territory holders tend to limit their contests to the exchange of signals (Temeles 1994): the "dear enemy" effect. In some species, neighbors go one step further and cooperate to maintain stability of ownership. Thus, in rock pipits (*Anthus petrosus*) familiar neighbors enter into an armistice when confronted with an intruder in border zones of their territories. Neighbors also support familiar territory owners in coordinated evictions of intruders inside the familiar neighbor's territory, whereas neighbors are not tolerated in adjacent territory in the absence of intruders (Elfström 1997). As noted above, owners benefit from this mutual help, because they avoid having to renegotiate the boundaries with a new and potentially stronger male (Getty 1987).

Indeed, neighborliness seems to benefit territory holders, as floaters and challengers are attracted to areas where territory owners are engaged in boundary disputes (Beletsky 1992).

Theoretically, dominance in hierarchies and territoriality are quite distinct. Dominance of territory owners is dependent on location but independent of the relationship they have with a rival (neighbors and floaters are equally unwelcome). By contrast, dominance in groups is independent of where the encounter takes place but depends on the identity of the opponent: A dominates B in all contexts. On an abstract level, however, dominance is simply a power relation that is contingent on a condition: location in the case of territoriality, individual identity in the case of dominance between group members. Not surprisingly, both systems may also be nested with dominance depending on both the identities of the individuals involved and additional factors such as relative location or ownership conventions (Packer & Pusey 1985; Hand 1986). In capuchins (*Cebus apella*), for instance, dominants often refrain from taking easily monopolizable food from subordinates (e.g., Thierry et al. 1989b). Kummer & Cords (1991) confirmed this finding experimentally by attaching ropes to food items: respect of ownership decreased as the length of the attached rope increased. Similarly, among hamadryas baboons (*Papio hamadryas*), who live in harems nested in larger, male-bonded units, a dominant male usually respects the bond between a subordinate male and "his" female (Bachmann & Kummer 1980).

Dominance in Groups: Hierarchies

In group-living animals, dyadic dominance-subordination relationships interconnect to form networks of relationships. These networks are called dominance hierarchies, or rank orders. Hierarchies exhibit a variety of interesting features. Thus, they can be either linear (A > B > C > D >

E > F), nonlinear or triangular (A > B and B > C, but C > A), or pyramidal (A > [B = C = D = F]) or even reflect a class system ([A + B] > [C = D + E + F]). Note that features of hierarchies such as linearity or pyramidality are relational properties of networks of dyads, rather than properties of individuals or dyads.

Because individuals showing certain behavioral tendencies can be selected for or against by natural selection, whereas groups with certain properties cannot (Williams 1966), features of hierarchies need to be explained at the level of individuals or dyads. This is not to say that features of the hierarchy cannot affect an individual's fitness or the likelihood of escalated aggression. But it is impossible for an individual to change the group's hierarchy so that everybody's fitness will be enhanced (Hawkes 1992). Nonetheless, it is interesting to examine ways in which overall aggression levels are affected by strategies of individuals that seem to be aimed at maintaining the stability of existing dominance relations.

Agonistic encounters among solitary animals are usually dyadic with power asymmetries depending on intrinsic factors such as body condition, weaponry, experience, alertness, and the like. By contrast, extrinsic power is derived from an individual's ability to obtain agonistic support from coalition partners. Extrinsic power thus comes into play only when three or more individuals can be potentially engaged in a conflict. A systematic influence of third parties on the power asymmetries among other conspecifics is contingent on the formation of individualized relationships as found among territory neighbors and, most prominently, among members of social groups with stable composition. We use *alliance* as the generic term for the relationship between two individuals cooperating in agonistic contexts, including short-term "coalitions" among less loyal partners.

In the following section we present support patterns as strategies employed by third parties to manipulate power relationships among other individuals, particularly the degree of dyadic power asymmetry and the stability of dominance relationships. We distinguish between *polarizing support*, which accentuates dyadic power asymmetries, and *leveling support* given to inferior individuals, thus compensating for dyadic power asymmetries (cf. Kuester & Preuschoft unpublished). Our focus is on some gregarious primates with their tight-knit social networks, because here alliances are of overwhelming importance, making individuals mutually dependent (e.g., Harcourt & de Waal 1992). This dependency has dramatic consequences on conflict management, since one party's inclination to escalate is bounded by the other's leverage power (see below). We address the conditions under which dominance and alliance formation within groups evolve and how the relative importance of dominance and team work is expected to shape dominance styles. Closing, we return to the issue of communication by discussing interspecific variation in status signaling in the light of differences in dominance styles (cf. de Waal 1989).

Group-Level Processes: The Role of Alliances

Polarizing Support: Accentuating and Maintaining Dominance Relationships If there were no maturational changes in strength, dominance ranks should be quite stable over time. Indeed, dominance relations may be self-reinforcing whenever dominance leads to access to limiting resources. Thus, relative to the subordinate, the dominant will inevitably acquire more of the limiting resource than the subordinate and thus gain in strength, size, or number of offspring who function as alliance partners. There are, however, additional, coalitionary processes that help to maintain the stability of dominance relations. Among these is providing aggressive support by joining dominant aggressors against their subordinate targets (T). In a *conservative alliance* both the supporter (S) and the aggressor (A) are higher ranking than the target (A and S > T). If the target ranks in between the supporter and the aggressor, this is called a *bridging alliance* (e.g., A > T > S).

A special type of bridging alliance triggers rank acquisition in matrilineal societies (Chapais 1992), characteristic for many cercopithecine primates in which adult females outrank all unrelated females dominated by their mothers and sisters (e.g., Kawamura 1965 [1958]). Often, the ranks of these matrilines remain stable over several generations (e.g., Samuels et al. 1987). However, an immature daughter of a high-ranking female (highborn female) first ranks below adult, physically stronger females who are dominated by her female kin (lowborn targets). The eventual rise in rank of a highborn female to a position right below her mother is mediated socially. Not surprisingly, kin support, especially by the mother, plays an important role when maturing females challenge lowborn females (Chapais 1992).

Remarkably, unrelated females also help maturing females to attain rank positions right below their mothers by providing agonistic support against lowborn target females (reviewed in Chapais 1992). While supporters, who occupy positions intermediate between the matrilines of the target and the supported female, may profit from their cooperation by eventually gaining a powerful friend, the benefits are much less straightforward when supporters dominate both the maturing female's matriline and the target (S > A's matriline > T > A). Moreover, high-ranking supporters shape the maturing female's targeting behavior by selectively providing support against females she is "bound" to outrank but withholding support when she targets females dominant to her own matriline.

This selective support minimizes the probability of overachievement by the maturing female by preventing upwardly mobile females from forming destabilizing alliances, while ultimately securing the supporter's dominance position. Why should dominant females care about the rank order among their subordinates? The only plausible explanation for this behavior pattern seems to be that females have a genuine interest in maintaining the stability of the rank order. It seems as if violation of that "traditional" order is likely to entail considerable costs associated with matrilineal upheavals, costs that are borne by all members of a group, not only those falling in rank (see below). Nonkin support in rank acquisition by cercopithecine females thus suggests that the benefits of social stability as provided by a clear dominance hierarchy outweigh the costs of seemingly "altruistic" support by dominants. The same argument for goal-directed maintenance of social stability may also explain xenophobic reactions to immigrants (e.g., Bernstein et al. 1974; Southwick et al. 1974): the presence of a new group member may effectively shatter the existing support networks.

Leveling Support: Compensating for Power Asymmetries
Third parties may protect inferior targets from aggression either by interfering aggressively in an ongoing conflict or by pacifying interventions (Petit & Thierry, Box 13.1). Regular interventions that result in a suppression of within-group conflicts or in defying relentless dominance exertion are expected when supporters would suffer costs should they permit the out-competing of subordinates. Such costs could ensue when group cohesion is jeopardized by the dispersing effects of aggression. Insufficient group cohesion may lead to an increased vulnerability to extragroup threats such as predators or rivaling social groups. Incidentally, documentation of elaborate pacifying interventions exists for Tonkean macaques (*Macaca*

tonkeana: Petit & Thierry 1994) and chimpanzees (*Pan troglodytes:* e.g., Boehm 1994)—species that are renowned for their tolerant social manners and, at least in the case of the chimpanzee, a strong need for cooperation among group members.

Similarly, in one-male groups the dominant male is expected to have a strong interest in the compatibility of the females with whom he is associated, since the monopolizability of a harem depends on its cohesiveness. Indeed, alpha males have been reported to "police" and effectively "control" the conflicts among "their" females (e.g., hamadryas baboons, *Papio hamadryas:* Kummer 1968; chacma baboons, *P. ursinus:* Hamilton & Bulger 1992; pigtail macaques, *Macaca nemestrina:* Oswald & Erwin 1976). Remarkably, these groups were all characterized by the absence of close female-female bonds and an established matrilineal rank system. Keeping conflicts between group members in check requires the intervener to rank higher than both target and aggressor. Alternatively, efficient leveling of power asymmetries among opponents is possible only when the combined forces of supporter and target exceed that of the aggressor.

In Barbary macaques (*Macaca sylvanus*) dominance is less pronounced among males than among females (Preuschoft et al. 1998), and this difference is also reflected in clearly different support strategies: female interventions in female-female conflicts are usually polarizing and conservative of the matrilineal rank system. Males intervene regularly in conflicts among other males and almost always support the inferior, thus effectively balancing dyadic power asymmetries (Kuester & Preuschoft unpublished). The resulting stalemates among male Barbary macaques are punctuated by severe and escalated aggression, whereas the stable dominance-subordination relationships among females coincide with a lack of escalated aggression (Preuschoft & Paul 2000).

Revolutionary Alliances and Violent Overthrows   Since the consequences of matrilineal rank accumulate over time, subordinates can be expected to rebel as such a system consistently puts them on the losing side. Lethal aggression can be expected if "the value of the future is close to zero" (Enquist & Leimar 1990, p. 1). Changes in the power relationships (e.g., Gouzoules 1980; Takahata 1991), in demography or group composition (e.g., Bernstein et al. 1974; Southwick et al. 1974), and thus potentially in support networks may trigger high levels of aggression and escalation. Experiments by Chapais and colleagues (summarized in Chapais 1992, 1995) show that competition for rank is a fundamental feature of primate societies with subordinates competing as conditional opportunists. They are opportunists in that they use any change in power asymmetry against any target, and their opportunism is conditional on power asymmetries becoming pronounced and being in stark contrast with the existing hierarchy. For example, in matrilineal rank systems subordinates rebel only when they receive support against dominant targets, either from individuals ranking above the target (bridging alliances) or from a large number of coalition partners ranking below it (revolutionary alliances). But even with this minimal risk strategy matrilineal overthrows are invariably associated with major escalated fights, severe injuries, trauma, and even deaths among the otherwise so restrained contestants. In the end, whole matrilines eventually switch ranks.

Matriline upheavals in the very large Arashiyama group of Japanese macaques (*Macaca fuscata*) occurred twice over a twelve-year period (Takahata 1991). In both cases, the dramatic reshuffling of ranks among matrilines precipitated a permanent fission of the group. Similarly, the overthrow of a high-ranking matriline in one of the Salem groups of Barbary macaques by all lower-ranking matrilines was the prelude to a three-year double fissioning process (Steuckardt 1991; Kuester & Paul 1997). As Takahata (1991) put it: "Disorder in the rank order may make it difficult for females to coexist" (p. 128).

In sum, dominance in groups seems to function as a conflict management device, preventing escalated competition by conventionalizing means and priority of access, thus allowing for peaceful coexistence of group members. Violence is, however, only suspended and may break out if subordinate individuals or alliances seize opportunities to reverse existing rank relationships.

The Socioecology of Power Asymmetries

As mentioned above, overt behavioral conflicts are expected only when animals contest access to clumped resources (Wrangham 1980; van Schaik 1989). When resources are dispersed, animals depending on them will disperse, too. However, when resources are scattered but individuals benefit from each other's presence, they will be gregarious, have few conflicts over resources and minute power asymmetries, and will entertain egalitarian relationships. By contrast, when vital resources occur in small defendable patches, contest competition arises between group members. Such within-group contest favors the evolution of power asymmetries and their stabilization by reliable support from relatives (e.g., Mitchell et al. 1991), and strongly despotic dominance relationships should ensue (van Schaik 1989). When the cooperation of subordinates is indispensable to dominants, dominance may be relaxed, because subordinates can exert leverage power by withholding cooperation (Vehrencamp 1983).

Leverage is the ability to withhold a commodity, independent of intrinsic power. An individual has leverage over another when that individual possesses something that the other desires/needs but cannot acquire by force (Hand 1986). Such a

situation might arise when large coalitions are needed so that subordinates have the option of withholding support or even joining rivaling teams. In this case the capacity to motivate group members to contribute to the common goal becomes vital. Unless cooperation can be enforced, animals are expected to "shop" for suitable cooperation partners, and weaker animals may opt to team up with partners who make the most "concessions." When individuals can choose partners in a market situation, potential partners are expected to engage in self-advertisement and bargaining (Noë & Hammerstein 1994, 1995). For instance, for their triadic interactions with infants, male Barbary macaques prefer infants with whom their male partner entertains an affiliative bond (Kuester & Paul, Box 5.2). In doing so males might advertise their capacity to restrain aggressiveness and their consideration of another male's interests. Similarly, the greeting ceremonies of baboons characteristically occur among individuals whose relationships are characterized by competitiveness and mutual tolerance or even support. Greeting functions to reassure partners that are uncertain or ambivalent about their relationship (Colmenares et al., Box 5.3). Reassurance is accomplished by demonstrations of trust and trustworthiness: partners make themselves vulnerable to each other. For example, among male Barbary macaques it is usually the dominant who presents to be mounted. As mounting is accompanied by ritualized biting by the mounter (Fig. 5.1), adopting the role of the mountee is risky, particularly when tension is high. Yet this is the typical situation in which these rituals are employed (S. Preuschoft unpublished data).

Hence, when subordinates have leverage power, dyadic relationships are not despotic. Instead, tolerant relationships arise, in which there is potentially a large discrepancy between signaled dominance status and the winning of conflicts of interest.

FIGURE 5.1. Male-male mounting with ritualized biting in Barbary macaques. Photograph by Signe Preuschoft.

Formal Indicators of Subordination and Dominance

De Waal (1977, 1986; de Waal & Luttrell 1985) noted that despite some inconsistencies in the direction of winning, certain displays are perfectly unidirectional. These signals were called indicators of the formal dominance relationship between two individuals (de Waal 1986). Thus, although particular conflicts are not always won by the same member of the dyad, the directionality of a formal dominance signal remains the same regardless of the situational context: neither variation in motivation nor proximity of other group members affects who emits the signal and who receives it. Together with recent findings in cognitive ethology (e.g., Hogue et al. 1996; Delius & Siemann 1998), this suggests that some species are capable of making transitive inferences and entertaining relational concepts such as dominance (e.g., Seyfarth & Cheney 1988; Preuschoft 1999).

BOX 5.2

The Use of Infants to Buffer Male Aggression

Jutta Kuester & Andreas Paul

Close relationships between males and small infants are rare in Old World monkeys (Whitten 1987). Notable exceptions are Barbary macaques (*Macaca sylvanus*: Lahiri & Southwick 1966; Deag & Crook 1971) and also Tibetan macaques (*M. thibetana*: Deng 1993; Ogawa 1995a, 1995b). Adult and adolescent males of these species carry, huddle, groom, and protect infants that may be only a few days old. Each male establishes close relationships with only a small number of infants and usually ignores all others. On the other hand, while some infants have no contacts with males at all, very "attractive" infants may spend up to 30 percent of the daytime in contact with different males during their first months of life.

Apart from interactions between one male and an infant, interactions in which two males are involved are very frequent. These triadic male-infant interactions can be initiated by the male that carries the infant or by the male without the infant. One male approaches the other and then both sit closely together and hold the infant between them like a bridge. They lift it up, lip smack or teeth chatter, and nuzzle the infant's body and genital area. Fully adult males usually leave each other again immediately after such an interaction. If one or both males are adolescent, grooming occurs more frequently. Hence, they maintain body contact for longer periods.

Three hypotheses have been proposed to explain why Barbary macaque males interact with infants.

The agonistic buffering hypothesis was the first and assumes that the males primarily benefit from these interactions: males use infants as social tools in order to stabilize or regulate their relationships with other males. Specifically, the presence of an infant should inhibit aggression (Deag & Crook

BOX 5.2 (continued)

1971; Deag 1980). If this hypothesis is true, mainly subordinate males should carry infants and initiate triadic interactions with dominant males. Moreover, the frequency of infant carrying and triadic interactions should increase in actual or potential aggressive situations.

The investment hypothesis emphasizes the dyadic male-infant interactions and assumes that the infants primarily benefit from their relationships with males: male care is essential for survival and future reproductive success of infants (Taub 1980). If this hypothesis is true, males should prefer related infants, and infants that have relationships with males should have a higher survival/reproductive success than infants without such relationships.

The mating effort hypothesis assumes again that males primarily benefit from their relationships with infants: males "care" for infants in order to signal to the mothers their paternal qualities, thereby increasing the chances to become a preferred mating partner of those females (Taub 1990). If this hypothesis is true, "caretakers" should become preferred sexual partners, and a male should become the father of the next younger siblings of the infants for which he cared.

We tested these hypotheses by using our large data set on dyadic and triadic male-infant interactions of more than 100 males and about 150 infants belonging to a free-ranging population of Barbary macaques living in a 16-hectare enclosure. Maternal kin relations were also known from observational data, and paternal kin relations were determined by DNA fingerprinting.

The last two hypotheses were not supported by our data (Kuester & Paul 1986; Paul et al. 1992, 1996). Males showed no preferences for maternally related infants, and they established close relationships with their own offspring as often as expected by chance (similar results were found for wild groups: Ménard et al. 1992). Furthermore, "caretakers" had no greater chances than "noncaretakers" to become a preferred mating partner of the infant's mother or to sire the next infant. We could not detect any effects of male-infant relationships on survival, rank, or reproductive success of the infants (Kuester & Paul 1988; Paul & Kuester 1988).

The agonistic buffering hypothesis was tested with quantitative data on dyadic and triadic male-infant inter-

Two adult male Barbary macaques interact in a friendly manner with an infant. Photograph by Andreas Paul.

actions, and many but not all results supported this hypothesis (Paul et al. 1996). We found a positive correlation between the time males spent in contact with an infant and the number of triadic interactions they initiated. This suggests that males interact with infants *in order* to interact with other males. The comparably long periods of dyadic contacts between males and infants are prerequisites for these interactions. Dyadic contacts also guarantee the familiarity between males and "their" infants that is necessary for successful interactions.

Also in accordance with the hypothesis, the majority of carrier-initiated triadic interactions (80 percent) were initiated by males subordinate to their recipients. As predicted, triadic male-infant interactions were also over-represented among males with small rank differences, presumably the rivals with the highest risk of escalation.

However, for those interactions initiated by non-carriers there was no relationship between contact time and number of interactions and no rank bias. This may be related with the comparably unstable rank relations between males and also with the high social power of older, usually low-ranking adult males that is based on the strong support they receive during conflicts (Kuester & Preuschoft unpublished). Under such varying social conditions, it is at least plausible that also (momentarily) dominant males initiate triadic male-infant interactions.

BOX 5.2 (continued)

Aggressions between mature males increased each day during the time of food provisioning, indicating that this was a "tense" situation. As predicted, triadic male-infant interactions increased, too. However, during the time of the year with the most tension, that is, the mating season, the frequency of triadic interactions did not increase.

Infants were no "perfect buffers" against aggression. In a study based on focal observations of males, 17 percent of the approaches initiated by carriers were terminated by aggression from the prospective recipient before contact was established. Despite this, the presence of an infant seems to be a safe way to establish close contacts, among adult males perhaps the only way. Grooming between adult males was rare. We observed only 3 grooming episodes in 96 hours of observation, and all episodes were preceded by triadic male-infant interactions. If one or both males were adolescent, grooming contacts could be established without an infant, but 85 percent of their grooming episodes were also preceded by triadic interactions. Another study found that infants increase the proximity between males in general. Males had more male neighbors in the presence of infants than without infants (Deag 1980).

Only 5 percent of the triadic male-infant interactions in our study were preceded by aggressions between the participants. This further suggests that the function of these interactions is primarily "buffering," that is, aggres-

sion prevention, and not the "repair" of a damaged relationship.

These findings suggest that the relationships between males and small infants and the triadic male-infant interactions in Barbary macaques are related to conflict management. Triadic interactions occur in tense situations, but mainly before potential conflicts escalate and very rarely thereafter. The interactions allow close body contact between males, but their intense focus on the infant during these contacts reduces the risk that a gesture, mimic, or eye contact may be perceived as aggression. Triadic male-infant interactions are probably a form of conflict management but not a form of conflict resolution, since the principal source of male conflicts, that is, access to fertile females, cannot be eliminated. Triadic interactions did not increase during the mating season, suggesting that conflict management by this means reaches its limits when male-male competition is really acute. While severity of aggression among males and injuries show a peak during the mating season (Kuester & Paul 1992), the aggression rate remains low (Kuester & Preuschoft unpublished). This is probably related to risk avoidance or suppression, owing to the strong support that recipients of aggression receive from other males. The relationship between male support and triadic male-infant interactions has still to be evaluated, however. Yet unexplored are the benefits females gain from allowing males to use their very young infants for own affairs.

Here, we slightly revise de Waal's (1986) terminology by distinguishing between formal indicators of subordination and formal indicators of dominance. Displays used to indicate submission or yielding in interactions can be labeled indicators of subordination if they are performed unidirectionally in a given dyad. Similarly, displays indicating a tendency for aggression or assertion are labeled indicators of dominance when they are exchanged unidirectionally in a dyad (Preuschoft 1999). The status indicators found by de

Waal were all indicators of subordination. A number of studies, however, have now documented the existence of unidirectional indicators of dominance.

Table 5.2 summarizes evidence on unidirectional agonistic displays for various primate species. The silent bared-teeth display signals subordination in longtail (*Macaca fascicularis*) and rhesus macaques (*M. mulatta*). Other species lack formal subordination signals but have formal indicators of dominance. Stumptail macaques (*M. arctoides*)

BOX 5.3

Greeting Ceremonies in Baboons and Hyenas

Fernando Colmenares, Heribert Hofer, & Marion L. East

In humans and many nonhuman animals as well, the meeting of two strangers or of two acquaintances after a temporary separation is typically initiated by the exchange of a certain class of behavioral patterns referred to as greetings (Barrows 1995). In addition to this general context, greeting interactions are also frequent in other situations in which the interactants need to coordinate their actions with each other but are uncertain about each other's behavioral response. Two such greeting-eliciting situations are competition for resources and cooperation. Here we examine the nature and function of greeting behavior, mainly in relation to the management and resolution of conflicts between individuals, drawing on detailed studies of greeting in baboons (*Papio* spp.: Colmenares 1990, 1991) and spotted hyenas (*Crocuta crocuta:* East et al. 1993).

The most commonly held view about the causation and function of greetings is that they reduce agonistic tendencies (i.e., fear or aggression) in the interactants when there is uncertainty about each other's motivations and hence about the intensity and quality of future interactions. This is because greetings (1) signal the performer's (at least initial) tendency to use nonagonistic behavior in the exchange, (2) tend to elicit a similarly nonagonistic response in the partner, and (3) facilitate the actual occurrence of "friendly" interactions between partners that, especially in the context of conflict, are deemed as necessary for repairing, reinforcing, or establishing valuable social relationships.

Although in baboons greeting interactions may occur between individuals of any age/sex/dominance class, those occurring between adult males are especially revealing of the diversity of facial, vocal, postural, and locomotory patterns potentially involved and the probable underlying motivation. Thus, male-male greetings typically consist of a sequence of three stages of approach/retreat patterns (Colmenares 1990). In stage I, the speed

of locomotion decreases as the approaching partners get closer and closer to each other. All along, they maintain eye contact and display lip smacking and ear flattening. In stage II, when they reach close proximity, partners momentarily stop locomotion, adopt a face-to-face, rump-to-face, or side-by-side body orientation and, in some cases, make brief contact with each other (e.g., touch or grasp the other's rear or penis). In this stage, the intensity of the nonagonistic signals displayed (e.g., lip smacking and ear flattening) reaches a peak. Finally, in stage III participants move away from each other, decreasing the speed of locomotion as they move farther apart. Exposure of vulnerable parts to the other partner is also characteristic of hyena greetings. They stand head to tail, lift the hind leg nearest to the partner, raise their tails, erect their penis (males) or pseudopenis (females), and sniff and lick each other's erect "penis," scent gland, and anus.

In both baboons and hyenas we find symmetrical or asymmetrical greetings depending on whether partners use the same or different patterns during the interaction (Colmenares 1990; East et al. 1993). Sometimes, greeting initiations are not responded to with other greeting patterns; these are considered unreciprocated greeting interactions (Colmenares 1990).

Greeting patterns may take place in several contexts. They are often displayed during conflict-provoking situations, such as the meeting of two group members after a temporary separation, the presence of some competition-eliciting attractive resource (e.g., space, food, mates, partners), or the maintenance of interactions that require mutual agreement between the partners in the fulfilling of roles (e.g., in play or grooming interactions). For example, hyenas frequently exchange greetings when they meet another clan member (Kruuk 1972; East et al. 1993). Male hamadryas baboons (*Papio hamadryas*) competing over females often greet each other before any overt agonistic signal has been exchanged (Kummer et al. 1974; Colmenares 1991). Juvenile baboons also use greeting during play episodes when the actions of one or both partners are getting very rough. They interrupt the play interaction, greet each other, and then resume play (Colmenares unpublished observations).

BOX 5.3 (continued)

Sequence of patterns exchanged during a greeting ceremony between male hamadryas baboons. (a) Male Aa (on the left) stops before male Am (on the right), orients the rear toward him, and lip smacks at him. Am responds by flattening the ears (with scalp retraction) and lip smacking at Aa. (b) Am moves the rear toward Aa so that they stand side by side and reaches out to touch Aa's rear. They both carry on lip smacking and ear flattening at each other. (c) Am keeps the rear presentation, while Aa moves toward Am and gently touches his genitalia. Photographs by Félix Zaragoza.

Greeting can also be used during ongoing aggressive conflicts. For example, a male baboon can direct greeting patterns (lip smacking, ear flattening, grasping of the rear, mounting, or embracing) toward a female that he has just attacked during herding. In addition, bystanders can direct greeting patterns toward combatants. For example, in one-male units of hamadryas baboons, the harem-holding male often interferes in quarrels among females. About one-fourth of these male interventions involve greeting patterns (e.g., embracing or mounting) directed at the target of aggression (Zaragoza & Colmenares unpublished).

Greeting also occurs soon after conflicts between former antagonists (i.e., reconciliation) or between antagonists and bystanders (East et al. 1993; Zaragoza & Colmenares unpublished). Thus, three major con-text-based types of greetings can be identified: pre-conflict, conflict, and post-conflict greetings. Reassurance, appeasement, and assessment of the partner's tendencies seem to be the most common functions of these greetings.

An analysis of 1,583 greetings of 20 adult male baboons showed that greetings were most common and most often unreciprocated when the partners had unclear dominance relationships with each other or when they were contesting the access to females already "owned" by one of them. Symmetrical greetings were most frequent when both partners were closely matched in fighting ability (Colmenares 1990, 1991). Similar findings have been obtained in other baboon studies (wild *Papio cynocephalus anubis*: Smuts & Watanabe 1990; captive *P. hamadryas*: Müller & Colmenares unpublished).

BOX 5.3 (continued)

In a detailed field study of 3,396 greetings of the Serengeti spotted hyenas, which live in female-dominated societies, greetings were exchanged mostly between adult individuals of similar rank (East et al. 1993). The exchange of symmetrical gestures is positively related to the degree of symmetry (or similarity) in dominance rank.

In hyenas, subordinate females greet dominants more often than the reverse. In hamadryas baboons, the direction of greetings between dominant, harem-holding males and subordinate, follower males is determined by the degree of asymmetry in their relationship. Typically, in the beginning of their relationship, the young follower approaches, presents its rear to, and allows itself to be grasped or mounted by the harem male. He also responds in this way when approached by the harem male. However, in later stages, when rivalry over females increases, the role of initiator is reversed; now it is the harem male who repeatedly directs greeting approaches to the follower, especially when the latter refuses to greet back (Kummer et al. 1974; Colmenares 1991).

In summary, the form of greeting exchanges, the contexts in which they occur most commonly (e.g., highly competitive situations), and the identity of the most frequent performers, that is, high rankers with potentially ambiguous dominance relationships (males in baboons and females in hyenas), suggest that greeting interactions are often driven by motivational ambivalence.

Although sometimes the approach of a greeting individual elicits an aggressive response from the recipient in both baboons and hyenas, a reduction in the recipient's agonistic motivation (i.e., fear or aggression) seems to be the most frequent immediate effect of greeting. Thus, the greeting of subordinates reduces the dominant partner's agonistic tendencies and increases the latter's social tolerance of the subordinates. The greeting of dominants, on the other hand, reduces the subordinate partner's fear and flight tendencies, facilitating social interaction and group cohesion. Therefore, in baboons and in hyenas, greeting is an important mechanism of conflict management.

Greeting ceremonies might also be interpreted as instances of negotiation (e.g., Colmenares 1990, 1991; Smuts & Watanabe 1990; de Waal 1996; Smith 1997). Greetings are common in contexts in which partners in a social relationship disagree about their roles and behavioral options. By greeting, individuals may be testing each other's interests, influencing each other's willingness to make concessions, and negotiating the shape and course of their mutual and interdependent relationship while avoiding aggressive escalation.

express formal dominance by a ritualized mock bite directed at a usually cooperative subordinate, whereas in Barbary macaques and gelada baboons (*Theropithecus gelada*) dominant status is manifested by a threat face. Yet systematic studies on other species, such as Tonkean and liontail (*M. silenus*) macaques failed to find unidirectional agonistic signals altogether.

The use of a formal status signal within a species is sometimes characteristic of only a specific category of dyads. For instance, among Barbary macaques the rounded-mouth threat (Fig. 5.2) is regularly exchanged among members of all demographic classes—but not among top-ranking adult males (Deag 1974; Preuschoft et al. 1998). Chimpanzees use a vocal-gestural signal of subordination consisting of repetitive pant-grunting and bowing toward the dominant. In this male-bonded species, pant-grunting is typically directed at dominant adult males, whereas female status relationships are more ambiguous (Nishida & Hiraiwa-Hasegawa 1987) and are rarely expressed in pant-grunting (Noë et al. 1980).

The evidence compiled in Table 5.2 suggests that there is an association between, on the one hand, the degree of power asymmetry and, on the other hand, the presence of formal indicators of subordination or of dominance. We will now pro-

TABLE 5.2
Unidirectional Status Indicators in Some Primate Species

Power Relations	Species	Signal	Status of Sender	Unidirectionality	% One-Way Relationships	Source
strongly asymmetrical	*Macaca mulatta*	silent bared-teeth	subordinate	.99	97	1
	Macaca fascicularis	silent bared-teeth	subordinate	1	99.5	2
	Macaca arctoides	mock bite	dominant	yes	95	3
	Macaca sylvanus	rounded-mouth	dominant	.99	96	4
	Theropithecus gelada	rounded-mouth	dominant	.97	97	5
more symmetrical	*Macaca silenus*	—	—	—	—	6
	Macaca tonkeana	—	—	—	—	7

Source: 1: de Waal & Luttrell 1985; 2: Preuschoft et al. 1995; 3: Demaria & Thierry 1990; 4: Deag 1974; 5: Reichler 1996; 6: Tennemann 1992; Beckmann 1997; 7: Thierry et al. 1989a; Preuschoft 1995.

Note: Unidirectionality is expressed by the directional consistency index (DC-index) that is calculated across all dyads and reflects the frequency with which the behavior occurred in its more frequent direction relative to the total number of times the behavior occurred. The DC-index ranges from 0 (perfect bidirectionality) to 1 (perfect unidirectionality; cf. van Hooff & Wensing 1987). "Yes": Demaria & Thierry contained no data matrices on which comparative calculations could be based.

One-way relationships are expressed as the percentage of dyads seen to exchange the display. Note that all signals were also significantly linear between dyads.

Degree of power asymmetry: species are classified according to the relationships in the resident sex, here all females. Assessment is based on comparative studies of competitive tactics, communication, conflict prevention, and resolution (Thierry 1985; de Waal 1989, 1996; Preuschoft 1995; Aureli et al. 1997).

FIGURE 5.2. A female Barbary macaque directs a rounded-mouth threat face at the playmate of her infant. Photograph by Signe Preuschoft.

pose a deductive argument to explain this pattern (Fig. 5.3).

Predictability of Power Asymmetries We can confidently assume that primates have rather accurate information about power relationships within their groups (e.g., Byrne & Whiten 1988). Therefore individuals should "formalize" (de Waal 1986) their power relationship whenever intrinsic power asymmetries are so pronounced, or support by third parties is provided so predictably, that a violent contest would be redundant and thus can be replaced by the unidirectional exchange of signals ("threat right," cf. van Rhijn & Vodegel 1980; see Fig. 5.3). When predictability is low, however, a considerable degree of probing and assessment is needed to manage conflicts. This occurs when intrinsic power asymmetry is low or when support patterns and efficiency are highly variable across situations. In such situations, we expect that ritualized agonistic signals not be exchanged unidirectionally, that is, dominance will be context-dependent, and there will be no formalization of dominance status (Fig. 5.3).

This framework may explain the variation in agonistic signaling documented in Table 5.2. In despotic dominance relationships, subordinates have virtually no leverage power. Subordinate individuals take the initiative and spontaneously signal their assessment of being competitively inferior, thus preventing further escalation and gaining peaceful coexistence in exchange for yielding. The advantage for subordinates is that they save the energy they would have to spend otherwise on withdrawal and avoidance and that they maintain or gain access to preferred sites or social partners. Preuschoft et al. (1995) show exactly this effect of the subordination signal in longtail macaques: time spent in affiliative contact declines during the two minutes before an individual shows the silent bared-teeth signal, suggesting that the display is a response to a deterioration in access to huddles or grooming clusters, but it increases again after the signal is given.

By contrast, in more tolerant dominance relationships, because of their mutual dependence, dominants do not habitually chase subordinates away from resources (e.g., Matsumura 1998). Consequently, inferior animals do not constantly have to solicit tolerance and will therefore not show spontaneous, unprovoked displays of subordination. In certain situations, such as mating or feeding, however, dominants are expected to insist in taking the priority that their greater power confers to them. Formal indicators of dominance allow dominant individuals to withhold potentially disruptive claims unless they can be certain that the costs of assertion are exceeded by the benefits gained.

Finally, if competitive outcomes are unpredictable, either because situational factors regularly override dominance (extremely tolerant) or because dominance is not established in the first place (egalitarian), all agonistic signals should be context-dependent and hence not unidirectional. It is possible that the Tonkean and liontail macaques are like this. In both species agonistic interactions regularly involve mild physical contact, recruitment signals, and third-party intervention (Beckmann 1997; Thierry, Chapter 6).

In sum, in despotic societies conflicts are usually prevented by spontaneous submission, and overt aggression is relatively rare but escalated. In more "tolerant" societies, subordinates are reluctant to offer submission in exchange for toleration, and dominants are forced to induce submission and compliance in subordinates actively by using formal threat signals (e.g., Reichler et al. 1997). The constant probing and calculated generosity so characteristic of more symmetrical power relations is also found in chimpanzees. In this species, however, dominance is not granted on the condition that an individual can induce subordination, but subordination is the key to equality: under the condition of formal subordination, dominants reward subordinates not only with tolerance but even by sharing food or matings (de Waal 1982). Although this mechanism appears to be present in some dyads of longtail macaques (de Waal 1977), full-fledged conditional reassurance (de Waal 1986) is probably a derived feature of chimpanzees, perhaps because it presupposes cognitive capacities that macaques do not possess.

Conclusion

We examined how dominance and communication of power asymmetries serve as devices to man-

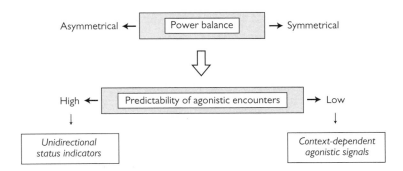

FIGURE 5.3. The degree of power asymmetry in a dyad results from intrinsic and extrinsic factors. When power is distributed rather asymmetrically, the predictability of winning is high; hence dominance status can be formalized by the unidirectional display of status indicators. When power is more balanced, the predictability of outcomes is low, and negotiations involve the exchange of context-dependent agonistic signals.

age the rate of aggression and the likelihood that conflicts will escalate into physical fighting. We argued that escalation is generally delayed as long as possible, even when strangers come into conflict, by stepwise assessment of the relative fighting power of the contestants. Escalation can also be avoided altogether by the use of signals that reliably reflect motivation and fighting ability or by arbitrary conventions. Among animals with individualized relationships in which dominance is established, fights are mostly replaced by the exchange of signals. Additional social processes in groups or territory neighborhoods, mainly involving third-party interventions, apparently function to maintain stability of power relations. Thus, an array of surprisingly sophisticated behaviors serve to avoid escalation of conflicts into damaging fights.

References

Archer, J., & Huntingford, F. 1994. Game theory models and escalation of animal fights. In: *The Dynamics of Aggression* (M. Potegal & J. F. Knutson, eds.), pp. 3–31. Hillsdale, N.J.: Lawrence Erlbaum Associates.

Aureli, F., Das, M., & Veenema, H. 1997. Differential kinship effect on reconciliation in three species of macaques (*Macaca fascicularis*, *M. fuscata* and *M. sylvanus*). *Journal of Comparative Psychology*, 111: 91–99.

Axelrod, R. 1997. *The Complexity of Cooperation.* Princeton: Princeton University Press.

Axelrod, R., & Hamilton, W. D. 1981. The evolution of cooperation. *Science*, 211: 1390–1396.

Bachmann, C., & Kummer, H. 1980. Male assessment of female choice in Hamadryas baboons. *Behavioral Ecology and Sociobiology*, 6: 315–321.

Barker, D. G., Murohy, J. B., & Smith, K. W. 1979. Social behaviour in a captive group of Indian pythons, *Python moluras*, with formation of linear social hierarchy. *Copeia*, 466–471.

Barrows, E. M. 1995. *Animal Behavior Desk Reference.* Boca Raton, Fla.: Chemical Rubber Company Press.

Beckmann, F. 1997. Vergleichende Untersuchungen zur Mimik bei Bartaffen (*Macaca silenus L.*). Diplomarbeit, Freie Universität Berlin.

Beletsky, L. D. 1992. Social stability and territory acquisition in birds. *Behaviour*, 123: 290–313.

Bernstein, I. S. 1981. Dominance: The baby and the bathwater. Incl. Open Peer Commentary. *Behavioral and Brain Sciences*, 4: 419–557.

Bernstein, I. S., Gordon, T. G., & Rose, R. M. 1974. Factors influencing the expression of aggression during introductions to rhesus monkey groups. In: *Primate Aggression, Territoriality, and Xenophobia: A Comparative Perspective* (R. L. Holloway, ed.), pp. 211–240. New York: Academic Press.

Boehm, C. 1994. Pacifying interventions at Arnhem Zoo and Gombe. In: *Chimpanzee Cultures* (R. W. Wrangham, W. C. McGrew, F. B. M. de Waal, &

P. G. Heltne, eds.), pp. 211–242. Cambridge: Harvard University Press.

Boerlijst, M. C., Nowak, M. A., & Sigmund, K. 1997. The logic of contrition. *Journal of Theoretical Biology*, 185: 281–293.

Byrne, R. 1995. *The Thinking Ape*. Oxford: Oxford University Press.

Byrne, R. W., & Whiten, A. 1988. *Machiavellian Intelligence: Social Expertise and the Evolution of Intellect in Monkeys, Apes, and Humans*. Oxford: Clarendon Press.

Caryl, P. G. 1979. Communication by agonistic displays: What can games theory contribute to ethology? *Behaviour*, 68: 137–169.

Caryl, P. G. 1981. Escalated fighting and the war of nerves: Games theory and animal combat. In: *Perspectives in Ethology*, Vol. 4 (P. P. G. Bateson & H. P. Klopfer, eds.), pp. 199–224. London: Plenum Press.

Caryl, P. G. 1982. Animal signals: A reply to Hinde. *Animal Behaviour*, 30: 240–244.

Chapais, B. 1992. The role of alliances in the social inheritance of rank among female primates. In: *Coalitions and Alliances in Humans and Other Animals* (A. Harcourt & F. B. M. de Waal, eds.), pp. 29–59. Oxford: Oxford University Press.

Chapais, B. 1995. Alliances as a means of competition in primates: Evolutionary, developmental, and cognitive aspects. *Yearbook of Physical Anthropology*, 38: 115–136.

Chase, I. D., Bartolomeo, C., & Dugatkin, L. A. 1994. Aggressive interactions and inter-contest interval: How long do winners keep winning? *Animal Behaviour*, 48: 393–400.

Chase, I. D. 1986. Explanations of hierarchy structure. *Animal Behaviour*, 34: 1265–1266.

Clutton-Brock, T. H., Guinness, F. E., & Albon, S. D. 1982. *Red Deer: Behaviour and Ecology of the Two Sexes*. Edinburgh: Edinburgh University Press.

Clutton-Brock, T. H., & Parker, G. A. 1995. Punishment in animal societies. *Nature*, 373: 209–216.

Colmenares, F. 1990. Greeting behaviour in male baboons, I: Communication, reciprocity and symmetry. *Behaviour*, 113: 81–116.

Colmenares, F. 1991. Greeting behaviour between male baboons: Oestrous females, rivalry and negotiation. *Animal Behaviour*, 41: 49–60.

Dawkins, R., & Krebs, J. R. 1978. Animal signals: Information or manipulation? In: *Behavioral Ecology: An Evolutionary Approach* (R. Krebs & N. B. Davies, eds.), pp. 282–309. Oxford: Blackwell.

Deag, J. M. 1974. A Study of the Social Behaviour and Ecology of the Wild Barbary Macaque, *Macaca sylvanus* L. Ph.D. diss., University of Bristol.

Deag, J. M. 1980. Interactions between males and unweaned Barbary macaques: Testing the agonistic buffering hypothesis. *Behaviour*, 75: 54–81.

Deag, J. M., & Crook, J. H. 1971. Social behaviour and "agonistic buffering" in the wild Barbary macaque *Macaca sylvana* L. *Folia primatologica*, 15: 183–200.

Delius, J. D., & Siemann, M. 1998. Transitive responding in animals and humans: Exaptation rather than adaptation? *Behavioural Processes*, 42: 107–137.

Demaria, C., & Thierry, B. 1990. Formal biting in stumptailed macaques (*Macaca arctoides*). *American Journal of Primatology*, 20: 133–140.

Deng, Z.-Y. 1993. Social development of infants of *Macaca thibetana* at Mount Emei, China. *Folia primatologica*, 60: 28–35.

de Waal, F. B. M. 1977. The organization of agonistic relationships within two captive groups of Java-Monkeys (*Macaca fascicularis*). *Zeitschrift für Tierpsychologie*, 44: 225–282.

de Waal, F. B. M. 1982. *Chimpanzee Politics*. New York: Harper & Row.

de Waal, F. B. M. 1986. The integration of dominance and social bonding in primates. *Quarterly Review of Biology*, 61: 459–479.

de Waal, F. B. M. 1989. Dominance "style" and primate social organization. In: *Comparative Socioecology* (V. Standon & R. A. Foley, eds.), pp. 243–264. Oxford: Blackwell.

de Waal, F. B. M. 1996. Conflict as negotiation. In: *Great Ape Societies* (W. C. McGrew, L. F. Marchant, & T. Nishida, eds.), pp. 159–172. Cambridge: Cambridge University Press.

de Waal, F. B. M., & Aureli, F. 1997. Conflict resolution and distress alleviation in monkeys and apes. In: *The Integrative Neurobiology of Affiliation* (S. C. Carter, I. I. Lederhendler, & B. Kirkpatrick, eds.), pp. 317–328. New York: New York Academy of Sciences.

de Waal, F. B. M., & Luttrell, L. 1985. The formal hierarchy of rhesus monkeys: An investigation of the bared-teeth display. *American Journal of Primatology*, 9: 73–85.

Dugatkin, L. A., & Reeve, H. D. 1998. *Game Theory and Animal Behavior*. Oxford: Oxford University Press.

Dunbar, R. I. M. 1988. *Primate Social Systems*. London: Croom and Helm.

East, M. L., Hofer, H., & Wickler, W. 1993. The erect "penis" is a flag of submission in a female-dominated society: Greetings in Serengeti spotted hyenas. *Behavioral Ecology and Sociobiology*, 33: 355–370.

Elfström, S. T. 1997. Fighting behaviour and strategy of rock pipits, *Anthus petrosus*, neighbours: Cooperative defence. *Animal Behaviour*, 54: 535–542.

Enquist, M., & Leimar, O. 1983. Evolution of fighting behaviour: Decision rules and assessment of relative strength. *Journal of Theoretical Biology*, 102: 387–410.

Enquist, M., & Leimar, O. 1990. The evolution of fatal fighting. *Animal Behaviour*, 39: 1–9.

Enquist, M., Leimar, O., Ljungberg, T., Malner, Y., & Segerdahl, N. 1990. A test of the sequential assessment game: Fighting in the cichlid fish *Nannacara anomala*. *Animal Behaviour*, 40: 1–14.

Getty, T. 1987. Dear enemies and the prisoner's dilemma: Why should territorial neighbours form defensive coalitions? *Animal Behaviour*, 27: 327–336.

Gouzoules, H. 1980. A description of genealogical rank changes in a troop of Japanese monkeys (*Macaca fuscata*). *Primates*, 21: 262–267.

Grafen, A. 1990. Biological signals as handicaps. *Journal of Theoretical Biology*, 144: 517–546.

Hamilton, W. J., & Bulger, J. 1992. Facultative expression of behavioral differences between one-male and multimale savanna baboon groups. *American Journal of Primatology*, 28: 61–71.

Hammerstein, P. 1981. The role of asymmetries in animal contests. *Animal Behaviour*, 29: 193–205.

Hand, J. L. 1986. Resolution of social conflicts: Dominance, egalitarianism, spheres of dominance, and game theory. *Quarterly Review of Biology*, 61: 201–220.

Hansen, A. J., & Rohwer, S. 1986. Coverable badges and resource defence in birds. *Animal Behaviour*, 34: 69–76.

Harcourt, A. H., & de Waal, F. B. M. 1992. *Coalitions and Alliances in Humans and Other Animals*. Oxford: Oxford University Press.

Hawkes, K. 1992. Sharing and collective action. In: *Evolutionary Ecology and Human Behavior* (E. A. Smith & B. Winterhalder, eds.), pp. 269–300. New York: Aldine de Gruyter.

Hinde, R. A. 1976. Interactions, relationships and social structure in non-human primates. *Man*, 11: 1–17.

Hinde, R. A. 1981. Animal signals: Ethological and games theory approaches are not incompatible. *Animal Behaviour*, 29: 535–542.

Hinde, R. A. 1985a. Expression and negotiation. In: *The Development of Expressive Behavior* (G. Zivin, ed.), pp. 103–116. Orlando, Fla.: Academic Press.

Hinde, R. A. 1985b. Was "The Expression of the Emotions" a misleading phrase? *Animal Behaviour*, 33: 985–992.

Hogue, M. E., Beaugrand, J. P., & Lagué, P. C. 1996. Coherent use of information by hens observing their former dominant defeating or being defeated by a stranger. *Behavioural Processes*, 38: 241–252.

Huntingford, F. A., & Turner, A. K. 1987. *Animal Conflict*. London: Chapman and Hall.

Karavanich, C., & Atema, J. 1998. Individual recognition and memory in lobster dominance. *Animal Behaviour*, 56: 1553–1560.

Kawamura, S. 1965 [1958]. Matriarchal social ranks in the Minoo-B troop: A study of the rank system of Japanese monkeys. In: *Japanese Monkeys* (K. Imanishi & S. Altmann, eds.), pp. 66–104. Atlanta: Emory University.

Krebs, J. R., & Dawkins, R. 1984. Animal signals: Mind reading and manipulation. In: *Behavioural Ecology* (J. R. Krebs & N. B. Davies, eds.), pp. 380–402. Oxford: Blackwell Scientific.

Kruuk, H. 1972. *The Spotted Hyaena*. Chicago: University of Chicago Press.

Kuester, J., & Paul, A. 1986. Male-infant relationships in semi-free-ranging Barbary macaques (*Macaca sylvanus*) of Affenberg Salem/FRG: Testing the "male care" hypothesis. *American Journal of Primatology*, 10: 315–327.

Kuester, J., & Paul, A. 1988. Rank relations of juvenile and subadult natal males of Barbary macaques (*Macaca sylvanus*) at Affenberg Salem. *Folia primatologica*, 51: 33–44.

Kuester, J., & Paul, A. 1992. Influence of male competition and female mate choice on male mating success in Barbary macaques (*Macaca sylvanus*). *Behaviour*, 120: 192–217.

Kuester, J., & Paul, A. 1997. Group fission in Barbary macaques (*Macaca sylvanus*) at Affenberg Salem. *International Journal of Primatology*, 18: 941–966.

Kummer, H. 1968. *Social Organization of Hamadryas Baboons*. Chicago: University of Chicago Press.

Kummer, H., & Cords, M. 1991. Cues of ownership in long-tailed macaques, *Macaca fascicularis*. *Animal Behaviour*, 42: 529–549.

Kummer, H., Götz, W., & Angst, W. 1974. Triadic differentiation: An inhibitory process protecting pair bonds in baboons. *Behaviour*, 49: 62–87.

Lahiri, R. K., & Southwick, C. H. 1966. Parental care in *Macaca sylvana*. *Folia primatologica*, 4: 257–264.

Lee, P. C. 1994. Social structure and evolution. In: *Behaviour and Evolution* (P. J. B. Slater & T. R. Halliday, eds.), pp. 266–303. Cambridge: Cambridge University Press.

Matsumura, S. 1998. Relaxed dominance relations among female Moor macaques (*Macaca maurus*) in their natural habitat, South Sulawesi, Indonesia. *Folia primatologica*, 69: 346–356.

Maynard Smith, J. 1982a. Do animals convey information about their intentions? *Journal of Theoretical Biology*, 97: 1–5.

Maynard Smith, J. 1982b. *Evolution and the Theory of Games*. Cambridge: Cambridge University Press.

Maynard Smith, J., & Harper, D. G. C. 1988. The evolution of aggression: Can selection generate variability? *Philosophical Transactions of the Royal Society*, ser. B, 319: 557–570.

Maynard Smith, J., & Parker, G. A. 1976. The logic of asymmetric contests. *Behaviour*, 24: 159–175.

Maynard Smith, J., & Price, G. R. 1973. The logic of animal conflict. *Nature*, 246: 15–18.

Ménard, N., Scheffrahn, W., Vallet, D., Zidane, C., & Reber, C. 1992. Application of blood protein electrophoresis and DNA fingerprinting to the analysis of paternity and social characteristics of wild Barbary macaques. In: *Paternity in Primates: Genetic Tests and Theories* (R. D. Martin, A. F. Dixson, & E. J. Wickings, eds.), pp. 155–174. Basel: Karger.

Metz, K. J., & Weatherhead, P. J. 1992. Seeing red: Uncovering coverable badges in red-winged blackbirds. *Animal Behaviour*, 43: 223–229.

Mitchell, C. L., Boinski, S., & van Schaik, C. P. 1991. Competitive regimes and female bonding in two species of squirrel monkeys (*Saimiri oerstedi* and *S. sciureus*). *Behavioral Ecology and Sociobiology*, 28: 55–60.

Nicholson, A. J. 1954. An outline of the dynamics of animal populations. *Australian Journal of Zoology*, 2: 9–65.

Nishida, T., & Hiraiwa-Hasegawa, M. 1987. Chimpanzees and bonobos: Cooperative relationships among males. In *Primate Societies* (B. B. Smuts, D. L. Cheney, R. M. Seyfarth, R. W. Wrangham, & T. T. Struhsaker, eds.), pp. 165–178. Chicago: University of Chicago Press.

Noë, R., & Hammerstein, P. 1994. Biological markets: Supply and demand determine the effect of partner choice in cooperation, mutualism and mating. *Behavioral Ecology and Sociobiology*, 35: 1–11.

Noë, R., & Hammerstein, P. 1995. Biological markets. *Trends in Ecology and Evolution*, 10: 336–339.

Noë, R., de Waal, F. B. M., & van Hooff, J. A. R. A. M. 1980. Types of dominance in a chimpanzee colony. *Folia primatologica*, 34: 90–110.

Nowak, M. A., & Sigmund, K. 1993. A strategy of win-stay, lose-shift that outperforms tit-for-tat in the Prisoner's Dilemma game. *Nature*, 364: 56–58.

Ogawa, H. 1995a. Bridging behavior and other affiliative interactions among male Tibetan macaques (*Macaca thibetana*). *International Journal of Primatology*, 16: 707–729.

Ogawa, H. 1995b. Recognition of social relationships in bridging behavior among Tibetan macaques (*Macaca thibetana*). *American Journal of Primatology*, 35: 305–310.

Okamoto, K. 1998. Punishment and apology in the iterated prisoner's dilemma. *Primate Research*, 14: 217–219 (in Japanese).

Oswald, M., & Erwin, J. 1976. Control of intragroup aggression by male pigtail monkeys (*Macaca nemestrina*). *Nature*, 262: 686–687.

Packer, C., & Pusey, A. 1985. Asymmetric contests in social mammals: Respect, manipulation and age-specific aspects. In: *Evolution: Essays in Honour of John Maynard Smith* (P. J. Greenwood, P. H. Harvey, & M. Slatkin, eds.), pp. 173–186. Cambridge: Cambridge University Press.

Pärt, T., & Qvarnström, A. 1997. Badge size in collared flycatchers predicts outcome of male competition over territories. *Animal Behaviour*, 54: 893–899.

Paul, A., & Kuester, J. 1988. Life history patterns of semi-free-ranging Barbary macaques (*Macaca sylvanus*) at Affenberg Salem (FRG). In: *Ecology and Behavior of Food-Enhanced Primate Groups* (J. E. Fa & C. H. Southwick, eds.), pp. 199–228. New York: Alan R. Liss.

Paul, A., Kuester, J., & Arnemann, J. 1992. DNA fingerprinting reveals that infant care by male Barbary macaques (*Macaca sylvanus*) is not paternal investment. *Folia primatologica*, 58: 93–98.

Paul, A., Kuester, J., & Arnemann, J. 1996. The sociobiology of male-infant interactions in Barbary macaques, *Macaca sylvanus*. *Animal Behaviour*, 51: 155–170.

Petit, O., & Thierry, B. 1994. Aggressive and peaceful interventions in conflicts in Tonkean macaques. *Animal Behaviour*, 48: 1427–1436.

Popp, J. L., & DeVore, I. 1979. Aggressive competition and social dominance theory: Synopsis. In: *The Great Apes* (D. Hamburg & E. McCown, eds.), pp. 316–338. Menlo Park, Calif.: Benjamin Cummings.

Preuschoft, S. 1992. "Laughter" and "smile" in Barbary macaques (*Macaca sylvanus*). *Ethology*, 91: 200–236.

Preuschoft, S., ed. 1995. *"Laughter" and "Smiling" in Macaques: An Evolutionary Perspective*. Utrecht: University of Utrecht.

Preuschoft, S. 1999. Are primates behaviorists? Formal dominance, cognition, and free-floating rationales. *Journal of Comparative Psychology*, 113: 1–5.

Preuschoft, S., & Paul, A. 2000. Dominance, egalitarianism and stalemate: An experimental approach to male-male competition in Barbary macaques. In:

Primate Males (P. M. Kappeler, ed.), pp. 205–216. Cambridge: Cambridge University Press.

Preuschoft, S., & Preuschoft, H. 1994. Primate nonverbal communication: Our communicatory heritage. In: *Origins of Semiosis* (W. Nöth, ed.), pp. 15–58. Berlin: Mouton de Gruyter.

Preuschoft, S., Gevers, E., & van Hooff, J. A. R. A. M. 1995. Functional differentiation in the affiliative facial displays of longtailed macaques (*Macaca fascicularis*). In: *"Laughter" and "Smiling" in Macaques: An Evolutionary Perspective* (S. Preuschoft, ed.), pp. 59–88. Utrecht: University of Utrecht.

Preuschoft, S., Paul, A., & Kuester, J. 1998. Dominance styles of female and male Barbary macaques. *Behaviour*, 135: 731–755.

Pusey, A. E., & Packer, C. 1997. The ecology of relationships. In: *Behavioural Ecology: An Evolutionary Approach*, 4th edn. (J. R. Krebs & N. B. Davies, eds.), pp. 254–283. London: Blackwell.

Reichler, S. 1996. Untersuchungen zum Dominanzstil bei Geladas, *Theropithecus gelada*. Diplomarbeit, Ruhr-Universität Bochum.

Reichler, S., Dücker, S., & Preuschoft, S. 1997. Agonistic behaviour and dominance style in gelada baboons (*Theropithecus gelada*). *Folia primatologica*, 69: 200.

Riechert, S. E. 1998. Game theory and animal contests. In: *Game Theory and Animal Behavior* (L. A. Dugatkin & H. K. Reeve, eds.), pp. 64–93. Oxford: Oxford University Press.

Rohwer, S. 1982. The evolution of reliable and unreliable badges of fighting ability. *American Zoologist*, 22: 531–546.

Samuels, A., Silk, J. B., & Altmann, J. 1987. Continuity and change in dominance relations among female baboons. *Animal Behaviour*, 35: 785–793.

Scott, J. P. 1958. *Aggression*. Chicago: University of Chicago Press.

Seyfarth, R. M., & Cheney, D. L. 1988. Do monkeys understand their relations? In: *Machiavellian Intelligence* (R. Byrne & A. Whiten, eds.), pp. 69–84. Oxford: Clarendon Press.

Slater, P. J. B. 1986. Individual differences and dominance hierarchies. *Animal Behaviour*, 34: 1264–1265.

Smith, W. J. 1997. The behavior of communicating, after twenty years. In: *Perspectives in Ethology*, Vol. 12: *Communication* (D. H. Owings, M. D. Beecher, & N. S. Thompson, eds.), pp. 7–53. New York: Plenum Press.

Smuts, B. B., & Watanabe, J. M. 1990. Social relationships and ritualized greetings in adult male baboons (*Papio cynocephalus anubis*). *International Journal of Primatology*, 11: 147–172.

Southwick, C. H., Siddiqi, M. F., Farooqui, M. Y., & Pal, B. C. 1974. Xenophobia among free-ranging rhesus groups in India. In: *Primate Aggression, Territoriality and Xenophobia* (R. L. Holloway, ed.), pp. 185–210. New York: Academic Press.

Steuckardt, A. 1991. Der Einfluß der Familiengröße auf das Beziehungsnetz adulter Berberaffenweibchen (*Macaca sylvanus L.* 1758). Diplomarbeit, Universität Göttingen.

Takahata, Y. 1991. Diachronic changes in the dominance relations of adult female Japanese monkeys of the Arashiyama B group. In: *The Monkeys of Arashiyama: Thirty-five Years of Research in Japan and the West* (L. M. Fedigan & P. J. Asquith, eds.), pp. 123–139. Albany: State University of New York Press.

Taub, D. M. 1980. Testing the "Agonistic Buffering" hypothesis I: The dynamics of participation in the triadic interactions. *Behavioral Ecology and Sociobiology*, 6: 187–197.

Taub, D. M. 1990. The functions of primate paternalism: A cross-species review. In: *Pedophilia: Biosocial Dimensions* (J. R. Feierman, ed.), pp. 338–377. New York: Springer.

Temeles, E. J. 1994. The role of neighbours in territorial systems: When are they "dear enemies"? *Animal Behaviour*, 47: 339–350.

Tennemann, A. 1992. Soziale Beziehungen und Verhaltensprofile untersucht an sechs Bartaffengruppen (*Macaca silenus L.*) unter verschiedenen Haltungsbedingungen. Diplomarbeit, Unversität Köln.

Thierry, B. 1985. Coadaptation des variables sociales: L'example des sytèmes sociaux des macaques. *Les Colloques de l'INRA*, 38: 91–100.

Thierry, B., Demaria, C., Preuschoft, S., & Desportes, C. 1989a. Structural convergence between silent bared-teeth display and relaxed open-mouth display in the Tonkean macaque (*Macaca tonkeana*). *Folia primatologica*, 52: 178–184.

Thierry, B., Wunderlich, D., & Gueth, C. 1989b. Possession and transfer of objects in a group of brown capuchins (*Cebus apella*). *Behaviour*, 110: 294–305.

Thompson, C. W., & Moore, M. C. 1991. Throat colour reliably signals status in male tree lizards, *Urosaurus ornatus*. *Animal Behaviour*, 42: 745–753.

van Hooff, J. A. R. A. M., & Wensing, J. A. B. 1987. Dominance and its behavioral measures in a captive wolf pack. In: *Man and Wolf* (H. Frank, ed.), pp. 219–252. Dordrecht: Junk.

van Rhijn, J. G. 1980. Communication by agonistic displays: A discussion. *Behaviour*, 74: 284–293.

van Rhijn, J. G., & Vodegel R. 1980. Being honest about one's intentions: An evolutionary strategy for animal conflicts. *Journal of Theoretical Biology*, 85: 623–641.

van Schaik, C. P. 1989. The ecology of social relationships amongst female primates. In: *Comparative Socioecology* (V. Standon & R. A. Foley, eds.), pp. 195–218. Oxford: Blackwell:

van Schaik, C. P. 1996. Social evolution in primates: The role of ecological factors and male behaviour. *Proceedings of the British Academy*, 88: 9–31.

Vehrencamp, S. 1983. A model for the evolution of despotic versus egalitarian societies. *Animal Behaviour*, 31: 667–682.

von Neumann, J., & Morgenstern, O. 1944. *Theory of Games and Economic Behavior*. Princeton: Princeton University Press.

Waas, J. R. 1991. Do little blue penguins signal their intentions during aggressive interactions with strangers? *Animal Behaviour*, 41: 375–382.

Wagner, W. E., Jr. 1992. Deceptive or honest signalling of fighting ability? A test of alternative hypotheses for the function of changes in call dominant frequency by male cricket frogs. *Animal Behaviour*, 44: 449–462.

Whitfield, D. P. 1987. Plumage variability, status signalling and individual recognition in avian flocks. *Trends in Ecology and Evolution*, 2: 408–415.

Whitten, P. L. 1987. Infants and adult males. In: *Primate Societies* (B. B. Smuts, D. L. Cheney, R. M. Seyfarth, R. W. Wrangham, & T. T. Strusaker, eds.), pp. 343–357. Chicago: University of Chicago Press.

Williams, G. C. 1966. *Adaptation and Natural Selection: A Critique of Some Current Evolutionary Thought.* Princeton: Princeton University Press.

Wrangham, R. W. 1980. An ecological model of female-bonded primate groups. *Behaviour,* 75: 262–300.

Zucker, N. 1994a. A dual status-signalling system: A matter of redundancy or differing roles? *Animal Behaviour,* 47: 15–22.

Zucker, N. 1994b. Social influence on the use of a modifiable status signal. *Animal Behaviour,* 48: 1317–1324.

CHAPTER 6

Covariation of Conflict Management Patterns across Macaque Species

Bernard Thierry

Researchers have found an astonishing diversity in conflict management patterns among non-human primates. By using the comparative method to study different social organizations, we may understand the processes underlying conflict management and, subsequently, the evolutionary determinants of social organizations. Typically, researchers resort to using ecological factors to explain patterns of competition and dominance (van Schaik 1989). Complex organizations, however, are systems of interrelated patterns. We should ask to what extent connections between patterns might act as constraints at the evolutionary level (Thierry 1997). In other words, are the social organizations that animals produce to cope with environmental constraints limitless in their form, or are social organizations limited to finite sets of forms by structural constraints that define

the possibilities open to them? In the latter case, social traits would evolve as sets of covariant traits, and we should find them together in "packages."

That traits of organisms covary through evolution is an empirical finding. A number of morphological and life history traits are linked by allometric relations, but whether social traits can be similarly correlated in nonhuman primates is more problematic (Harvey et al. 1987; Martins 1996). Although social groups are interconnected wholes that may display some autonomy, they have no central unit directing behavior or homeostasis. Producers of social organizations are individuals, and these individuals are autonomous reproductive agents who follow their own strategies. It may be held that organizational features are constrained only by conditions of external requirements. On the other hand, it is also possible that behavior is not infinitely plastic and that individuals in groups are not independent players. Kummer (1971) captured this point when he wrote, "If a species succeeds in adapting by introducing a structural change compatible with its preexisting structure, it thereby alters a component of its social, behavioral, or morphological organization. The result may be disharmony within the organization, so that other components must be altered to accommodate the main change and to reestablish a functioning entity. Such secondary adjustment also must be within the capacity of the heritage" (p. 91).

The best evidence for the clustering of traits relating to conflict management, dominance, and nepotism comes from the comparative study of macaques. In what follows, I review the evidence that supports the existence of covariant sets of traits within macaque social organizations. I start with a brief presentation of the phyletic relations of macaques and their socioecology. I then examine the meaning of covariation in macaques and its possible existence in other primate species. Finally, I consider the origin of covariation, asking specifically whether it may be attributed to ecological and phylogenetic factors.

The Macaque Genus

Most macaques are found in forests or semiopen habitats in southern and eastern Asia, although one species does live in the forests of northwestern Africa (Fooden 1982; Table 6.1). Macaques are mainly frugivorous and semiterrestrial. The degree of frugivory and arboreality, however, is quite variable according to species. Diet includes leaves, buds, seeds, and insects. Macaques may be found in very different habitats from tropical to temperate regions: rain, deciduous, or coniferous forests and also riverine refuges, mangrove swamps, semideserts, or areas of human settlement (Fooden

1982; Richard et al. 1989). A broad range in habitat variation is observed not only between species but also within several species.

Current knowledge regarding the biology, evolution, and behavior of macaques is especially advanced. The various macaque species can be arranged in several phyletic groups using a classification system based on morphological and molecular markers (Fooden 1976; Delson 1980; Hoelzer & Melnick 1996). There are three main groups: the *fascicularis* group, containing four species (longtail, rhesus, Japanese, and Taiwan macaques); the *sinica* group, which also contains four species (toque, bonnet, Assamese, and Tibetan macaques); and the *silenus* group, which includes liontail and pigtail macaques and the seven species inhabiting the island of Sulawesi. There are also two monotypic species groups: one for the Barbary macaque, which is close to the *silenus* group, and another for the stumptail macaque, which is close to the *sinica* group. The genus *Macaca* is thus derived from three distinct phyletic lineages, *silenus-sylvanus*, *sinica-arctoides*, and *fascicularis* (Table 6.1). These lineages correspond to three successive radiations in Asia. Most of the differentiation that produced extant species occurred during the Pleistocene and the Holocene (Delson 1980; Eudey 1980; Hoelzer & Melnick 1996).

Macaques form large multimale-multifemale groups. The sex ratio among adult macaques in these groups is biased in favor of females. Males migrate when reaching sexual maturity, and they may transfer between groups several times in their lives (Pusey & Packer 1987). Consequently, their social links and dominance ranks fluctuate. In contrast, most females remain for the course of their lives in their natal groups. Females form matrilineal subgroups where kin-related females maintain preferential bonds and support one another during conflicts. Owing to alliances among kin, the dominance status of females is quite stable

TABLE 6.1

Phyletic Lineages and Geographic Distribution of Macaque Species

Species	Distribution
FASCICULARIS LINEAGE	
longtail macaque (*M. fascicularis*)	Indochinese peninsula, Indonesia, Philippines
rhesus macaque (*M. mulatta*)	Continental South and East Asia
Japanese macaque (*M. fuscata*)	Japan
Taiwan macaque (*M. cyclopis*)	Taiwan
SINICA-ARCTOIDES LINEAGE	
toque macaque (*M. sinica*)	Sri Lanka
bonnet macaque (*M. radiata*)	South and West India
Assamese macaque (*M. assamensis*)	Continental Southeast Asia
Tibetan macaque (*M. thibetana*)	East and Central China
stumptail macaque (*M. arctoides*)	South China, Indochinese peninsula
SILENUS-SYLVANUS LINEAGE	
liontail macaque (*M. silenus*)	Southwest India
pigtail macaque (*M. nemestrina*)	Indochinese peninsula, Sumatra, Borneo
crested macaque (*M. nigra*)	North Sulawesi
Gorontalo macaque (*M. nigrescens*)	North Sulawesi
Heck's macaque (*M. hecki*)	North Sulawesi
Tonkean macaque (*M. tonkeana*)	Central Sulawesi
moor macaque (*M. maurus*)	Southwest Sulawesi
booted macaque (*M. ochreata*)	Southeast Sulawesi
Muna-Butung macaque (*M. brunnescens*)	Southeast Sulawesi
Barbary macaque (*M. sylvanus*)	Algeria, Morocco

Source: Fooden 1976, 1982.

and depends on the matrilines to which they belong (Kawamura 1965 [1958]; Paul & Kuester 1987; Datta 1992). Despite sharing these basic organizational features, however, macaque species do display broad interspecific variation with regard to patterns of aggression, reconciliation, dominance, nepotism, socialization, and temperament, as detailed below.

Evidence for Covariation

Aggression and Reconciliation

The degree of asymmetry in conflicts is especially variable among macaques. In rhesus and Japanese macaques, contests are highly unidirectional: the target of aggression generally flees or submits, and severe biting is not rare. Post-conflict friendly

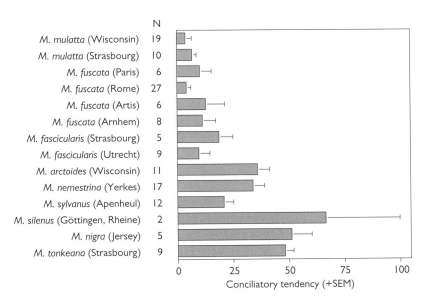

FIGURE 6.1. Conciliatory tendencies across nine species of macaques (from Thierry et al. 1997). The conciliatory tendencies are calculated according to Veenema (Box 2.1) on fourteen captive populations of macaques (original data from de Waal & Luttrell 1989; Demaria & Thierry 1992; Abegg et al. 1996; Castles et al. 1996; Aureli et al. 1997; Petit et al. 1997; Schino et al. 1998). To standardize the data, only dyadic conflicts occurring between pairs of unrelated females (older than 3.5 years) and between pairs of females and unrelated juveniles (between 1.5 and 3.5 years of age) were entered into the analysis. N refers to the number of focal females in each group. SEM = standard error of the mean.

reunions between previous opponents (i.e., reconciliation) are not frequent (Thierry 1986b, 1990a; de Waal & Luttrell 1989; Aureli et al. 1993; Butovskaya 1993; Chaffin et al. 1995; Petit et al. 1997; Schino et al. 1998). The conciliatory tendencies measured among unrelated individuals consistently rated between 4 and 12 percent in six different groups of rhesus and Japanese macaques (Fig. 6.1). This picture departs from that observed in the Tonkean, moor, and crested macaques—all species from the Sulawesi island—in which a majority of aggressive acts induce protests or counterattacks from the targets (Fig. 6.2a). The intensity of aggression is low, and biting is neither frequent nor injurious. Measures of conciliatory tendency yield high values: reconciliation tends to occur as much as 50 percent of the time among unrelated partners (Bernstein et al. 1983; Thierry 1986b; Matsumura 1991, 1996; Petit et al. 1997; Thierry et al. 1997; Fig. 6.1).

Other macaque species are located intermediately between the rhesus and Japanese macaques and the Sulawesi macaques. According to aggression and reconciliation patterns, longtail and pigtail macaques are more similar to rhesus and Japanese macaques (Angst 1975; Thierry 1986b; Judge 1991; Castles et al. 1996; Aureli et al. 1997), whereas stumptail and liontail macaques tend toward the Sulawesi macaques (de Waal & Luttrell 1989; Butovskaya 1993; Abegg et al. 1996; Abegg 1998; Fig. 6.1). As the above patterns indicate, we can arrange macaque species along a four-grade scale according to conflict management patterns and other behavioral traits (see below; Table 6.2). Although patterns of aggression of Barbary macaques have not been compared with those of other species (but see Preuschoft et al. 1998), their conciliatory tendencies suggest they should be placed in the third grade (Aureli et al. 1997; Fig. 6.1).

Correlations between reconciliation rates, aggression intensity, and asymmetry may be explained by proximate mechanisms. Competitors' readiness to fight for resources depends on the risk incurred by doing so (Popp & DeVore 1979). If the risk of injury is high, the best tactic for the target of aggression is to avoid the opponent rather than to counterattack. On the other hand, when targets can easily retaliate, initial aggressors risk becoming the recipients of sharp and dangerous attacks. Whereas unidirectional contests and increased risk of injury inhibit the occurrence of affiliative contacts between opponents, weak asymmetry and uncertainty about outcomes may create room for negotiation (Silk 1997). By facilitating information exchange, conciliatory behaviors may prevent conflict and improve social relationships (de Waal 1986; Thierry, 1986b; de Waal & Luttrell 1989; Petit et al. 1997). Additionally, the occurrence of conciliatory behaviors might subsequently affect aggression levels because group members are less willing to jeopardize valuable relationships.

Covariation is not limited to conflict resolution mechanisms, such as reconciliation, but it includes other forms of conflict management. Species from the third and fourth grades display high rates of specific affiliative contacts such as clasps and embraces (Thierry 1984; de Waal 1989; de Waal & Luttrell 1989; Matsumura 1991; Abegg 1998). They are also characterized by the development of special behaviors that may reduce social tension and facilitate social contact. In stumptail macaques, a ritualized soft biting performed by higher-ranking individuals toward subordinates may bring about positive interactions and the end of conflicts (de Waal 1989; Demaria & Thierry 1990). In Barbary macaques, adult males use infants as buffers to facilitate approach and affiliation (Deag 1980; Kuester & Paul, Box 5.2). In Sulawesi and liontail macaques, individuals regularly intervene in, and occasionally terminate, conflicts by appeasing one of the opponents (Abegg 1998; Petit & Thierry, Box 13.1). Such patterns are basically absent in species from the first grade (rhesus and Japanese macaques). The use of infants in triadic interactions can occur in longtail macaques,

albeit at a low rate (de Waal et al. 1976; Thierry personal observation). Clasping interactions also occur at a low rate in the second grade (longtail macaques: Thierry 1985a; pigtail macaques: Castles et al. 1996; Maestripieri 1996).

Displays of dominance and submission serve as conflict management devices (Preuschoft & van Schaik, Chapter 5). Patterns of submission, like the other social traits described above, also vary consistently along the scale (Table 6.2). In rhesus and Japanese macaques, for example, subordinates retract the lips and expose the teeth to express submission (de Waal & Luttrell 1985; Preuschoft 1995). By using this facial expression outside the context of conflict, individuals formally acknowledge their lower status relative to higher-ranking conspecifics. A similar display, the silent bared-teeth display, is also observed in species from the fourth grade, but in these species it does not have a submissive function (Fig. 6.2b); instead, it signals the sender's peaceful intentions and serves to initiate affiliative interactions (Dixson 1977; Thierry et al. 1989; Petit & Thierry 1992). In the second and third grades, the silent bared-teeth display is mainly submissive as it is in the first grade (Bertrand 1969; Angst 1975; de Waal et al. 1976; Preuschoft 1995; Maestripieri 1996). In Barbary macaques, however, it sometimes occurs in affiliative contexts, and in liontail macaques its meaning is often positive—albeit submissive—in a significant proportion of interactions (Preuschoft 1995; Abegg 1998).

Species from the third and fourth grades exhibit yet another facial expression, the teeth-chattering display. The teeth-chattering display is characterized by vertical retraction of the lips and by repeated opening and closing of the mouth (Fig. 6.2c). Teeth chattering is virtually absent in the first and fourth grades. Its meaning is similar to that of the bared-teeth display notwithstanding some functional variation. For example, it is an

FIGURE 6.2a–c. Typical behaviors in macaques. (a) A dominant male is challenged by a juvenile female and her mother during bidirectional contest in Tonkean macaques (photograph by B. Thierry). (b) An adult male directs an affiliative silent bared-teeth display toward a juvenile in crested macaques (photograph by O. Petit). (c) Silent bared-teeth and teeth chattering in bonnet macaques (photograph by C. Abegg).

affiliative signal in stumptail and Barbary macaques and an exclusively submissive signal in longtail macaques (Bertrand 1969; Preuschoft 1995). In sum, the degree of asymmetry in conflict management is related to the differential development of submissive and reassurance behaviors in the macaque species.

Dominance and Kinship Networks

The species-specific style of social relationships also appears to covary with the above patterns (Table 6.2). The dominance gradient is the steepest in the first grade, in that in species of this grade, social life is governed by rigid hierarchies (Kawamura 1965 [1958]; Sade 1972; Kurland 1977; de Waal 1989, 1991). Power asymmetry determines who may interact with whom. It affects how an individual chooses partners for proximity, affiliation, or play and whether the distribution of choices is skewed in favor of higher-ranking individuals. Group members compete for access to the strongest allies, limiting the number of partners available to subordinate individuals. Subordinate individuals may be inhibited from approaching or

TABLE 6.2
Tentative Scaling of Macaque Social Organizations

1st Grade	2nd Grade	3rd Grade	4th Grade
rhesus macaque	longtail macaque	stumptail macaque	Tonkean macaque
Japanese macaque	pigtail macaque	Barbary macaque	moor macaque
(Taiwan macaque)		liontail macaque	crested macaque
		bonnet macaque	(Gorontalo macaque)
		(toque macaque)	(Heck's macaque)
		(Tibetan macaque)	(booted macaque)
		(Assamese macaque)	(Muna-Butung macaque)

Note: Species are ordered mainly based on their conciliatory tendency and social tolerance, which increase from the left (1st grade) to the right (4th grade), and based on their asymmetry of contests, dominance gradient, and kin bias, which decrease from the left to the right. For the least-known species (indicated in parentheses), location on grade is predicted from only a few behavioral traits (see text).

contacting higher-ranking individuals because of the possibility, and subsequent cost, of attack. This picture differs from that observed in Sulawesi macaques, in which the dominance gradient is substantially less severe, although dominance ranks are stable. In these species, status differences do not hinder contacts between group members and have little effect on grooming distribution (Thierry et al. 1990, 1994; Matsumura 1991; Petit et al. 1992).

The social relationships of rhesus and Japanese macaques are marked by strong kinship networks. Females have a high preference for relatives, and strict rules of inheritance determine the acquisition of dominance rank within matrilines (Kawamura 1965 [1958]; Sade 1972; Kurland 1977; Datta 1992). Females achieve, by adulthood, a dominance rank just below their mothers and rarely outrank them. Because females choose to support their youngest relatives in contests, dominance ranks are ordered inversely to age within matrilines: younger daughters dominate their elder sisters. By applying this rule of rank inheritance

(i.e., youngest ascendancy), it is usually possible to predict the dominance status of a female at adulthood from her birth rank and mother's status.

The influence of kinship on competition and proximity among partners is still important in other grades. The degree to which females prefer relatives, however, in affiliative contact, social grooming, and support in conflicts is less pronounced in the third grade than in the first two grades (de Waal & Luttrell 1989; Butovskaya 1993; Aureli et al. 1997). Kin bias is still weaker in Sulawesi macaques, and it is not possible to characterize the social structure using only patterns of spatial distribution in these species (Thierry et al. 1990, 1994; Matsumura & Okamoto 1997). We lack detailed data about female rank acquisition in many macaques. We know that in longtail macaques (second grade), rank acquisition follows the rules described in the first grade, although rank reversals between mothers and daughters are not uncommon (Angst 1975). In Barbary macaques (third grade), older females are often outranked by their daughters, and females are usually subordinate to their older sisters (Paul & Kuester 1987). In Tonkean macaques (fourth grade), rank reversal of mothers and daughters is common, and therefore the rule of youngest ascendancy is not relevant (Thierry unpublished data).

Using computerized models, it is possible to show that the degree of dominance asymmetry may be related to the level of aggression intensity (Hemelrijk 1999). It is understandable that power inequality between conspecifics is related to the asymmetry observed in competition and submission patterns (de Waal 1986; Thierry 1990b; Preuschoft & van Schaik, Chapter 5). Determining the relationship between dominance and nepotism is, however, less straightforward. I proposed that the occurrence of frequent coalitions creates a link between degree of nepotism and dominance asymmetry (Thierry 1990b). When most alliances involve relatives, the dominance status of individuals depends primarily on the power of the kin subgroup to which they belong. This increases rank differences between nonrelatives and further develops kin alliances, generating group structures based on strong hierarchies. Conversely, when kin bias is less pronounced, coalitions involving nonrelatives are more common, dominance appears more dependent on individual attributes, and the individual retains some degree of freedom with regard to power networks. Dominance relationships remain balanced among group members, and close ties exist even between nonrelatives. Recent comparisons between macaque species support the view that dominance gradient and nepotism level among macaque females are connected by positive feedback (Aureli et al. 1997; Thierry et al. 1997).

Males, on the other hand, do not depend on their kin to the same extent as females because they break their social ties when emigrating from their natal troop. Variation in intermale tolerance among species may explain differences in male emigration patterns among grades. Whereas male rhesus and Japanese macaques typically migrate from their natal groups when they are three to five years of age, first emigration occurs some years later in other grades. Consistent differences are observed between species in the number of migrating males of all ages and the duration of solitary phases, which seem especially long in the first two grades (Mehlman 1986; Pusey & Packer 1987; Oi 1990; but see Paul & Kuester 1985). Interestingly, these two grades are also characterized by low levels of intermale tolerance (Caldecott 1986; Hill 1994).

Socialization and Temperament

Maternal behavior also appears to covary with tolerance and asymmetry in social relationships. Except for the highest-ranking females, mothers in the first two grades are quite protective of infants. They frequently retrieve their infants, restricting their interactions mostly to relatives. The amount of care by females other than the mother is therefore limited. Mothers belonging to species from the other two grades are quite permissive; most females in the group handle and carry infants from an early age (Kaufman & Rosenblum 1969; Kurland 1977; Hiraiwa 1981; Thierry 1985b; Small 1990; Mason et al. 1993; Maestripieri 1994).

Covariation between mothering behaviors and dominance patterns may be explained by the level of protection needed by infants in a given social environment. Mothers living in strict hierarchies are very restrictive. In more tolerant macaque societies, females are more confident, reflected by the degree to which they allow their offspring to move about unrestricted. Another index that reflects the level of tolerance during socialization is the rate of immatures' interference in mounts. So far, various hypotheses have been proposed to explain why immatures approach mating pairs and

BOX 6.1

Physiological Correlates of Individual Dominance Style

Robert Sapolsky

A number of primatologists have pioneered quantitative approaches to the study of personality differences—consistency in temperament, responsivity, patterns of social affiliation, or the social roles played among primates (Smuts 1985; Goodall 1986; Clarke & Boinski 1995). In their wake, some primatologists have examined the physiological correlates of personality. Are they a cause or consequence of the personality differences? Do they carry health or selection advantages? Do they provide insight as to consequences for conflict management of the differing coping styles of individuals with differing personalities?

This approach is derived from studies examining physiological correlates of rank. It is worth reviewing those studies to appreciate the personality correlates better.

Early studies of rank and physiology produced a seemingly consistent picture that was logical in the context of stress (Sapolsky 1998). Studies of various captive Old World monkey species, squirrel monkeys (*Saimiri sciureus*), tree shrews (*Tupaia belangeri*), and wild baboons (*Papio anubis*) suggested that subordinate individuals of both sexes suffer the most stress, have chronically overactive stress responses, and are more vulnerable to stress-related disease. Such individuals were reported to have elevated basal glucocorticoid levels. Glucocorticoids are hormones secreted in response to stress; although essential for dealing with an acute crisis, an excess is pathogenic. Furthermore, subordinates are hypertensive, have a sluggish cardiovascular response to stress and a sluggish recovery, and have suppressed levels of "good" (HDL) cholesterol, fewer circulating lymphocytes, and more risk of atherosclerosis. Finally, subordinate individuals' reproductive capacities are diminished. Subordinate females are more likely to be anovulatory, and although subordinate males do not have lower basal testosterone levels, those levels are the most readily suppressed by stress. When physiological measures could be taken before and after hierarchical formation in captive groups, it was clear that rank differences preceded physiological differences (Sapolsky 1993b).

For the same stressor, a stress response or a stress-related disease is more likely among individuals with no sense of control predictability, no outlets for frustration, or no social support (Levine et al. 1989; Sapolsky 1998). The physical and psychological stressors associated with subordinance among these species were readily interpreted in this context. For a subordinate, even daily life is stressful, chronically activating the stress response.

Recently, this basic set of results has been modified as we have learned more about the effects of the social context in which individuals live:

- The rank/physiology relationship is influenced by quality of life in a particular group. Thus, the maladaptive endocrine features of subordinance are less severe in macaque groups with high rates of reconciliation (Gust et al. 1993) or low rates of redirected aggression in baboon troops (Sapolsky 1986). Moreover, during rare periods of social instability when the hierarchy reorganizes, dominant animals, amid the unpredictable, tense shifts of ranks, now have the chronically activated stress responses (Sapolsky 1993b).

- The relationship is also influenced by quality of life in a particular species. For most macaques and baboons, subordinance is imposed from above and involves high rates of received redirected aggression and rising in the hierarchy by force. In contrast, among marmosets and tamarins, subordinates are usually younger siblings or offspring of the dominant female, free from redirected aggression, aiding in cooperative breeding, and ultimately rising in rank by waiting their turn. Among these latter species, dominant animals have the highest basal glucocorticoid levels (Abbott et al. 1997).

- Personal experience modifies the relationship between rank and physiology. The individual rates at which animals are subject to social stressors or have access to social support can predict endocrine profiles (Sapolsky 1993a).

BOX 6.1 (continued)

Thus, there is no single physiological profile of dominance; rather, the profile is modified by the rate of exposure to stressors and the availability of coping mechanisms.

The personality studies extend this view by showing that physiology is also influenced by an individual's style of reaction to stressors and sources of coping. For example, there are stable personality differences in macaques in their reactivity to novelty. Such traits appear early in life, and animals whose behaviors are most disrupted by novelty hypersecrete glucocorticoids and are more at risk for atherosclerosis (Manuck et al. 1995; Suomi 1997).

My students and I have identified two clusters of personality traits that predict low basal glucocorticoid levels among wild male baboons, after controlling for rank (Sapolsky & Ray 1989; Ray & Sapolsky 1992; Virgin & Sapolsky 1997). These traits predict hormone levels better than does social rank and reflect the psychological modifiers of stress physiology.

Low basal glucocorticoid levels occur among males with a cluster of traits concerning male-male competition. These include the greatest tendency to (1) distinguish between threatening and neutral interactions with rivals (as assessed by transitional probabilities of behavior—is the mere presence of a rival as likely to disrupt ongoing behavior as is an overt threat by that rival?); (2) initiate fights when being threatened with high intensity; (3) distinguish between winning and losing a fight (again, as assessed by transitional probabilities of behavior—does the likelihood of resuming a prior, ongoing behavior differ between winning and losing?); (4) redirect aggression after losing a fight.

Broadly, these "low glucocorticoid" animals distinguish between true stressors and minor events, exert some control over the former, distinguish between good and bad outcomes, and redirect aggression if the latter. This agrees with the known protective effects of control, predictability and having outlets. Moreover, these reactivity traits are stable over time, and "low glucocorticoid" males have longer tenures in the dominant cohort.

A second, independent cluster of traits also predicts low basal glucocorticoid levels. These "low glucocorti-

A male baboon (right side) bites a female as redirected aggression after being harassed by another male (left side). Photograph by Robert Sapolsky.

coid" males have high rates of grooming, being groomed, sitting in contact, and playing with infants. These findings agree with an extensive literature demonstrating the ability of social affiliation to lower glucocorticoid levels and facets of the stress response and to have beneficial health effects (see Sapolsky 1998).

What relevance do these individual dominance styles and their physiological consequences have for controlling aggression and for conflict resolution? The message is mixed. "Low glucocorticoid" males with greater amounts of social contact and affiliation and less provocability by rivals may decrease overall levels of group conflict. These are healthier individuals who, of considerable importance, probably have greater cumulative reproductive success (because they have atypically frequent affiliative relationships with females when past their male-male competitive peak) (Sapolsky 1996).

In contrast, low glucocorticoid levels also occur among males most prone toward redirected aggression. These beneficial health consequences probably reflect the fact that redirected aggression releases frustration (Levine et al. 1989) and can decrease the likelihood of repeated attack by the former aggressor (possibly by diverting his attention to the new victim, or even by eliciting his joining in the attack as a form of reconciliation) (Aureli & van Schaik 1991).

It is not clear whether the selective advantages for a temperament prone toward social affiliation and low provocability outweigh those for an individual prone toward redirected aggression. Nonetheless, the occurrence of any advantages for the redirection strategy highlights a problem that bedevils those concerned with fostering cooperation and altruism. In the abstract, it is obvious that a strategy that decreases conflict and controls the negative consequences of aggression for an individual need not have a positive impact on all other group members. Nonetheless, it is far from desirable to live in a social group in which the predominant form of stress management consists of avoiding ulcers by giving them.

direct affiliative or ambivalent behaviors toward them. Regardless of the function of such interferences, it appears that young individuals may enter into proximity of mating pairs only if males tolerate them (Thierry 1986a). Interferences by immatures are frequent in species from the third and fourth grades (Dixson 1977; Niemeyer & Chamove 1983; Thierry 1986a; Kumar 1987; Kuester & Paul 1989; Matsumura 1995). They are, however, rare in longtail macaques (de Benedictis 1973; Gore 1986) and virtually absent in pigtail, rhesus, and Japanese macaques.

Social behavior, whether aggressive, tolerant, or maternal, has emotional basis (Mason et al. 1993; Aureli & Smucny, Chapter 10; Sapolsky, Box 6.1). Consistent differences are found between macaque species in response to stress and novelty when measured by arousal, alarm, and exploration behaviors, corticosteroid levels, or heart rates. Rhesus and longtail macaques are less explorative with regard to their physical environment when compared with liontail or Tonkean macaques (Thierry et al. 1994; Clarke & Boinski 1995). As a general rule, species from the third and fourth grades are less easily aroused than are species from the first and second grades (de Waal 1989; Clarke & Boinski 1995).

The existence of species-specific temperaments suggests that individual physiological and psycho-logical characters are linked (Mendoza & Mason 1989; Clarke & Boinski 1995; Sapolsky, Box 6.1). For instance, there is a genetic component for sensitivity of the autonomic nervous system or rates of serotoninergic activity, both traits being correlated with reactivity thresholds and aggression levels (Higley & Linnoila 1997). Correlations also arise at the phenotype level through developmental processes (Thierry 1997). The individual's characteristics are influenced by conspecifics through socialization. The experience of individuals growing up in highly hierarchical and nepotistic social structures is different from that of others who develop in more tolerant and open societies. Consequently, the two subsets of individuals should acquire different temperaments. Temperament may influence behavioral style, which itself can be subject to intergenerational transmission and the subsequent perpetuation of a set of developmental pathways (Berman 1990; de Waal 1996).

Generality of Covariation

Intraspecific Variation

Direct interspecific comparisons have been made only for a limited sample of species, groups, and traits. The variability inherent to individual behavior and the relativity of our measurements call for caution. Rating species along a discrete and bipolar scale is inevitably reducing. My classification of macaque species in four grades according to sets of traits is a first attempt. Although it is easy to group those species that fit into either the first or the fourth grade, it is more difficult to group intermediate species. There may be inconsistencies for some traits. Each grade should be considered a modal site on a continuous scale. Although each species is assigned to one grade, a more accurate model would represent the various study populations of each species using a cluster of points centered on one grade and would allow for overlap with other clusters centered on neighboring grades.

Broad intraspecific variation presents a real challenge when looking for interspecific variation. Questions remain about how strongly social traits are interconnected (Castles et al. 1996; Bernstein & Cooper 1998). Average traits vary within populations of the same species according to factors such as age, sex, kinship, or population density (e.g., Bernstein & Ehardt 1985; Thierry 1990a; Judge & de Waal 1997; Schino et al. 1998). Additionally, the assessment of traits makes sense only in stable conditions. Social instability creates a temporary dissociation of traits. In periods of rank reversals, for instance, relationships are dramatically altered. Intense aggression is then temporarily associated with bidirectional social interactions (de Waal 1986; Demaria & Thierry 1990).

Better knowledge about the extent of intraspecific variation of each trait would improve the assessment of covariation. We are aware of some examples of intraspecific variation in conciliatory patterns that are relevant to interspecific comparisons. When young rhesus macaques cohabit with young stumptail macaques, the conciliatory tendencies of the former are increased threefold, stabilizing at a level comparable to the rates typical of the latter species (de Waal & Johanowicz 1993).

Conciliatory tendencies are also influenced by kinship bonds. In species of the first two grades, reconciliation rates between kin are much higher than those between unrelated individuals. Therefore, the reconciliatory rates between kin in the first two grades may approach those found in the species of the last two grades, in which differences between kin and nonkin are less pronounced (Thierry 1990a; Demaria & Thierry 1992; Castles et al. 1996; Aureli et al. 1997).

Most of the previously retained traits are the result of "transactions," that is, they require both the expression of behavior by one individual and a subsequent response by another (Mason et al. 1993). Transactions grasp the nature of social interchanges better than mere behavior frequencies and duration, which are more sensitive to environmental influences. Comparisons between studies are performed using ratios and qualitative differences. Actually, the variation range of transactions remains limited (e.g., for conflict, reconciliation, and mother-offspring interactions: Berman 1980; Johnson & Southwick 1987; Demaria & Thierry 1990; Aureli 1992; Call et al. 1996; Judge & de Waal 1997; Abegg 1998; Schino et al. 1998).

Some traits such as emigration patterns may be more affected by ecological and demographical conditions than others, making it harder to demonstrate true interspecific differences. Demography affects social relationships by determining the number of partners available to each individual. Lack of competition and support by kin allies are likely to lower dominance asymmetries and prevent the emergence of the youngest ascendancy rule (Datta 1992; Hill & Okayasu 1996). In groups in which there are few related individuals, reconciliation with unrelated individuals may occur more frequently than in more typical groups (Aureli unpublished data). Average tolerance level among males and females may be related to the number of available adult partners (Hill 1994).

Intermale tolerance is usually less pronounced in the first two grades. In small groups, however, the number of partners is limited; even rhesus and Japanese macaque males may then exchange relatively high rates of affiliative behaviors (Hill 1994).

Another important point is that covariation may occur at the intraspecific level. Within-group covariation was shown for reconciliation rates, aggression intensity, and bidirectionality of contests by comparing kin and nonkin interactions in Japanese macaques (Thierry 1990a). Bites were found to be particularly frequent between individuals with marked dominance asymmetries in rhesus macaques (de Waal 1991). In addition, variations in the distribution of friendly behavior across group members were associated with differences in conciliatory tendency of two groups of pigtail macaques (Castles et al. 1996). Yet intraspecific covariation may be difficult to demonstrate in the absence of dramatic between-group differences, because of noise induced by demographical and environmental factors, and the possible occurrence of floor or ceiling effects likely to prevent directional changes. For example, when a species is characterized by a low conciliatory tendency, a stronger asymmetry in conflicts could not lead to a significant decrease in reconciliation rates.

Predicting Social Traits and Species Classification

The four-grade scale provides an operational framework that allows one to generate falsifiable hypotheses. In species such as bonnet macaques, aggression and reconciliation have not been quantitatively compared. But their patterns of temperament, submission, dominance, nepotism, affiliation, socialization, female acquisition of rank, and male emigration consistently place them into the third grade (Kaufman & Rosenblum 1969; Simonds 1973; Caine & Mitchell 1980; Silk et al. 1981; Silk

& Samuels 1984; Mason et al. 1993; see Fig. 6.2c). One can therefore predict that rates of reconciliation should be relatively high in bonnet macaques. We may similarly expect relatively low rates of severe biting in Barbary macaques, a high degree of mother permissiveness in crested macaques, and delayed male migration in moor macaques.

In the least-known species, a few traits may be used to predict others. If the little-known Sulawesi macaques share an affiliative bared-teeth display with the other species originating from the same island (Thierry et al. 1994), they should also share the other distinctive traits of the fourth grade (Table 6.2). From the combination of teeth chattering, infant use by males, and juveniles' interference in matings, Tibetan macaques most likely belong in the third grade (Deng 1993; Xiong 1993; Ogawa 1995; Zhao 1996). The same patterns most likely hold for Assamese macaques (Bernstein & Cooper 1998; T. Wangchuk personal communication), thus placing them in the third grade as well. Another species showing teeth chattering, the toque macaque, should be set in the third grade judging from the development of embrace gestures, age at male migration, and weak influence of mother's rank on dominance relationships between immatures (Dittus 1975, 1977; Baker-Dittus 1985; Table 6.2). Last, the submissive bared-teeth display together with the absence of teeth chattering and the absence of juveniles' interference in matings (Y. Wu personal communication) suggests that the Taiwan macaque should be grouped with species in the first grade (Table 6.2).

Seemingly Uncoupled Traits

Any significant deviation from the scale should be closely examined. Traits that do not covary like others reveal the action of additional constraints. A first case is provided by the unusually high levels of infant handling displayed by adult males in

Barbary and Tibetan macaques. Both species contrast with others from same or other grades with regard to this behavior, which cannot be explained by mating or paternal investment (Deng 1993; Ogawa 1995; Kuester & Paul, Box 5.2). It may not be due only to chance that both species share another peculiar behavior: adult males use particular infants to facilitate approach of one another and to help regulate their social relationships (Deag 1980; Ogawa 1995; Kuester & Paul, Box 5.2). It was hypothesized that infant use derives from infant handling (Taub 1980; Zhao 1996). Yet the converse might also be true. Infant use does not occur in a vacuum; it is embedded in a set of affiliative interactions occurring between adult males and infants. Exploitation of infants as social tools is most likely an adaptation allowing negotiation among males. Actual handling by males would, in this case, be the by-product. Both behaviors are based on one and the same character: a strong attraction toward infants. This kind of reasoning may explain why infant handling by males is not very developed in other species. Interestingly, in stumptail macaques, infant use sometimes occurs, and, correspondingly, males may handle infants (Gouzoules 1975; Estrada & Sandoval 1977). There are also hints that male Assamese macaques display high levels of infant handling and exploitation (Bernstein & Cooper 1998), providing a test case for the existence of a link between both features.

If specific modes of conflict management induce the emergence of additional traits as secondary effects, external constraints may strongly affect other traits as well. This appears to be the case for patterns of male reproductive competition, which do not fit the four-grade classification. Macaque species differ markedly in the duration of the associations occurring between adult males and females approaching ovulation. In some species, mate guarding by the male may last several days

(e.g., Tonkean, liontail, and pigtail macaques). In others, the male follows and mates with the female for only a few hours (e.g., rhesus and Japanese macaques) or even a few minutes (e.g., Barbary macaques) (Caldecott 1986). Recent studies indicate that seasonality is the key factor influencing the mating tactics available to males. Species living in temperate regions experience seasonal mating. When several females enter estrus simultaneously, they cannot be monopolized by one male. It follows that dominance rank and paternity rate are only slightly correlated. In contrast, reproduction takes place year-round in tropical species. There is generally no more than one female in estrus at a time in a group, so the dominant male is in a position to control her during the fertile period. Consequently, the probability of a male's paternity is closely correlated with his dominance rank (Oi 1996; Thierry et al. 1996; Paul 1997). This entails the following paradox: in nonseasonal species with limited dominance asymmetry (Tonkean and stumptail macaques: third and fourth grades), social rank has more influence on the reproductive success of males than in species in which hierarchical differences are marked but where females' fertility is synchronous (rhesus and Japanese macaques: first grade).

Extension to Other Nonhuman Primates

Analyses of the degree of covariation should be carried out in closely related species in which differences in the formalization of dominance relationships have been found—for example, in squirrel monkeys (*Saimiri sciureus* vs. *S. oerstedi*: Mitchell et al. 1991), capuchin monkeys (*Cebus apella* vs. *C. albifrons*: Janson 1986), chimpanzees (*Pan troglodytes* vs. *P. paniscus*: de Waal 1989, Furuichi & Ihobe 1994), and lemurs (*Lemur catta* vs. *Eulemur fulvus*: Pereira & Kappeler, Box 15.2).

Some taxonomic groups, such as baboons and mangabeys (*Papio* and *Cercocebus*), are similar to macaques. The patterns of conflict management and nepotism of baboons and mangabeys suggest that they may belong to the second and third grades on the macaque scale (see Ehardt 1988; Pereira 1988; Gust & Gordon 1993; Petit et al. 1997; Colmenares et al., Box 5.3). Guenons belong to another related genus (*Cercopithecus*). Most guenon species are characterized by high-intensity aggression along with little expression of appeasement behaviors and weak nepotism (e.g., Rowell 1971; Kaplan 1987; Rowell et al. 1991). Direct comparisons between guenons, baboons, and mangabeys would provide an opportunity to test covariation as found among macaques.

Any extension of comparisons outside the genus *Macaca* should be made with caution. Homology of traits between species belonging to different taxonomic groups is not warranted. Absence or addition of a single trait may induce nonlinear effects, thereby modifying relationships between traits. As previously mentioned, rates of kin coalitions may have profound consequences on dominance gradient between group members (Thierry 1990b). In several species with a social organization different from that of macaques, weak asymmetry in contests and dominance is associated with elevated percentages of biting together with low rates of appeasement and kin support (e.g., *Alouatta seniculus*: Crockett 1984; *Cercopithecus* spp.: Rowell 1971; Kaplan 1987; *Gorilla gorilla*: Watts 1994).

Origin of Covariation

Ecological Determinants

The empirical finding that macaque social organizations represent covariant sets of characters indicates that these organizations belong to a finite set of possible forms. This raises the question about the nature of the factors that shape them. Given the correlated occurrence of traits, one crude ecological factor cannot be expected to tune each

of them separately. Nonetheless, environmental factors might act on some key features, which would work as pacemakers for the whole organization. Other traits would be shaped by correlated responses.

According to a first model, resource distribution would determine intermale competition (Caldecott 1986). In poor habitats, reproductive requirements should drive females to be selective with males; thus creating antagonism among them, females reduce the number of males within breeding groups. In high-quality habitats, good food conditions should generate more tolerance among males. As any male could have fathered any offspring, such a situation would favor male parental investment and low degrees of kin bias. This reasoning, however, fails to produce a consistent classification of macaque social organizations. As previously mentioned, mating tactics and rates of infant handling displayed by males do not covary with levels of aggression and nepotism. This model does not match the four-grade scale that classifies the differences in social traits across macaques known so far.

According to a second model, macaques may be divided into two categories: *weed* species, which develop in close association with human settlements, and *nonweed* species, which inhabit undisturbed forests (Richard et al. 1989). It is assumed that life in association with people requires some particular temperamental qualities such as being explorative and aggressive. Results show, however, that these two temperamental qualities are not correlated among macaques (see above). In addition, this ecological classification is not consistent with the social one; for instance, rhesus and bonnet macaques (first and third grades) are considered weeds, while Japanese and stumptail macaques (respectively from the first and third grades, too) are classified as nonweeds. In any case, the distinction between weed and nonweed species does not help explain the four-grade scale.

A more general socioecological model explains the patterns of competition observed among nonhuman primates by distinguishing *despotic* and *egalitarian* species (van Schaik 1989). The basic argument is that animals live in groups for protection from predators and that group living, in turn, induces competition between individuals and groups. In mainly frugivorous species such as macaques, females should constitute kin-bonded coalitions to face overt competition that arises for resources within and between groups. If predation risks are high, the costs of leaving the group are elevated for subordinates. As a consequence, dominants take the lion's share, and the relationships between unrelated group members are despotic. On the contrary, when predation risks are low, subordinates are not forced to remain in the group. Dominants, in this case, benefit from subordinates' cooperation against external threats and so must accept a relatively equal exploitation of resources. This condition would produce rather egalitarian relationships even among nonkin.

This socioecological model accounts for the contrasted "dominance styles" observed among macaques. The weaker kin bias and the lower power asymmetry found in the last two grades may be explained by high levels of between-group competition and the consequent need for communal defense (de Waal & Luttrell 1989; Aureli et al. 1997). We still lack, however, detailed data concerning levels of predation and intergroup competition from wild populations of most macaques. There is some information that appears inconsistent with the model's predictions. For instance, liontail macaques exhibit relaxed dominance relationships even though they live in habitats containing large predators (Abegg et al. 1996). In addition, no evidence for higher intergroup competition in the species of the last two grades has been provided yet (e.g., Matsumura 1998).

Phylogenetic Inertia

Phylogeny was found to influence primate social organization if taxa and behavior are classified in broad categories (di Fiore & Rendall 1994). This influence is expected to remain substantial even at a finer-grained level (Martins 1996). When examining the four-grade social scale from the point of view of macaque phyletic radiations (Table 6.1), it appears that the *fascicularis* lineage is located in the first and second grades. The *sinica-arctoides* lineage is located in the third grade. The *silenus-sylvanus* lineage belongs primarily in the third and fourth grades, except for pigtail macaques that are in the second grade (Table 6.2). The location of each lineage is *mostly* restricted to one or two grades in the scale. A fair degree of phylogenetic inertia is thus revealed in the social patterns of macaques. Pigtail macaques, however, depart somewhat from other members of their lineage. Perhaps they diverged more quickly for some unknown reason. Or perhaps they are actually more closely related to the *fascicularis* lineage as indicated by some phyletic reconstructions (Purvis 1995). In any case, the various traits of the social organization of pigtail macaques are consistently clustered, as are the traits characteristic of any macaque species.

By limiting the changes possible to societies, interconnections between traits act as constraints that channel evolutionary processes (Thierry 1990b, 1997). Interconnections may occur at the genome or organism level, as exemplified by the increasing evidence for species-specific temperaments. They may also stem from the social organization itself, as suggested by the interplay of conflict management, dominance, and nepotism patterns. The good match of each of the three macaque lineages with the grades of the tentative classification suggests that the core of the species-specific systems of interconnections underwent limited changes during the past several hundred thousand years. Ecological conditions may have changed several times during this period. Actually, macaque populations have experienced more than once a switch between warm and temperate climate during recent geological time (Eudey 1980; Richard et al. 1989). Such a change can quickly alter reproductive seasonality, which in turn influences the conditions of male mating competition and consequently the genetic structure of populations.

Conclusion

The covariation of traits produced specific sets of social mechanisms to deal with intragroup conflict. Rhesus macaques, for example, manage competition by respecting strict rules of submission and dominance, whereas Tonkean macaques regulate conflict by retaliating and reconciling frequently. Each social organization is the product of history and environment, but only some sets of social traits are within the realms of possi-

bility. These sets appear quite stable at the evolutionary time scale of macaques. We still need to determine to what extent genetic flows within populations depend on the mechanisms of competition regulation particular to each species.

At present, the ecological preferences of macaques appear independent of their phylogeny (Fooden 1982; Richard et al. 1989). Future field studies should aim to assess to what extent the contrasting behavioral styles of macaques, and other species, arise from constraints internal to societies (Thierry 1990b) or represent specific solutions to environmental problems, as predicted by socioecological theories (de Waal & Luttrell 1989; van Schaik 1989). The issue, however, should not be limited to a dichotomous question about the respective roles of phylogeny and adaptation in the evolution of primates. By studying the genetic and phenotypic processes underlying covariation (Thierry 1997), we may reach a deeper understanding of the developmental processes responsible for the interconnections between conflict management patterns and the other traits of social organizations.

References

Abbott, D., Saltzman, W., Schultz-Darken, N., & Smith, T. 1997. Specific neuroendocrine mechanisms not involving generalized stress mediate social regulation of female reproduction in cooperatively breeding marmoset monkeys. In: The Integrative Neurobiology of Affiliation (C. S. Carter, I. I. Ledenhendler, & B. Kirkpatrick, eds.), pp. 114–128. New York: Annals of the New York Academy of Sciences.

Abegg, C. 1998. Constance et Variabilité des Comportements Sociaux en Fonction de l'Environnement chez Deux Espèces de Macaques (Macaca fuscata, Macaca silenus). Ph.D. diss., Université de Strasbourg.

Abegg, C., Thierry, B., & Kaumanns, W. 1996. Reconciliation in three groups of lion-tailed macaques. International Journal of Primatology, 17: 803–816.

Angst, W. 1975. Basic data and concepts on the social organization of Macaca fascicularis. In: Primate Behavior, Vol. 4 (L. A. Rosenblum, ed.), pp. 325–388. New York: Academic Press.

Aureli, F. 1992. Post-conflict behaviour among wild long-tailed macaques (Macaca fascicularis). Behavioral Ecology and Sociobiology, 31: 329–337.

Aureli, F., & van Schaik, C. 1991. Post-conflict behaviour in long-tailed macaques (Macaca fascicularis): II. Coping with the uncertainty. Ethology, 84: 101–114.

Aureli, F., Veenema, H. C., van Panthaleon van Eck, C. J., & van Hooff, J. A. R. A. M. 1993. Reconciliation, consolation, and redirection in Japanese macaques (Macaca fuscata). Behaviour, 124: 1–21.

Aureli, F., Das, M., & Veenema, H. C. 1997. Differential kinship effect on reconciliation in three species of macaques (*Macaca fascicularis*, *M. fuscata*, and *M. sylvanus*). *Journal of Comparative Psychology*, 111: 91–99.

Baker-Dittus, A. 1985. Infant and juvenile-directed care behaviors in adult toque macaques, *Macaca sinica*. Ph.D. diss., University of Maryland.

Berman, C. M. 1980. Mother-infant relationships among free-ranging rhesus monkeys on Cayo Santiago: A comparison with captive pairs. *Animal Behaviour*, 28: 860–873.

Berman, C. M. 1990. Intergenerational transmission of maternal rejection rates among free-ranging rhesus monkeys. *Animal Behaviour*, 39: 329–337.

Bernstein, I. S., & Cooper, M. A. 1998. Ambiguities in the behavior of Assamese macaques. *American Journal of Primatology*, 45: 170–171.

Bernstein, I. S., & Ehardt, C. L. 1985. Age-sex class differences in the expression of agonistic behaviors in rhesus monkey (*Macaca mulatta*) groups. *Journal of Comparative Psychology*, 99: 115–132.

Bernstein, I. S., Williams, L., & Ramsay, M. 1983. The expression of aggression in Old World monkeys. *International Journal of Primatology*, 4: 113–125.

Bertrand, M. 1969. *The Behavioral Repertoire of the Stumptail Macaque*. Basel: Karger.

Butovskaya, M. 1993. Kinship and different dominance styles in groups of three species of the genus *Macaca* (*M. arctoides*, *M. mulatta*, *M. fascicularis*). *Folia primatologica*, 60: 210–224.

Caine, N., & Mitchell, G. 1980. Species differences in the interest shown in infants by juvenile female macaques (*Macaca radiata* and *M. mulatta*). *International Journal of Primatology*, 1: 323–332.

Caldecott, J. O. 1986. Mating patterns, societies and the ecogeography of macaques. *Animal Behaviour*, 34: 208–220.

Call, J., Judge, P. G., & de Waal, F. B. M. 1996. Influence of kinship and spatial density on reconciliation and grooming in rhesus monkeys. *American Journal of Primatology*, 39: 35–45.

Castles, D., Aureli, F., & de Waal, F. B. M. 1996. Variation in conciliation tendency and relationship quality across groups of pigtail macaques. *Animal Behaviour*, 52: 389–403.

Chaffin, C. L., Friedlen, K., & de Waal, F. B. M. 1995. Dominance style of Japanese macaques compared with rhesus and stumptail macaques. *American Journal of Primatology*, 35: 103–116.

Clarke, S. A., & Boinski, S. 1995. Temperament in nonhuman primates. *American Journal of Primatology*, 37: 103–125.

Crockett, C. M. 1984. Emigration by female red howler monkeys and the case for female competition. In: *Female Primates: Studies by Women Primatologists* (M. Small, ed.), pp. 159–173. New York: Alan Liss.

Datta, S. B. 1992. Effects of availability of allies on female dominance structure. In: *Coalitions and Alliances in Humans and Other Animals* (A. H. Harcourt & F. B. M. de Waal, eds.), pp. 61–82. Oxford: Oxford University Press.

Deag, J. M. 1980. Interactions between males and unweaned Barbary macaques: Testing the agonistic buffering hypothesis. *Behaviour*, 75: 54–81.

de Benedictis, T. 1973. The behavior of young primates during adult copulations: Observations of a *Macaca irus* colony. *American Anthropologist*, 75: 1469–1484.

Delson, E. 1980. Fossil macaques, phyletic relationships and a scenario of deployment. In: *The Macaques Studies in Ecology, Behavior, and Evolution* (D. G. Lindburg, ed.), pp. 10–30. New York: Van Nostrand Rheinhold.

Demaria, C., & Thierry, B. 1989. Lack of effects of environmental changes on agonistic behavior patterns in a captive group of stumptailed macaques (*Macaca arctoides*). *Aggressive Behavior*, 15: 353–360.

Demaria, C., & Thierry, B. 1990. Formal biting in stumptailed macaques (*Macaca arctoides*). *American Journal of Primatology*, 20: 133–140.

Demaria, C., & Thierry, B. 1992. The ability to reconcile in Tonkean and rhesus macaques. *Abstracts of the XIV Congress of International Primatological Society*, p. 101. Strasbourg: SICOP.

Deng, Z. Y. 1993. Social development of infants of *Macaca thibetana* at Mount Emei, China. *Folia primatologica*, 60: 28–35.

de Waal, F. B. M. 1986. The integration of dominance and social bonding in primates. *Quarterly Review of Biology*, 61: 459–479.

de Waal, F. B. M. 1989. *Peacemaking among Primates*. Cambridge: Harvard University Press.

de Waal, F. B. M. 1991. Rank distance as a central feature of rhesus monkey social organization: A sociometric analysis. *Animal Behaviour*, 41: 383–395.

de Waal, F. B. M. 1996. Macaque social culture: Development and perpetuation of affiliative networks. *Journal of Comparative Psychology*, 110: 147–154.

de Waal, F. B. M., & Johanowicz, D. L. 1993. Modification of reconciliation behavior through social experience: An experiment with two macaque species. *Child Development*, 64: 897–908.

de Waal, F. B. M., & Luttrell, L. M. 1985. The formal hierarchy of rhesus macaques: An investigation of the bared-teeth display. *American Journal of Primatology*, 9: 73–85.

de Waal, F. B. M., & Luttrell, L. M. 1989. Toward a comparative socioecology of the genus *Macaca*: Different dominance styles in rhesus and stumptailed macaques. *American Journal of Primatology*, 19: 83–109.

de Waal, F. B. M., van Hooff, J. A. R. A. M., & Netto, W. J. 1976. An ethological analysis of types of agonistic interaction in a captive group of Java-monkeys (*Macaca fascicularis*). *Primates*, 17: 257–290.

di Fiore, A., & Rendall, D. 1994. Evolution of social organization: A reappraisal for primates by using phylogenetic methods. *Proceedings of the National Academy of Sciences of the USA*, 91: 9941–9945.

Dittus, W. P. J. 1975. Population dynamics of the toque monkey, *Macaca sinica*. In: *Socioecology and Psychology of Primates* (R. H. Tuttle, ed.), pp. 125–151. The Hague: Mouton.

Dittus, W. P. J. 1977. The social regulation of population density and age-sex distribution in the toque monkey. *Behaviour*, 63: 281–322.

Dixson, A. F. 1977. Observations on the displays, menstrual cycles and sexual behaviour of the "black ape" of Celebes (*Macaca nigra*). *Journal of Zoology*, 182: 63–84.

Ehardt, C. L. 1988. Absence of strongly kin-preferential behavior by adult female sooty mangabeys (*Cercocebus atys*). *American Journal of Physical Anthropology*, 76: 233–243.

Estrada, A., & Sandoval, J. M. 1977. Social relations in a free-ranging troop of stumptail macaques (*Macaca arctoides*): Male-care behaviour I. *Primates*, 18: 793–813.

Eudey, A. A. 1980. Pleistocene glacial phenomena and the evolution of Asian macaques. In: *The Macaques: Studies in Ecology, Behavior, and Evolution* (D. G. Lindburg, ed.), pp. 52–83. New York: Van Nostrand Rheinhold.

Fooden, J. 1976. Provisional classification and key to living species of macaques (Primates: *Macaca*). *Folia primatologica*, 25: 225–236.

Fooden, J. 1982. Ecogeographic segregation of macaque species. *Primates*, 23: 574–579.

Furuichi, T., & Ihobe, H. 1994. Variation in male relationships in bonobos and chimpanzees. *Behaviour*, 130: 211–228.

Goodall, J. 1986. *The Chimpanzees of Gombe*. Cambridge: Harvard University Press, Belknap Press.

Gore, M. A. 1986. Mother-offspring conflict and interference at mother's mating in *Macaca fascicularis*. *Primates*, 27: 205–214.

Gouzoules, H. 1975. Maternal rank and early social interactions of infant stumptail macaques, *Macaca arctoides*. *Primates*, 16: 405–418.

Gust, D., & Gordon, T. 1993. Conflict resolution in sooty mangabeys. *Animal Behaviour*, 46: 685–694.

Gust, D., Gordon, T., Hambright, K., & Wilson, M. 1993. Relationship between social factors and pituitary-adrenocortical activity in female rhesus monkeys. *Hormones and Behavior*, 27: 318–324.

Harvey, P. H., Martin, R. D., & Clutton-Brock, T. H. 1987. Life histories in comparative perspective. In: *Primate Societies* (B. B. Smuts, D. L. Cheney, R. M. Seyfarth, R. W. Wrangham, & T. T. Struhsaker, eds.), pp. 181–196. Chicago: University of Chicago Press.

Hemelrijk, C. K. 1999. An individual-oriented model on the emergence of despotic and egalitarian societies. *Proceedings of the Royal Society, London*, B, 266: 361–369.

Higley, J. D., & Linnoila, M. 1997. Low central nervous system serotoninergic activity is traitlike and correlates with impulsive behavior. *Annals of the New York Academy of Sciences*, 836: 39–56.

Hill, D. A. 1994. Affiliative behaviour between adult males of the genus *Macaca*. *Behaviour*, 130: 293–308.

Hill, D. A., & Okayasu, N. 1996. Determinants of dominance among female macaques: Nepotism, demography and danger. In: *Evolution and Ecology of Macaque Societies* (J. E. Fa & D. G. Lindburg, eds.), pp. 459–472. Cambridge: Cambridge University Press.

Hiraiwa, M. 1981. Maternal and alloparental care in a troop of free-ranging Japanese monkeys. *Primates*, 22: 309–329.

Hoelzer, G. A., & Melnick, D. J. 1996. Evolutionary relationships of the macaques. In: *Evolution and Ecology of Macaque Societies* (J. E. Fa & D. G. Lindburg, eds.), pp. 3–39. Cambridge: Cambridge University Press.

Janson, C. H. 1986. The mating system as a determinant of social evolution in capuchin monkeys (*Cebus*). In: *Primate Ecology and Conservation* (J. G. Else & P. C. Lee, eds.), pp.169–179. Cambridge: Cambridge University Press.

Johnson, R. L., & Southwick, C. H. 1987. Ecological constraints on the development of infant independence in rhesus. *American Journal of Primatology*, 13: 103–118.

Judge, P. G. 1991. Dyadic and triadic reconciliation in pigtail macaques (*Macaca nemestrina*). *American Journal of Primatology*, 23: 225–237.

Judge, P. G., & de Waal, F. B. M. 1997. Rhesus monkey behaviour under diverse population densities: Coping with long-term crowding. *Animal Behaviour*, 54: 643–662.

Kaplan, J. R. 1987. Dominance and affiliation in the Cercopithecini and Papionini: A comparative examination. In: *Comparative Behavior of African Monkeys* (E. L. Zucker, ed.), pp. 127–150. New York: Alan Liss.

Kaufman, I. C., & Rosenblum, L. A. 1969. The waning of the mother-infant bond in two species of macaque. In: *Determinants of the Mother-Infant Bond in Two Species of Macaque* (B. M. Foss, ed.), pp. 41–59. London: Methuen.

Kawamura, S. 1965 [1958]. Matriarchal social ranks in the Minoo-B troop: A study of the rank system of Japanese macaques. *Primates*, 1: 149–156. Translation in: *Japanese Monkeys* (S. A. Altmann, ed.), pp. 105–112. Atlanta: Emory University Press.

Kuester, J., & Paul, A. 1989. Reproductive strategies of subadult Barbary macaque males at Affenberg Salem. In: *The Sociobiology of Sexual and Reproductive Strategies* (A. E. Rasa, C. Vogel, & E. Voland, eds.), pp. 93–109. London: Chapman Hall.

Kumar, A. 1987. The ecology and population dynamics of the lion-tailed macaque (*Macaca silenus*) in South India. Ph.D. diss., Cambridge University.

Kummer, H. 1971. *Primate Societies*. Chicago: Aldine.

Kurland, J. A. 1977. *Kin Selection in the Japanese Monkey*. Basel: Karger.

Levine, S., Coe, C., & Wiener, S. 1989. The psycho-neuroendocrinology of stress: A psychobiological perspective. In: *Psychoneuroendocrinology* (S. Levine & R. Brush, eds.), pp. 181–204. New York: Academic Press.

Maestripieri, D. 1994. Mother-infant relationships in three species of macaques (*Macaca mulatta, M. nemestrina, M. arctoides*). II. The social environment. *Behaviour*, 131: 97–113.

Maestripieri, D. 1996. Gestural communication and its cognitive implications in pigtail macaques (*Macaca nemestrina*). *Behaviour*, 133: 997–1022.

Manuck, S., Marsland, A., Kaplan, J., & Williams, J. 1995. The pathogenicity of behavior and its neuroendocrine mediation: An example from coronary artery disease. *Psychosomatic Medicine*, 57: 215–221.

Martins, E. 1996. *Phylogenies and the Comparative Method in Animal Behavior*. New York: Oxford University Press.

Mason, W. A., Long, D. D., & Mendoza, S. P. 1993. Temperament and mother-infant conflict in macaques: A transactional analysis. In: *Primate Social Conflict* (W. A. Mason & S. P. Mendoza, eds.), pp. 205–227. New York: State University of New York Press.

Matsumura, S. 1991. A preliminary report on the ecology and social behavior of moor macaques (*Macaca maurus*) in Sulawesi, Indonesia. *Kyoto University Overseas Research Report Studies in Asian Non-human Primates*, 8: 27–41.

Matsumura, S. 1995. Affiliative mounting interference in *Macaca maurus. Kyoto University Overseas Research Report Studies in Asian Non-human Primates*, 9: 1–5.

Matsumura, S. 1996. Postconflict affiliative contacts between former opponents among wild moor macaques (*Macaca maurus*). *American Journal of Primatology*, 38: 211–219.

Matsumura, S. 1998. Relaxed dominance relations among female moor macaques (*Macaca maurus*) in their natural habitat, South Sulawesi, Indonesia. *Folia primatologica*, 69: 346–356.

Matsumura, S., & Okamoto, K. 1997. Factors affecting proximity among members of a wild group of moor macaques during feeding, moving, and resting. *International Journal of Primatology*, 18: 929–940.

Mehlman, P. 1986. Male intergroup mobility in a wild population of the Barbary macaque (*Macaca sylvanus*), Ghomaran Rif Mountains, Morocco. *American Journal of Primatology*, 10: 67–81.

Mendoza, S. P., & Mason, W. A. 1989. Primate relationships: Social dispositions and physiological responses. In: *Perspectives in Primate Biology*, Vol. 2 (P. K. Seth & S. Seth, eds.), pp. 129–143. New Delhi: Today & Tomorrow.

Mitchell, C. L., Boinski, S., & van Schaik, C. P. 1991. Competitive regimes and female bonding in two species of squirrel monkeys (*Saimiri oerstedi* and *S. sciureus*). *Behavioral Ecology and Sociobiology*, 28: 55–60.

Niemeyer, C. L., & Chamove, A. S. 1983. Motivation of harassment of matings in stumptailed macaques. *Behaviour*, 87: 298–323.

Ogawa, H. 1995. Bridging behavior and other affiliative interactions among male Tibetan macaques (*Macaca thibetana*). *International Journal of Primatology*, 16: 707–727.

Oi, T. 1990. Population organization of wild pig-tailed macaques (*Macaca nemestrina nemestrina*) in West Sumatra. *Primates*, 31: 15–31.

Oi, T. 1996. Sexual behaviour and mating system of the wild pig-tailed macaque in West Sumatra. In: *Evolution and Ecology of Macaque Societies* (J. E. Fa & D. G. Lindburg, eds.), pp. 342–368. Cambridge: Cambridge University Press.

Paul, A. 1997. Breeding seasonality affects the association between dominance and reproductive success in non-human primates. *Folia primatologica*, 68: 344–349.

Paul, A., & Kuester, J. 1985. Intergroup transfer and incest avoidance in semifree-ranging Barbary macaques (*Macaca sylvanus*) at Salem (FRG). *American Journal of Primatology*, 8: 317–322.

Paul, A., & Kuester, J. 1987. Dominance, kinship and reproductive value in female Barbary macaques (*Macaca sylvanus*). *Behavioral Ecology and Sociobiology*, 21: 323–331.

Pereira, M. E. 1988. Agonistic interactions of juvenile savanna baboons. I. Fundamental features. *Ethology*, 79: 195–217.

Petit, O., & Thierry, B. 1992. Affiliative function of the silent bared-teeth display in moor macaques (*Macaca maurus*): Further evidence for the particular status of Sulawesi macaques. *International Journal of Primatology*, 13: 97–105.

Petit, O., Desportes, C., & Thierry, B. 1992. Differential probability of "coproduction" in two species of macaque (*Macaca tonkeana, M. mulatta*). *Ethology*, 90: 107–120.

Petit, O., Abegg, C., & Thierry, B. 1997. A comparative study of aggression and conciliation in three cercopithecine monkeys (*Macaca fuscata, Macaca nigra, Papio papio*). *Behaviour*, 134: 415–431.

Popp, J. L., & DeVore, I. 1979. Aggressive competition and social dominance theory: Synopsis. In: *The Great Apes* (D. A. Hamburg & E. R. McCown, eds.), pp. 317–338. Menlo Park, Calif.: Benjamin/Cummings.

Preuschoft, S. 1995. "Laughter" and "smiling" in macaques. Ph.D. diss., Rijksuniversiteit Utrecht.

Preuschoft, S., Paul, A., & Kuester, J. 1998. Dominance styles of female and male Barbary macaques (*Macaca sylvanus*). *Behaviour*, 135: 731–755.

Purvis, A. 1995. A composite estimate of primate phylogeny. *Philosophical Transactions of the Royal Society of London*, 348: 405–421.

Pusey, A. E., & Packer, C. 1987. Dispersal and philopatry. In: *Primate Societies* (B. B. Smuts, D. L. Cheney, R. M. Seyfarth, R. W. Wrangham, & T. T.

Struhsaker, eds.), pp. 250–266. Chicago: University of Chicago Press.

Ray, J., & Sapolsky, R. 1992. Styles of male social behavior and their endocrine correlates among high-ranking baboons. *American Journal of Primatology*, 28: 231–240.

Richard, A., Goldstein, S. J., & Dewar, R. E. 1989. Weed macaques: The evolutionary implications of macaque feeding ecology. *International Journal of Primatology*, 10: 569–594.

Rowell, T. E. 1971. Organization of caged groups of *Cercopithecus* monkeys. *Animal Behaviour*, 19: 625–645.

Rowell, T. E., Wilson, C., & Cords, M. 1991. Reciprocity and partner preference in grooming of female blue monkeys. *International Journal of Primatology*, 12: 319–336.

Sade, D. S. 1972. A longitudinal study of social behavior in rhesus monkeys. In: *The Functional and Evolutionary Biology of Primates* (R. H. Tuttle, ed.), pp. 378–398. Chicago: Aldine.

Sapolsky, R. 1986. Endocrine and behavioral correlates of drought in the wild baboon. *American Journal of Primatology*, 11: 217–221.

Sapolsky, R. 1993a. Endocrinology alfresco: Psycho-endocrine studies of wild baboons. *Recent Progress in Hormone Research*, 48: 437–462.

Sapolsky, R. 1993b. The physiology of dominance in stable versus unstable social hierarchies. In: *Primate Social Conflict* (W. Mason & S. Mendoza, eds.), pp. 171–189. New York: State University of New York Press.

Sapolsky, R. 1996. Why should an aged male baboon transfer troops? *American Journal of Primatology*, 39: 149–155.

Sapolsky, R. 1998. *Why Zebras Don't Get Ulcers: A Guide to Stress, Stress-Related Diseases and Coping*, 2nd edn. New York: W. H. Freeman.

Sapolsky, R., & Ray, J. 1989. Styles of dominance and their physiological correlates among wild baboons. *American Journal of Primatology*, 18: 1–12.

Schino, G., Rosati, L., & Aureli, F. 1998. Intragroup variation in conciliatory tendencies in captive Japanese macaques. *Behaviour*, 135: 897–912.

Silk, J. B. 1997. The function of peaceful post-conflict contacts among primates. *Primates*, 38: 265–279.

Silk, J. B., & Samuels, A. 1984. Triadic interactions among *Macaca radiata*: Passports and buffers. *American Journal of Primatology*, 6: 373–376.

Silk, J. B., Samuels, A., & Rodman, P. S. 1981. Hierarchical organization of female *Macaca radiata* in captivity. *Primates*, 22: 84–95.

Simonds, P. E. 1973. Outcast males and social structure among bonnet macaques (*Macaca radiata*). *American Journal of Physical Anthropology*, 38: 599–604.

Small, M. F. 1990. Alloparental behaviour in Barbary macaques, *Macaca sylvanus*. *Animal Behaviour*, 39: 297–306.

Smuts, B. 1985. *Sex and Friendship in Baboons*. New York: Hawthorne, Aldine Press.

Suomi, S. 1997. Early determinants of behaviour: Evidence from primate studies. *British Medical Bulletin*, 53: 270–286.

Taub, D. M. 1980. Testing the "agonistic buffering" hypothesis. *Behavioral Ecology and Sociobiology*, 6: 187–197.

Thierry, B. 1984. Clasping behaviour in *Macaca tonkeana*. *Behaviour*, 89: 1–28.

Thierry, B. 1985a. Le comportement d'étreinte dans un groupe de macaques de Java (*Macaca fascicularis*). *Biology of Behaviour*, 10: 23–30.

Thierry, B. 1985b. Social development in three species of macaque (*Macaca mulatta, M. fascicularis, M. tonkeana*): A preliminary report on the first ten weeks of life. *Behavioural Processes*, 11: 89–95.

Thierry, B. 1986a. Affiliative interference in mounts in a group of Tonkean macaques (*Macaca tonkeana*). *American Journal of Primatology*, 14: 89–97.

Thierry, B. 1986b. A comparative study of aggression and response to aggression in three species of macaque. In: *Primate Ontogeny, Cognition, and Social Behaviour* (J. G. Else & P. C. Lee, eds.), pp. 307–313. Cambridge: Cambridge University Press.

Thierry, B. 1990a. L'état d'équilibre entre comportements agonistiques chez un groupe de macaques japonais (*Macaca fuscata*). *Comptes-rendus de l'Académie des Sciences de Paris*, 310 III: 35–40.

Thierry, B. 1990b. Feedback loop between kinship and dominance: The macaque model. *Journal of Theoretical Biology*, 145: 511–521.

Thierry, B. 1997. Adaptation and self-organization in primate societies. *Diogenes*, 180: 39–71.

Thierry, B., Demaria, C., Preuschoft, S., & Desportes, C. 1989. Structural convergence between silent bared-teeth display and relaxed open-mouth display in the Tonkean macaque (*Macaca tonkeana*). *Folia primatologica*, 52: 178–184.

Thierry, B., Gauthier, C., & Peignot, P. 1990. Social grooming in Tonkean macaques (*Macaca tonkeana*). *International Journal of Primatology*, 11: 357–375.

Thierry, B., Anderson, J. R., Demaria, C., Desportes, C., & Petit, O. 1994. Tonkean macaque behaviour from the perspective of the evolution of Sulawesi macaques. In: *Current Primatology*, Vol. 2 (J. J. Roeder, B. Thierry, J. R. Anderson, & N. Herrenschmidt, eds.), pp. 103–117. Strasbourg: Université Louis Pasteur.

Thierry, B., Heistermann, M., Aujard, F., & Hodges, J. K. 1996. Long-term data on basic parameters and evaluation of endocrine, morphological and behavioral measures for monitoring reproductive status in a group of semifree ranging Tonkean macaques (*Macaca tonkeana*). *American Journal of Primatology*, 39: 47–62.

Thierry, B., Aureli, F., de Waal, F. B. M., & Petit, O. 1997. Variation in reconciliation patterns and social organization across nine species of macaques. *Advances in Ethology*, 32: S39.

van Schaik, C. P. 1989. The ecology of social relationships amongst female primates. In: *Comparative Socioecology* (V. Standen & R. A. Foley, eds.), pp. 195–218. Oxford: Blackwell.

Virgin, C., & Sapolsky, R. 1997. Styles of male social behavior and their endocrine correlates among low-ranking baboons. *American Journal of Primatology*, 42: 25–34.

Watts, D. P. 1994. Agonistic relationships between female mountain gorillas (*Gorilla gorilla beringei*). *Behavioral Ecology and Sociobiology*, 34: 347–358.

Xiong, C. P. 1993. Sexual harassment of copulation in Tibetan monkeys (*Macaca thibetana*). *Acta Theriologica Sinica*, 13: 172–180.

Zhao, Q. K. 1996. Male-infant-male interactions in Tibetan macaques. *Primates*, 37: 135–143.

CHAPTER 7

Coping with Crowded Conditions

Peter G. Judge

Introduction

Research on nonhuman primate behavior that occurs following aggression has flourished while relatively little attention has been directed toward behavioral mechanisms that may prevent fights from occurring. Nevertheless, many of the responses that are thought to resolve conflicts following aggression (e.g., allogrooming, huddling, touching, submissive displays, and appeasement gestures) occur routinely throughout the day. If these responses limit aggression, decrease the likelihood of further escalation, reduce social tension, and restore disrupted relationships after fights (de Waal, Chapter 2), might the same responses serve similar functions outside overtly aggressive contexts? For example, if a conflict-provoking situation arises between two animals and one grooms the other, then aggression may be less likely than when no affiliative behavior is exchanged.

One method for evaluating such conflict management is to observe primate groups in contexts known to increase social tension and the potential for aggression. If submission, tension-regulating behavior, and friendly responses increase in such competitive situations, then animals may be attempting to reduce escalation of conflict. Two such contexts have been investigated in primate groups: scheduled feeding and crowding. Scheduled feeding of provisioned groups produces a tense situation in anticipation of competition for food and does elicit attempts to reduce conflict beforehand (Koyama, Box 7.1). Crowding produces another situation in which increased competition for space produces a context in which aggression may be reduced through behavioral mechanisms. With respect to crowding, two main questions arise: Does crowding increase aggression, and do animals modify behavior under crowded conditions to reduce the potential for conflict? The answer to both questions is a qualified yes, but in order to understand the current thinking concerning coping with crowding, it is necessary to trace the origins of the research.

Background and Popular Conceptions of Crowding Research

Interest in the influence of high density on behavior grew out of research on crowding as a natural stressor (Christian 1961). Endocrine reactions in crowded rodent populations resulted in decreased reproductive capacity and increased mortality, thereby acting as a natural population control. Similar studies explored the behavior of rodents during crowding and implicated increased aggression as one of the possible stressors linked to increased mortality (Southwick 1955; Myers 1966). In his classic experiment on rats (*Rattus norvegicus*), Calhoun (1962b) opened floodgates when he concluded that overpopulation produced a "behavioral sink" into social pathology that included high aggression, cannibalism, hypersexuality, social withdrawal, and other abnormal behavior.

Calhoun's paper helped fuel existing concerns over human overpopulation, urbanization, inner-city violence, and their consequences to physical and mental health. His conclusions were quickly incorporated into the popular literature and extrapolated directly to humans (Russell & Russell 1968; Morris 1969; Ardrey 1970). The outburst of editorializing that followed firmly established that crowding had a negative influence (Gad 1973). Amid the furor, however, more pragmatic critics argued that conclusions were based on very little data; results from animal studies were inconsistent; and there was no firm basis for extrapolating results of animal studies to humans (Freedman 1973; Gad

---

BOX 7.1

Conflict Prevention before Feeding

Nicola F. Koyama

The majority of conflict management studies address how individuals deal either with existing conflicts or with the effects of conflicts. Here I focus on the strategies individuals use to reduce the likelihood of conflict before it erupts. These pre-conflict behaviors are, cognitively, more demanding than post-conflict strategies as they imply that individuals possess the ability to anticipate future conflict.

Anecdotal evidence (Goodall 1986, p. 570; de Waal 1989c, p. 38; Byrne 1995, pp. 154–158), observational evidence of the transport of tools (Boesch & Boesch 1983), and experimental evidence (Rensch & Doehl 1967; Doehl 1968) strongly suggest that apes are indeed capable of planning and foresight. However, planning, a goal-directed sequence of behavior, does require an understanding of causality distinct from basic associative learning (Visalberghi & Tomasello 1998). The experimental evidence for an understanding of the cause-effect relations involved in a tool task is tentative for chimpanzees and lacking in monkeys and has yet to be fully corroborated. There is evidence, however, to suggest that a group of cells in the cerebral cortex of Japanese macaques (*Macaca fuscata*) is exclusively related to planning motor tasks (Tanji & Shima 1994), and in humans it is the frontal cortex that mediates planning (reviewed in Owen 1997). It is not difficult to see how the ability to anticipate future events could have considerable survival value, conferring a selective advantage on those individuals capable of planning. Moreover, given the evolution of the primate brain, we should expect to find precursors of such behaviors in apes and monkeys.

A type of conflict that can be anticipated, because of its predictability, is that occurring over food. Both in the wild and in captivity, a clumped food distribution by its very nature elicits stronger

BOX 7.1 (continued)

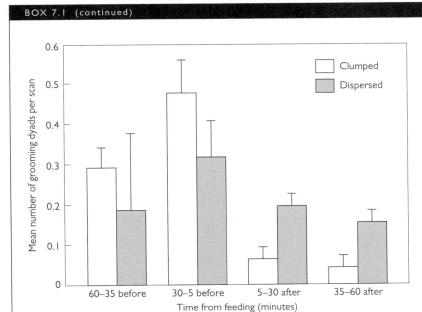

The mean number of grooming dyads per five-minute scan in the two half-hour blocks before (n = 15) and after (n = 12) clumped feeds and before (n = 7) and after (n = 9) dispersed feeds in captive chimpanzees. The proportion of dyads that groomed during the pre-feed hour is significantly greater than the postfeed hour for clumped, but not dispersed, feeding times. After Koyama & Dunbar (1996).

interindividual competition than dispersed food that cannot easily be monopolized (e.g., Boccia et al. 1988; Aureli et al. 1992; Sterck & Steenbeck 1997). If individuals are able to anticipate these periods of increased competition, they may attempt preemptively to reduce the likelihood of conflict by employing tension-reduction strategies.

In nonhuman primates, sociopositive interactions have been reported to increase in prefeeding situations. In chimpanzees (*Pan troglodytes*) affiliative contacts exchanged during celebrations, shortly before food provisioning, have been found to reduce food-related aggression significantly (de Waal 1992). In addition, sexual behavior in bonobos (*Pan paniscus*) occurs in a variety of tense situations including feeding time (de Waal 1989a) and is thought to "dissolve inter-individual tension," increasing tolerance and facilitating food sharing (Kuroda 1980).

If individuals are indeed able to anticipate future episodes of conflict and attempt to prevent them, we would expect to see the following patterns. First, individuals should show a *selective* increase in appeasement behavior before a predictable period of conflict, for example, scheduled feeding time. Second, this increase

should be more marked before feeding on clumped food than dispersed food. Third, attempts to increase tolerance during feeding should especially be found in those species that tolerate proximity during feeding (e.g., stumptail macaques, *Macaca arctoides*) or engage in food sharing (e.g., chimpanzees, bonobos), or both.

Affiliative interactions, such as grooming, have long been thought to function as tension-reduction strategies (Terry 1970; Schino et al. 1988). An increase in grooming rates prior to feeding time has been reported in captive bonobos (de Waal 1987) when prefeeding grooming rates were found to be above baseline but not significantly so. Further, provisioning with monopolizable food increases social tension among juvenile rhesus macaques (*Macaca mulatta*), and grooming increases following the provision of monopolizable but not dispersed food (de Waal 1984).

Two captive studies have conducted detailed tests of the above predictions. A selective increase in affiliative contacts was noted in captive stumptail macaques when alpha males were particularly attractive (received most competition to groom) and received the most grooming during the pre-mealtime period (four hours) when compared with the post-mealtime period (four

BOX 7.1 (continued)

hours) (Mayagoitia et al. 1993). Thus, the increase in grooming rates was directed toward the most powerful individuals in the group.

In captive chimpanzees grooming and proximity rates increased sharply in the hour prior to feeding time, and, more striking, these elevated prefeeding levels of grooming and proximity were only significantly higher than postfeeding levels before clumped feeding times but not before dispersed feeding times. Feeding time location allowed the chimpanzees to anticipate the nature of the food supply. The food was dispersed around the outdoor enclosure, whereas it was left clumped in a pile when given in the indoor quarters. In contrast to stumptail macaques, chimpanzees preferred to associate significantly more with their normal grooming partners and close kin, and not with the highest ranking, during the prefeeding period (Koyama & Dunbar 1996). Furthermore, chimpanzees were in proximity during clumped feeding times with those individuals they had spent more time near and had groomed more often before feeding time, suggesting that they were attempting to increase tolerance to facilitate co-feeding. That this selective increase in affiliative tactics was more marked before clumped feeding times but not before dispersed feeding times strongly supports the hypothesis that chimpanzees are able to anticipate future conflict and plan their behavior accordingly. Another study supports the selectivity of the partner: chimpanzees increased food sharing only with partners that had groomed them in the two hours prior to feeding (de Waal 1997).

In non-primate species also, clumped food elicits an increase in aggression (e.g., snowshoe hare, *Lepus americanus*: Ferron 1993; red deer, *Cervus elaphus*: Schmidt et al. 1998; Indian blackbucks, *Antilope cervicapra*: Gan-

slosser personal communication). Although no studies in the non-primate literature have investigated affiliation tactics prior to feeding, an increase in nonaggressive and sociopositive interactions (e.g., approach, naso-nasal contact, rubbing head on other's body, sniffing body) *during* clumped, but not dispersed, feeding sessions has been reported in captive Bongo antelopes, *Taurotragus euryceros* (Ganslosser & Brunner 1997). In contrast to primates, however, there was no significant difference in aggressive interactions between clumped and dispersed conditions, suggesting either that the increase in sociopositive interactions effectively reduces food-related aggression or that there may be little need for prefeeding tension-reduction mechanisms in this species.

In addition to post-conflict reconciliation (the focus of many chapters in this volume), nonhuman primates also use other, more powerful pre-conflict methods to maintain peace and harmony within their group. They adopt specific strategies to dissipate tension and reduce conflict before it erupts, suggesting that they possess the ability to anticipate these events and plan their behavior accordingly. With only a few systematic investigations available it is not possible to speculate on whether any phylogenetic trend exists and to what extent both apes and monkeys attempt to manipulate the pre-conflict situation. The scarcity of non-primate studies makes it difficult to state whether such conflict-prevention mechanisms exist in non-primate "contest-type" feeders. Research on wild populations is needed to determine the relevance of pre-conflict tension-reduction mechanisms under natural conditions. Given that clumped food increases competition and aggression in the wild, it is highly likely that similar tension-reduction and conflict-prevention mechanisms do exist in wild populations.

1973). Further, high density did correlate with increased pathology (e.g., Schmitt 1966; Galle et al. 1972), but causal relationships were not convincing because other relevant variables, such as income level and ethnicity, were intercorrelated with density and pathology. When these intervening variables were accounted for mathemati-

cally, density alone could have little or no relationship with pathology (Freedman et al. 1975). Despite indeterminate evidence, the axiom that high density had a negative influence on behavior, particularly aggression, persisted.

Concentrated study of the effects of density on human and nonhuman primates originated in this atmosphere, and early descriptions of crowding in nonhuman primates were similar to the harrowing scenario already described for rats. Based primarily on observations of primates held in zoo exhibits, crowding was said to produce extreme violence, a high level of mortality, and a struggle for survival (Russell & Russell 1968). The sensationalistic language made popular in the 1960s persisted even into the late 1970s as crowding in a small group of baboons was seen to produce "social disintegration," "individual pathology," and even a "ghetto riot" (Elton 1979). As systematic research progressed and methodology was refined, however, conflicting evidence accumulated, and definitive claims within the scientific community declined.

Below I summarize investigations into the relationship between density and aggression in nonhuman primates. I include a brief description of the methodologies used to study density effects and list some confounds associated with the research. I then review the evidence for coping behavior under crowded conditions in nonhuman primates. I also present a summary of density and aggression studies in non-primates. Last, I present a general summary of human density research and cite selected studies of human behavior under high density that are analogous to nonhuman primate studies.

Nonhuman Primates and Density

Density and Aggression

Investigation of the relationship between density and aggression in primates can be grouped roughly into three periods: initial investigations in the 1960s; development and refinements in the 1970s and 1980s; and more comprehensive studies in the 1990s. Many of the early investigations that described a relationship between density and nonhuman primate behavior were not studies of crowding per se but were conducted to assess whether behavior observed in captive primate groups was representative of that observed in wild populations (e.g., Kummer & Kurt 1965; Alexander & Bowers 1967; Rowell 1967). The general conclusions were that the behavioral repertoire of animals in captive groups was essentially the same as that in wild groups with just a few responses being unique to the wild or captivity. The main difference between conditions was a general increase in almost all social behavior in captivity (Kummer & Kurt 1965; Rowell 1967), particularly aggressive behavior (Rowell 1967). Since the captive conditions were more crowded than feral groups, many of the behavioral differences observed were attributed to differences in density.

Concurrent investigations of primates that focused specifically on the relationship between density, crowding, and aggression were also reporting a positive correlation between density and aggression. High-density city-dwelling rhesus monkeys (*Macaca mulatta*) were more aggressive than less crowded forest-dwelling monkeys (Southwick 1969); captive rhesus monkeys increased aggression when they were crowded into half their compound (Southwick 1967); and urban rhesus monkeys were more aggressive and more successful competitors than rural monkeys when the two were paired in experimental tests (Singh 1968). Japanese macaques (*Macaca fuscata*) also showed increased aggressiveness under captive or crowded conditions (Alexander & Bowers 1967; Alexander & Roth 1971).

These early studies of primates seemed to support a "density/aggression" relationship similar to that described in rats (Calhoun 1962b): higher density or crowding produces higher aggressiveness. Subsequent studies of primates have produced inconsistent results, however. Approximately half of these studies (15 of 29) find more aggressiveness under higher density, while others find no relationship or even less aggression at higher density (Table 7.1).

TABLE 7.1
Investigations into the Relationship between Density and Behavior in Nonhuman Primates

Design	Duration of High Density	Species	Change in Behavior under Higher Density			Source
			Aggression	Submission	Affiliation	
WITHIN-GROUP COMPARISONS	2–3 hours	*Macaca fascicularis*	More	—	Less[a]	Aureli et al. 1995
	1–9 hours	*Macaca mulatta*	More	More	Less[a]	Judge & de Waal 1993
	1–5 days	*Pan troglodytes*	Less	Less	Less[b]	Aureli & de Waal 1997
	4–6 days	*Macaca fuscata*	More	More	—	Alexander & Roth 1971
	7 days	*Macaca mulatta*	More	—	—	Southwick 1967
	1 month	*Macaca mulatta*	Less	—	Less	Novak & Drewsen 1989
	1 month	*Cercopithecus aethiops*	No effect	—	More	McGuire et al. 1978
	5 months	*Papio anubis*	More	More	—	Elton & Anderson 1977
	4.5–6 months	*Pan troglodytes*	More	More	More	Nieuwenhuijsen & de Waal 1982
	6 months	*Macaca mulatta*	More	—	—	Boyce et al. 1998
	1 to 2 years	*Macaca fuscata*	No effect	—	—	Eaton et al. 1981
BETWEEN-GROUP COMPARISONS	1 month	*Macaca mulatta*	More	—	—	Southwick 1969
	2 months	*Macaca mulatta*	Less	—	More	Drickamer 1973
	6 months	*Cercopithecus aethiops*	No effect	—	More	McGuire et al. 1983
	6 years	*Macaca mulatta*	No effect	More	More[c]	Novak et al. 1992
	lifelong	*Macaca mulatta*	More	More	More	Judge & de Waal 1997
	unspecified	*Papio hamadryas*	More	More	No effect	Kummer & Kurt 1965

INCONCLUSIVE STUDIES

Design	Duration of High Density	Species	Aggression	Submission	Affiliation	Source
WITHIN-GROUP COMPARISONS	20 minutes	*Macaca nemestrina*	Less[d]	—	—	Anderson et al. 1977
	1–3 weeks	*Macaca mulatta*	More[e]	—	—	Southwick 1967
	4 weeks	*Galago senegalensis*	No effect[f]	—	No effect	Nash & Chilton 1986
	2–6 months	*Macaca arctoides*	More[f]	More	No effect	Demaria & Thierry 1989
	10 months	*Macaca mulatta*	Less[e]	—	—	Bercovitch & Lebron 1991
	variable	*Cercopithecus aethiops*	No effect[e]	More	—	McGuire et al. 1978
BETWEEN-GROUP COMPARISONS	variable	*Macaca nemestrina*	No effect[e]	—	—	Dazey et al. 1977
	variable	*Macaca nemestrina*	More[e]	—	—	Erwin & Erwin 1976
	variable	*Papio anubis*	More[e]	More	More	Rowell 1967
	9 months	*Macaca fuscata*	More[f]	—	—	Alexander & Bowers 1967
	lifelong	*Macaca mulatta*	Less[f]	—	More	Marriott 1988
	lifelong	*Macaca mulatta*	No effect[f]	—	—	de Waal 1989b

Some of the inconsistency is attributable to the numerous confounding variables investigators have identified over the past 35 years that must be controlled or taken into consideration when studying the effects of density. The confounds are inherent in designs used to study the two basic types of density: spatial density and social density. In a *spatial density* study, the number of animals under investigation is held constant while the amount of available space is varied. Usually a within-group investigation is conducted in which the amount of space available to a group is either increased or decreased and behavioral changes are assessed (e.g., Southwick 1967). Between-group spatial density designs compare the behavior of similar-sized groups living in environments with different amounts of available space (e.g., Judge & de Waal 1997). In a *social density* study, the number of animals within a group is varied while the amount of space available to a group is held constant. Within-group investigations of social density include adding or removing animals from a group (e.g., Southwick 1969; Bercovitch & Lebron 1991) and allowing a population to grow naturally within a fixed amount of space (e.g., Calhoun 1962b; Eaton et al. 1981). Between-group social density investigations involve comparing different-sized groups living in the same amount of space (e.g., Erwin & Erwin 1976). In each type of investiga-

tion, higher density corresponds to less average space available to each animal.

Confounds arise from moving groups between smaller and larger enclosures, restricting groups within their usual enclosure, and comparing groups from different environments. Simply relocating animals to an unfamiliar enclosure can influence social behavior (Nash & Chilton 1986) and can cause increases in some forms of aggression that are not related to density (Alexander & Roth 1971). If the physical attributes of the different environments being compared are not equivalent, then differences in configuration, escape opportunities, and places of concealment may influence aggression rates independent of density (Erwin et al. 1976; Anderson et al. 1977). Differences in the age/sex composition between comparison groups (Kummer & Kurt 1965; Rowell 1967) and amount of provisioning (see Asquith 1989 for review) may also influence aggressive rates and confound density comparisons. Finally, manipulating a group's social density by adding or removing individuals is so disruptive to the social relationships of an established primate group that valid conclusions cannot be drawn concerning density (Southwick 1967).

This partial list of potential confounds should not create the impression that the study of density is a hopeless endeavor. Progress from earlier

---

Note: "Duration of high density" in "between-group comparisons" refers to the amount of time that the highest-density groups had been living under their density condition at the start of a study. Values usually represent the amount of time between the formation of a captive group and the start of behavioral observations. "Variable" crowding durations were cases in which group sizes were manipulated so frequently that duration at a highest density could not be assessed. The general conclusions listed concerning "More" or "Less" of a particular behavior category under higher density represent my, or the author's, general conclusions from what are sometimes complex series of analyses involving several behavior categories. "More" aggression was indicated if increases in any intensity of aggression occurred (heavy, mild, or both) or if only one age/sex class increased aggression. Letter notations provide more details or describe confounds.
[a]Less grooming occurred, but more huddling occurred.
[b]Less grooming occurred.
[c]More grooming occurred.
[d]Confounded because of the configuration of the testing apparatus: males controlled female aggression in the smaller compartment.
[e]Confounded because of manipulations of group composition.
[f]No statistical tests were used to draw conclusions.

work has allowed later investigators to identify and control confounding variables, conduct more exacting investigations, and make more accurate interpretations. However, if inconclusive studies with acknowledged confounds, apparent confounds, or no statistical tests are excluded, results are still inconsistent: only 59 percent of the remaining studies support a density/aggression relationship (10 of the 17 in the upper portion of Table 7.1).

The inconsistency across studies might be due to differences in methodology or design, but there does not appear to be a unifying pattern among those presented in Table 7.1. Both within-group and between-group comparisons produce inconsistent results, and studies with shorter exposure to high density produce results as inconsistent as those with longer exposure. Perhaps the most consistent results come from studies in which groups are crowded into a restricted space for a week or less, observed, and then released. Aggression usually increases during this brief exposure to high density (four of the first five studies listed in Table 7.1). The one exception is an experiment with chimpanzees (*Pan troglodytes*) rather than macaques in which aggression decreased during crowding (Aureli & de Waal 1997).

Such species differences in reactions to crowding may also account for inconsistent results across studies, but reliable conclusions must be available within a particular species before comparisons can be made between species. Too few studies have been conducted on single species to make generalizations, however. Rhesus macaques are the most extensively investigated primate (47 percent of studies in the upper portion of Table 7.1), yet even in this pugnacious species there is no consistent relationship between increased density and aggression. Only five of eight rhesus studies support a density/aggression relationship.

Attempts to interpret such conflicting results may seem futile, but the problem may actually be an oversimplified expectation that all individuals in a primate group will show a generalized increase in all forms of aggression. Primate aggression is expressed in a range of intensities that may serve a variety of functions. Among studies in which more intense aggression, such as bites, chases, and other physical attacks, was distinguished from milder forms of aggression, such as threats, some responses were influenced by density, but others were not. Observed changes at higher density include increases in heavy and not mild aggression (Alexander & Roth 1971; Erwin 1979); increases in mild and not heavy aggression (Judge & de Waal 1993; Aureli et al. 1995); and increases in both heavy and mild aggression (Judge & de Waal 1997).

Sex differences in response to higher density also occur. Judge & de Waal (1997) found that increases in heavy and mild aggression observed at higher density were only by adult females; adult males showed no increases in any forms of aggression with higher density. During a five-year longitudinal study of increasing social density, adult female aggression decreased as density increased, while there was no effect on adult male aggression (Eaton et al. 1981). In contrast, adult male, and not female, Japanese macaques increased aggression during a crowding experiment (Alexander & Roth 1971).

This variability illustrates the difficulty in making generalizations concerning a group as a whole. De Waal (1989b) has emphasized that it may be naive to expect a simple relationship between density and aggression in primates because primates have complex and flexible social behavior and advanced cognitive abilities. They do not react passively to changing situations, such as increases in density, but may actively change their behavior to cope with adverse situations.

Coping Strategies

De Waal (1989b) pointed out that primates have evolved behavioral mechanisms to control aggression and reduce conflict. For example, dominance relationships are formed that moderate the amount of aggression between two individuals (Preuschoft & van Schaik, Chapter 5). Primates also use less intense aggressive signals, such as threats, to enforce a dominance relationship rather than full-scale attack. Primates have developed formal submissive displays that clearly signal subordinate status and may preclude aggressive enforcement of a dominance relationship. For example, the bared-teeth display in rhesus macaques (de Waal & Luttrell 1985) and submissive greetings in chimpanzees (Noë et al. 1980) are unidirectional responses that correspond to a dominance hierarchy and also occur outside aggressive contexts. Primates also use tension-reducing mechanisms to ease stressful social situations. For example, allogrooming is known to have a calming, tension-reducing effect on recipients (Boccia et al. 1989; Aureli et al. 1999), and primates allogroom more during tense situations (de Waal 1984; Schino et al. 1988). Finally, former opponents exchange affiliative responses shortly after fights to reconcile aggression and restore disrupted relationships (de Waal, Chapter 2).

De Waal (1989b) proposed that primates increase these behavioral responses under conditions of high density to reduce tension and the increased risk of aggression. This "coping model" states that aggression may increase somewhat at higher density, but submissive responses and friendly responses, particularly tension-reduction responses such as allogrooming, will also increase at higher density. These behavioral modifications may take long periods of time, even generations, to develop, as animals learn to control their aggressive responses.

The original evidence for a coping model was a spatial density experiment conducted before the model was formally proposed (Nieuwenhuijsen & de Waal 1982). The behavior of a captive group of chimpanzees was compared when they were in a large outdoor compound and when they were confined to much smaller indoor quarters during the cold winter months. Chimpanzees were found to be just as sociable or solitary in the confined space as in the outdoor compound (as measured by proximity indexes); however, formal submissive greetings and allogrooming increased in the higher density condition. Mild, but not severe, aggression also increased during crowding. The study implied that animals modified their behavior to reduce the increased risk of escalated aggression.

Earlier studies of density and aggression that also reported rates of submission, allogrooming, and other affiliative behavior provide retrospective support for the coping model (Table 7.1). Rowell (1967) reported a general increase in all categories of social behavior in captive compared with wild baboons (*Papio anubis*) but said the most dramatic increase was in dominance/submissive responses. Captive vervets (*Cercopithecus aethiops*) living at higher density did not increase aggression rates but did increase affiliative or submissive behavior (McGuire et al. 1978; McGuire et al. 1983). Marriott (1988) found that captive free-ranging rhesus macaques living under high density had half as many aggressive encounters and performed twice as much social grooming as a wild group.

Short-Term Crowding Investigations of the coping model since its inception incorporate recent knowledge of conflict resolution and tension reduction into the hypotheses tested. These more comprehensive studies examine many categories of social behavior and assess overall patterns of increase and decrease among different classes of dyadic relationships in a group. In one of the first tests of the model, rhesus macaques were periodically crowded into the familiar indoor section of their indoor/outdoor enclosure (Judge & de Waal 1993). Behavior during crowding was then compared with behavior in the outdoor section. Monkeys showed no increases in severe aggression when crowded but did show increases in mild aggression (threats and rough behavior) that dispersed animals in the confined area without escalation (Fig. 7.1). Monkeys avoided dominant animals more and exaggerated conformity to the established dominance hierarchy by increasing formal submissive displays. Allogrooming

FIGURE 7.1. Threat display by an adult rhesus female. Photograph by Frans B. M. de Waal.

decreased during crowding, but animals increased huddling up with their kin apparently to "lay low" and avoid confrontation.

The pattern of changes in several behavior categories during crowding seemed to contribute to the single objective of decreasing the number of interactions and avoiding conflicts. The increases in only mild aggression under crowded conditions might even be considered a form of restraint by aggressors and thus a type of conflict management. We therefore called the constellation of changes a *strategy* and in this specific case a *conflict-avoidance* strategy. Although use of the term strategy to categorize a group of potentially unrelated behavior changes may be unjustified, an independent study of a different species of macaque found virtually the same complex pattern of behavior changes during short-term crowding (Aureli et al. 1995). The consistency of results was impressive considering there were distinct differences in experimental conditions and sampling procedures between the two studies.

Chimpanzees also appeared to use a behavioral strategy when they were temporarily crowded into the indoor portion of their indoor/outdoor enclosure (Aureli & de Waal 1997). While crowded into less than half of their usual living space, animals decreased allogrooming, submissive greeting, and aggressive behavior, and only juvenile play increased. The behavioral changes were interpreted as an *inhibition* strategy whereby adult animals reduced all forms of behavior to decrease social interaction and the possibility for conflict. It is notable that chimpanzee aggression *decreased* during confinement, unlike macaques, who increased mild aggression during crowding (Judge & de Waal 1993; Aureli et al. 1995). Results indicate that different species may have unique reactions to short-term crowding and that more than one strategy may be used in response to increased density.

Long-Term Crowding   Although interesting findings, conflict-avoidance and inhibition strategies were not the specific pattern of response predicted by the coping model. The coping model emphasized an active attempt to reduce tension, but allo-grooming decreased in all of these short-term crowding studies. Animals appeared to avoid conflict rather than actively attempt to reduce conflict. These results did not contradict the coping model as it was originally proposed, however, because an original prediction of the coping model was that an active *conflict-reduction* lifestyle (or strategy) may take long periods of time to develop (de Waal 1989b). Results from short-term crowding experiments may not assess the capacity of primates to adopt a conflict-reduction pattern because animals are not given enough time to adjust to the new conditions. Indeed, when chimpanzees were crowded for longer periods, they adopted a tension-reduction pattern that included increased allogrooming and submissive greetings (Nieuwenhuijsen & de Waal 1982) rather than the inhibition pattern of shorter-term crowding (Aureli & de Waal 1997).

Tests of the coping model using groups that have lived under high-density conditions for long periods are generally supportive of a tension-reduction strategy. Novak et al. (1992) compared groups of rhesus macaques that had lived under the same indoor caging conditions for the first two to three years of life but then were transferred to environments of differing density for the next several years. Two groups lived in indoor enclosures while another group was moved to a much larger five-acre outdoor compound. The high-density indoor groups groomed one another more often and performed more dominance/submissive displays than the lower-density outdoor group, but the high-density groups were not more aggressive than the low-density group.

In what is probably the most comprehensive test of the coping model to date, Judge & de Waal (1997) compared several groups of rhesus macaques living under a wide range of densities. The living conditions compared were relatively high density indoor/outdoor enclosures, medium-density outdoor corrals, and a relatively low density free-ranging island population. The high-density condition was more than two thousand times more crowded than the low-density condition.

Trend analyses of adult behavior showed that as density increased there was an increase in all categories of behavior examined (heavy aggression, mild aggression, submissive teeth-baring displays, and allogrooming). The result implies that the reduced interindividual distances brought about by higher density produce an overall increase in all types of social interaction. Not all animals showed a generalized increase in behavior, however. Adult males increased allogrooming as density increased but not aggression (Fig. 7.2a), while adult females increased both allogrooming and aggression (Fig. 7.2b).

When we analyzed the behavior of males and females toward specific classes of partners, particular strategies were evident. As density increased, adult female interactions toward adult males showed the tension-reduction pattern described by the coping model. Females did not increase any forms of aggression against males as density increased but did increase formal submissive displays and allogrooming (Fig. 7.3).

As density increased, adult males did not show increased aggression toward any partner class tested (adult males, adult females, or juveniles). Either the density variable had no influence on male aggression, or males inhibited aggressive responses at higher density. Adult male behavior toward adult males did not show density-related trends in the other behavior categories tested (allogrooming and bared-teeth displays). Males,

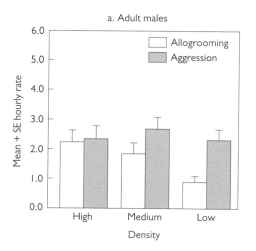

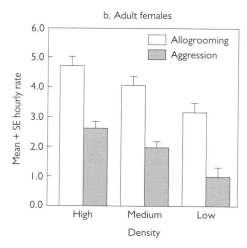

FIGURE 7.2a–b. Allogrooming and aggression by (a) adult male and (b) adult female rhesus macaques to all group members under three levels of population density. The measure provided is mean hourly rate per subject plus the standard error. After Judge & de Waal (1997).

FIGURE 7.3. An adult rhesus female grooms an adult male. Photograph by Frans B. M. de Waal.

however, increased allogrooming with the other members of the group (adult females and juveniles) as density increased. The behavior of adult males toward the female matrilines in their groups could therefore be considered a tension-reduction strategy: they did not increase aggression with these partners but did increase the affiliative behavior of allogrooming at higher density.

Adult female behavior toward other adult females and their immature offspring differed depending on whether the animals were kin or nonkin. As density increased, adult females increased both heavy and mild aggression against their kin but did not increase formal submissive displays or allogrooming. Although this pattern reflects a density/aggression relationship and indicates that crowding has negative consequences on relationships among kin, the unameliorated aggression did not produce the debilitating effect that a density/aggression relationship might imply. Grooming among kin was equivalent at all density conditions, and huddling between adult females and their kin increased as density increased, indicating that relationships with kin remained intact.

Heavy and mild aggression increased between adult females and nonkin as density increased, but adult females also increased formal displays and grooming with nonkin partners. Thus, higher density disturbed adult female relationships with both kin and nonkin partners, but with nonkin partners high density also increased tension-reduction responses. The increases in submissive displays and grooming observed between nonkin females at higher density were not simply immediate reactions to the increased aggression between them. The increases occurred outside aggressive contexts. Since the displays and grooming were not contiguous with aggression, they provide evidence that animals attempted to cope with the potential for aggression and not the consequences of ongoing aggression.

Attempts by adult females to reduce conflict with nonkin under the conditions of heightened antagonism present at high density are not surprising considering the rigid matrilineally based structure of rhesus macaque society. Entire matrilines typically dominate the next-lowest-ranking matriline in a group with very little overlap so that matrilines form a linear dominance hierarchy (Sade 1967; de Waal & Luttrell 1986). Rhesus macaques are quite nepotistic during aggressive interactions, and when two animals from different matrilines fight, animals often join in on the side of their relatives (Kaplan 1978; Bernstein & Ehardt 1985). Increased allogrooming and increased formal expression of established dominance relationships between matrilines at higher density may help moderate potential escalation.

Judge & de Waal's (1997) study may provide a confusing picture of the relationship between density and behavior, since increased density had a variety of influences on different relationships. However, several conclusions can be drawn from formal tests of the coping model and earlier studies. First, primate behavior is definitely influenced

by density as changes occur under different density conditions. Second, aggressive behavior is not the only response influenced by density. A wide range of social behavior including submissive displays and affiliative responses are also affected. Third, results still show that there is no simple relationship between density and behavior in primates. Results differ across species, experimental situations, and individuals within a group. Such variation presents a challenge and indicates a promising area for understanding the behavioral flexibility and the adaptive capacity of nonhuman primates. Fourth, in some cases, the patterns of change adopted appear to be strategies for avoiding or reducing conflict and maintaining peaceful relations during crowding. Not all studies produce clear indications of such coping strategies, however (e.g., Call et al. 1996). Fifth, there is evidence that primates adopt different strategies (e.g., inhibition, conflict avoidance, or tension reduction) depending on the duration of crowding, the species studied, and the individual characteristics of the interacting partners (i.e., the age, sex, and kinship of particular dyads).

Finally, the coping model (de Waal 1989b) predicted that reconciliation would increase at higher density because animals would be more motivated to restore disrupted relationships under more crowded conditions. However, no studies that have investigated the relationship between density and reconciliation have found a relationship. Relative levels of reconciliation remain the same during short-term crowding (Judge & de Waal 1993; Aureli et al. 1995), and intergroup comparisons testing reconciliation rates at various densities have found no differences across conditions (Aureli 1992; Call et al. 1996; Judge & de Waal 1997). Reconciliation may be a process that is dependent on the context of the current confrontation or the value of the relationship to opponents (Cords & Thurnheer 1993; Castles et al. 1996; Cords & Aureli, Chapter 9), and not influenced by a general external variable such as density.

Coping Mechanisms in Non-Primates

The coping model emphasized nonhuman primates because they are known to exhibit conflict-resolution mechanisms, flexibility of behavior, and advanced cognitive abilities (de Waal 1989c). However, non-primates perform submissive displays, affiliative behavior, and tension-reducing responses that may also increase at higher density in order to cope with crowding. Rasa (1979) obtained results consistent with the coping model in a spatial density experiment in which a captive family group of dwarf mongooses (*Helogale undulata rufula*) was restricted to one-fifth of their usual enclosure. Aggression initially doubled but returned to baseline rates. Greetings, a submissive response that inhibits aggression (Rasa 1977), increased and was sustained at elevated levels throughout the experiment. Allogrooming and a related integrative behavior (allomarking) increased. As in primates, crowding had differential effects on particular relationships. Further, relationships that were disrupted, strained, or strengthened during crowding were those that might have been predicted within dwarf mongoose society in which an alpha male and female pair attempt to maintain reproductive exclusivity over subordinate males and females. Rasa (1979) concludes that "the first response to crowding is an almost immediate rise in intragroup aggression which appears to be effectively 'blocked' by a parallel rise in submissive and integrative behavior" (p. 325). Although litters born during crowding were cannibalized and two animals died as a result of aggression-induced stress, Rasa suggests that effective behavioral coping mechanisms prevented a complete dissolution of society.

Mongoose responses to crowding indicate that species other than primates modify behavior under high density and that the behavioral changes may be adaptive strategies to manage conflict. Unfortunately, Rasa's experiment may be the only non-primate study that has observed the full range of a species' behavior in adequate detail to draw conclusions concerning coping behavior and density. An ethogram must be compiled that is detailed enough to suggest the possible function of particular responses (e.g., affiliation, appeasement, submission, threat), and a species must exhibit behavioral mechanisms that reduce conflict or tension before coping can be tested under various density conditions (cf. Schino, Chapter 11).

A density/aggression relationship can be evaluated to some extent in non-primates because aggressive behavior has been examined at varying densities. Non-primate studies are subject to some of the same confounds described in primate studies, however (Archer 1970). For example, xenophobic reactions to strangers and disruptions of established social relationships can have a more pronounced influence on the behavior of non-primates than increased density (e.g., Thiessen 1963; Wolfe & Summerlin 1968; Lobb & McCain 1978). In addition, few aggressive trends reported are based on systematic collection of behavioral data or statistical comparisons. Many reported differences are based on impressions, anecdotes, or examination of means.

Increased aggression at higher densities has been reported across a wide range of non-primate species including rabbits (*Oryctolagus cuniculus*: Myers 1966), birds (*Gallus gallus*: Flickinger 1966), and fish (*Lepomis gibbosus*: Erickson 1967; Table 7.2). The relationship is not universal, however, since higher density does not always lead to increased aggression. Considering that the popular concept of a density/aggression relationship originated from investigations of non-primates, surprisingly

TABLE 7.2
Investigations into the Relationship between Density and Behavior in Non-Primates

Design	Duration of High Density	Common Name	Species	Change in Behavior under Higher Density			Source
				Aggression	Submission	Affiliation	
WITHIN-GROUP COMPARISONS	8 days	Rat	*Rattus* sp.	More[a]	—	—	Lobb & McCain 1978
	6 weeks	Sunfish	*Lepomis gibbosus*	More	—	—	Erickson 1967
	4 months	Mouse	*Mus muculus*	No effect	—	—	Lloyd & Christian 1967
	5 months	Mongoose	*Helogale undulata*	More	More	More	Rasa 1979
	6 months	Mouse	*Mus muculus*	More	—	—	Southwick 1955
	1 year	Rat	*Rattus norvegicus*	More	—	—	Calhoun 1962a, b
	1–2 years	Vole	*Microtus pennsylvanicus*	More	—	—	Louch 1956
	2 years	Rabbit	*Oryctolagus cuniculus*	More	—	—	Myers 1966
BETWEEN-GROUP COMPARISONS	5 minutes	Pigeon	*Columba* sp.	More	—	—	Willis 1966
	15 minutes	Mouse	*Mus* sp.	No effect[b]	—	—	Thiessen 1966
	2–56 days	Cuttlefish	*Sepia officinalis*	More	More	—	Boal et al. 1999
	3–12 weeks	Chicken	*Gallus gallus*	Less	—	—	Craig et al. 1969
	2.5–7.5 months	Chicken	*Gallus gallus*	More[c]	—	—	Flickinger 1966
	4–9 months	Deermouse	*Peromyscus maniculatus*	No effect	—	—	Terman 1974
	6 months	Mouse	*Mus muculus*	No effect	—	—	Southwick 1955
	9 months	Mouse	*Mus muculus*	Less	—	—	Rowe et al. 1964
	10 months–1 year	Vole	*Microtus agrestis*	More	—	—	Clarke 1955

Note: "Duration of high density" in "between-group comparisons" refers to the amount of time that the highest-density groups were exposed to high density. The general conclusions listed concerning the effect of higher density in a particular behavior category represent the full gamut of inference from probability testing to general impressions. Since so few studies were available for inclusion in the table, inconclusive studies without statistical tests were not differentiated from more rigorous investigations. Letter notations provide more details or describe confounds.

[a]Confounded because increased aggression was attributed to the social disruption of combining strange groups of rats rather than increased social density.

[b]Confounded because aggression in this short-term study was assessed on newly formed groups of male strangers.

[c]The author concluded that density had "no effect" on aggression, but examination of group means indicates higher density had an effect as strong as or stronger than those reported as "more" in other studies.

few studies have actually examined the relation-ship, and fewer still have found support for the relationship (11 of 17 studies in Table 7.2).

Interestingly, within-group studies show fairly consistent support for a density/aggression rela-tionship in non-primates (seven of eight studies in Table 7.2), but between-group studies do not. The difference occurs because single groups tend to increase aggressive behavior as density increases, but high-density groups do not neces-sarily have higher aggression rates than other lower-density groups (Southwick 1955). Some high-density groups can be rather peaceful while some low-density groups can be quite combative. The individual characteristics and aggressive ten-dencies of animals that compose populations can influence aggression more so than increasing population density. Even the results of Calhoun's (1962b) influential rat studies were dictated by the unique behavior of a few individuals. The infa-mous "behavioral sinks" developed when a few dominant adult males established breeding terri-tories in quarter sections of the compartmental-ized pens used in the experiments. The remainder of the colony became restricted to single com-partments. In colonies in which males did not establish territories or did so in a manner that did not restrict the rest of the colony, no "behavioral sinks" developed (Calhoun 1962b). This outcome of increased density is rarely cited.

Density, Crowding, and Human Behavior

Humans are certainly influenced by density and the experience of crowding, but the relationship is not well understood. Models proposing the influence of density (see review and synthesis by Baum & Paulus 1987) typically depict density and many other variables (e.g., physical setting, indi-vidual differences in reactions to crowding, cul-tural experience) as contributing to a subjective

experience of crowding. Density produces the feeling of crowding by acting on other mediating variables such as perceived control and overstim-ulation (Baum & Paulus 1987; Evans & Lepore 1992). Perceived crowding produces stress. Stress may be mitigated by coping mechanisms. Suc-cessful or unsuccessful coping at high density then affects behavior (e.g., aggressiveness), phys-iology (e.g., pathology), and cognitive processes (e.g., learned helplessness). Actual density (people per unit space) becomes one component of a complex sequential process.

Studies usually test only one of many vari-ables that may be operating at a single stage in the sequential process. Thus, the general model described has been pieced together from diverse investigations, many of which focus on a specific human application. For example, high density appears to have a negative influence on the per-formance of complex cognitive tasks (Paulus et al. 1976). Such findings are relevant for providing optimum environments for school performance and worker productivity and indicate the impor-tance of understanding the effects of density. However, for all its potential application, the model is not well tested. Perhaps the most neglected stage of the model includes the use of coping mechanisms by humans when crowding is perceived as stressful (Baum & Paulus 1987). Cognitive coping mechanisms have been hypoth-esized (Sherrod 1974; Baron & Rodin 1978), and behavioral changes have been interpreted as coping responses (Loo 1972), but a coping pro-cess in humans is virtually uninvestigated.

Comparisons between humans and other ani-mals are limited because human density studies usually focus on cognitive outcomes that cannot be inferred in other animals (Baron & Needel 1980). Typically, self-reports and "paper and pencil" assessments that measure cognitive and affective states are correlated to density conditions. For example, reported feelings of group hostility might be compared in larger and smaller rooms (Evans 1979). Comparisons between the overt behavior of humans during crowding and that of other animals are also difficult because some of the interactions of interest (e.g., physical aggression) are rare in humans, especially in a laboratory setting. Investigators must sometimes design procedures that assess aggressive behavior indirectly. For example, Freedman et al. (1972) assessed aggressive tendency at different densities using mock jury trials and games that could involve competition or cooperation (prisoner's dilemma). "Aggression" was scored as the severity of jury sentences or number of competitive versus cooperative responses. Conclusions drawn from such indirect assessments may not be generalizable as aggression in other animals.

In order to make comparisons between human and nonhuman primates, I have evaluated only data from human studies that report either social interactions or nonverbal behavior during crowding. Although a strictly ethological approach is the most realistic means of comparison, it is limited because behavioral data are frustratingly absent from a vast majority of the hundreds of density studies conducted on humans. Further, most human studies reporting behavioral data are restricted to observations of young children in a classroom setting, and evidence suggests that the behavioral reactions of young children to high density diminish with age (Loo & Smetana 1978). High density can have distinct influences on the affective states of older children with little or no apparent influence on behavior (Loo & Smetana 1978). A strictly ethological approach overlooks such cognitive and affective reactions to high density.

The behavioral reactions of humans to crowding are variable (Table 7.3), but the most common result is an increase in aggression (60 percent of studies: 9 of 15) and a decrease in affiliative interactions under higher density (67 percent of studies: 12 of 18). Aggression may include physical aggression, angry expressions, rough play, and stealing toys (Hutt & Vaizey 1966; McGrew 1972; Ginsburg et al. 1977; Loo 1978). Reduced affiliation may include decreases in social proximity, friendly behavior, general social activity, and increases in solitary behavior (McGrew 1970, 1972; Loo 1972, 1978; Sundstrom 1975). The decrease in social interaction has generally been interpreted as a *withdrawal* strategy to cope with the increased social stimulation at higher density (Baum & Paulus 1987). Remarkably, the pattern of response is quite similar to the *conflict-avoidance* strategy observed in macaques under short-term crowding (Judge & de Waal 1993; Aureli et al. 1995).

Aggression and withdrawal should not be considered the typical reaction of humans to crowding, however. Of 12 studies that assessed both aggression and affiliation, only 4 found both increased aggression and decreased affiliation (Table 7.3). Further, individuals differ in their responses to crowding. Increased aggression during crowding appears to be more likely in males than in females (Rohe & Patterson 1974; Loo 1978; Loo & Kennelly 1979), particularly in isosexual groups (Freedman et al. 1972). Individual personality traits may also influence reactions to crowding. Loo (1978), for example, found that hyperactive children were even more active under crowded conditions and exhibited nervous pacing. Such differential effects preclude describing a singular human pattern of response to crowding.

Human studies also typically assess only short-term effects of density because ethical considerations prohibit subjecting humans to prolonged experimental crowding. Exceptions include experiments designed to assess the reactions of groups to specific conditions of extended confinement

TABLE 7.3

Selected Investigations into the Relationship between Density and Behavior in Humans

Design	Duration of High Density	Age Class of Subjects	Change in Behavior under Higher Density			Source
			Aggression	Submission	Affiliation	
WITHIN-GROUP COMPARISONS	30 minutes	Children	More	—	More	McGrew 1972
	30 minutes	Children	More	—	No effect	McGrew 1972
	30 minutes	Children	—	—	Less	McGrew 1970
	30 minutes	Children	—	—	Less	McGrew 1970
	40 minutes	Children	More	—	—	Ginsburg et al. 1977
	45 minutes	Children	More	—	—	Rohe & Patterson 1974
	45 minutes	Children	No effect	—	No effect	Smith & Connolly 1977
	45 minutes	Children	No effect[a]	—	No effect	Smith & Connolly 1977
	1 hour	Children	More	More	Less	Loo 1978
	hours[b]	Children	More	—	Less	Hutt & Vaizey 1966
	hours[b]	Children	No effect	—	Less	Liddell & Kruger 1987
BETWEEN-GROUP COMPARISONS	30 minutes	Collegians	—	—	Less	Sundstrom 1975
	48 minutes	Children	Less	No effect	Less	Loo 1972
	1 hour	Children	More	No effect	Less	Loo 1978
	1 hour	Children	More	—	Less	Loo & Kennelly 1979
	1 hour	Collegians	No effect[c]	—	More[d]	Stokols et al. 1973
	1 hour	Children	No effect	—	No effect	Loo & Smetana 1978
	4 hours	Teenagers	More[e]	—	—	Freedman et al. 1972
	3.5 months	Collegians	—	—	Less	Baum & Davis 1980
	7 months	Collegians	—	—	Less	Baum & Valins 1979
	8 months	Collegians	—	—	Less	Evans & Lepore 1993

Note: "Duration of high density" in "between-group comparisons" refers to the amount of time that the highest-density groups were exposed to high density. Affiliation in human studies included any type of nonaggressive social interaction. Letter notations provide more details of results.
[a]One of four groups showed increased aggression under high density.
[b]Sessions of "free play" during attendance at care facilities.
[c]Hostile comments were considered aggression.
[d]Laughter was considered affiliation.
[e]Aggression was measured by mock jury verdicts and a competitive game. Only males increased aggression; females showed no effect.

such as simulations of fallout shelters, isolated military duty, and space flight (see review by Smith 1969). As in most human studies, changes in affect, cognition, and task performance are emphasized more so than overt social behavior, and behavior assessments are mostly anecdotal. Nevertheless, studies have reported that hostility and interpersonal conflicts increase and that people tend to withdraw from social interaction as long-term confinement progresses (Smith 1969), a pattern similar to some short-term crowding experiments. An inherent confound in such confinement studies is that subjects are aware that confinement is temporary and that useful data are being generated from their involvement. Although confinement sometimes lasted for months, awareness of the circumstances of the situation may render such studies more comparable to short-term crowding studies.

More naturalistic studies of the long-term effects of density examine human populations already living under high density (e.g., urban residents, prisoners, and students in college dormitories). Investigators often use a correlational approach to assess the relationship between a measure of population density and behavior. For example, various measures of density have been positively correlated to categories of behavior analogous to aggression: violent crimes (McCarthy et al. 1975); prison incarcerations (Schmitt 1966); civil unrest (Welch & Booth 1974); juvenile delinquency (Schmitt 1966; Galle et al. 1972; Levy & Herzog 1974); aggressive offenses (Levy & Herzog 1974); and prison misconduct (Megargee 1977; Nacci et al. 1977). Although these studies appear to support a density/aggression relationship in humans, some studies find no correlation between high density and aggression (e.g., murder: Bagley 1989). Further, there are acknowledged limitations to such correlational studies (see reviews by Fischer et al. 1975 and Epstein & Baum 1978). Apparent relationships between density and indexes of aggressiveness can often be accounted for by intervening variables that are also correlated with density, such as socioeconomic status, education, and ethnicity (Gillis 1974; Freedman et al. 1975). Although correlational studies suggest a relationship, they do not demonstrate direct causal links between density and aggressiveness and therefore provide no basis to conclude that long-term crowding increases aggression in humans.

More direct investigations of long-term crowding indicate that humans may avoid or withdraw from social interaction in response to chronic crowding (see the last three studies listed in Table 7.3). For example, college dormitory residents sharing common facilities with many people spent less time socializing and kept the doors to their rooms closed more often than less crowded students (Baum & Valins 1979; Baum & Davis 1980). Again, such withdrawal has been interpreted as a coping strategy to reduce the stress of increased social interaction. In contrast, nonhuman primates that have lived at high density for long periods increase all types of social interaction (Judge & de Waal 1997). Perhaps the high-density environments of nonhuman primates do not provide secluded locations for individuals to isolate themselves from interactions with others. Or perhaps nonhuman primate groups are so firmly established with such long histories of valuable relationships that they take a more active and direct role in mediating social interaction.

Baum & Paulus (1987) note that investigation into the effects of human crowding subsided after a proliferation of research in the 1970s, partially because many questions were answered. The effects of high density were found to be less overwhelming than originally proposed, and the effects were not necessarily aversive for all individuals. Research also decreased because density studies were subsumed into the general category of stress research, in which crowding was lumped together as one of several environmental stressors. Many aspects of density effects remain to be studied, however, and recent investigations focusing on the cognitive and emotional consequences of household crowding, rather than regional density, appear to be revitalizing interest in human density research (e.g., Evans & Lepore 1993; Fuller et al. 1996). Further, human behavioral reactions to crowding have yet to be studied in detail. Much like post-conflict reconciliation, work has proceeded on nonhuman primates while similar work on humans is rare (Verbeek et al., Chapter 3). Although selective, the human results described appear to show some behavioral similarities between humans and other animals. Freedman (1979) argues that we should not assume that the reactions of human and nonhuman primates to crowding differ until data indicate otherwise.

Conclusions and Future Directions

High density increases the potential for social conflict, and individuals change their behavior under high density in ways an observer would predict if they were attempting to reduce or neutralize the risk of aggression. The changes in behavior imply the use of conflict management strategies, but results are only correlational. Gross changes in behavior are measured and related to density, but the cause-and-effect relationships behind the behavioral changes are unknown. For example, when conflict management behavior increases at higher density, are the responses conciliatory reactions to the increased aggression occurring at high density, or are they attempts to forestall aggression before it is instigated? Judge & de Waal's (1997) observations of increased allogrooming and submissive displays in nonaggressive contexts at high density suggest the latter possibility, but answering such questions will require more precise experimental testing than a correlational approach. The next step is to examine behavior at an individual level, rather than a group level, and conduct temporal analyses to determine the consequences of behavior. For example, are individuals that adopt particular coping strategies more or less likely to become involved in aggressive interactions than those that do not? Answers would allow us to infer the actual function of coping behavior rather than imply function indirectly with such labels

as *conflict-avoidance strategy, tension-reduction strategy,* and *inhibition strategy.*

High density also produces social tension and stress, and de Waal (1989b) has predicted that animals will increase tension-reduction behavior for relief, but this aspect of density and coping has not been investigated thoroughly. Allogrooming, a tension-reduction response (Terry 1970; Schino et al. 1988), sometimes increases under conditions of long-term crowding, but actual reduction of tension is implied, not tested. Research showing that scratching, an easily recorded self-directed behavior, is a reliable indicator of tension and anxiety (Maestripieri et al. 1992; Schino et al. 1996) and recent research that relates physiological measures of stress, such as heart rate, directly with ongoing behavior (Aureli et al. 1999; Aureli & Smucny, Chapter 10) provide unlimited prospects in this area: coping behavior can be directly related to proposed emotional outcomes. For example, Aureli & de Waal (1997) found that chimpanzees used a behavioral inhibition strategy to reduce the possibility of aggression while undergoing short-term crowding, but behavioral indicators of anxiety increased. They conclude that the inhibition strategy may be useful for a short term, but a more active tension-reduction strategy would have to be used for the long term in order to be adaptive. Measurements of tension will enable testing such hypotheses. Emotional assessments of nonhuman primates will also provide more latitude for comparisons with the many human studies that assess emotional reactions to crowding.

Finally, as a conflict management paradigm the coping model (de Waal 1989b) originally proposed that individuals take active measures, such as increased allogrooming and formal submissive displays, to reduce social tension and conflict under high density. Subsequent tests of the model have shown that reactions to crowding are complex and may include different strategies, some of which are not active but constrained responses to crowding (e.g., an inhibition strategy). Perhaps the concept of the coping model should be expanded to represent any general means to curtail conflict at high density that includes

several different strategies. Results are not entirely consistent, and the obtrusive effects of confounding variables have not been overcome, but patterns are beginning to emerge. A vast area of exploration exists for testing hypotheses and corroborating current conclusions.

References

Alexander, B. K., & Bowers, M. J. 1967. The social structure of the Oregon troop of Japanese macaques. *Primates*, 8: 333–340.

Alexander, B. K., & Roth, E. M. 1971. The effects of acute crowding on aggressive behavior of Japanese monkeys. *Behaviour*, 39: 73–89.

Anderson, B., Erwin, N., Flynn, D., Lewis, L., & Erwin, J. 1977. Effects of short-term crowding on aggression in captive groups of pigtail monkeys (*Macaca nemestrina*). *Aggressive Behavior*, 3: 33–46.

Archer, J. 1970. Effects of population density on behaviour in rodents. In: *Social Behavior of Birds and Mammals: Essays on the Social Ethology of Animals and Man* (J. H. Crook, ed.), pp. 169–210. New York: Academic Press.

Ardrey, R. 1970. *The Social Contract.* New York: Atheneum.

Asquith, P. J. 1989. Provisioning and the study of free-ranging primates: History, effects, and prospects. *Yearbook of Physical Anthropology*, 32: 129–158.

Aureli, F. 1992. Post-conflict behaviour among wild long-tailed macaques (*Macaca fascicularis*). *Behavioral Ecology and Sociobiology*, 31: 329–337.

Aureli, F., & de Waal, F. B. M. 1997. Inhibition of social behavior in chimpanzees under high-density conditions. *American Journal of Primatology*, 41: 213–228.

Aureli, F., Cordischi, C., Cozzolino, R., & Scucchi, S. 1992. Agonistic tactics in competition for grooming and feeding among Japanese macaques. *Folia primatologica*, 58: 150–154.

Aureli, F., van Panthaleon van Eck, C. J., & Veenema, H. C. 1995. Long-tailed macaques avoid conflicts during short-term crowding. *Aggressive Behavior*, 21: 113–122.

Aureli, F., Preston, S. D., & de Waal, F. B. M. 1999. Heart rate responses to social interactions in free-

moving rhesus macaques: A pilot study. *Journal of Comparative Psychology*, 113: 59–65.

Bagley, C. 1989. Urban crowding and the murder rate in Bombay, India. *Perceptual and Motor Skills*, 69: 1241–1242.

Baron, R. M., & Needel, S. P. 1980. Toward an understanding of the differences in the responses of humans and other animals to density. *Psychological Review*, 87: 320–326.

Baron, R. M., & Rodin, S. 1978. Personal control as a mediator of crowding. In: *Advances in Environmental Psychology* (A. Baum, J. E. Singer, & S. Valins, eds.), pp. 145–190. Hillsdale, N.J.: Lawrence Erlbaum Associates.

Baum, A., & Davis, G. E. 1980. Reducing the stress of high-density living: An architectural intervention. *Journal of Personality and Social Psychology*, 38: 471–481.

Baum, A., & Paulus, P. B. 1987. Crowding. In: *Handbook of Environmental Psychology* (D. Stokols & I. Altman, eds.), pp. 533–570. New York: Wiley.

Baum, A., & Valins, S. 1979. Architectural mediation of residential density and control: Crowding and the regulation of social contact. In: *Advances in Experimental Social Psychology*, Vol. 12 (L. Berkowitz, ed.), pp. 131–175. New York: Academic Press.

Bercovitch, F. B., & Lebron, M. R. 1991. Impact of artificial fissioning and social networks on levels of aggression and affiliation in primates. *Aggressive Behavior*, 17: 17–25.

Bernstein, I. S., & Ehardt, C. L. 1985. Intragroup agonistic behavior in rhesus monkeys *Macaca mulatta*. *International Journal of Primatology*, 6: 209–226.

Boal, J. G., Hylton, R. A., Gonzalez, S. A., & Hanlon, R. T. 1999. Effects of crowding on the social behavior of cuttlefish (*Sepia officinalis*). *Contemporary Topics*, 38: 49–55.

Boccia, M. L., Laudenslager, M., & Reite, M. 1988. Food distribution, dominance and aggressive behaviors in Bonnet macaques. *American Journal of Primatology*, 16: 123–130.

Boccia, M. L., Reite, M., & Laudenslager, M. 1989. On the physiology of grooming in a pigtail macaque. *Physiology and Behavior*, 45: 667–670.

Boesch, C., & Boesch, H. 1983. Optimisation of nut-cracking with natural hammers by wild chimpanzees. *Behaviour*, 83: 265–286.

Boyce, W. T., O'Neill-Wagner, P., Price, C. S., Haines, M., & Suomi, S. J. 1998. Crowding stress and violent injuries among behaviorally inhibited rhesus macaques. *Health Psychology*, 17: 285–289.

Byrne, R. 1995. *The Thinking Ape*. Oxford: Oxford University Press.

Calhoun, J. B. 1962a. A "behavioral sink." In: *Roots of Behavior* (E. L. Bliss, ed.), pp. 295–315. New York: Harper.

Calhoun, J. B. 1962b. Population density and social pathology. *Scientific American*, 206: 139–148.

Call, J., Judge, P. G., & de Waal, F. B. M. 1996. Influence of kinship and spatial density on reconciliation and grooming in rhesus monkeys. *American Journal of Primatology*, 39: 35–45.

Castles, D. L., Aureli, F., & de Waal, F. B. M. 1996. Variation in conciliatory tendency and relationship quality across groups of pigtail macaques. *Animal Behaviour*, 52: 389–403.

Christian, J. J. 1961. Phenomena associated with population density. *Proceedings of the National Academy of Sciences of the USA*, 47: 428–449.

Clarke, J. R. 1955. Influence of numbers on reproduction and survival in two experimental vole populations. *Proceedings of the Royal Society of London: Series B*, 144: 68–85.

Cords, M., & Thurnheer, S. 1993. Reconciling with valuable partners by long-tailed macaques. *Ethology*, 93: 315–325.

Craig, J. V., Biswas, D. K., & Guhl, A. M. 1969. Agonistic behavior influenced by strangeness, crowding and heredity in female domestic fowl (*Gallus gallus*). *Animal Behaviour*, 17: 498–506.

Dazey, J., Kuyk, K., Oswald, M., Martenson, J., & Erwin, J. 1977. Effects of group composition on agonistic behavior of captive pigtail macaques, *Macaca nemestrina*. *American Journal of Physical Anthropology*, 46: 73–76.

Demaria, C., & Thierry, B. 1989. Lack of effects of environmental changes on agonistic behavior patterns in a stabilizing group of stumptailed macaques (*Macaca arctoides*). *Aggressive Behavior*, 15: 353–360.

de Waal, F. B. M. 1984. Coping with social tension: Sex differences in the effect of food provision to small rhesus monkey groups. *Animal Behaviour*, 32: 765–773.

de Waal, F. B. M. 1987. Tension regulation and non-reproductive functions of sex in captive bonobos (*Pan paniscus*). *National Geographic Research*, 3: 318–335.

de Waal, F. B. M. 1989a. Behavioral contrasts between bonobo and chimpanzee. In: *Understanding Chimpanzees* (P. G. Heltne & L. A. Marquardt, eds.), pp. 154–175. Cambridge, Mass.: Harvard University Press.

de Waal, F. B. M. 1989b. The myth of a simple relation between space and aggression in captive primates. *Zoo Biology Supplement*, 1: 141–148.

de Waal, F. B. M. 1989c. *Peacemaking among Primates*. Cambridge: Harvard University Press.

de Waal, F. B. M. 1992. Appeasement, celebration and food sharing in the two *Pan* species. In: *Topics in Primatology*, Vol. 1: *Human Origin* (T. Nishida, W. C. McGrew, P. R. Marler, M. Pickford, & F. B. M. de Waal, eds.), pp. 37–50. Tokyo: University of Tokyo Press.

de Waal, F. B. M. 1997. The chimpanzee's service economy: Food for grooming. *Evolution and Human Behavior*, 18: 375–386.

de Waal, F. B. M., & Luttrell, L. M. 1985. The formal hierarchy of rhesus monkeys: An investigation of the bared-teeth display. *American Journal of Primatology*, 9: 73–85.

de Waal, F. B. M., & Luttrell, L. M. 1986. The similarity principle underlying social bonding among female rhesus monkeys. *Folia primatologica*, 46: 215–234.

Doehl, J. 1968. Ueber die faehigkeit einer Schimpansin, Umwege mit selbstaendigen Zwischenzielen zu ueberblicken. *Zeitschrift für Tierpsychologie*, 25: 89–103.

Drickamer, L. C. 1973. Semi-natural and enclosed groups of *Macaca mulatta*: A behavioral comparison. *American Journal of Physical Anthropology*, 39: 249–254.

Eaton, G. G., Modahl, K. B., & Johnson, D. F. 1981. Aggressive behavior in a confined troop of Japanese macaques: Effects of density, season, and gender. *Aggressive Behavior*, 7: 145–164.

Elton, R. H., & Anderson, B. V. 1977. The social behavior of a group of baboons (*Papio anubis*) under artificial crowding. *Primates*, 18: 225–234.

Elton, R. H. 1979. Baboon behavior under crowded conditions. In: *Captivity and Behavior* (J. Erwin, T. L. Maple, & G. Mitchell, eds.), pp. 125–138. New York: Van Nostrand Rheinhold.

Epstein, Y. M., & Baum, A. 1978. Crowding: Methods of study. In: *Human Response to Crowding* (A. Baum & Y. M. Epstein, eds.), pp. 141–164. Hillsdale, N.J.: Lawrence Erlbaum Associates.

Erickson, J. C. 1967. Social structure, territoriality and stress reactions in the sunfish. *Physiological Zoology*, 40: 40–48.

Erwin, J. 1979. Aggression in captive macaques: Interaction of social and spatial factors. In: *Captivity and Behavior* (J. Erwin, T. L. Maple, & G. Mitchell, eds.), pp. 139–171. New York: Van Nostrand Rheinhold.

Erwin, J., Anderson, B., Erwin, N., Lewis, L., & Flynn, D. 1976. Aggression in captive groups of pigtail monkeys: Effects of provision and cover. *Perception and Motor Skills*, 42: 319–324.

Erwin, N., & Erwin, J. 1976. Social density and aggression in captive groups of pigtail monkeys (*Macaca nemestrina*). *Applied Animal Ethology*, 2: 265–269.

Evans, G. W. 1979. Behavioral and physiological consequences of crowding in humans. *Journal of Applied Social Psychology*, 9: 27–46.

Evans, G. W., & Lepore, S. J. 1992. Conceptual and analytic issues in crowding research. *Journal of Environmental Psychology*, 12: 163–173.

Evans, G. W., & Lepore, S. J. 1993. Household crowding and social support: A quasiexperimental analysis. *Journal of Personality and Social Psychology*, 65: 308–316.

Ferron, J. 1993. How do population-density and food-supply influence social behavior in the snowshoe hare (*Lepus americanus*). *Canadian Journal of Zoology*, 71: 1084–1089.

Fischer, C. S., Baldassare, M., & Ofshe, R. J. 1975. Crowding studies and urban life: A critical review. *Journal of the American Institute of Planners*, 41: 406–418.

Flickinger, G. L. 1966. Response of the testes to social interaction among grouped chickens. *General and Comparative Endocrinology*, 6: 89–98.

Freedman, J. L. 1973. The effects of population density on humans. In: *Psychological Perspectives on Population* (J. T. Fawcett, ed.), pp. 209–238. New York: Basic Books.

Freedman, J. L. 1979. Reconciling apparent differences between the responses of humans and other animals to crowding. *Psychological Review*, 86: 80–85.

Freedman, J. L., Levy, A., Buchanan, R., & Price, J. 1972. Crowding and human aggressiveness. *Journal of Experimental Social Psychology*, 8: 528–548.

Freedman, J. L., Heshka, S., & Levy, A. 1975. Population density and pathology: Is there a relationship? *Journal of Experimental Social Psychology*, 11: 539–552.

Fuller, T. D., Edwards, J. N., Vorakitphokatorn, S., & Sermsri, S. 1996. Chronic stress and psychological well-being: Evidence from Thailand on household crowding. *Social Science and Medicine*, 42: 265–280.

Gad, G. 1973. "Crowding" and "pathologies": Some critical remarks. *Canadian Geographer*, 17: 373–390.

Galle, O. R., Gove, W. R., & McPherson, J. M. 1972. Population density and pathology: What are the relations for man? *Science*, 176: 23–30.

Ganslosser, U., & Brunner, C. 1997. Influence of food distribution on behavior in captive bongos, *Taurotragus euryceros*: An experimental investigation. *Zoo Biology*, 16: 237–245.

Gillis, A. R. 1974. Population density and social pathology: The case of building type, social allowance and juvenile delinquency. *Social Forces*, 53: 306–314.

Ginsburg, H. J., Pollman, V. A., Wauson, M. S., & Hope, M. L. 1977. Variation of aggressive interaction among male elementary school children as a function of changes in spatial density. *Environmental Psychology and Nonverbal Behavior*, 2: 67–75.

Goodall, J. 1986. *The Chimpanzees of Gombe.* Cambridge: Harvard University Press, Belknap Press.

Hutt, C., & Vaizey, M. 1966. Differential effects of group density on social behavior. *Nature*, 209: 1371–1372.

Judge, P. G., & de Waal, F. B. M. 1993. Conflict avoidance among rhesus monkeys: Coping with short-term crowding. *Animal Behaviour*, 46: 221–232.

Judge, P. G., & de Waal, F. B. M. 1997. Rhesus monkey behaviour under diverse population densities: Coping with long-term crowding. *Animal Behaviour*, 54: 643–662.

Kaplan, J. R. 1978. Fight interference and altruism in rhesus monkeys. *American Journal of Physical Anthropology*, 49: 241–249.

Koyama, N. F., & Dunbar, R. I. M. 1996. Anticipation of conflict by chimpanzees. *Primates*, 37: 79–86.

Kummer, H., & Kurt, F. 1965. A comparison of social behavior in captive and wild hamadryas baboons. In: *The Baboon in Medical Research* (H. Vagtborg, ed.), pp. 65–80. Austin: University of Texas Press.

Kuroda, S. 1980. Social behavior of the pygmy chimpanzees. *Primates*, 21: 181–197.

Levy, L., & Herzog, A. N. 1974. Effects of population density and crowding on health and social adaptation in the Netherlands. *Journal of Health and Social Behavior*, 15: 228–240.

Liddell, C., & Kruger, P. 1987. Activity and social behavior in a South African township nursery: Some effects of crowding. *Merrill-Palmer Quarterly*, 33: 195–211.

Lloyd, J. A., & Christian, J. J. 1967. Relationship of activity and aggression to density in two confined populations of house mice (*Mus musculus*). *Journal of Mammalogy*, 48: 262–269.

Lobb, M., & McCain, G. 1978. Population density and nonaggressive competition. *Animal Learning and Behavior*, 6: 98–105.

Loo, C. M. 1972. The effects of spatial density on the social behavior of children. *Journal of Applied Social Psychology*, 2: 372–381.

Loo, C. M. 1978. Density, crowding, and preschool children. In: *Human Response to Crowding* (A. Baum & Y. M. Epstein, eds.), pp. 371–388. Hillsdale, N.J.: Lawrence Erlbaum Associates.

Loo, C. M., & Kennelly, D. 1979. Social density: Its effects on behaviors and perceptions of preschoolers. *Environmental Psychology and Nonverbal Behavior*, 3: 131–146.

Loo, C. M., & Smetana, J. 1978. The effects of crowding on the behavior and perception of 10-year-old boys. *Environmental Psychology and Nonverbal Behavior*, 2: 226–249.

Louch, C. D. 1956. Adrenocortical activity in relation to the density and dynamics of three confined populations of *Microtus pennsylvanicus*. *Ecology*, 37: 701–713.

Maestripieri, D., Schino, G., Aureli, F., & Troisi, A. 1992. A modest proposal: Displacement activities as an indicator of emotions in primates. *Animal Behaviour*, 44: 967–979.

Marriott, B. M. 1988. Time budgets of rhesus monkeys (*Macaca mulatta*) in a forest habitat in Nepal and on Cayo Santiago. In: *Ecology and Behavior of Food-Enhanced Primate Groups* (J. E. Fa & C. H. Southwick, eds.), pp. 125–149. New York: Liss Publishing Company.

Mayagoitia, L., Santillan-Doherty, A. M., Lopez-Vergara, L., & Mondragon-Ceballos, R. 1993. Affiliation tactics prior to a period of competition in captive groups of stumptail macaques. *Ethology, Ecology and Evolution*, 4: 435–446.

McCarthy, J. D., Galle, O. R., & Zimmern, W. 1975. Population density, social structure, and interpersonal violence. *American Behavioral Scientist*, 18: 771–789.

McGrew, P. L. 1970. Social and spatial density effects on spacing behavior in preschool children. *Journal of Child Psychology and Psychiatry*, 11: 197–205.

McGrew, W. C. 1972. *An Ethological Study of Children's Behavior.* New York: Academic Press.

McGuire, M. T., Cole, S. R., & Crookshank, C. 1978. Effects of social and spatial density changes in *Cercopithecus aethiops sabaeus*. *Primates*, 19: 615–631.

McGuire, M. T., Raleigh, M. J., & Johnson, C. 1983. Social dominance in adult male vervet monkeys:

General consideration. *Social Science Information*, 22: 89–123.

Megargee, E. I. 1977. The association of population density, reduced space, and uncomfortable temperatures with misconduct in a prison community. *American Journal of Community Psychology*, 5: 289–298.

Morris, D. 1969. *The Human Zoo*. New York: McGraw-Hill.

Myers, K. 1966. The effects of density on sociality and health in mammals. *Proceedings of the Ecological Society Australia*, 1: 40–64.

Nacci, P. L., Teitelbaum, H. E., & Prather, J. 1977. Population density and inmate misconduct rates in the federal prison system. *Federal Probation*, June, 26–31.

Nash, L. T., & Chilton, S. 1986. Space or novelty?: Effects of altered cage size on *Galago* behavior. *American Journal of Primatology*, 10: 37–49.

Nieuwenhuijsen, K., & de Waal, F. B. M. 1982. Effects of spatial crowding on social behavior in a chimpanzee colony. *Zoo Biology*, 1: 5–28.

Noë, R., de Waal, F. B. M., & van Hooff, J. A. R. A. M. 1980. Types of dominance in a chimpanzee colony. *Folia primatologica*, 34: 90–110.

Novak, M. A., & Drewsen, K. H. 1989. Enriching the lives of captive primates: Issues and problems. In: *Housing, Care and Psychological Wellbeing of Captive and Laboratory Primates* (E. F. Segal, ed.), pp. 161–182. Park Ridge, N.J.: Noyes Publications.

Novak, M. A., O'Neill, P., & Suomi, S. J. 1992. Adjustments and adaptations to indoor and outdoor environments: Continuity and change in young adult rhesus monkeys. *American Journal of Primatology*, 28: 125–138.

Owen, A. M. 1997. Cognitive planning in humans: Neuropsychological, neuroanatomical and neuropharmacological perspectives. *Progress in Neurobiology*, 53: 431–450.

Paulus, P. B., Annis, A. B., Seta, J. J., Schkade, J. K., & Matthews, R. W. 1976. Density does affect task performance. *Journal of Personality and Social Psychology*, 34: 248–253.

Rasa, O. A. E. 1977. The ethology and sociobiology of the dwarf mongoose (*Helogale undulata rufula*). *Zeitschrift für Tierpsychologie*, 43: 337–406.

Rasa, O. A. E. 1979. The effects of crowding on the social relationships and behaviour of the dwarf mongoose (*Helogale undulata rufula*). *Zeitschrift für Tierpsychologie*, 49: 317–329.

Rensch, B., & Doehl, J. 1967. Spontanes Oeffnen verschiedener Kistenverschluesse durch einen Schimpansen. *Zeitschrift für Tierpsychologie*, 24: 476–489.

Rohe, W., & Patterson, A. H. 1974. The effects of varied levels of resources and density on behavior in a day care center. In: *Man-Environment Interactions* (D. H. Carson, ed.), pp. 161–171. Washington, D.C.: Environmental Design Research Association.

Rowe, F. P., Taylor, E. J., & Chudley, A. H. J. 1964. The effect of crowding on reproduction of the housemouse (*Mus musculus* L.) living in corn-ricks. *Journal of Animal Ecology*, 33: 477–484.

Rowell, T. E. 1967. A quantitative comparison of the behaviour of a wild and a caged baboon group. *Animal Behaviour*, 15: 499–509.

Russell, C., & Russell, W. M. S. 1968. *Violence, Monkeys and Man*. London: Macmillan.

Sade, D. S. 1967. Determinants of dominance in a group of free-ranging rhesus monkeys. In: *Social Communication among Primates* (S. A. Altmann, ed.), pp. 99–114. Chicago: University of Chicago Press.

Schino, G., Scucchi, S., Maestripieri, D., & Turillazzi, P. G. 1988. Allogrooming as a tension-reduction mechanism: A behavioral approach. *American Journal of Primatology*, 16: 43–50.

Schino, G., Perretta, G., Taglioni, A. M., Monaco, V., & Troisi, A. 1996. Primate displacement activities as an ethopharmacological model of anxiety. *Anxiety*, 2: 186–191.

Schmidt, K. T., Seivwright, L. J., Hoi, H., & Staines, B. W. 1998. The effect of depletion and predictability of distinct food patches on the timing of aggression in red deer stags. *Ecography*, 21: 415–422.

Schmitt, R. C. 1966. Density, health, and social disorganization. *Journal of the American Institute of Planners*, 32: 38–40.

Sherrod, D. R. 1974. Crowding, perceived control and behavioral aftereffects. *Journal of Applied Social Psychology*, 4: 171–186.

Singh, S. D. 1968. Social interactions between the rural and urban monkeys (*Macaca mulatta*). *Primates*, 9: 69–74.

Smith, P. K., & Connolly, K. J. 1977. Social and aggressive behavior in preschool children as a function of crowding. *Social Science Information*, 16: 601–620.

Smith, S. 1969. Studies of small groups in confinement. In: *Sensory Deprivation: Fifteen Years of Research* (J. P. Zubek, ed.), pp. 374–403. New York: Appleton-Century-Crofts.

Southwick, C. H. 1955. Regulatory mechanisms of house mouse populations: Social behavior affecting litter survival. *Ecology*, 36: 627–634.

Southwick, C. H. 1967. An experimental study of intragroup agonistic behavior in rhesus monkeys (*Macaca mulatta*). *Behaviour*, 28: 182–209.

Southwick, C. H. 1969. Aggressive behaviour of rhesus monkeys in natural and captive groups. In: *Aggressive Behavior* (S. Garattini & E. B. Sigg, eds.), pp. 32–43. New York: Wiley.

Sterck, E. H. M., & Steenbeck, R. 1997. Female dominance relationships and food competition in the sympatric Thomas langur and long-tailed macaque. *Behaviour*, 134: 749–774.

Stokols, D., Rall, M., Pinner, B., & Schopler, J. 1973. Physical, social, and personal determinants of the perception of crowding. *Environment and Behavior*, 4: 87–115.

Sundstrom, E. 1975. An experimental study of crowding: Effects of room size, intrusion, and goal blocking on nonverbal behavior, self-disclosure, and self-reported stress. *Journal of Personality and Social Psychology*, 32: 645–654.

Tanji, J., & Shima, K. 1994. Role for supplementary motor area cells in planning several movements ahead. *Nature*, 371: 413–416.

Terman, C. R. 1974. Behavioral factors associated with cessation of growth of laboratory populations of prairie deermice. *Research in Population Ecology*, 15: 138–147.

Terry, R. 1970. Primate grooming as a tension reduction mechanism. *Journal of Psychology*, 76: 129–136.

Thiessen, D. D. 1963. Varying sensitivity of C57BL/Crg1 mice to grouping. *Science*, 141: 827–828.

Thiessen, D. D. 1966. Role of physical injury in the physiological effects of population density in mice. *Journal of Comparative and Physiological Psychology*, 62: 322–324.

Visalberghi, E., & Tomasello, M. 1998. Primate causal understanding in the physical and psychological domains. *Behavioural Processes*, 42: 189–203.

Welch, S., & Booth, A. 1974. Crowding as a factor in political aggression: Theoretical aspects and an analysis of some crossnational data. *Social Science Information*, 13: 151–162.

Willis, F. N. J. 1966. Fighting in pigeons relative to available space. *Psychonomic Science*, 4: 315–316.

Wolfe, J. L., & Summerlin, C. T. 1968. Agonistic behavior in organized and disorganized cotton rat populations. *Science*, 160: 98–99.

CHAPTER 8

The Peacefulness of Cooperatively Breeding Primates

Colleen M. Schaffner & Nancy G. Caine

Many researchers have provided information that supports the impression that callitrichids (tamarins and marmosets) lead a pacific group life. Callitrichids are highly cooperative: all individuals within a group share in a variety of tasks that include searching for and consuming food, antipredator detection and mobbing, defense of the home territory, and shared care of offspring. These well-documented characteristics suggest that marmosets and tamarins are not only cooperative but also nonaggressive and that family life is peaceful. However, high levels of cooperation are not necessarily exclusive of aggression. In our chapter we present data on the rates and forms of aggression in three species, representing three of the four callitrichid genera. We also present an analysis of post-conflict behavior that sheds light on the consequences of aggression. Finally, we explore the role of reproductive inhibition as a proximate mechanism for maintaining low aggression and in turn facilitating cooperation in callitrichids.

Taxonomic Overview of the Callitrichids

The primate family Callitrichidae is composed of approximately 30 distinct species within four genera (Rylands et al. 1993). Callitrichids include the diminutive pygmy marmosets (*Cebuella*); the *Callithrix* marmosets, which include 10 to 12 distinct species; the *Saguinus* tamarins, which include 12 recognized species (Fig. 8.1), and four species of lion tamarins (*Leontopithecus*). Although the callitrichid group represents a large number of species, there are a variety of taxonomic characteristics that distinguish them from other New World primates.

FIGURE 8.1. A family group of captive red-bellied tamarins (*Saguinus labiatus*) emerging from the nesting box. The group consists of six members: a breeding pair and their four adult offspring. Photograph by Nancy G. Caine.

Physical features shared by all marmosets and tamarins include a small body size, clawlike nails, specialized scent glands, a vomeronasal organ, and birth of twins (Eisenberg 1978; Mittermeier et al. 1988). In addition, distinct traits distinguish marmosets from tamarins. The tamarins, *Leontopithecus* and *Saguinus*, have a larger body size and are frugivorous and insectivorous, only occasionally exploiting exudate sources (Kleiman 1978; Snowdon & Soini 1988). The marmosets, *Callithrix* and *Cebuella*, are smaller bodied and rely on gum exudates as a stable food source (Soini 1988; Stevenson & Rylands 1988). Furthermore, the marmosets have developed special dentition and gut morphology for extracting and digesting gums (Ferrari et al. 1993).

Callitrichids have a remarkably flexible social structure (Savage & Baker 1996). Originally, based on a few field studies (Dawson 1978; Izawa 1978; Neyman 1978) and data from captive groups (Epple 1975; Kleiman 1978), tamarins and marmosets were thought to be strictly monogamous, with breeding limited to a single male and female. In the past decade, the picture has become progressively more complex. Reports from the field and captivity indicate that callitrichid groups exhibit a range of mating patterns including monogamy, polyandry, polygyny, and polygynandry. However, two common threads exist among all groups: the presence of relatively long lasting sociosexual bonds among breeding males and females (Goldizen 1989; Baker et al. 1993; Digby & Ferrari 1994) and cooperative social relationships (cf. Garber 1997).

Behavioral Indications of Cooperation

Marmosets and tamarins share a variety of behavioral traits that warrant the characterization "cooperative." For example, they travel in their extended family groups to forage (Garber 1997), take turns at feeding sites (Addington 1998), and share food. In captivity, all individuals will descend on food dishes when introduced to the cage and readily tolerate the removal of food directly from their mouth and hands by all group members (Wied's marmosets, *C. kuhli*: Schaffner personal observation). In addition, adult lion tamarins use vocal cues to promote food transfers to their youngest offspring (Brown & Mack 1978; Ferrari 1987). This cooperative nature is not limited to food-related activities. At night, individuals pack together in the roosting site with arms and legs intertwined (Snowdon & Soini 1988;

Caine et al. 1992; Schaffner personal observation). Furthermore, red-bellied tamarins (S. labiatus) display coordinated antipredator surveillance, in which individuals take turns serving as sentinel (Caine 1986; Buchanan-Smith & Hardie 1997).

Perhaps the hallmark of callitrichid cooperation is the shared rearing of young, in which everyone assists in carrying and provisioning infants. Like other species of birds and mammals that breed cooperatively (see Solomon & French 1997), mature offspring delay dispersal from their home territory, aid the breeding individuals (most often their parents) in rearing infants, and postpone independent breeding well beyond reproductive maturity (French 1997). The essential nature of shared infant care is evident in the extent to which fathers and maturing offspring carry infants (Fig. 8.2). For example, in saddle-back tamarins

FIGURE 8.2. A Wied's marmoset father (*Callithrix kuhli*) carrying his two-month-old twin daughters. Photograph by Jeffrey E. Fite.

(S. fuscicollis), adult males carry offspring twice as often as mothers, and nonreproductive members carry offspring 25 percent of the time (Goldizen 1987a). Cooperative care of young may, in fact, be essential to infant survival. Goldizen et al. (1996) maintain that it is nearly impossible for pairs of saddle-back tamarins to rear offspring successfully in the wild without the assistance of other group members.

Not only do marmosets and tamarins cooperate in a variety of tasks, but all group members affiliate and maintain contact with one another at high rates. Goldizen (1989) found that wild saddle-back tamarins engage in allogrooming activities in approximately 10 percent of sampling intervals and are nearly always within two meters of at least one group member. Captive pairs of pied tamarins (S. bicolor) spend more than 25 percent of observations sitting in contact with each other (Wormell 1994), and similar rates are reported for captive cotton-top tamarins (S. oedipus: Price 1992). Systematic behavioral observations of Wied's marmosets in family groups indicate that individuals are either sitting next to group mates or engaged in social activity approximately 30 percent of the time (Stevenson & Poole 1976; Schaffner unpublished data). Tamarins and pygmy marmosets also rely on monitoring vocalizations to maintain contact with group members throughout the day (Pola & Snowdon 1975; Caine & Stevens 1990). Whether sitting in proximity, engaging in direct social interactions, or maintaining associations via monitoring calls, group mates nearly constantly engage in affiliative contact with one another.

Low Aggression and Peaceful Coexistence

Cooperative behavior and aggression are not necessarily mutually exclusive, but egalitarian social order and low rates of aggression may facilitate the level of cooperation present in marmoset and

TABLE 8.1

The Frequency and Intensity of Aggression Observed in
Wied's Marmosets, Pygmy Marmosets, and Red-Bellied Tamarins

Species	Number of Groups Observed	Total Hours Observed	Rate of Bouts per Hour per Individual	% of Bouts Limited to Visual or Vocal Threats	% of Bouts Potentially Injurious
Red-bellied tamarins[1]	4	149	0.22	90	3
Wied's marmosets[2]	4	29	0.16	83	13
Pygmy marmosets[3]	10	16	0.92	50	4

Source: The data presented here were collected as parts of larger studies of social behavior by 1: Schaffner 1991, 2: Smith et al. 1998, and 3: Addington 1998. All occurrences of aggressive behaviors were scored (Altmann 1974).

tamarin groups (cf. Caine 1993). A variety of researchers drawing on observations from the field and captivity have commented on the mildness and relatively rare incidence of aggression among group mates. In cotton-top and lion tamarins aggression usually takes the form of visual threats and vocalizations without physical contact (Price & Hannah 1983). Coates & Poole (1983) report that aggression in red-bellied tamarins is brief and typically consists of facial and vocal threats. Note that Coates & Poole (1983) and Price & Hannah (1983) make little mention of physical contact in their descriptions of aggression.

Studies that have specifically investigated aggression in callitrichids focus primarily on experimental conditions (e.g., intruder studies: French & Snowdon 1981; Anzenberger 1985; Harrison & Tardif 1988; French & Inglett 1989; Schaffner & French 1997) and specific situations such as sibling fights (Sutcliffe & Poole 1984; Smith & French 1997) and group expulsions (Kleiman 1979; Inglett et al. 1989), in which aggression involves chasing, attempted attacks, biting, and tumble fights. Group expulsions often involved severe aggression, but violent expulsions have not been observed in the wild and may be limited to captive situations in which animals cannot disperse. There is little documentation of the frequency and form of "everyday" aggression in undisturbed family groups. In the following section, we present data on the levels and forms of aggression in captive undisturbed groups of red-bellied tamarins, pygmy marmosets, and Wied's marmosets.

Aggression in Undisturbed Groups

Ten red-bellied tamarins, 16 Wied's marmosets, and 18 pygmy marmosets contributed to the data (Table 8.1). For red-bellied tamarins and Wied's marmosets, aggressive bouts occurred at a rate of once every five hours per individual. Rates for pygmy marmosets were higher, approximately one time per hour per individual. The rate at which bouts of aggression were potentially injurious was extremely low. For red-bellied tamarins, nearly all the aggressive bouts were restricted to a vocal chattering threat that was occasionally followed by chases and lunges; however, in only one case was the aggression potentially injurious (i.e., tumble fighting and uninhibited biting). In Wied's marmosets, the majority of aggression consisted of visual threats, which involved the flattening of ear tufts and squinting of the eyes. Potentially injurious aggression occurred only once. For pygmy marmosets, approximately half of aggressive bouts were restricted to visual or vocal displays, whereas half of the aggression involved physical contact. However, only a handful of bouts could have resulted in serious injury; most of the contact aggression consisted of mild pushing and cuffing with the hand. For none of the species did the expression of aggression differ as a function of sex or age class, and most of the animals were equally likely to be either the actors or recipients of aggression.

Overall, the rates of aggression reported above are within the range of aggressive bouts reported in other studies of captive and free-ranging callitrichids (Epple 1975; Box & Morris 1980; Coates & Poole 1983; Price & Hannah 1983; Heymann 1996). Our group sizes were small (two to six individuals) but are well within the limits that are typical for free-ranging tamarin and marmoset species (Goldizen 1987b; Ferrari & Lopes Ferrari 1989; Ferrari & Rylands 1994). It is likely that as group size increases the frequency of aggressive bouts per hour increases, but we suspect that the rate of aggression per individual per hour remains constant.

Contrasting rates of aggressive interactions in callitrichids with other primate species reveal that marmosets and tamarins exhibit aggressive behavior 3 to 20 times less often than macaque species, depending on the species comparison (de Waal & Luttrell 1989; Castles et al. 1996). In pygmy marmosets aggression occurs two to four times less often than in rhesus macaques (*Macaca mulatta*) and stumptail macaques (*M. arctoides*) but occurs at a higher rate than reported for pigtail macaques (*M. nemestrina*). However, when comparing the rate at which potentially injurious aggression occurs, pygmy marmosets are seven to eight times less likely than pigtail macaques to exhibit severe forms of aggression (Castles et al. 1996). The rates of aggression we reported above are comparable to aggression exhibited in other species that are typically thought to be nonaggressive (e.g., patas

monkeys, *Erythrocebus patas:* Loy et al. 1993; howler monkeys, *Aloutta palliata:* Zucker & Clarke 1998). Most of what we know about callitrichid aggression comes from captive groups; rates of aggression in wild marmosets and tamarins are probably lower than rates of their captive counterparts and might, in fact, be lower than those reported for the vast majority of primates species (for another example of low aggression in New World primates, see Strier et al., Box 15.1).

Reconciliation

Over the past 20 years, primatologists have become increasingly interested in the ways that monkeys and apes cope with the potential disruptions in social relationships that might follow aggressive encounters. Reconciliation, a friendly reunion of former opponents following a conflict, has been studied in a variety of primate species, particularly macaques, baboons, and apes (de Waal, Chapter 2). The presumed function of reconciliation varies, depending on the species and the authors' interpretation (Kappeler & van Schaik 1992; Cords & Aureli 1996; Silk 1996). Most scientists agree that the function of reconciliation is to repair damage between former combatants, particularly in cases in which relationships are considered "valuable" (de Waal & Aureli 1997; Cords & Aureli, Chapter 9; but see Silk, Box 9.1).

Unlike other primate species that have been subjects of reconciliation research to date, marmosets and tamarins live in extended family groups in which cooperation is the defining feature of group life (Caine 1993; Garber 1997). All callitrichids are highly cooperative, and reliable status relationships, if they exist at all apart from breeding priorities, are difficult to detect (Box & Morris 1980; Knox & Sade 1991). As described above, callitrichids depend on one another for vigilance and mobbing, safety while sleeping, infant care, and foraging success; if reconciliation is the way that primates repair the negative consequences of aggression in order to retain valuable social bonds, then callitrichids should reconcile frequently. However, our informal observations of one species, the red-bellied tamarins, did not generate the impression that tamarins reconcile after aggression. Consequently, we designed a study of reconciliation in red-bellied tamarins living in social groups. The results, presented below, lead us to conclude that the strong cohesive and cooperative tendencies of callitrichids somehow obviate the need for reconciliation following aggression.

We observed 10 red-bellied tamarins in four groups for 149 hours during which 77 postconflict observations were made following aggressive interactions. The two combatants were monitored for 20 minutes following the conflict (PC) and for 20 minutes on the next possible observation day at the same time of day to provide matched-control (MC) data following de Waal & Yoshihara's (1983) procedure. We scored the frequency and duration of the following interactions between the two opponents: approach, move away, in contact, allogroom, and stay (stay occurred when the individual remained stationary following an approach by its combatant). Statistical analyses first compared PC and MC data across all pairs of combatants. Then, we limited our analyses to conflicts between breeding individuals, as they are perhaps the most valuable partners (Kleiman 1977; Evans 1983; Savage et al. 1988; Goldizen 1989).

When all combatant pairs were examined, we found no evidence of reconciliation following aggression. Nor did we find evidence of reconciliation when we restricted our analyses to cases of aggression involving only pair mates (i.e., breeding males and females). Analyses of minutes to first contact, time spent in contact, duration of grooming, and approaches revealed no difference for breeding pairs between PC and MC observa-

tions. Finally, when we examined each opponent dyad individually, one potentially interesting finding emerged. Although no single breeding pair showed evidence of reconciliation, one dyad did spend more time in contact following conflict than during MC observations. Two males that resided together with a single unrelated female (polyandrous grouping) spent more time in contact following conflict than they did during MC observations. Future research should examine the possibility that the nature of competition and cooperation among polyandrous males may require restoration of the relationship because their bonds may be less "secure" than other callitrichid relationships. Alternatively, the "attraction" between polyandrous males may reflect a subtle form of mate guarding in which the males maintain close proximity to each other while in the company of the female.

One of the important aspects of data used to assess reconciliation behavior involves the time it takes for individuals to make contact with each other after a fight. Individuals who are reconciling should make contact with each other more quickly than they do during matched-control observations. As shown in Figure 8.3a, the tamarins made contact with each other in the first minute after a fight. However, they also made contact with each other in the first minute of matched-control observations. Indeed, the tamarins were often in contact at the point when PC and MC observations began, reflecting the tamarins' general tendency to affiliate with one another. How can we see an increased tendency to affiliate after a fight (if there is such a tendency) when the tamarins are so often near each other to begin with? One might argue that looking at those cases in which the combatants were *not* in contact when observations began might reveal differences in how quickly tamarins come together (i.e., reconcile) after a fight, compared with control periods. However,

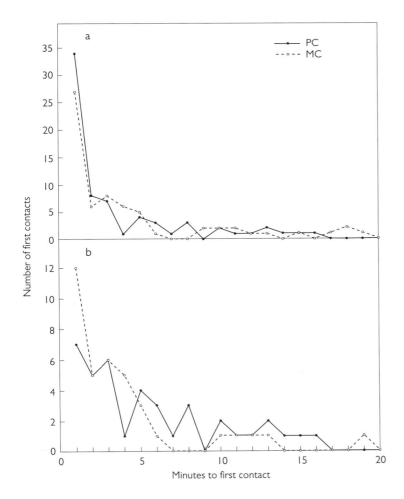

FIGURE 8.3. The number of times in which the first affiliative contact occurred between former opponents in each of the first 20-minute intervals. In both the post-conflict (PC) periods and matched-control (MC) periods, former opponents made affiliative contact or were already in affiliative contact most frequently in the first minute of observation, reflecting the affiliative nature of tamarin social groups. Panel a includes all PC-MC events, whereas b is limited to conflicts in which opponent pairs were not in contact at the start of either the PC or the MC periods.

as shown in Figure 8.3b, focusing on "initially dispersed" cases (N = 45) once again revealed no evidence of reconciling following an altercation.

What accounts for the lack of reconciliation between tamarins following an aggressive interaction? The aggressive interactions did not appear to disrupt the bonds between combatants. In fact, it was extremely difficult to determine if the aggression had any effect on opponents, as the behavior preceding aggression often resumed immediately following the termination of conflict (e.g., eating next to each other, grooming). Perhaps relationships formed in response to the overriding needs of

callitrichids to cooperate for the sake of avoiding predation and caring for infants reduce the extent to which those relationships are vulnerable to disruption from "everyday" aggression. The social tolerance and cooperative relationships exhibited by tamarins may generate sufficient social "security" that relationships simply do not require reaffirmation following an altercation (see Cords & Aureli, Chapter 9). It may be the case that by their virtually constant cooperative interactions, callitrichids may obviate the need for reconciliation altogether.

Our conclusion that aggression does not disrupt social relationships is not meant to suggest that callitrichids do not actively establish, reinforce, or maintain social bonds. Schaffner & French (1997) have shown that Wied's marmoset pairs approach each other more following exposure to unfamiliar intruders; and Shepherd & French (1999) have shown that separation from the pair mate increases calling and results in higher levels of sociosexual behavior following reunion. In addition, common marmosets (*C. jaccus*) alter their behavior in the presence of unfamiliar potential mates as a function of social status. Unpaired females solicit and engage in sexual behavior with novel males, whereas paired females do not (Anzenberger 1985). It appears, however, that aggression of the sort that accounts for the great majority of antagonism in tamarins and marmosets does not threaten relationships between group mates. It remains to be seen if the rare but more severe instances of aggression, sometimes seen in captive callitrichids, in which one individual relentlessly and systematically pursues and fights with another group mate, generate any reconciliatory activity. However, this form of aggression is used to evict individuals from the group and therefore is not likely to be followed by efforts to repair the relationship. In fact, following severe conflicts it is extremely difficult to reintroduce ejected individuals into the group (see Inglett et al. 1989).

Reproductive Inhibition: A Role in Reducing Aggression?

One proximate factor that may contribute heavily to the low levels of aggression in callitrichids is the inhibition of reproduction among nonbreeding individuals. As mentioned previously, adult daughters and sons delay breeding beyond the age of reproductive competence (see French 1997 for review). For *Callithrix* marmosets (Barrett et al. 1990) and *Saguinus* tamarins (Epple & Katz 1984) there is strong evidence that reproductive inhibition is maintained through olfactory cues on the part of breeding females, whereas in lion tamarins inhibition is communicated via behavioral mechanisms (French 1997). Reproductive inhibition likely arose for multiple reasons. Inhibition may serve to (1) increase successful rearing of offspring by promoting singular breeding in females and making available "helpers" to mitigate the high costs of reproduction that characterize callitrichids; (2) provide opportunities for kin-selected cooperative infant care by individuals who have limited breeding opportunities; and (3) provide nonbreeding individuals with the opportunity to gain experience essential for the survival of their own offspring (French et al. 1996b). In addition, predation pressure may have contributed to the adaptive significance of cooperative breeding and the reproductive inhibition that supports it (Caine 1993; Tardif 1994). While carrying infants, captive individuals occupy the safest places in enclosures (Price 1991) and significantly reduce time devoted to foraging (Goldizen 1987a; Caine 1996). Multiple individuals are necessary to carry young if adults are unwilling to risk foraging while carrying infants. Likewise, emigrating may mean that the individual finds itself temporarily alone, without the support of multiple group mates. Because this condition is probably extremely dangerous for callitrichids (Caine 1993), individuals may delay breeding in exchange for the relative safety of membership in an established social group.

In sum, selection pressures relating directly (e.g., via mitigation of energetic costs) and indirectly (e.g., via predation pressures) to reproductive success probably led to reproductive inhibition in callitrichids; the proximate mechanisms promoting the inhibition involve reproductive behavior and physiology. Reproductive inhibition maintains adult sons and daughters in a reproductively and behaviorally juvenile state. Under conditions of reduced mating competition, aggression should be less likely. If this contention is correct, there should be evidence of increased aggression when animals are released from inhibition (or fail to become inhibited) within contexts in which intrasexual competition is potentially high. As described below, there is some evidence that this is the case.

Reproductive Inhibition Ceases— Aggression Increases

In *Saguinus* tamarins, reproductive cycling is usually limited to a single female. Savage et al. (1997) reported on a case in wild tamarins in which a cotton-top tamarin daughter's ovulatory cycling and subsequent pregnancy led to increased aggressive interactions between her and her mother, the breeding female. Ultimately the daughter was peripheralized and left the group. Although cotton-top tamarins are only rarely aggressive, when nonbreeding daughters begin to cycle, rates of aggression among group members increase, particularly between the cycling daughter and the breeding female (Savage personal communication).

In captive groups of *Callithrix* marmosets, daughters often cycle within the family group, but these cycles are characterized by progesterone insufficiency in the luteal phase, and levels of aggression from mother to daughter do not change (Saltzman et al. 1997; Smith et al. 1997; Schaffner personal observation). However, aggressive altercations among siblings increase in response to the onset of cycling by one or both of those siblings. On occasion, older daughters evict younger siblings when physiological puberty is attained (common marmosets: Hubrecht 1989; Geoffroy's marmosets, *C. geoffroyi*: Caine personal observation; Wied's marmosets: Smith unpublished data). Although ovulatory cycling in daughters appears to·go unnoticed by the breeding female, daughters who give birth to young within their natal group are prone to aggressive interactions with

their mothers, and in extreme cases, infanticide, probably by the breeding female, occurs (Digby 1995a). Furthermore, Digby (1995b) reported that a daughter breeding in her family group emigrated shortly after her infant died, suggesting possible expulsion from the group.

There is very little information pertaining to the sociosexual relationships of pygmy marmosets. At this time, information regarding severe forms of aggression, that is, group expulsions and sibling fights, is limited to casual observations of researchers and colony managers. Addington (personal communication) reports that group expulsions occur, on occasion, particularly in larger groups. It is not known whether the expulsions and sibling fights are associated with changes in the reproductive status of maturing offspring. However, Carlson et al. (1997) did not report mother-daughter aggression when a daughter cycled within her family group. In light of preliminary findings regarding the extent of physiological inhibition in pygmy marmosets (Carlson et al. 1997) and other similarities that exist among *Callithrix* and *Cebuella* (e.g., reliance on gums and smaller home ranges), it is likely that pygmy marmosets exhibit patterns of aggression similar to those reported for *Callithrix* marmosets when breakthroughs in inhibition occur.

In lion tamarins behavioral inhibition but not physiological inhibition occurs. All daughters cycle in their natal groups at about 16 months of age, and the cycles are physiologically indistinguishable from the cycles of their mothers (French & Stribley 1987; French et al. 1989). In wild groups, Deitz & Baker (1993) discovered 11 instances of two females reproducing in 35 different groups. Interestingly, there are more reports of aggression, sibling fights, and parent-offspring conflicts in lion tamarins than in the other callitrichid genera, and a third of female-female altercations in captive lion tamarins are deadly (Kleiman 1979;

Inglett et al. 1989). It may be that behavioral control of breeding rights is less effective than physiological control in terms of limiting ovulation and pregnancy and reducing aggression within groups.

Although physiological inhibition seems to play a role in reducing aggression, it cannot be the only causal factor. When *Saguinus* tamarin females cycle in the presence of the breeding female, extreme aggression is very likely. But *Callithrix* females are able to cycle in the presence of the breeding female without consequent aggression. The degree of relatedness among members of *Callithrix* groups is probably higher than it is among tamarin groups because of stability of membership over time (Ferrari & Lopes Ferrari 1989; Digby & Barreto 1993; Ferrari & Digby 1996). If this is true, kin-selected tolerance may play a role in reducing aggression among cycling *Callithrix* females. However, this does not explain the fact that, as stated above, *Callithrix* siblings may become aggressive to one another when one or both of them cycles, and infanticide has been reported in *Callithrix* females. High degrees of relatedness within groups also means that the only mating options available to *Callithrix* daughters are fathers and brothers; inbreeding avoidance may limit competition behaviorally, thus reducing aggression when females cycle. However, the many ways in which reproductive inhibition interacts with other aspects of callitrichid social behavior and ecology to modulate aggression are yet to be understood. Much more data are needed to verify and understand the circumstances under which reproductive inhibition relates to aggression in marmosets and tamarins.

Conclusions and Future Research

Cooperation and peaceful relationships are ubiquitous within callitrichid groups, permeating every aspect of group life, including foraging and feeding, antipredator behavior, sleeping, and infant care. We were able to document in three different species from three different genera that aggression is infrequent and almost always mild. Furthermore, our data from red-bellied tamarins fail to reveal reconciliation following aggression. If reconciliation is necessary to repair temporarily disrupted relationships, we must conclude that everyday sorts of aggression in undisturbed family groups of tamarins do not lead to such disruptions.

We proposed that reproductive inhibition, particularly physiological inhibition, is a key proximate mechanism for maintaining the peaceful and cooperative life of callitrichids. There is indeed some evidence for an association between severe aggression and failures of reproductive inhibition. Less effective reproductive inhibition appears to be associated with more intragroup aggression. But more detailed data from more species, particularly *Callithrix* marmosets, *Saguinus* tamarins, and pygmy marmosets, on the rate of aggression and its relationship to failures of inhibition are needed. In lion tamarins, systematic data on the form of aggression in intact family groups would be valuable. If our hypothesis is correct and reproductive inhibition reduces aggression within groups, it is likely that baseline rates of aggression, as well as the frequency of potentially injurious aggression, are higher in lion tamarin families than in other callitrichid genera. In addition, in the absence of physiological inhibition of breeding, lion tamarins may rely on other proximate mechanisms to maintain peaceful bonds.

Our discussion of the relationship between physiological inhibition and aggression is limited to females: there are no data to address the role of reproductive inhibition in regulating aggressive behavior among males. Male social behavior is an area of special interest because of the frequent reports of polyandrous groups in wild and captive callitrichids (Garber et al. 1984; Goldizen 1987a; Soini 1987; Price & McGrew 1991; Rothe & Koenig 1991; Baker et al. 1993; Schaffner 1996). French et al. (1996a) found in Wied's marmosets that nonbreeding adult males, housed in their family groups, have significantly lower levels of urinary testosterone than their fathers. Furthermore, as reported for *Saguinus* tamarin females, male marmosets and tamarins pursue copulations only if removed from their natal group and placed with an unfamiliar female (saddle-back tamarins: Epple & Katz 1980; common marmosets: Abbott 1984). Nonbreeding common marmoset males also fail to copulate with unfamiliar females if visual contact is maintained with the family (Anzenberger 1985). These findings indicate that as in females, there is physiological or behavioral reproductive inhibition, or both, in males, which may account for low rates of male-male aggression.

We are beginning to understand the mechanisms underlying the peaceful life of marmosets and tamarins. Reproductive inhibition and nonaggressive cooperative lifestyle have been two very common themes in callitrichid research. However, these two areas of research have not overlapped. Ours has been an initial attempt to establish an association between the cooperative behavior and inhibition research to explain the low level of aggressiveness in callitrichids. We are optimistic that with more detailed data we will be able to disclose the secret of conflict management in these peaceful societies.

References

Abbott, D. H. 1984. Behavioral and physiological suppression of fertility in subordinate marmoset monkeys. *American Journal of Primatology*, 6: 169–186.

Addington, R. L. 1998. Pygmy marmosets (*Cebuella pygmaea*) are egalitarian social foragers. *American Journal of Primatology*, 45: 161.

Altmann, J. 1974. Observational study of behavior: Sampling methods. *Behaviour*, 49: 227–267.

Anzenberger, G. 1985. How stranger encounters of common marmosets (*Callithrix jacchus jacchus*) are influenced by family members: The quality of behavior. *Folia primatologica*, 45: 202–224.

Baker, A. J., Dietz, J. M., & Kleiman, D. G. 1993. Behavioural evidence for monopolization of paternity in multi-male groups of golden lion tamarins. *Animal Behaviour*, 46: 1091–1103.

Barrett, J., Abbott, D. H., & George, J. M. 1990. Extension of reproductive suppression by pheromonal cues in subordinate female marmoset monkeys, *Callithrix jacchus. Journal of Reproduction and Fertility*, 90: 411–418.

Box H., & Morris, J. M. 1980. Behavioural observations on captive pairs of wild caught tamarins (*Saguinus mystax*). *Primates*, 21: 53–65.

Brown, K., & Mack, D. S. 1978. Food sharing among captive *Leontopithecus rosalia. Folia primatologica*, 29: 268–290.

Buchanan-Smith, H. M., & Hardie, S. M. 1997. Tamarins mixed-species groups: An evaluation of a combined captive and field approach. *Folia primatologica*, 68: 272–286.

Caine, N. G. 1986. Visual monitoring of threatening objects by captive tamarins (*Saguinus labiatus*). *American Journal of Primatology*, 10: 1–8.

Caine, N. G. 1993. Flexibility and co-operation as unifying themes in *Saguinus* social organization and behaviour: The role of predation pressures. In: *Marmosets and Tamarins: Systematic, Behaviour, and Ecology* (A. B. Rylands, ed.), pp. 200–219. Oxford: Oxford University Press.

Caine, N. G. 1996. Foraging for animal prey by outdoor groups of Geoffroy's marmosets (*Callithrix geoffroyi*). *International Journal of Primatology*, 17: 933–945.

Caine, N. G., & Stevens C. 1990. Evidence for a "monitoring call" in red-bellied tamarins. *American Journal of Primatology*, 22: 251–262.

Caine, N. G., Potter, M. P., & Mayer, K. E. 1992. Sleeping site selection by captive tamarins (*Saguinus labiatus*). *Ethology*, 90: 63–71.

Carlson, A. A., Ziegler, T. E., & Snowdon, C. T. 1997. Ovarian function of pygmy marmoset daughters (*Cebuella pygmaea*) in intact and motherless families. *American Journal of Primatology*, 43: 340–347.

Castles, D. L., Aureli, F., & de Waal, F. B. M. 1996. Variation in conciliatory tendency and relationship quality across groups of pigtail macaques. *Animal Behaviour*, 52: 389–403.

Coates, A., & Poole, T. B. 1983. The behavior of the callitrichid monkey, *Saguinus labiatus labiatus*, in the laboratory. *International Journal of Primatology*, 4: 339–369.

Cords, M., & Aureli, F. 1996. Reasons for reconciling. *Evolutionary Anthropology*, 5: 42–45.

Dawson, G. A. 1978. Composition and stability of social groups of the tamarin, *Saguinus oedipus geoffroyi*, in Panama: Ecological and behavioral implications. In: *The Biology and Conservation of the Callitrichidae* (D. G. Kleiman, ed.), pp. 23–37. Washington, D.C.: Smithsonian Institution Press.

Deitz, J. M., & Baker, A. J. 1993. Polygyny and female reproductive success in golden lion tamarins (*Leontopithecus rosalia*). *Animal Behaviour*, 46: 1067–1078.

de Waal, F. B. M., & Aureli, F. 1997. Conflict resolution and distress alleviation in monkeys and apes. In: *The Integrative Neurobiology of Affiliation* (C. S. Carter, I. I. Lenderhendler, & B. Kirkpatrick, eds.), pp. 317–328. New York: Annals of the New York Academy of Sciences.

de Waal, F. B. M., & Luttrell, L. M. 1989. Toward a comparative socioecology of the Genus *Macaca*: Different dominance styles in rhesus and stumptail monkeys. *American Journal of Primatology*, 19: 83–109.

de Waal, F. B. M., & Yoshihara, D. 1983. Reconciliation and redirected affection in rhesus monkeys. *Behaviour*, 85: 145–224.

Digby, L. J. 1995a. Infant care, infanticide, and female reproductive strategies in polygynous groups of common marmosets (*Callithrix jacchus*). *Behavioral Ecology and Sociobiology*, 37: 51–61.

Digby, L. J. 1995b. Social organization in a wild population of *Callithrix jacchus*: II. Intra-group social behavior. *Primates*, 36: 361–375.

Digby, L. J., & Barreto, C. E. 1993. Social organization in a wild population of *Callithrix jacchus*: Group Composition and Dynamics. *Folia primatologica*, 61: 123–134.

Digby, L. J., & Ferrari, S. F. 1994. Multiple breeding females in free-ranging groups of *Callithrix jacchus*. *International Journal of Primatology*, 15: 389–397.

Eisenberg, J. F. 1978. Comparative ecology and reproduction of New World monkeys. In: *The Biology and Conservation of the Callitrichidae* (D. G. Kleiman, ed.), pp. 13–22. Washington, D.C.: Smithsonian Institution Press.

Epple, G. 1975. The behavior of the marmoset monkeys (Callitrichidae). In: *Primate Behavior*, Vol. 4 (L. A. Rosenblum, ed.), pp. 195–239. New York: Academic Press.

Epple, G., & Katz, Y. 1980. Social influences on first reproductive success and related behaviors in the saddle-back tamarins (*Saguinus fuscicollis, Callitrichidae*) *International Journal of Primatology*, 1: 171–183.

Epple, G., & Katz, Y. 1984. Social influences on estrogen excretion and ovarian cyclicity in saddle back tamarins (*Saguinus fuscicollis*). *American Journal of Primatology*, 6: 215–227.

Evans, S. 1983. The pair-bond of the common marmoset, *Callithrix jacchus jacchus*: An experimental investigation. *Animal Behaviour*, 31: 651–658.

Ferrari, S. F. 1987. Food transfer in a wild marmoset group. *Folia primatologica*, 48: 203–206.

Ferrari, S. F., & Digby, L. J. 1996. Wild *Callithrix* groups: Stable extended families? *American Journal of Primatology*, 38: 19–27.

Ferrari, S. F., & Lopes Ferrari, M. A. 1989. A re-evaluation of the social organisation of the Callitrichidae, with reference to the ecological differences between genera. *Folia primatologica*, 52: 132–147.

Ferrari, S. F., & Rylands, A. B. 1994. Activity budgets and differential visibility in field studies of three marmosets (*Callithrix* spp.). *Folia primatologica*, 63: 78–83.

Ferrari, S. F., Lopes, M. A., & Krause, E. A. K. 1993. Gut morphology of *Callithrix nigriceps* and *Saguinus labiatus* from western Brazilian Amazonia. *American Journal of Physical Anthropology*, 90: 487–493.

French, J. A. 1997. Proximate regulation of singular breeding in callitrichid primates. In: *Cooperative Breeding in Mammals* (N. G. Solomon & J. A. French, eds.), pp. 40–81. Cambridge: Cambridge University Press.

French, J. A., & Inglett, B. J. 1989. Female-female aggression and male indifference in response to unfamiliar intruder in lion tamarins. *Animal Behaviour*, 37: 487–497.

French, J. A., & Snowdon, C. T. 1981. Sexual dimorphism in response to unfamiliar intruders in the tamarin, *Saguinus oedipus*. *Animal Behaviour*, 29: 822–829.

French, J. A., & Stribley, J. A. 1987. Synchronization of ovarian cycles within and between social groups in the golden lion tamarin (*Leontopithecus rosalia*). *American Journal of Primatology*, 34: 115–132.

French, J. A., Inglett, B. J., & Dethlefs, T. 1989. The reproductive status of non-breeding group members in captive golden lion tamarin social groups. *American Journal of Primatology*, 18: 73–86.

French, J. A., Smith, T. E., & Schaffner, C. M. 1996a. Sources of variation in the contexts and mechanisms of suppression in callitrichid primates. Invited paper presented at the 26th Congress of the International Primatological Society and the 19th Conference of the American Society of Primatologists, August 11–16, Madison, Wisc.

French, J. A., Brewer, K. J., Schaffner, C. M., Schalley, J., Hightower-Merritt, D. L., Smith, T. E., & Bell, S. M. 1996b. Urinary steroid and gonadotropin excretion across the reproductive cycle in female Wied's black tufted-ear marmosets (*Callithrix kuhli*). *American Journal of Primatology*, 40: 231–246.

Garber, P. A. 1997. One for all and breeding for one: Cooperation and competition as a tamarin reproductive strategy. *Evolutionary Anthropology*, 7: 187–199.

Garber, P. A., Moya, L., & Malago, C. A. 1984. A preliminary field study of the moustached tamarin monkey (*Saguinus mystax*) in northeastern Peru: Questions concerned with the evolution of a communal breeding system. *Folia primatologica*, 42: 17–32.

Goldizen, A. W. 1987a. Facultative polyandry and the role of infant-carrying in wild saddle-back tamarins (*Saguinus fuscicollis*). *Behavioral Ecology and Sociobiology*, 20: 99–109.

Goldizen, A. W. 1987b. Tamarins and marmosets: Communal care of offspring. In: *Primate Societies* (B. B. Smuts, D. L. Cheney, R. M. Seyfarth, R. W. Wrangham, & T. T. Struhsaker, eds.), pp. 34–43. Chicago: Chicago University Press.

Goldizen, A. W. 1989. Social relationships in a cooperatively polyandrous group of tamarins (*Saguinus fuscicollis*). *Behavioral Ecology and Sociobiology*, 24: 79–89.

Goldizen, A. W., Mendelson, J., & Terborgh, J. 1996. Saddle-back tamarin (*Saguinus fuscicollis*) reproductive strategies: Evidence from a thirteen-year study of a marked population. *American Journal of Primatology*, 38: 57–84.

Harrison, M. L., & Tardif, S. D. 1988. Kin preference in marmosets and tamarins: *Saguinus oedipus* and *Callithrix jacchus* (Callitrichidae, Primates). *American Journal of Physical Anthropology*, 77: 377–384.

Heymann, E. W. 1996. Social behavior of wild moustached tamarins, *Saguinus mystax*, at the Estación Biológica Wuebrada Blanco, Peruvian Amazonia. *American Journal of Primatology*, 38: 101–113.

Hubrecht, R. C. 1989. Fertility of daughters in common marmoset (*Callithrix jacchus jacchus*) family groups. *Primates*, 30: 423–432.

Inglett, B. J., French, J. A., Simmons, L. G., & Vires, K. W. 1989. Dynamics of intrafamily aggression and social reintegration in lion tamarins. *Zoo Biology*, 8: 67–78.

Izawa, K. 1978. A field study of the ecology and behavior of the black-mantled tamarin (*Saguinus nigricollis*). *Primates*, 19: 241–274.

Kappeler, P. M., & van Schaik, C. P. 1992. Methodological and evolutionary aspects of reconciliation among primates. *Ethology*, 92: 51–69.

Kleiman, D. G. 1977. Monogamy in mammals. *Quarterly Review of Biology*, 52: 39–69.

Kleiman, D. G. 1978. Characteristics of reproduction and sociosexual interactions in pairs of lion tamarins (*Leontopithecus rosalia*) during the reproductive cycle. In: *Biology and Conservation of the Callitrichidae* (D. G.

Kleiman, ed.), pp. 181–190. Washington, D.C.: Smithsonian Institution Press.

Kleiman, D. G. 1979. Parent-offspring conflict and sibling competition in a monogamous primate. *American Naturalist*, 194: 753–760.

Knox, K. L., & Sade, D. S. 1991. Social behavior of the emperor tamarin in captivity: Components of agonistic display and the agonistic network. *International Journal of Primatology*, 12: 439–480.

Loy, J., Argo, B., Nestell, G., Vallett, S., & Wanamaker, G. 1993. A reanalysis of patas monkeys' "grimace and gecker" display and a discussion of their lack of formal dominance. *International Journal of Primatology*, 14: 879–893.

Mittermeier, R. A., Rylands, A. B., & Coimbra-Fihlo, A. F. 1988. Systematic: Species and subspecies. In: *Ecology and Behavior of Neotropical Primates* (R. A. Mittermeier, A. B. Rylands, A. F. Coimbra-Filho, & G. A. B. Fonseca, eds.), pp. 13–75. Washington, D.C.: World Wildlife Fund.

Neyman, P. F. 1978. Aspects of the ecology and social organization of free-ranging cotton-top tamarins (*Saguinus oedipus*) and the conservation status of the species. In: *Biology and Conservation of the Callitrichidae* (D. G. Kleiman, ed.), pp. 39–71. Washington, D.C.: Smithsonian Institution Press.

Peres, C. A. 1989. Costs and benefits of territorial defense in wild golden lion tamarins, *Leontopithecus rosalia*. *Behavioral Ecology and Sociobiology*, 25: 227–233.

Pola, Y. V., & Snowdon, C. T. 1975. The vocalizations of pygmy marmosets (*Cebuella pygmaea*). *Animal Behaviour*, 23: 826–842.

Price, E. C. 1991. The costs of infant carrying in captive cotton-top tamarins. *American Journal of Primatology*, 26: 23–33.

Price, E. C. 1992. Changes in the activity of captive cotton-top tamarins (*Saguinus oedipus*) over the breeding cycle. *Primates*, 33: 99–106.

Price, E. C., & Hannah, A. C. 1983. A preliminary comparison of groups structure in the golden lion tamarin, *Leontopithecus r. rosalia*, and the cotton-top tamarin, *Saguinus oedipus*. *Dodo, Journal of the Jersey Wildlife Preservation Trust*, 20: 36–48.

Price, E. C., & McGrew, W. C. 1991. Departures from monogamy: Survey and synthesis from colonies of cotton-top tamarins. *Folia primatologica*, 57: 16–27.

Rothe, H., & Koenig, A. 1991. Variability of social organization in captive common marmosets (*Callithrix jacchus*). *Folia primatologica*, 57: 28–33.

Rylands, A. B., Coimbra-Fihlo, A. F., & Mittermeier, R. A. 1993. Systematics, geographic distribution, and some notes on the conservation status of the Callitrichidae. In: *Marmosets and Tamarins: Systematic, Behaviour, and Ecology* (A. B. Rylands, ed.), pp. 11–77. Oxford: Oxford University Press.

Saltzman, W., Severin, J. M., Schultz-Darken, N. J., & Abbott, D. H. 1997. Behavioral and social correlates of escape from suppression of ovulation in female common marmosets housed with the natal family. *American Journal of Primatology*, 41: 1–22.

Savage, A., & Baker, A. J. 1996. Callitrichid social structure and mating system: Evidence from field studies. *American Journal of Primatology*, 38: 1–4.

Savage, A., Ziegler, T. E., & Snowdon, C. T. 1988. Sociosexual development, pair bond formation, and mechanisms of fertility suppression in female cotton-top tamarins (*Saguinus oedipus oedipus*). *American Journal of Primatology*, 14: 345–359.

Savage, A., Shideler, S. E., Soto, L. H., Causado, J., Giraldo, L. H., Lasley, B. L., & Snowdon, C. T. 1997. Reproductive events of wild cotton-top tamarins (*Saguinus oedipus*) in Colombia. *American Journal of Primatology*, 43: 347–356.

Schaffner, C. M. 1991. Aggression and post-conflict behavior in red-bellied tamarins (*Saguinus labiatus*). Master's thesis, Bucknell University.

Schaffner, C. M. 1996. Social and endocrine factors in the establishment and maintenance of sociosexual relationships in Wied's black-tufted ear marmoset (*Callithrix kuhli*). Ph.D. diss., University of Nebraska.

Schaffner, C. M., & French, J. A. 1997. Group size and aggression: "Recruitment incentives" in a cooperatively breeding primate. *Animal Behaviour*, 54: 171–180.

Shepherd, R. E., & French, J. A. 1999. Comparative analysis of sociality in lion tamarins (*Leontopithecus rosalia*) and marmosets (*Callithrix kuli*): Responses to separation from long term pairmates. *Journal of Comparative Psychology*, 113: 24–32.

Silk, J. B. 1996. Why do primates reconcile? *Evolutionary Anthropology*, 5: 39–42.

Smith, T. E., & French, J. A. 1997. Social and reproductive condition modulates urinary cortisol excretion in black tufted-ear marmosets (*Callithrix kubli*). *American Journal of Primatology*, 42: 253–267.

Smith, T. E., Schaffner, C. M., & French, J. A. 1997. Social and developmental influences on reproductive function in female Wied's black tufted-ear marmosets (*Callithrix kubli*). *Hormones and Behavior*, 31: 159–168.

Smith, T. E., McGreer-Whitworth, B., & French, J. A. 1998. Close proximity of the heterosexual partner reduces the physiological and behavioral consequences of novel-cage housing in black-tufted ear marmosets (*Callithrix kubli*). *Hormones and Behavior*, 34: 211–222.

Snowdon, C. T., & Soini, P. 1988. The tamarins, genus *Saguinus*. In: *Ecology and Behavior of Neotropical Primates* (R. A. Mittermeier, A. B. Rylands, A. Coimbra-Filho, & G. A. B. Fonseca, eds.), pp. 223–298. Washington, D.C.: World Wildlife Fund.

Soini, P. 1987. Sociosexual behavior of a free-ranging *Cebuella pygmaea* (Callitrichidae: Platyrrhini) troop during postpartum estrus of its reproductive female. *American Journal of Primatology*, 13: 223–230.

Soini, P. 1988. The pygmy marmoset, genus *Cebuella*. In: *Ecology and Behavior of Neotropical Primates* (R. A. Mittermeier, A. B. Rylands, A. Coimbra-Filho, &

G. A. B. Fonseca, eds.), pp. 79–129. Washington, D.C.: World Wildlife Fund.

Solomon, N. G., & French, J. A. 1997. *Cooperative Breeding in Mammals*. Cambridge: Cambridge University Press.

Stevenson, M. F., & Poole, T. B. 1976. An ethogram of the common marmoset *Callithrix jacchus jacchus*. *Animal Behaviour*, 24: 428–451.

Stevenson, M. F., & Rylands, A. B. 1988. The marmosets, Genus *Callithrix*. In: *Ecology and Behavior of Neotropical Primates* (R. A. Mittermeier, A. B. Rylands, A. Coimbra-Filho, & G. A. B. Fonseca, eds.), pp. 131–222. Washington, D.C.: World Wildlife Fund.

Sutcliffe, A. G., & Poole, T. B. 1984. Intragroup agonistic behavior in captive groups of the common marmoset *Callithrix jacchus jacchus*. *International Journal of Primatology*, 5: 473–489.

Tardif, S. D. 1994. Relative energetic cost of infant care in small-bodied neotropic primates and its relation to infant-care patterns. *American Journal of Primatology*, 34: 133–144.

Wormell, D. 1994. Relationships in male-female pairs of pied tamarins *Saguinus bicolor bicolor*. *Dodo, Journal of the Jersey Wildlife Preservation Trust*, 30: 70–79.

Ziegler, T. E., Bridson, W. E., Snowdon, C. T., & Eman, S. 1987. The endocrinology of puberty and reproductive functioning in female cotton-top tamarins (*Saguinus oedipus*) under varying social conditions. *Biology of Reproduction*, 37: 618–627.

Zucker, E. L., & Clarke, M. R. 1998. Agonistic and affiliative relationships of adult female howlers (*Alouatta palliata*) in Costa Rica over a 4-year period. *International Journal of Primatology*, 19: 443–449.

PART III

Repairing the Damage

Introduction

Conflict management uses many different tools, but perhaps the most readily observable is the effort of former adversaries to come together and "undo" the damage to their relationship following open conflict. In humans, this can be accomplished through verbal apology and compensatory gifts, and in animals through the well-studied process of reconciliation, that is, post-conflict friendly interaction between former opponents. This process allows parties who recently engaged in hostilities to resume cooperation on which survival may depend. It is therefore assumed that reconciliation is most typical of cooperative, or valuable, relationships.

Cords and Aureli explore this variable in Chapter 9 and add to the current models by proposing two further components of social relationships—security and compatibility—as factors affecting the chance that relationship repair will take place. They propose various ways to measure these three relationship variables. The independent evaluation of each variable is the first step toward a complete definition and subsequent testing of this three-way model.

In Box 9.1, **Silk** proposes what she sees as an alternative view of the relationship repair function of reconciliation. She suggests that the effect of post-conflict reunions may be short lasting only, namely, a signaling of the ceasing of hostilities. Future work on short-term versus long-term effects may resolve the issue raised or perhaps indicate that both are at work because long-term effects commonly depend on an accumulation of short-term ones. The concept of social relationship assumes such an accumulation.

Call presents data in Box 9.2 about the way in which distance is regulated in two macaque species following a fight. Distance regulation is more subtle than reconciliation by means of affiliative contact, such as grooming, but may be just as effective and meaningful. His study confirms dramatic species differences and warns against matched-control designs that ignore the distance dimension.

In Chapter 10, **Aureli and Smucny** point out that behind the observable behavior of conflict and its aftermath reside emotions of anxiety, attraction, and continued hostility that mediate the events observed, including reconciliation and possibly other forms of conflict management. One gauge of these emotions is physiological measurement such as heart rate or cortisol level, but there exist also behavioral indicators of arousal and anxiety that provide interesting insights into the "state of mind" induced by conflict. This raises the possibility that signals exchanged in the reconciliation context serve to manipulate the emotions of the other. The reassuring effect of one such signal among wild baboons is addressed by **Cheney and Seyfarth** in Box 10.1: if a dominant individual grunts at another after a fight, this facilitates subsequent friendly interaction between them.

Weaver and de Waal report in Box 10.2 that the nature of the mother-offspring relationship is a good predictor of how young monkeys will engage in post-conflict interactions with unrelated adults. Both the way these adults react to the youngsters after conflict and the way the youngsters themselves react to these adults mirror the security or insecurity of the mother-offspring attachment.

In Chapter 11 by **Schino**, and Boxes 11.1–11.3 by **Rowell**, **Samuels and Flaherty**, and **Hofer and East**, the scarce literature on conflict resolution in non-primate species, and the possible reason for this scarcity, are reviewed. From the beginning, primatologists have argued that there is no reason why reconciliation should be limited to chimpanzees, in which it was first discovered. And indeed, the phenomenon has been demonstrated in widely different primate species (Appendix A). Rowell, however, points out how the concept of reconciliation did not fit the motivational models of early ethologists. Together with certain biases in evolutionary models (de Waal, Chapter 2; Matsumura & Okamoto, Box 5.1), and the selective reading habits of scientists pointed out by Schino, this may have led to an underappreciation of the widespread importance of conflict resolution. The result is that we know extremely little about reconciliation in non-primate species. What is presented by these authors is extremely encouraging, however, as it indicates existence of the phenomenon in animals ranging from dolphins to hyenas and from goats to monkeys.

The same techniques used to demonstrate reconciliation in animals can be used to investigate how children make peace. In a cross-cultural and developmental perspective that nicely complements that of Fry (Chapter 16) and Park and Enright (Box 17.1), **Butovskaya, Verbeek, Ljungberg, and Lunardini** summarize in Chapter 12 a large amount of data on children in five separate geographic regions in the world. This is the first such approach to spontaneous conflict resolution in children, and the results are informative. For example, some conciliatory gestures (e.g., object offers, hugs) were found in all human children, whereas others (e.g., ritualized peacemaking rhymes) were restricted to specific cultures. The great variation in both the tendency and method of conflict resolution—even though the phenomenon itself is universal—may provide a fertile ground for theorizing about how the general society affects the way we learn to handle conflict.

In Box 12.1 **Potegal** explores one important context in which learning takes place, that of distress about parental control over resources during a period in which, over most of human history, weaning took place. The great similarity between children's temper tantrums and those of young primates, and the similarity in response by the caregiver (initial distancing but eventual affiliation), suggest a shared developmental underpinning of reconciliation. It also illustrates again the power of the valuable-relationship hypothesis, as the parent is central in the life of the child, and hence every conflict needs to be followed by repair in order to preserve a relationship on which the youngster's survival and the parents' reproduction depend.

CHAPTER 9

Reconciliation and Relationship Qualities

Marina Cords & Filippo Aureli

When conflict has erupted and subsided, former opponents have the option of reconciling their differences. In nonhuman animals, such reconciliations have been documented systematically in many species of primates (Appendix A; de Waal, Chapter 2), and more anecdotal evidence exists for other mammals as well (Schino, Chapter 11). In these animals, reconciliation takes the form of a friendly encounter between former opponents and typically occurs just a few minutes after the end of the behavioral conflict (Kappeler & van Schaik 1992). Both observational and experimental studies have documented the function of these friendly reunions in reestablishing characteristic levels of tolerance in the dyad (Cords 1992), reducing the likelihood of further attacks (Aureli & van Schaik 1991; Cords 1992) and relieving signs of anxiety enhanced by the conflict (Aureli

& Smucny, Chapter 10; Cheney & Seyfarth, Box 10.1). Human children participate in similar post-conflict reunions but also engage in conciliatory behavior while terminating conflict (Butovskaya et al., Chapter 12; Verbeek et al., Chapter 3). Conciliatory behavior in children may include friendly interactions like hand-holding, embracing, kissing, or simple resumption of joint play. Children may also interact verbally, making cooperative propositions (e.g., "You can help me build this house") or offering to share objects (even those not immediately available, e.g., "I will bring you my new toy"; Sackin & Thelen 1984; Butovskaya et al., Chapter 12). When conflicts end in these conciliatory ways, children are more likely to interact subsequently in a friendly manner (Sackin & Thelen 1984; Vespo & Caplan 1993; Verbeek et al., Chapter 3). This pattern suggests that conciliatory behavior of human children restores tolerance and maintains social relationships, as it does in other primates (Cords & Killen 1998).

Although former opponents have the option of reconciling after their conflict, they may not actually do so. In fact, empirical studies of nonhuman primates have shown that the proportion of aggressive conflicts followed by friendly reunion is usually lower than 100 percent, with exact figures varying by the group or species studied or the particular dyads within a group (Appendix A). Given that reconciliatory reunions have beneficial consequences for the individuals involved, it may seem surprising that they do not occur more often. One must remember, however, that attempts to reconcile may also be costly. Approaching a former opponent who was just previously aggressively motivated may be risky because aggression can be renewed (e.g., Aureli & van Schaik 1991; Cords 1992), and cases of "false reconciliation" have been reported in which one former opponent is invited to approach the other only to be attacked at the last moment (de Waal 1986b; Cheney & Seyfarth 1990, p. 195). If such costs are associated with post-conflict reunions, one would expect them to occur only when the benefits of reconciling exceed the costs. Thus explaining the occurrence of reconciliation becomes a matter of assessing the relative costs and benefits of engaging in this behavior.

In this chapter, we focus on understanding variation in the likelihood of reconciliation from the perspective that its function is to repair the relationship between the two opponents (de Waal & van Roosmalen 1979; Cords & Aureli 1996; see Silk, Box 9.1, for another perspective). The nature of the opponents' relationship is thus central to an evaluation of the benefits of reconciling but is also likely to influence the costs. We consider qualities of the social relationship between the two oppo-

nents, as well as individual characteristics of each of them, that might influence the tendency to reconcile. Whether a friendly reunion occurs after a particular conflict will depend on the interaction of the animals' tendencies to come together and the social and physical situation in which they find themselves.

Some of our ideas began their development in earlier publications (Cords 1988; Cords & Aureli 1993; Cords & Thurnheer 1993). In following this earlier work, we have chosen to focus on three particular qualities of social relationships (Table 9.1): the *value* and *security* of relations with a social partner relate especially to the benefits of reconciling conflicts with that partner, while the *compatibility* of partners relates most to the costs of reconciling. We explain these terms immediately below and summarize empirical evidence of the relevance of each of them. Recognizing the limited nature of this evidence, and hoping to spur further research, we then suggest ways in which these qualities might be measured empirically and so related to

TABLE 9.1

Relationship Qualities That
May Influence Conciliatory Tendency

Value	What the subject gains from her or his relationship with a partner, which depends on what the partner has to offer, how willing she or he is to offer it, and how accessible a partner she or he is
Security	The perceived probability that the relationship with the partner will change, which relates to the consistency of the partner's behavioral responses
Compatibility	The general tenor of social interactions in a dyad, which may result from both the temperament of the partners and their shared history of social exchanges

conciliatory tendency. Our intention is not to present a comprehensive model that explains variation in conciliatory tendencies: our more modest objective is simply to take the first step of identifying factors, related to social relationships, that may influence that tendency and suggesting ways that such influence can be verified. A second step would involve gathering (more) data on the influence of these factors. Only with these data in hand would it be possible to move on to the third step of determining how these relationship qualities interact with one another, and with contextual factors, to produce post-conflict outcomes.

Three Key Relationship Qualities and Conciliatory Tendency

Our view of social relationships follows from the pioneering work of Hinde (1979) and Kummer (1978). Hinde described a relationship in terms of the content, patterning, and quality of social interactions between two individuals. He recognized that earlier interactions could influence later ones, so that any pair of individuals would establish a unique history of interactions and thus a unique relationship. Kummer added a functional perspective to Hinde's descriptive view by emphasizing the way in which an animal that interacts socially with another individual can influence the likelihood that its partner behaves in a way from which it benefits. Kummer viewed relationships as social investments, and interactions as ways of maximizing gain (or perhaps minimizing loss; see Cords 1997) from those investments. Some interactions also represent the relationship "at work," in the sense that one or both partners are benefiting directly from the interaction.

From Kummer's (1978) view follows the notion that social partners have a certain value to those with whom they interact. He described the components of value as behavioral tendencies (how

---

BOX 9.1

The Function of Peaceful Post-Conflict Interactions

An Alternate View

Joan B. Silk

For most of the contributors to this volume and for many of the readers of these contributions, the function of peaceful post-conflict interactions is well established. Peaceful interactions after conflicts provide a means of resolving conflicts and repairing relationships that have been damaged by conflict (de Waal & van Roosmalen 1979). However, I believe that there are theoretical (Silk 1997) and empirical reasons to question this conclusion (Silk 1996). My goal here is to outline an alternative explanation for peaceful post-conflict behavior.

Discussions of the function of post-conflict behaviors are complicated by the terms we use to describe these interactions. This is because *reconciliation* is a functional label, like affiliation or aggression, not a descriptive one, like grooming or biting (Silk 1998). However, I believe that important questions about the function of these interactions remain unresolved. Thus, I use the term *peaceful post-conflict interactions* to refer to nonaggressive interactions between former opponents that take place in the minutes that follow conflicts.

I have recently proposed an alternative explanation for the evolution of peaceful post-conflict interactions (Silk 1996, 1997). I suggest that peaceful post-conflict interactions are honest signals that indicate that the conflict is over and that the actor's intentions are now benign. The logic underlying this hypothesis goes like this. After conflicts, even conflicts that have clearly decided outcomes, there may be some uncertainty about whether the conflict will continue (Aureli et al. 1989; Aureli & van Schaik 1991). This uncertainty makes it potentially dangerous or stressful for former adversaries to approach each other or reestablish peaceful contact. Both victims and aggressors exhibit behavioral

BOX 9.1 (continued)

signs of stress, for example, scratching and other forms of self-directed behavior, in this situation (Aureli & Smucny, Chapter 10).

Some behaviors may be honest signals that indicate that the conflict is over and the actor's intentions are peaceful or benign. A good example of these kinds of signals comes from observations of conflicts among free-ranging female baboons (*Papio cynocephalus ursinus*) in the Okavango Delta of Botswana. After conflicts, females often approached their former opponents and grunted to them. These grunts facilitated nonaggressive interactions between former opponents after conflicts (Silk et al. 1996; Cheney & Seyfarth 1997; Cheney & Seyfarth, Box 10.1) and alleviated the victim's concern about becoming the target of renewed aggression from her former aggressor (Cheney et al. 1995). Thus, grunts seem to be honest signals of the actor's intention to behave peacefully. Signals of benign intent can evolve, even when there is some conflict of interest between participants, and seem to play an important role in many social contexts among primates (Silk et al. in press).

Cords & Aureli (1996) have suggested that the benign intent hypothesis is simply a proximate explanation for how contact between former opponents is restored, while the relationship-repair hypothesis provides an ultimate explanation for why former opponents are motivated to repair relationships after conflict occurs. According to this view, a female baboon's grunt signals her intention to behave peacefully toward former opponents, and this enables the female to interact with her former opponent in the immediate aftermath of conflict. However, a female's "decision" to grunt to any particular opponent is based on the benefits that the female will ultimately derive from repairing and preserving her relationship with her former opponent.

Thus, the following question arises: Does the benign intent hypothesis provide a sufficient and complete explanation for the occurrence of peaceful post-conflict behavior in primates? I believe that it does. My intuition is based on my reading of the literature on reconciliation and my observations of post-conflict behavior among baboons. I outline two points that support this view below.

In the Okavango Delta, female baboons were significantly more likely to give peaceful post-conflict signals

An adult female baboon sits beside a female with a newborn infant and grunts to the mother. Then, the female gently handles the infant while the mother watches. This sequence of events is typical of many interactions among adult female baboons. Grunts seem to function as signals of the actor's intention to behave benignly and facilitate nonaggressive interactions. Photograph by Joan B. Silk.

when their former opponent had a young infant than when she did not (Silk et al. 1996). Moreover, the rate of peaceful post-conflict signals closely tracked the rate of infant handling as infants matured. In this group, females directed peaceful post-conflict signals selectively to mothers of newborn infants because they were motivated to interact with their infants, not because they were attempting to repair and preserve valued long-term social bonds. Although Cords & Aureli (1996) have disputed the relevance of this observation, I remain convinced that it supports the idea that peaceful post-conflict signals are not always used to repair and preserve social bonds.

The second observation that contributes to my intuition is that high rates of reconciliation often occur among kin (reviewed by de Waal & Aureli 1996; Silk et al. 1996). The relationship-repair model predicts that peaceful post-conflict signals should occur most often within dyads that value their relationships *because* conflict dam-

BOX 9.1 (continued)

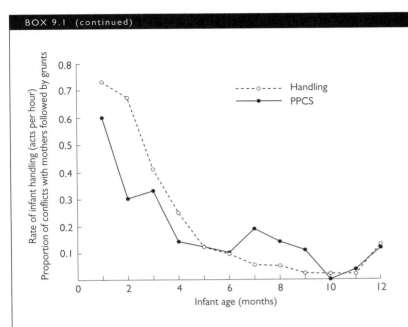

As infants grew older, the rate of infant handling (acts per hour) by other adult female baboons declined sharply. The proportion of all conflicts with mothers that were followed by peaceful post-conflict signals (PPCS) also declined as infants matured (Silk, Cheney, & Seyfarth unpublished data).

ages social bonds. According to Cords (1988), kin, whose relationships are valuable but resilient, should have little need to reconcile because their relationships are unlikely to be disrupted by conflict. The security of social bonds does not figure in recent writings about the pattern of peaceful post-conflict behavior (e.g., de Waal & Aureli 1996), but I believe that Cords's original logic was cogent. Therefore, evidence of high rates of peaceful post-conflict behavior among kin seems to fit the benign intent hypothesis better than the relationship-repair hypothesis.

Although we have studied peaceful post-conflict behavior in dozens of species (Appendix A), we do not yet have enough information about its form, pattern, and function to draw firm conclusions about the selective forces that have shaped its evolution. Predictions derived from these two hypotheses need to be tested empirically, and the logic underlying these hypotheses needs to be scrutinized further. However, this debate will be productive if it stimulates researchers to give more careful attention to why and how nonhuman primates resolve conflicts and to address functional questions explicitly.

likely is the partner to act in a way that benefits the subject), availability (how accessible is the partner for such beneficial interaction), and qualities of the partner (such as social status, reproductive condition, knowledge, or skills) that make interaction with that partner beneficial. The behavioral tendencies of relevance are likely to depend on the type of organism: studies of nonhuman primates have emphasized the tendencies to tolerate others near resources, to support them

in aggressive encounters with third parties, to protect them against external threats, to facilitate access to food or social resources, and to be a willing mate (Cords 1997; van Schaik & Aureli, Chapter 15).

No two partners are likely to be equally valuable to a given individual, as they may differ along any of the three dimensions that define value. For the same reason, two individuals are unlikely to value each other to the same degree; in terms of value, a social relationship is very likely to be asymmetrical.

The value of a social partner should be a critical factor that influences the likelihood of reconciliation (de Waal 1986a, 1989; Kappeler & van Schaik 1992). When a social partner is more valuable, restoring amicable relations with that individual is more important. Restoring amicable relations is less important with a partner who is not valuable. There are reasons, however, why reconciliation might still occur with partners of little or no value: it may be advantageous to avoid escalated conflict with anyone because of the risks and uncertainties associated with such conflict, and post-conflict reunions may contribute to the development of valuable relationships that are not yet established. We would nevertheless expect reconciliation to occur less often with partners of little or no value because disturbance of relationships with more valuable partners entails a larger loss of benefits.

Although the value of a social partner is likely to be a primary factor influencing the likelihood of reconciling after aggression, the security of the relationship may be an important secondary factor. By security we refer to the confidence that an individual has that its relationship with a partner is firmly established and thus unlikely to change suddenly. One's relationship with a partner is secure if the partner's behavioral tendencies and dispositions (Kummer 1978) are predictable. We would expect that when social partners are valuable, relationships that are less secure are in greater need of repair after aggressive conflict. Relations with unpredictable but valuable partners often require more nurturing care generally: post-conflict reconciliation reconfirms mutual interest.

Security, like value, may not be symmetrically experienced and assessed by social partners. Two individuals may not show the same degree of predictability in response to each other. Differences in response consistency could result from differences in personality, social priorities, or the availability of alternative partners (cf. Noë et al. 1991 on market effects).

Although the value of the social partner and the security of the relationship with that partner should influence the benefits of reconciling after aggression, compatibility may be a third factor that influences primarily the costs of reconciliation. By compatibility, we refer to the general tenor of social relations within a dyad, which may be a product of both individual characteristics such as temperament and a shared history of nonantagonistic interactions in a variety of contexts, not just a post-conflict one. Compatibility is important insofar as it influences accessibility of the social partner. When the members of a dyad are in the habit of interacting in nonantagonistic ways in many contexts, it may be easier for them to engage in a post-conflict friendly reunion because this is the sort of interaction they usually have with each other, and so it is a familiar course of action. Given this history, the result of approaching a former opponent is less risky.

Compatibility is directly related to the shared experience of social partners and so would be the same for both partners in an absolute sense. These partners may differ, however, in their perspectives on the compatibility of their relationship, depending on how it compares with the compatibilities of other relationships they have. What is a relatively compatible relationship for one partner may be a relatively incompatible relationship for the other.

Evidence That Relationship Qualities Influence Reconciliation Rates

In this section, we will briefly review the evidence from published studies, mostly of nonhuman primates, that is relevant to the three factors outlined above. Four points need mentioning at the outset. First, as we have noted, the value, security, and compatibility of a dyadic relationship may not be the same for both members of the dyad. Since it takes two animals to engage in a post-conflict reunion, the chances of reconciliation actually occurring depend on the perspectives of both opponents and are increased when both partners find their mutual relationship more valuable, less secure, and characterized by higher levels of compatibility. Few studies, however, have focused on the potentially differing perspectives of social partners. Second, variation in conciliatory tendency can occur on many scales—within a dyad over time, between dyads in a single group, between groups of a single species, and also between species. Studies of nonhuman primates have emphasized the between-dyads and between-species comparisons, and our review necessarily reflects this focus. Third, measuring reconciliatory tendencies is not straightforward. Veenema et al. (1994; Veenema, Box 2.1) have discussed measurement techniques, but their methodological refinements were published after some of the data we review and were not incorporated in all studies published later. It is therefore possible that some of the results that have been published will need modification in the future. Fourth, no observational study has singled out the effect of any one of the three variables that we consider (i.e., value, security, and compatibility). Although the authors usually choose to emphasize one factor over another in their interpretations, it is quite likely that there is some covariance in these factors and that observed effects may have multiple causes.

Value

Support for the influence of relationship value on reconciliation behavior comes from both experimental and observational studies. Using an exper-

imental approach, Cords & Thurnheer (1993) measured the reconciliatory tendency in seven dyads of longtail macaques (*Macaca fascicularis*) as a function of the value of the social partner, which was determined by the circumstances of the experiment. In the baseline phase, each of the monkeys in a pair could gain access to coveted food independently. The value of the social partner was increased in the experimental phase by making access to food contingent on the joint presence of the partner; now, each monkey "needed" its partner to get the food. Tests of reconciliation tendency took place immediately before the monkeys were released to this feeding apparatus, in an adjacent section of the mesh enclosure. Aggressive conflicts were provoked, and the animals were given time to make a friendly reunion. As predicted, reunions after aggression were much more likely when the partner's value was enhanced in the experimental phase (Fig. 9.1), and alternative explanations of this increase (e.g., Chalmeau et al. 1997) could be ruled out.

Further evidence that relationship value influences conciliatory tendencies comes from observational studies of various primate taxa, in which the frequency of post-conflict reunions varies among dyads of different types. In macaques, higher rates of post-conflict reunions are typically found in dyads (namely kin) most likely to form alliances (reviewed in Aureli et al. 1997). Similarly, in gorillas (*Gorilla gorilla*) reconciliation occurs only in dyads (namely those involving an adult female and an adult male) in which alliance formation occurs and has significant fitness consequences (in that male gorillas defend females and young against other males, who may be infanticidal; Watts 1995). In chimpanzees (*Pan troglodytes*), males, but not females, are important alliance partners that support each other in intra- and intercommunity competition: males are also more likely to reconcile after aggressive conflict than are females

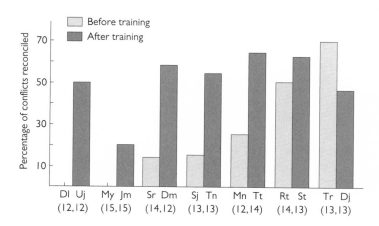

FIGURE 9.1. Results from Cords & Thurnheer's (1993) experiment showing that longtail macaques increase the likelihood of reconciling conflicts when their relationship is more valuable. The percentage of conflicts reconciled is shown for seven dyads when food could be acquired independently (before training, partner less valuable) and when the partner was needed to access food (after training, partner more valuable). A tick mark above the x-axis indicates a baseline rate of zero. Under the identification codes for each dyad, the numbers of conflicts in the baseline and experimental phases are given in parentheses. The median reconciliation rate increased more than threefold from the baseline to the experimental phase.

(de Waal 1986c; Goodall 1986). Sex differences in reconciliation tendencies among juvenile longtail macaques who had conflicts with unrelated adult females from their natal groups may also relate to the value of relationships with these partners. Juvenile females reconciled more often than juvenile males with unrelated adult female opponents (Cords & Aureli 1993). In this female-philopatric species, only juvenile females will have long-term relationships with unrelated adult females; juvenile males should have less incentive to maintain relationships with unrelated adult females. The lack of post-conflict reunions among ringtailed lemurs (*Lemur catta*) also supports the importance of relationship value as an influence on reconciliation (Kappeler 1993). In this species, interactions within dyads are either almost exclusively affiliative or almost exclusively antagonistic. Whereas the affiliative dyads have only rare opportunities to reconcile because conflicts are rare, the antagonistic dyads have opportunity but no reason to reconcile, since neither partner is likely to benefit from exclusively antagonistic interactions (see also Pereira & Kappeler, Box 15.2).

Among human children, relationships with friends are important to development (Hartup et al. 1988; Bukowski et al. 1996), and children more often use constructive kinds of conflict termination strategies with friends than with nonfriends, reflecting the higher value of relationships with friends (Cords & Killen 1998; Verbeek et al., Chapter 3). Post-conflict reunion in children appears not to occur more often among friends, however (Butovskaya et al., Chapter 12; Verbeek et al., Chapter 3), perhaps because human children use conciliatory conflict termination more often than post-conflict reunion to repair their social relationships.

Only one study has produced results that seem to contradict the importance of relationship value. The frequency of post-conflict reunion by juvenile longtail macaques was not related to the frequency with which their opponents supported them as coalition partners in the months around the time of the conflict (Cords & Aureli 1993). Cords & Aureli offered two possible explanations for their results. First, asymmetries in the value of partners may have limited the mutual interest in reconciliation: because of their small size, support behavior is needed especially by juveniles, but juveniles cannot offer very effective support to their partners in return. Second, juveniles are often supported by high-ranking adults, with whom compatibility may be low. Low compatibility may have prevented post-conflict reunions.

Intergroup variation in reconciliatory tendencies has also been interpreted in light of relationship value. Castles et al. (1996) found that members of a long-established group of pigtail macaques (*Macaca nemestrina*) had post-conflict reunions roughly twice as often as members of a newly formed group. The higher reunion level coincided with a social network in which individuals had fewer but more intense affiliative bonds, and reunion occurred especially often between individuals with these very strong affiliative ties. The authors reasoned that in a society in which affiliative relationships are stronger, each of those relationships must be more valuable.

Call et al. (1996) also compared reconciliatory tendencies in two groups and emphasized the importance of relationship value in explaining intergroup differences. Whereas individuals in a group of rhesus macaques (*Macaca mulatta*) living at low density followed the typical rhesus pattern of more post-conflict reunions with kin, those in another group living at high density had reunions as often with kin as with nonkin. These different patterns, which were also seen in grooming behavior, were interpreted as a strategic allocation of peacemaking behavior that reflected a heightened need for maintaining friendly relations between matrilines in relatively crowded conditions.

Interspecific differences in the frequency of post-conflict reunions have also been related to the value of social relationships. Among macaques there are species, such as rhesus, Japanese (*M. fuscata*), and longtail macaques, that typically have a low frequency of post-conflict reunions and that are more likely to have post-conflict reunions with kin than with nonkin. In other species, such as stumptail (*M. arctoides*), Barbary (*M. sylvanus*), and Tonkean macaques (*M. tonkeana*), post-conflict reunions occur frequently, and just as often between related and unrelated opponents (Thierry 1986; de Waal & Ren 1988; Aureli et al. 1997; Thierry,

Chapter 6). It has been suggested that in these latter species, cooperation by all group members against predators, other groups, or infanticidal males is more important than in the former species (de Waal & Luttrell 1989; Aureli et al. 1997; van Schaik & Aureli, Chapter 15). These suggestions will need to be confirmed with more natural history information for the species concerned, especially those with frequent and non-kin-biased reconciliation, which have been little studied in the wild.

Security

Security has not been explicitly studied as an independent variable that might influence the tendency to reconcile. However, some results suggest that it may be important. For example, when juvenile longtail macaques were studied in a temporary subgroup of same-sex peers, both males and females were more likely to have post-conflict reunions with an unrelated peer than with a related one (Cords 1988; Cords & Aureli 1993). This pattern could be interpreted in terms of the lower security of relationships among nonkin, given that these young animals have basically friendly and valuable relationships with all or most of their peers, regardless of kinship status.

The notion of security has figured more prominently in psychological studies of attachment than in ethological studies of social behavior. However, we are aware of only a few published studies that examined how attachment "styles" relate to the resolution of conflicts between the attached individuals. Kobak & Hazan (1991) and Simpson et al. (1996) related the attachment styles of human romantic partners to aspects of conflict resolution, as measured in laboratory negotiation sessions that involved real relationship problems identified by the subjects. In this research paradigm, all subjects discuss the problem (i.e., attempt a resolution) because they have been instructed to do so. The nature of the pair mates' discussions can then be rated, and their own reactions to these discussions can be reported. The results showed that less secure (i.e., more ambivalent) attachment styles coincide with less effective means of conflict resolution. Pistole (1989) found generally similar results using a questionnaire, rather than observations, to assess styles of conflict resolution among adult romantic partners. Subjects whose attachment to their partners was less secure (i.e., "anxious/ambivalent") scored lower on their ability to compromise, and to find a mutually satisfactory solution to conflict, than subjects whose attachments were "secure." Unfortunately, a comparison of these measures, which concern the nature of verbal resolution strategies, and the measures of conciliatory tendency typically used by primatologists, which concern how often post-conflict reunions occur, is not straightforward. Although the human studies do suggest that relationship security may influence how conflict resolution occurs, they do not reveal whether spontaneous conflict resolution is more likely to be attempted among insecurely attached opponents.

Compatibility

Evidence for compatibility as an important influence on tendency to reconcile comes from the many studies that looked at reconciliation tendency as a function of social bonding or friendliness. In studies of nonhuman primates, social bonding is usually measured by the frequency or duration of affiliative interaction, which is sometimes the very same kind of behavior used in post-conflict reunions. Measures of reconciliatory tendency do control for differential baseline rates of affiliation (Veenema, Box 2.1), so that higher reunion rates in friendly dyads would not just be a consequence of generally more frequent affiliation: these dyads would have more post-conflict affilia-

tive interaction on top of their higher baseline rates of affiliation.

Within groups of nonhuman primates, dyads that are more friendly reconcile more often (de Waal & Yoshihara 1983; de Waal & Ren 1988; Aureli et al. 1989; Cords & Aureli 1993; Watts 1995; Castles et al. 1996; Schino et al. 1998; Call et al. 1999). Groups of monkeys with more dyads that are very friendly also show higher reconciliation than other groups (Call et al. 1996; Castles et al. 1996). Species that are characterized by higher rates of affiliative interaction have higher conciliatory tendencies (de Waal & Luttrell 1989; Thierry, Chapter 6).

Affiliation and friendship have also been related to conflict resolution in dyads of human children. Despite some differences in how affiliation is measured (friendship in children may be self-reported), the results of these studies generally concur with those of nonhuman primates, showing that friendship facilitates conciliatory outcomes (specifically terminations) of conflict (Hartup et al. 1988; Butovskaya et al., Chapter 12; Verbeek et al., Chapter 3).

Measuring Relationship Qualities

Of the three relationship qualities we have considered, value and compatibility seem to be most supported by empirical evidence, while security has been little researched. Although the current evidence is a definite beginning, further study is needed to evaluate conclusively the importance of these relationship qualities and their relative roles in influencing reconciliation behavior. We recommend two general strategies for such research. First, multiple ways of operationalizing each of these qualities need to be developed. To the degree that different kinds of measurement of a particular factor relate similarly to reconciliation tendency, the case for that factor's influence is strengthened, and measurements that clearly differentiate between value, security, and compatibility can be chosen for future work. When we have the right way to measure value, security, and compatibility, it will be easier to avoid ambiguity in the interpretation of results. Some existing results are open to more than one interpretation and thus cannot be taken as unequivocal support for the effect of only one factor. For example, we cited the difference in the rates of post-conflict reunion in two groups of pigtail macaques (Castles et al. 1996) as evidence of the role of both value and compatibility. In this case, what the researchers actually measured was compatibility; however, they interpreted their results in terms of value. Their results would provide clearer support for the influence of value on reconciliation tendencies if value had been more directly measured. The second general research strategy we recommend follows from the first. The different operationalizations of value, security, and compatibility should be used collectively to assess the influence of these factors in the same study. This will allow researchers to study empirically the interactions among the factors.

In light of these recommendations, we now turn to ways of studying relationship value, security, and compatibility.

Measuring Value

How might one measure the value of a social partner? One possibility is to assume that the researchers can judge the relative value of different social partners by observing the sorts of behavioral interactions in which they engage. This is the approach most often taken by primatologists doing observational studies (see above). Frequent coalition partners are often assumed to be especially valuable partners, although adjustments for

the differential effectiveness of their support are not always made. Some researchers use broader categories of affiliative behavior to assess partner value. We caution against conflating the notion of value and general affiliation (see, for example, Kappeler and van Schaik's [1992] "good" relationships). Partners with whom one is especially friendly are often, but not necessarily, very valuable. For clear-cut operationalizations, assessments of value should be based on behavioral exchanges with clear fitness consequences, whereas measures of friendliness should be used for assessments of compatibility (see below).

An experimental situation may allow the researcher to be more certain that the value of partners is correctly assessed. For example, in their experiment on macaques described above, Cords & Thurnheer (1993) controlled the value of the social partner by adjusting the circumstances in which the subject could feed.

A researcher's inferences about who makes a valuable partner are likely to be fairly accurate for carefully and thoroughly studied species, but there is always a danger of overlooking currencies of value that are important to the animals. Furthermore, researchers are seldom able to demonstrate a direct link between the performance of some social act, thought to be valuable, and a change in lifetime reproductive success: many leaps of logic are typically involved in moving from statements such as "A tends to defend B when B is under attack" to "B is likely to have a higher reproductive output over its lifetime because of A's support." We would need to know not only the direct consequences of A's support in a single act but also how often and in what contexts such acts of support occur, whether they influence the receiver's competitive ability in later encounters (in which the mere presence of the former supporter may influence the outcome; Chapais 1992), and whether animals deprived of such support resort to alternative, but perhaps equally effective, strategies to achieve their lifetime reproductive output. Because such knowledge is difficult to attain, researchers in the real world always use at least some common sense and intuition in judging the value of social partners.

Another approach to measuring value might take some of the burden off the researcher's judgment. The behavior of the subjects could be used to gauge the value that they themselves attach to different social partners. Although the animal's assessment of value may differ from that of an observer, it is the animal's assessment that most directly influences its (conciliatory) behavior. If we assume that an animal is motivated to be in the company of a valuable social partner, one way to study relationship value from the subject's perspective is to give the subject the opportunity to choose its partners. To hold constant various factors that might influence the exercising of preference in a complex naturalistic social environment, choice tests in the laboratory would seem the best option. For example, Mason (1971) used such choice tests to study social preferences between opposite-sex individuals in titi monkeys (*Callicebus moloch*). Both male and female titi monkeys preferred a familiar pair mate over a stranger or an empty cage, and pair mates are known to maintain close bonds and jointly to defend their common range in the wild.

Some of the difficulties in interpreting preference tests are discussed by Dawkins (1980). First, preferences may not be constant: how valuable a particular social partner is may depend on the context, such as time of year, occurrence of social events, or the testing environment itself. A choice test measures the animal's preferences at a particular time and in a particular context: generalization to other times and contexts may not be warranted. Second, the value of the partner may not be the only basis for choice: the security of the relationship and the compatibility of the partner may also influence the choice and would need to be ruled out as influential factors.

Another way to access the animal's perspective on its partner's value is to measure its responses to changes in proximity (cf. de Waal 1986c, p. 463). For example, if separated from a social partner, how hard will the subject work to be reunited? Observational studies might examine how often and how soon a subject that has been displaced from its partner (e.g., by the arrival of a third party, or by an alarm response) attempts to regain proximity or social contact. If it has been displaced by a dominant third party, for example, does it wait for this individual to leave, or does it risk aggression from the dominant and immediately return to its partner? If it waits, does it show any behavioral signs of distress (see below)?

Experimental studies might stimulate a subject to approach a social partner, perhaps by creating a stressful situation in which contact with the partner could relieve distress. Various obstacles to access to the partner might then be set up, to see how much the subject is willing to endure to reach its partner. By eliminating a complex social background, captive study might better allow clearer interpretations of behavior in terms of specific dyadic relationships.

Separation-reunion paradigms are another way to measure a subject's response to changes in proximity under experimentally controlled conditions. If prevented from being with the partner, how distressed does the partner become? When partners have been separated, how do they respond to reunion? Research designs such as these have long been used in studies of attachment, especially between mothers and infants but also between peers. Variable responses to separation and reunion have been attributed to characteristics of the dyad's social relationship before the manipulation, although other factors (such as age-sex class, duration of separation, and the separation environment) are also involved (Mineka & Suomi 1978). In the attachment context, the value of a social partner is recognized in its ability to provide protection (Ainsworth et al. 1978); young animals especially are highly motivated to form attachments for their own safety, and when they perceive themselves to be at risk, they react strongly when separated from attachment figures. Although a dramatic protest-despair reaction may be less marked in older individuals whose attachments are less strong (Mineka & Suomi 1978), we think that even among adults there are likely to be detectable signs of distress upon separation, and relief from distress upon reunion, when an individual values its relationship with a particular partner. Indeed, Cubicciotti & Mason (1975) showed such effects in adult titi and squirrel monkey (*Saimiri sciureus*) pair mates that were separated from each other for only 90 minutes and then reunited. Similarly, adult humans with close relationships often report emotional reactions (such as loneliness, anxiety, or even despair) to physical separation (Feeny 1998).

The measures of affect used in separation and reunion studies of nonverbal animals can be both behavioral and physiological, and again the literature on mother-infant attachment in nonhuman primates provides many examples. Several investigators have identified a set of "disturbance" behaviors, which are rarely observed during baseline conditions but more common when animals are distressed (Laudenslager et al. 1990; Maestripieri et al. 1992). Such behavior includes startles, body shakes, self-scratching, stereotypies, slouched posture, and certain vocalizations. Behavior that is less limited to stressful situations may also change in occurrence and indicate distress: for example, the amount of time spent moving (or conversely, resting) is often used as an index of arousal in maternal separation studies (e.g., Hinde & Spencer-

Booth 1971; Cubicciotti & Mason 1975; Laudenslager et al. 1990).

Physiological measurement of distress at separation, or relief at reunion, is another possibility. Heart rate telemetry (e.g., Cubicciotti & Mason 1975), challenge tests for immunocompetence (e.g., Laudenslager et al. 1990), and measurement of adrenocortical activity (e.g., Smotherman et al. 1977; Coe et al. 1978; Mendoza & Mason 1986) or autonomic nervous system stimulation (Dettling et al. 1998) have all been used to evaluate a subject's response to separation from an attachment figure. Sometimes these physiological measures indicate effects of separation that could not be detected by concomitant behavioral study (e.g., Coe et al. 1978) or might be misinterpreted if only behavior was monitored (Levine et al. 1987). One significant disadvantage of using physiological measures, however, is that they often require invasive handling of the animals; they are therefore difficult to apply to field situations.

Using manipulations of proximity in the context we propose, namely, to allow the animals themselves to indicate how much they value different partners, will require some preliminary investigations. First, the measures used must be "good" measures (Martin & Bateson 1993), sensitive to an interdyadic level of variation. Some studies of maternal separation indicate such sensitivity in both behavioral and physiological measures: for example, Laudenslager et al. (1990) found variation in disturbance behaviors among macaque infants whose mothers had been removed from their group, and this behavioral variation among infants related closely to variation in immunocompetence. Adjustment of the length of separation, and the context of both separation and reunion, may be necessary to achieve good measurement.

Second, the measures must be the "right" measures (Martin & Bateson 1993). Although for our purposes the subject's valuation of its preseparation relationship is the most interesting factor that affects its response to separation, other factors may also influence that response, including individually varying aspects of temperament. These additional factors would have to be identified and evaluated and perhaps factored out analytically.

Third, these approaches are likely to be most effective when proximity and separation of partners are experimentally manipulated, as this allows for the most complete control of potentially confounding variables. Such manipulation is not always an option in studies of wild populations, but even here there may be naturally occurring separations or reunions that provide some insight into a subject's view of its partner (e.g., spatial displacements). Even an analysis of the situations in which separations do and do not occur might shed light on the animals' perspective on their social partners if it reveals the contexts in which they avoid or tolerate separation or respond in different ways to reunion.

Measuring Security

We believe there are at least two possible operationalizations of relationship security that should be investigated. As in the case of partner value, these operationalizations either derive from the researcher's assessment of behavioral patterns that "should" indicate predictability (and thus contribute to security), or rely on variation in the subject's behavior that may reflect its own assessment of such predictability and security.

To measure predictability directly, one might consider a partner's response to social overtures by the subject: for example, what proportion of approaches to the partner (Fig. 9.2) are acknowledged by a neutral or friendly response, and what proportion are ignored, followed by signs of tension (such as self-scratching: Maestripieri et al. 1992, or heart rate increase: Aureli et al. 1999), or even rebuffed? We would expect a less secure relationship to be characterized by more variable

FIGURE 9.2. In a group of pigtail macaques, an adult male approaches another adult male. It is at this stage of the interaction that we expect the behavioral responses of the approached individual to provide insight on its perception of the security of its relationship with the approaching individual. Responses could range from avoidance to friendly contact, or the approach could be ignored, rebuffed, or followed by signs of tension. Photograph by Filippo Aureli.

responses to approach and more signs of tension in one or both partners during an approach.

To measure security from the perspective of the subject, we have to assume that the subject's behavior varies according to the security of its relationship with its partner. We might expect, for example, the subject to be more hesitant in its efforts to approach and interact with its partner the less secure its relationship seems. The speed and directness of approach, whether signals of appeasement or friendly intent are given before or concurrently with approach, and whether there are signs of tension (such as self-scratching) in the approaching individual would all be potential measures of such hesitation. One might consider other contexts as well: for example, if the subject is the victim in an aggressive conflict with a third party, how readily does it request support from the partner? Nonhuman primates use gestures and vocalizations to enlist such support. The latency and frequency of these gestures could indicate how certain the subject is of the response of its partner.

The same sort of observations might be made in an experimental setting, in which the dyad is isolated from other group members. As long as the subjects are familiar with this arrangement, such an approach would allow better control over the social context of interaction between partners, so that one could be more certain that the subject's response to the partner is not influenced by the behavior or merely the presence of other individuals.

Additional manipulations could also be carried out: for example, if the subject was stressed by an aversive stimulus or a novel situation, and therefore motivated to approach its partner to relieve its anxiety, one could see how hesitant such approaches were and how consistently they were tolerated or responded to by the partner. Studies of attachment in humans have used mother-infant separations and reunions as a way of assessing the security of the attachment bond. The separation is itself a stressful experience for the infant, and the infant's response to separation, and especially reunion with the mother, is characteristically different for infants with secure versus ambivalent attachments (Ainsworth et al. 1978). Analogous results have been reported for nonhuman primates (e.g., Dettling et al. 1998). Studying the dynamics of social interaction during times of stress might offer yet another way to assess the security of relationships with partners that are able to relieve that stress.

Measuring Compatibility

Of the three relationship qualities that may influence reconciliatory tendencies, compatibility is perhaps the most straightforwardly measured. One should examine, outside a post-conflict context, the frequency and nature of behavioral acts like those used by the species in post-conflict reunions (see Call, Box 9.2, for discussion of these acts). If

BOX 9.2

Distance Regulation in Macaques
A Form of Implicit Reconciliation?

Josep Call

Several studies have shown that affiliative contact between opponents (i.e., reconciliation) increases after conflicts in a number of primate species (Kappeler & van Schaik 1992; de Waal 1993; Appendix A). Typically, these affiliative contacts involve behaviors such as allogrooming, touching, or other species-specific behaviors (e.g., hold-bottom). Although reconciliation is widespread among primates, its rate remains relatively low in most species. Even the most conciliatory species make contact after only 40 percent or less of conflicts, and some species show less than 10 percent of interopponent contacts (Thierry, Chapter 6). One possible explanation for this low level of reconciliation is that species may reconcile using alternative behaviors that do not involve physical friendly contact between individuals.

One behavior that may serve a conciliatory function is interopponent distance regulation. York & Rowell (1988) investigated reconciliation in patas monkeys (*Erythrocebus patas*) and noted that opponents tended to remain in close proximity after conflicts (see also Petit & Thierry 1994). Furthermore, Cords (1993) showed that distance reduction between opponents was equivalent to using contact behaviors in restoring the tolerance between former opponents in longtail macaques (*Macaca fascicularis*). Thus, one potentially important issue in studying conflict resolution is how individuals regulate their distance to former opponents after conflicts. We conducted a study to investigate how rhesus (*Macaca mulatta*) and stumptail macaque (*Macaca arctoides*) opponents, which differ in their amount of reconciliation measured by affiliative contacts (de Waal & Ren 1988), regulated their interindividual distance after conflicts. In particular,

BOX 9.2 (continued)

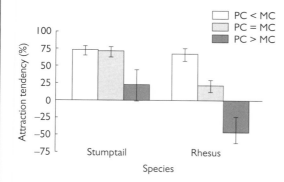

Mean attraction tendency as a function of initial interopponent distance in PC and MC periods. Positive values indicate smaller (minimum) interopponent distances in PC compared with MC periods, whereas negative values indicate smaller (minimum) interopponent distances in MC compared with PC periods. A zero value indicates that the minimum distance attained during PC is equal to that of MC periods.

we examined what proportion of conflicts resulted in a distance reduction as opposed to a distance increase between opponents regardless of whether there was any bodily contact between opponents. Spatial attraction between opponents (as opposed to dispersion) may serve a conciliatory function by signaling the end of aggressive intentions.

Both species were represented by multimale, multifemale groups housed in large outdoor compounds at the Yerkes Primate Center, Lawrenceville, Georgia (see Call et al. 1996, 1999, for further details). Ten-minute focal animal observations were conducted during post-conflict (PC) and matched-control (MC) periods based on the PC/MC method (de Waal & Yoshihara 1983) in which affiliative contacts between former opponents were recorded, along with interopponent distance at the beginning of each focal observation and every time it changed during the observation. Three categories of distance were distinguished: near (0–2 m), medium (2–5 m), and far (more than 5 m). A total of 561 and 251 PC-MC opponent pairs were obtained for rhesus and stumptail macaques, respectively.

Conciliatory tendency estimates the likelihood of post-conflict affiliative contacts between opponents, and

it is defined as attracted minus dispersed pairs divided by the total number of opponents pairs (Veenema et al. 1994). In the present study, an analogous measure called attraction tendency (AT) was employed to estimate the likelihood of post-conflict interopponent distance reductions. This new measure also involved subtracting dispersed from attracted pairs and dividing by the total number of opponent pairs, but attracted pairs were defined as those in which the *minimum* interopponent distance during the PC was smaller than the *minimum* interopponent distance during the MC period. A particular pair was also defined as attracted if the minimum interopponent distance was the same in the PC and MC periods but was attained earlier in the PC relative to the MC period. On the other hand, dispersed pairs were defined as those pairs in which the *minimum* interopponent distance during the PC was greater than the *minimum* interopponent distance in the MC period. A particular pair was also defined as dispersed if the minimum interopponent distance was the same in the PC and MC periods but was attained earlier in the MC than the PC period. Neutral pairs were those in which both the minimum interopponent distance and the time of attainment were the same in the PC and MC periods. A positive AT reflects attraction between opponents, whereas a negative AT reflects dispersion. All AT estimates were calculated at the individual level, and here we present the means of individual values.

Overall, both macaque species adopted significantly shorter distances to their opponents in the PC compared with the MC period (rhesus: AT = 26 percent; stumptail: AT = 70 percent). These results produced higher levels of attraction compared with those obtained using affiliative body contact as an indicator of reconciliation, which produced a conciliatory tendency of 7 percent and 35 percent in rhesus and stumptail macaques, respectively (Call et al. 1996, 1999). Using interopponent distance as opposed to affiliative contacts also dramatically reduced the number of neutral pairs obtained in both species. In rhesus macaques neutral pairs decreased from 88 percent to 19 percent when interindividual distance rather than affiliative contacts were used, whereas in stumptail macaques neutral pairs decreased from 55 percent to 10 percent.

BOX 9.2 (continued)

One possible confounding factor in this analysis based on distance is that for most PC-MC pairs the initial interopponent distance in PC and MC periods was not equal. The figure presents the mean attraction tendency for each species as a function of the initial distance in PC and MC periods. In both species, the closer opponents were at the beginning of the PC relative to the beginning of the MC period, the more likely they were to show a high AT. Call (in press) proposed various methods to control for the effect of unequal interopponent distances at the beginning of the PC and MC observations. One such method consists of restricting analyses to conflicts with equal initial interopponent distances in PC and MC periods. Focusing only on those PC-MC pairs with equal initial interopponent distance confirmed that both species reached shorter interopponent distances in PC compared with MC periods, and, more important, it indicated that interopponent distance was reduced after conflicts (i.e., AT > 0 in the figure). Moreover, those distance reductions often brought opponents within two meters or closer in both species. In particular, 69 percent of the attracted pairs in rhesus macaques resulted from opponents being within two meters or closer even though only 18 percent of those pairs were at that distance at the beginning of the PC period. This figure was even more pronounced in stumptail macaque opponents who were within two meters or less in 95 percent of the attracted pairs, although they were already within that distance in 43 percent of these pairs when the PC

observation started. Similar results were obtained when only those opponent pairs with equal distance in PC and MC periods were considered.

In summary, both rhesus and stumptail macaques reduced the interopponent distance after conflicts. The majority of these reductions brought opponents within two meters or less of each other, even when only those opponent pairs with identical initial interopponent distances in both PC and MC periods were considered. The attraction tendency based on distance reduction was higher than the conciliatory tendency based on affiliative contacts for both species as a result of a reduction of neutral pairs. This reduction is important because it offers a more dynamic picture of post-conflict situations, since even though relatively few affiliative contacts occurred between opponents, the majority of conflicts involved some form of post-conflict management in the form of distance regulation. Future studies should focus on the relation between interopponent distance regulation and the occurrence of affiliative contacts. An intriguing possibility is that former opponents may use distance reduction to probe the likelihood of additional attacks, in an attempt to reduce the uncertainty of continued aggression. Perhaps distance reduction could even have a function analogous to that of post-conflict affiliative contacts. If this is the case, this would explain why levels of contact affiliation are relatively low after fights, since interopponent distance reduction may be sufficient as a conciliatory behavior.

special acts are used (only) in a reconciliatory context, as is the case in some nonhuman primates, it would not be possible to study performances of these types of interactions outside the post-conflict context; instead, one might examine a broad set of friendly interactions, such as approaches followed by affiliation or neutral behavior versus those followed by aggression or spatial displacement.

A composite measure of friendliness was used in our study of juvenile longtail macaques (Cords & Aureli 1993). We assessed compatibility in terms

of the amount of time each juvenile spent in close proximity to a partner and the likelihood that the juvenile received aggression relative to the amount of time spent together. In these two measures, the top quartile of partners was clearly differentiated from the bottom three quartiles. Compatible partners for each juvenile were those whose scores for close proximity to that juvenile were in the top quartile and whose rates of aggression to the juvenile (relative to association time) were in the bottom three quartiles.

Individual Characteristics

We have so far emphasized qualities of the dyadic social relationship that influence the tendency to reconcile after aggression, but other factors may be at work, including characteristics of the individuals involved. Certain individuals might be more conciliatory than others; whether they can exercise their greater conciliatory tendency will depend on the identity of their former opponent.

From studies of nonhuman primates it is not yet known whether individuals differ in their conciliatory tendencies. Both empirical and analytical methods might be used to discover if such individual differences exist. Empirical evaluation might involve presenting different subjects with the opportunity to reconcile after aggression with particular opponents and seeing whether these opportunities are used. Variations on this general idea could include measurements other than the occurrence (or nonoccurrence) of post-conflict reunions. One might test subjects to see how hard they would work to gain an opportunity to reconcile—for example, to be admitted to an enclosure containing the former opponent. One might also measure how distressed the subject is if prevented from reconciling. Whatever the metric, these comparisons might be difficult in practice because one would have to ensure that the different subjects' relationships with their partners were equally valuable, secure, and compatible. Holding the identity of the opponent constant in interindividual comparisons would not necessarily guarantee such congruence. Confidence in the measurement of these relationship qualities would therefore be necessary.

An analytical approach would use individual identity as one factor in a multivariate analysis of conflict outcomes in which other factors, such as the relationship qualities discussed above, are also included, or at least randomized in the sample. A potential difficulty here might be in gathering sufficient data for all individuals in the group. Observational studies typically do not include aggressive conflicts among all group members, and certain dyads tend to dominate the data set. Experimentally staged conflicts, in which opponent identity is controlled, might offer one solution to this difficulty (Cords 1994).

If individual differences are found, new questions will emerge as to their causes. Both temperament (Kagan 1989; Clarke & Boinski 1995) and early experience might influence individual tendencies to reconcile. A role for early experience is suggested by de Waal & Johanowicz's (1993) experiment in which juvenile rhesus macaques, housed for five months with more conciliatory juvenile stumptail macaques, increased the proportion of fights that were reconciled threefold. Another study has shown that the type of attachment that juvenile capuchin monkeys (*Cebus apella*) develop with their mothers affects their conciliatory tendencies following conflicts with other group members (Weaver & de Waal, Box 10.2). Similarly, differences in the quality of attachment between human children and their mothers influence the type of conflict-resolution strategies that children employ in conflicts with one another (Park & Waters 1989).

Conclusion

Although our consideration of factors that influence the tendency to reconcile derives primarily from research on nonhuman primates, there is no reason to think that the factors we have identified as important would not apply as well to other animals whose societies are also characterized by the formation of long-lasting, individualized relationships. At present, however, we know comparatively little about reconciliation in these other species, and even less about variation in its occurrence and the causes of that variation (Schino, Chapter 11). Some of the ideas we have discussed—especially the notions of value and security—are touched on in the human attachment literature, as we have noted above. However, as far as we have been able to discern, the human attachment literature has only infrequently addressed issues relating to conflict resolution and, at least in the case of adults, reports methods of study, including laboratory exercises and questionnaires, that differ considerably from the observational methods in naturalistic settings used on nonhuman animals and some human children. Furthermore, the human attachment literature tends to emphasize variation in the style of conflict resolution, rather than the more basic questions of whether such resolution is attempted or is successful, so comparisons between human and nonhuman primates are difficult.

Given that the evidence currently available is subject to ambiguous interpretation, there is a strong need for clear-cut operationalizations of the various factors that may influence reconciliation and for investigations of the relative role of each factor.

We have suggested several ways in which some of the relationship qualities that influence conciliatory tendencies might be measured. To study how important these qualities are, they must be related to variation in reunion frequency. Another way to assess their effect would be to take advantage of the asymmetries that are likely to exist in these various factors, since two opponents are unlikely to experience their mutual relationship as equally valuable, secure, or compatible. To the extent that such asymmetries exist and can be detected (Aureli & Smucny, Chapter 10), one might expect differences in the degree to which one partner or the other tries to initiate post-conflict reconciliation. Unfortunately, however, attempts to reconcile are difficult to quan-

tify. Although one partner's approach to another for a friendly reunion may be easily recognized, eye contact or subtle shifts of posture or orientation may have preceded the approach and would be easily missed under typical observation conditions. Failed or aborted approaches (i.e., those not followed by a friendly reunion) might not even be recognized as attempts at reconciliation. Furthermore, individuals may not even try to reconcile if they expect such an attempt to be unsuccessful. For all these reasons, we are not optimistic that researchers will be able to detect effects of asymmetries in the partners' perception of their relationship on the initiative to reconcile. Instead, these asymmetries will have to be related to the number of reconciliations actually achieved.

Variation in the tendency to reconcile has been reported at different scales, from dyad to dyad within a group, between groups of the same species, and between species. The degree to which the relationship qualities we have discussed explain some or all of these scales of variation is a question for further empirical study. On the whole, there is more evidence at present for the role of some of these qualities—especially value and compatibility—in explaining variation in reconciliation tendencies among dyads in a group. It is possible, however, that larger-scale differences would simply result from group- or species-level variation in demography and social organization, insofar as these factors determine how many social partners an individual has, and in which contexts they interact. In addition, larger-scale differences, especially between species with disjunct gene pools, may result from species differences in temperament (Clarke & Boinski 1995; Thierry, Chapter 6), even though within each species, factors such as value, security, and compatibility influence variation among dyads in similar ways.

References

Ainsworth, M. D. S., Blehar, M. C., Waters, E., & Wall, S. 1978. *Patterns of Attachment.* Hillsdale, N.J.: Lawrence Erlbaum Associates.

Aureli, F., & van Schaik, C. P. 1991. Post-conflict behaviour in long-tailed macaques (*Macaca fascicularis*): II. Coping with the uncertainty. *Ethology,* 89: 101–114.

Aureli, F., van Schaik, C. P., & van Hooff, J. A. R. A. M. 1989. Functional aspects of reconciliation among captive long-tailed macaques. *American Journal of Primatology,* 19: 39–51.

Aureli, F., Das, M., & Veenema, H. C. 1997. Differential kinship effect on reconciliation in three species of macaques (*Macaca fascicularis, M. fuscata,* and *M. sylvanus*). *Journal of Comparative Psychology,* 111: 91–99.

Aureli, F., Preston, S. D., & de Waal, F. B. M. 1999. Heart rate responses to social interactions in free-moving rhesus macaques: A pilot study. *Journal of Comparative Psychology,* 113: 59–65.

Bukowski, W. M., Newcomb, A. F., & Hartup, W. W. 1996. *The Company They Keep: Friendship in Childhood and Adolescence.* Cambridge: Cambridge University Press.

Call, J. In press. The effect of inter-opponent distance on the assessment of reconciliation. *Primates.*

Call, J., Judge, P. G., & de Waal, F. B. M. 1996. Influence of kinship and spatial density on reconciliation and grooming in rhesus monkeys. *American Journal of Primatology,* 39: 35–45.

Call, J., Aureli, F., & de Waal, F. B. M. 1999. Reconciliation patterns among stumptail macaques: A multivariate approach. *Animal Behaviour,* 58: 165–172.

Castles, D. L., Aureli, F., & de Waal, F. B. M. 1996. Variation in conciliatory tendency and relationship quality across groups of pigtail macaques. *Animal Behaviour,* 52: 389–402.

Chalmeau, R., Visalberghi, E., & Gallo, A. 1997. Capuchin monkeys, *Cebus apella,* fail to understand a cooperative task. *Animal Behaviour,* 54: 1215–1225.

Chapais, B. 1992. The role of alliances in social inheritance of rank among female primates. In: *Coalitions and Alliances in Humans and Other Animals* (A. H. Harcourt & F. B. M. de Waal, eds.), pp. 29–59. Oxford: Oxford University Press.

Cheney, D. L., & Seyfarth, R. M. 1990. *How Monkeys See the World.* Chicago: University of Chicago Press.

Cheney, D. L., & Seyfarth, R. M. 1997. Reconciliatory grunts by dominant female baboons influence victims' behavior. *Animal Behaviour,* 54: 409–418.

Cheney, D. L., Seyfarth, R. M., & Silk, J. B. 1995. The role of grunts in reconciling opponents and facilitating interactions among adult female baboons. *Animal Behaviour,* 50: 249–257.

Clarke, A. S., & Boinski, S. 1995. Temperament in nonhuman primates. *American Journal of Primatology,* 37: 103–125.

Coe, C., Mendoza, S. P., Smotherman, W. D., & Levine, S. 1978. Mother-infant attachment in the squirrel monkey: Adrenal response to separation. *Behavioral Biology,* 22: 256–263.

Cords, M. 1988. Resolution of aggressive conflicts by immature long-tailed macaques *Macaca fascicularis. Animal Behaviour,* 36: 1124–1135.

Cords, M. 1992. Post-conflict reunions and reconciliation in long-tailed macaques. *Animal Behaviour,* 44: 57–61.

Cords, M. 1993. On operationally defining reconciliation. *American Journal of Primatology,* 29: 255–267.

Cords, M. 1994. Experimental approaches to the study of primate conflict resolution. In: *Current Primatology,* Vol. 2: *Social Development, Learning and Behaviour* (J. J. Roeder, B. Thierry, J. R. Anderson, & N. Herrenschmidt, eds.), pp. 127–136. Strasbourg: Université Louis Pasteur.

Cords, M. 1997. Friendships, alliances, reciprocity and repair. In: *Machiavellian Intelligence II: Extensions and Evaluations* (A. Whiten & R. Byrne, eds.), pp. 24–49. Cambridge: Cambridge University Press.

Cords, M., & Aureli, F. 1993. Patterns of reconciliation among juvenile long-tailed macaques. In: *Juvenile Primates: Life History, Development and Behavior* (M. E. Pereira & L. A. Fairbanks, eds.), pp. 271–284. New York: Oxford University Press.

Cords, M., & Aureli, F. 1996. Reasons for reconciling. *Evolutionary Anthropology,* 5: 42–45.

Cords, M., & Killen, M. 1998. Conflict resolution in human and non-human primates. In: *Piaget, Evolution, and Development* (J. Langer & M. Killen, eds.), pp. 193–218. Mahwah, N.J.: Lawrence Erlbaum Associates.

Cords, M., & Thurnheer, S. 1993. Reconciliation with valuable partners by long-tailed macaques. *Ethology*, 93: 315–325.

Cubicciotti, D. D., III, & Mason, W. A. 1975. Comparative studies of social behavior in *Callicebus* and *Saimiri*: Male-female emotional attachment. *Behavioral Biology*, 16: 185–197.

Dawkins, M. S. 1980. *Animal Suffering: The Science of Animal Welfare*. New York: Chapman and Hall.

Dettling, A., Pryce, C. R., Martin, R. D., & Dobeli, M. 1998. Physiological responses to parental separation and a strange situation are related to parental care received in juvenile Goeldi's monkeys (*Callimico goeldii*). *Developmental Psychobiology*, 33: 21–31.

de Waal, F. B. M. 1986a. Conflict resolution in monkeys and apes. In: *Primates: The Road to Self-sustaining Populations* (K. Benirschke, ed.), pp. 341–350. New York: Springer-Verlag.

de Waal, F. B. M. 1986b. Deception in the natural communication of chimpanzees. In: *Deception: Perspectives on Human and Nonhuman Deceit* (R. W. Mitchell & N. S. Thompson, eds.), pp. 221–244. Albany: State University of New York Press.

de Waal, F. B. M. 1986c. The integration of dominance and social bonding in primates. *Quarterly Review of Biology*, 61: 459–479.

de Waal, F. B. M. 1989. Dominance "style" and primate social organization. In: *Comparative Socioecology: The Behavioural Ecology of Humans and Other Mammals* (V. Standen & R. A. Foley, eds.), pp. 243–263. Oxford: Blackwell Scientific Publications.

de Waal, F. B. M. 1993. Reconciliation among primates: A review of empirical evidence and unresolved issues. In: *Primate Social Conflict* (W. A. Mason & S. P. Mendoza, eds.), pp. 111–144. Albany: State University of New York Press.

de Waal, F. B. M., & Aureli, F. 1996. Consolation, reconciliation, and a possible cognitive difference between macaques and chimpanzees. In: *Reaching into Thought: The Minds of the Great Apes* (A. E. Russon, K. A. Bard, & S. T. Parker, eds.), pp. 80–110. Cambridge: Cambridge University Press.

de Waal, F. B. M., & Johanowicz, D. L. 1993. Modification of reconciliation behavior through social experience: An experiment with two macaque species. *Child Development*, 64: 897–908.

de Waal, F. B. M., & Luttrell, L. M. 1989. Towards a comparative socioecology of the genus *Macaca*: Different dominance styles in rhesus and stumptail monkeys. *American Journal of Primatology*, 19: 83–110.

de Waal, F. B. M., & Ren, R. 1988. Comparison of the reconciliation of stumptail and rhesus macaques. *Ethology*, 78: 129–142.

de Waal, F. B. M., & van Roosmalen, A. 1979. Reconciliation and consolation among chimpanzees. *Behavioral Ecology and Sociobiology*, 5: 55–66.

de Waal, F. B. M., & Yoshihara, D. 1983. Reconciliation and redirected affection in rhesus monkeys. *Behaviour*, 85: 224–241.

Feeny, J. A. 1998. Adult attachment and relationship-centered anxiety. In: *Attachment Theory and Close Relationships* (J. A. Simpson & W. S. Rholes, eds.), pp. 189–218. New York: Guilford Press.

Goodall, J. 1986. *The Chimpanzees of Gombe*. Cambridge: Harvard University Press, Belknap Press.

Hartup, W. W., Laursen, B., Stewart, M. I., & Eastenson, A. 1988. Conflict and the friendship relations of young children. *Child Development*, 59: 1590–1600.

Hinde, R. A. 1979. *Towards Understanding Relationships*. London: Academic Press.

Hinde, R. A., & Spencer-Booth, Y. 1971. Effects of brief separations from mothers on rhesus monkeys. *Science*, 173: 111–118.

Kagan, J. 1989. Temperamental contributions to social behavior. *American Psychologist*, 44: 668–674.

Kappeler, P. M. 1993. Reconciliation and post-conflict behavior in ringtailed lemurs, *Lemur catta*, and red-fronted lemurs, *Eulemur fulvus rufus. Animal Behaviour*, 45: 901–915.

Kappeler, P. M., & van Schaik, C. P. 1992. Methodological and evolutionary aspects of reconciliation among primates. *Ethology*, 92: 51–69.

Kobak, R. R., & Hazan, C. 1991. Attachment in marriage: Effects of security and accuracy of working models. *Journal of Personality and Social Psychology*, 60: 861–869.

Kummer, H. 1978. On the value of social relationships to non-human primates. *Social Science Information*, 17: 687–705.

Laudenslager, M. L., Held, P. E., Boccia, M. L., Reite, M., & Cohen, J. J. 1990. Behavioral and immunological consequences of brief mother-infant separation: A species comparison. *Developmental Psychobiology*, 23: 247–264.

Levine, S., Wiener, S. G., Coe, C. L., Bayart, F. E. S., & Hayashi, K. T. 1987. Primate vocalization: A psychobiological approach. *Child Development*, 58: 1420–1430.

Maestripieri, D., Schino, G., Aureli, F., & Troisi, A. 1992. A modest proposal: Displacement activities as an indicator of emotions in primates. *Animal Behaviour*, 44: 967–979.

Martin, P., & Bateson, P. 1993. *Measuring Behaviour*. Cambridge: Cambridge University Press.

Mason, W. A. 1971. Field and laboratory studies of social organization in *Saimiri* and *Callicebus*. In: *Primate Behavior* (L. A. Rosenblum, ed.), pp. 107–137. New York: Academic Press.

Mendoza, S. P., & Mason, W. A. 1986. Contrasting responses to intruders and to involuntary separation by monogamous and polygynous New World monkeys. *Physiology and Behavior*, 38: 795–801.

Mineka, S., & Suomi, S. J. 1978. Social separation in monkeys. *Psychological Bulletin*, 85: 1376–1400.

Noë, R., van Schaik, C. P., & van Hooff, J. A. R. A. M. 1991. The market effect: An explanation for pay-off asymmetries among collaborating animals. *Ethology*, 87: 97–118.

Park, K. A., & Waters, E. 1989. Security of attachment and preschool friendships. *Child Development*, 60: 1976–1081.

Petit, O., & Thierry, B. 1994. Aggressive and peaceful interventions in conflicts in Tonkean macaques. *Animal Behaviour*, 48: 1427–1436.

Pistole, M. C. 1989. Attachment in adult romantic relationships: Style of conflict resolution and relationship satisfaction. *Journal of Social and Personal Relationships*, 6: 505–510.

Sackin, S., & Thelen, E. 1984. An ethological study of peaceful associative outcomes to conflict in preschool children. *Child Development*, 55: 1098–1102.

Schino, G., Rosati, L., & Aureli, F. 1998. Intragroup variation in conciliatory tendencies in captive Japanese macaques. *Behaviour*, 135: 897–912.

Silk, J. B. 1996. Why do primates reconcile? *Evolutionary Anthropology*, 5: 39–42.

Silk, J. B. 1997. The function of peaceful post-conflict contact among primates. *Primates*, 38: 265–279.

Silk, J. B. 1998. Making amends: Adaptive perspectives on conflict remediation in monkeys, apes, and humans. *Human Nature*, 9: 341–368.

Silk, J. B., Cheney, D. L., & Seyfarth, R. M. 1996. The form and function of post-conflict interactions between female baboons. *Animal Behaviour*, 52: 259–268.

Silk, J. B., Kaldor, E., & Boyd, R. In press. Cheap talk when interests conflict. *Animal Behaviour*.

Simpson, J. A., Rholes, W. S., & Phillips, D. 1996. Conflict in close relationships: An attachment perspective. *Journal of Personality and Social Psychology*, 71: 899–914.

Smotherman, W. P., Hunt, L. E., McGinnis, L. M., & Levine, S. 1977. Mother-infant separation in group-living rhesus macaques: A hormonal analysis. *Developmental Psychobiology*, 12: 211–217.

Thierry, B. 1986. A comparative study of aggression and response to aggression in three species of macaque. In: *Primate Ontogeny, Cognition and Social Behavior* (J. G. Else & P. C. Lee, eds.), pp. 307–313. Cambridge: Cambridge University Press.

Veenema, H. C., Das, M., & Aureli, F. 1994. Methodological improvements for the study of reconciliation. *Behavioural Processes*, 31: 29–38.

Vespo, J. E., & Caplan, M. 1993. Preschoolers' differential conflict behavior with friends and acquaintances. *Early Education and Development*, 4: 45–53.

Watts, D. P. 1995. Post-conflict social events in wild mountain gorillas (Mammalia, Hominoidea). I. Social interactions between opponents. *Ethology*, 100: 139–157.

York, A. D., & Rowell, T. E. 1988. Reconciliation following aggression in patas monkeys, *Erythrocebus patas*. *Animal Behaviour*, 36: 502–509.

CHAPTER 10

The Role of Emotion in Conflict and Conflict Resolution

Filippo Aureli & Darlene Smucny

Introduction

Emotions traditionally have been described as human subjective experiences and considered inaccessible to scientific investigation in animals. There have been, however, promising recent developments in the comparative study of emotion. The growing emphasis on the mediating role of emotions in the human literature (Panksepp 1989; Frijda 1994; Rolls 1995) is paralleled by new perspectives in animal research that consider emotions as mediators between an animal's perception of the social and physical environment and its behavioral responses (Crook 1989; Lott 1991; Whiten 1996). For example, emotion has been proposed to mediate intraspecific variation in gregariousness in response to different levels of predation pressure (Lott 1991): if anxiety is reduced by proximity with conspecifics, then aggregation is more likely to occur when risk of predation is high.

Can the perspective of emotional mediation increase our understanding of the dynamics of animal conflict resolution? Much is known about the patterns and possible function of *reconciliation*, that is, a friendly reunion between former opponents following a conflict, in nonhuman primates (de Waal, Chapter 2; Cords & Aureli, Chapter 9; Appendix A), but we do not know what is at the basis of the high flexibility of this behavior. For example, the same individual reconciles more often with certain group members (e.g., kin) than with others (e.g., Aureli et al. 1997); pairs of individuals increase their reconciliation frequency after the value of their relationship has increased (Cords & Thurnheer 1993); juveniles learn from juveniles of another species to reconcile at higher rates (de Waal & Johanowicz 1993). Can this flexibility be explained by variation in the underlying emotions?

Recent research has documented changes in anxiety during conflict and its resolution (see below for more details). Anxiety levels of both opponents are elevated following conflict between baboons (*Papio anubis* and *P. cynocephalus ursinus*) and macaques (*Macaca fascicularis* and *M. sylvanus*), and their friendly reunions reduce anxiety at baseline levels (Aureli & van Schaik 1991; Aureli 1997; Cheney & Seyfarth 1997; Castles & Whiten 1998; Das et al. 1998). If post-conflict anxiety depends on the characteristics of the conflict and the quality of the relationship between the opponents, can post-conflict anxiety somehow mediate the occurrence of reconciliation? Can differential anxiety under-lie the flexibility of this behavior?

We shall address these questions by embracing MacLean's (1952) view of emotions as brain func-tions involved in maintaining individual survival. His evolutionary approach to brain anatomy led him to point out that emotions are not uniquely human traits because most of the brain structures involved in emotions are essentially the same in all mammals and perhaps in all vertebrates. Escap-ing from danger, for example, is a basic survival response for all animals. The *conscious* feeling of fear, which humans may experience, is not needed to link the sensory input of detecting danger with the quick response of finding protection. *Uncon-scious* emotional mechanisms mediate the percep-tion of danger with the appropriate response in both humans and other animals. MacLean's con-ceptual framework is still at the basis of the cur-rent evolutionary view of emotion (LeDoux 1996), and significant advances in neurophysiology have supported this view (Panksepp 1989; Brothers 1990; Levenson 1992; Davidson & Sutton 1995). Although all researchers may not agree on a general, operational definition of *emotion*, many believe that "the term emotion should be right-fully used to designate a collection of responses triggered from parts of the brain to the body, and

from parts of the brain to other parts of the brain, using both neural and humoral routes. The end results of the collection of such responses is an emotional state" (Damasio 1998, p. 84).

The aim of our chapter is to examine how the concept of emotion may provide insight into the interpretation of conflict dynamics and conflict resolution. Emotion is used as an intervening vari-able between the internal and external factors leading to interindividual conflict of interest and the array of potential responses characterizing conflict and its aftermath (Fig. 10.1). Different internal and external stimuli produce different emotions of variable intensities, which in turn contribute to different behavioral and physiolog-ical responses. We evaluate whether anxiety and other emotions play a mediating role in conflict management by regulating the occurrence of spe-cific social interactions. We especially focus on the interactions that prevent conflict of interest from escalating to aggressive encounters (pre-escalation phase) and on the interactions that bring resolution after escalation (post-escalation phase).

We start our chapter by describing some behav-ioral and physiological measures used to study emotions. Then we review the evidence for emo-tional responses during the pre-escalation and post-escalation phases. We then examine whether emotion may be the underlying mechanism explaining the flexibility of conflict resolution, especially flexibility related to the quality of rela-tionships between opponents. We conclude our chapter with some considerations about how the perspective of emotion as a mediating variable may be useful to predict outcomes of social conflict as well as other social interactions.

Indicators of Emotion

Emotion is multiply determined as the result of previous experiences, internal states, external con-

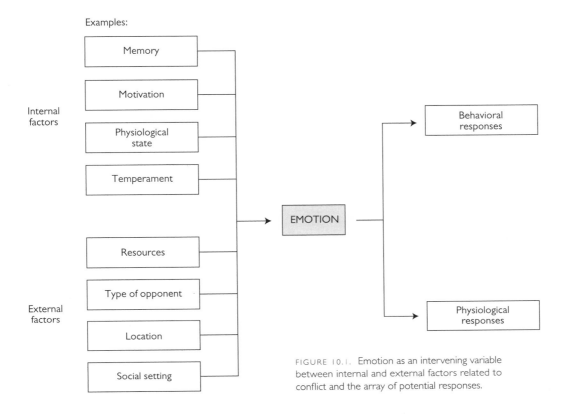

FIGURE 10.1. Emotion as an intervening variable between internal and external factors related to conflict and the array of potential responses.

ditions, and behavioral and physiological changes (Cacioppo et al. 1993). To investigate emotional responses, scientists need to monitor closely these changes and the conditions under which they occur. We briefly review here some methods used to quantify the behavioral and physiological responses during emotional experiences, focusing on methods relevant to the investigation of emotional changes in the pre- and post-escalation phases.

Behavioral Indicators

Darwin's *Expression of the Emotion in Man and Animals* (1872) provided the framework for all evolutionary approaches to emotion. Darwin was convinced that animals and humans express emotion in similar ways. Their facial and postural expressions convey the underlying emotion, thereby increasing the chances of survival and ultimately fitness.

Facial expressions are among the emotional indicators primarily studied in humans and other primates (in other taxa postural expressions are probably more revealing). Even though there is no consensus among emotion theorists, many investigators believe that different emotions are characterized by distinctive facial muscle responses (Ekman & Oster 1979). Researchers have also investigated vocal expressions of emotion. For example, throughout various taxa, species-specific distress calls appear to increase in frequency during social separation (e.g., Newman 1991; Panksepp et al. 1997).

Another important class of indicators of emotion is self-directed behavior, such as self-touch, self-scratching, and self-grooming. Apart from the hygienic function of self-directed behaviors, recent ethological studies on nonhuman primates have documented their occurrence in situations of uncertainty, social tension, or impending danger (Maestripieri et al. 1992). Similarly, high levels of self-directed behavior are characteristic of humans experiencing anxiety (Waxer 1977). Furthermore, pharmacological manipulation of stress and anxiety is followed by corresponding changes in the rates of self-scratching and self-grooming in rodents and monkeys (Moody et al. 1988; Schino et al. 1996).

Hence, facial, postural, and vocal expressions together with self-directed behavior can be easily used to examine emotional responses in the pre- and post-escalation phases.

Physiological Indicators

Even though there is still open debate about the correlation and causation between physiological processes and emotional expression (Levenson 1992; Cacioppo et al. 1993; Davidson & Sutton 1995; LeDoux 1996), many measures have been used to investigate the physiological changes associated with emotional experiences.

The monitoring of peripheral physiological parameters generally relies on relatively noninvasive techniques. Common techniques include the measurement of physiological functions such as heart rate, blood pressure, hand temperature, skin resistance, muscle tension, and hormonal levels. For example, in the classical work by Ekman et al. (1983), several measures were used to determine whether human "positive" and "negative" emotions could be differentiated on the basis of physiological responses.

Recent technical improvements in biotelemetry have allowed scientists to monitor physiological parameters, such as heart rate, in freely moving animals. This technique has enabled the study of predicted emotional responses to naturally occurring interactions, such as the proximity with group members or the receipt of allogrooming, in baboons and macaques living in social groups (Smith et al. 1986; Boccia et al. 1989; Aureli et al. 1999).

Various physiological events are characteristic of the "stress response" (Levine et al. 1989; Sapolsky 1998; von Holst 1998) that results in the rapid mobilization of energy: heart rate, blood pressure, and breathing rate all increase to transport nutrients and oxygen where needed. At the same time, other processes, such as digestion, growth, and reproduction, are inhibited to save energy. These responses are adaptive as they enable individuals to take immediate actions in threatening situations. Measures of variation in hormones and neurotransmitters associated with the stress response (e.g., cortisol, norepinephrine) have provided biological support for individual differences in emotional profiles of children, baboons, and macaques (Kagan et al. 1988; Rasmussen & Suomi 1989; Sapolsky, Box 6.1).

Scientists employ various methods to explore the role of central nervous system structures and different neurotransmitters in the regulation of emotion. These methods are typically invasive and include electrical stimulations, chemical manipulations, cerebrospinal fluid sampling, and lesions of specific brain areas. Through these techniques, much has been learned, for example, about the role of the amygdala in emotion regulation (LeDoux 1996). Noninvasive methods have recently become available to study the role of the central nervous system in emotion regulation. Activity of specific brain areas can be successfully investigated with electroencephalography, or EEG (e.g., Davidson

1995); positron emission tomography, or PET (e.g., Ingvar 1997); and functional magnetic resonance imaging, or fMRI (e.g., Kalin et al. 1997).

Emotion and Conflict

Most studies on animal conflict focus on aggression; thus our review of emotional responses to interindividual conflict is obviously biased. We do not include studies that use experimental paradigms to induce aggression, such as electroshocks, isolation, or predatory attacks, because the elicited responses appear to differ from those during naturally occurring conflict between conspecifics (Brain & Haug 1992). The emotional bases for the involvement in aggressive interactions have been discussed by various authors (e.g., Zillmann 1979; Averill 1982; Blanchard 1984; Berkowitz 1992; Potegal 1994). Here we focus on aspects that can highlight emotional mediation of conflict management and resolution.

The Pre-Escalation Phase

It is not easy to discern the early phase of conflict of interest between two individuals, especially when the conflict does not lead to overt expression, such as aggression or avoidance. It is possible, however, to identify situations that are potentially conflict-enhancing when similar situations have previously led to escalated conflict or when the circumstances are likely to provoke a conflict of interest, for example, competition for a food source.

Proximity to a dominant group member may be potentially risky because reduced interindividual distance usually precedes aggression. Physiological and behavioral evidence suggests that under these circumstances subordinate individuals may be tense or anxious (Fig. 10.2). The approach of a dominant individual produces an increase in heart rate in middle-ranking rhesus macaques (*Macaca*

FIGURE 10.2. Self-scratching by a female pigtail macaque as a sign of anxiety while in proximity with a higher-ranking female. Photograph by Michael Seres.

mulatta), whereas the approach of kin or subordinate individuals does not (Aureli et al. 1999). Similarly, longtail macaque females increase their rate of self-grooming and self-scratching when they are in proximity with a dominant male (Troisi & Schino 1987; Pavani et al. 1991). In the same species, treatment with anxiety-eliciting drugs is accompanied by increased fear responses to the approach of dominant males (Vellucci et al. 1986; Schino et al. 1996).

Behavioral indicators of anxiety also increase in frequency under other conditions that are associated with high risk of aggressive conflict. In captive group-living chimpanzees (*Pan troglodytes*) vocalizations of neighboring groups increase the likelihood of intragroup aggression and, as a consequence, are followed by higher rates of self-scratching (Baker & Aureli 1997). Rhesus macaque mothers scratch themselves and monitor others more often when their infant is not in contact with them, a situation that facilitates infant harassment by others (Maestripieri 1993). Furthermore, scratching rates of middle-ranking rhesus macaques increase during feeding tests in which clumped preferred food is initially monopolized by more dominant individuals (Diezinger & Anderson 1986).

Facial, postural, and vocal displays are often exchanged in situations of potential conflict (Preuschoft & van Schaik, Chapter 5). These displays are likely both the expression of the underlying emotion of the sender and signals for negotiation between the sender and the receiver (Hinde 1985). In other words, a display could communicate the sender's conditional tendency to act: a facial display may be the equivalent of saying, "If you don't stop, I will attack you." In addition, specific vocalizations are emitted in primate species in which other group members are likely to join aggressive encounters. Apart from providing information about the conflict and the type of opponent, screams are a reflection of the caller's emotional state (Rowell & Hinde 1962; Gouzoules et al. 1995). Thus, the expression of emotion at the early stage of conflict has a communicative function and provides information to the opponent and to third parties, which likely regulates subsequent behavior and prevents escalation.

Differences in display intensity can determine the winner of a conflict without further escalation, as in encounters between red stags (*Cervus elaphus*), in which the winner is often the one displaying the higher roaring rate and roaring exchanges are rarely followed by physical combat (Clutton-Brock & Albon 1979). In female Barbary macaques, the "round-mouth threat face" facilitates low-risk outcomes during competition over food (Preuschoft et al. 1998), and the "silent bared-teeth display," which is typically accompanied by submissive body movements, reduces the probability of attack from the receiver (Preuschoft 1992). Thus, emo-

tional expressions are effective in managing conflict of interest at its early stages.

The tendency either to communicate successfully or to escalate could depend on individual emotional dispositions or temperaments that often have a physiological basis. Aggressive and fearful expressions are regulated by complex interactions between internal and external factors (e.g., Huntingford & Turner 1987; Archer 1988; Berkowitz 1992) and are mediated by a range of physiological mechanisms (e.g., Gray 1987; Potegal 1994; LeDoux 1996). Although understanding the specific role of physiological mechanisms in the control of emotional expression is certainly important for the study of conflict management, our knowledge is as yet incomplete.

For example, animal and human studies reveal an influence of testosterone and serotonin on aggressive tendencies, but the emerging picture is far from simple. Although in a wide range of vertebrate species testosterone appears to facilitate aggression, the effects of past experience and social factors are often more prominent under stable social conditions than the influence of testosterone levels (Archer 1991; Rubinov & Schmidt 1996). Recent research has shown that measures of central testosterone concentrations (rather than plasma or salivary concentrations) combined with measures of serotonin could provide new insight (Higley et al. 1996). Testosterone may influence aggressive tendencies, and serotonin may regulate the threshold and intensity of the behavioral expression of aggression. Individuals with low serotonin levels exhibit impaired impulse control and a low threshold to escalate when facing potential conflict (Soubriè 1986; Higley et al. 1996).

Since central serotonin levels are stable over long periods and across experimental settings, they are considered "traitlike" (Higley & Linnoila 1997). The actions of serotonin are not limited to aggressive behavior and are likely influenced by other neurotransmitter and hormonal systems (Berman et al. 1997). In nonhuman primates, central serotonin levels affect aspects of social behaviors, such as dominance, affiliation, and sexual activity (Higley & Linnoila 1997). It is likely that central serotonin more directly affects information processing and decision making, which in turn influence the expression of aggression, among other behaviors (Spoont 1992). Hence, although the action could be indirect, individuals with different serotonin levels may differ in their response styles during the pre-escalation phase and in their likelihood to escalate.

The Post-Escalation Phase

When the overt manifestations of conflict (e.g., aggressive behavior) are over, the dispute may be ended but may nevertheless have long-lasting negative consequences for the participants as well as for their relationship. Furthermore, the conflict of interest still may persist, and new problems may arise, especially for the loser of the dispute. Emotional responses in the post-escalation phase reflect the opponents' perception of the dispute outcome; these responses may mediate actions for conflict resolution or minimally for coping with the current situation.

Winners and losers of the conflict are likely to experience different emotions in the aftermath of a dispute. A clear example is shown in Arsenio & Killen's (1996) study of children. After the dispute, children who were targets during the conflict displayed expressions of anger and sadness, whereas conflict initiators showed more happiness. The difference in emotional responses was likely due to the conflict outcome: in most cases initiators obtained the disputed object.

Territorial disputes occur naturally in the wild in many species, and when staged in laboratories, they result in marked physiological differences between winners and losers. For example, the

levels of hormones associated with the stress response, such as glucocorticoids, are higher in losers than in winners after dyadic confrontations in a variety of species (e.g., swordtail fish, *Xiphophorus helleri*: Hannes et al. 1984; guinea pigs, *Cavia aperea*: Haemisch 1990; tree shrews, *Tupaia belangeri*: von Holst 1998). In rats (*Rattus norvegicus*) differences between winners and losers may persist for several hours, even when the two opponents are separated from each other after the encounter (Schuurman 1980). In addition, these differences depend on the social setting and experience (Sachser & Lick 1991) and are not a mere reflection of differential physical activity (Sapolsky 1993). Interestingly, hostile behavior during marital conflict in otherwise happy human couples produces changes in blood pressure, immunological function, and hormonal concentration typical of the stress response (Kiecolt-Glaser et al. 1993; Malarkey et al. 1994).

These physiological changes appear to have an adaptive function. When losers do not leave the scene, as in most experimental settings, the long-term increase in glucocorticoids is likely to suppress aggressive behavior, thereby avoiding repeated involvement in potentially damaging fights (Brain 1980; Leshner 1983). In addition, the increase in glucocorticoids may facilitate submissive behavior, an adaptive response to prevent further attack by the opponent (Leshner 1980).

Losers do not necessarily react passively to the experience of a defeat. In tree shrews, some losers adopt a more active coping style in the aftermath of a fight (von Holst 1998). Variation in coping styles is also present in group-living species, in which dominance-subordinance relationships provide benefits to both partners by reducing disputes about priority every time there is potential competition (Preuschoft & van Schaik, Chapter 5). Even if overt aggression is reduced by establishing dominance relationships, conflicts of interest are usually not resolved in favor of subordinate individuals. In addition, subordinates have less control over the social environment and are strongly constrained in their actions. Accordingly, dominant and subordinate individuals differ in their physiological states (Sapolsky 1998; von Holst 1998). However, many social factors and personal experiences affect the relation between dominance rank and physiology; instead of a dominant/subordinate dichotomy, variation exists in behavioral coping styles and physiological responses within both dominant and subordinate individuals (Sapolsky, Box 6.1).

The drastic physiological responses associated with territorial disputes are not typical of the aftermath of everyday conflicts in well-established groups. In fact, even when conflicts of interests between group members escalate, the majority of conflicts are of low intensity (e.g., Bernstein & Ehardt 1985). Anxiety is the likely emotion to be experienced by participants of intragroup aggressive conflicts.

Physiological and behavioral measures of baboons and macaques support this view. For instance, the heart rate of both opponents remains elevated after the end of an aggressive interaction (Smith et al. 1986; Smucny et al. 1997) and takes longer to return to pre-conflict levels in the recipient of aggression (Smucny et al. 1997). Self-scratching rates are elevated during the post-conflict phase compared with control periods in both the recipient of aggression (Aureli et al. 1989; Aureli & van Schaik 1991; Aureli 1992; Castles & Whiten 1998) and the aggressor (Aureli 1997; Castles & Whiten 1998; Das et al. 1998). These physiological and behavioral responses can be interpreted as indications of anxiety attributed to the uncertainty of the post-conflict situation. As a consequence of the aggressive conflict, the recipient of aggression cannot be sure it will be tolerated around resources by the former aggressor (Cords 1992), and it is more likely to be attacked again (Aureli & van Schaik 1991; Aureli 1992; Cords 1992; Watts 1995; Silk et al. 1996; Castles & Whiten 1998), whereas the aggressor may lose the support of its previous opponent for future cooperative actions.

Emotion and Conflict Management

The behavioral and physiological responses to the different phases of a conflict are adaptations to critical situations and are likely to mediate conflict management and mitigate the negative consequences of aggressive interactions. For instance, the exchange of threat and submissive displays communicates the emotional disposition of the sender and its conditional actions (cf. Hinde 1985). These exchanges are likely to settle the conflict without the need for aggressive escalation (Preuschoft & van Schaik, Chapter 5). Similarly, the increase of glucocorticoids in recipients of aggression possibly facilitates the occurrence of submission (Leshner 1980).

Elevation in glucocorticoid levels represents an adaptive, short-term stress response. If the situation continues to be perceived as threatening or uncertain, the stress response will remain activated. This is not usually the case for solitary or semisolitary species because under natural conditions most of their conflicts are related to territorial disputes, and the common outcome is the withdrawal of the loser. In group-living animals, continuation of the threat or uncertainty is more likely because individuals do not tend to leave the group as a consequence of escalated conflicts (Aureli 1992). Prolonged activation of the stress response, however, has deleterious, possibly fatal, health consequences (Henry & Stephens 1977; Sapolsky 1998; von Holst 1998). Hence, mechanisms to cope with these negative consequences and to reduce risks and uncertainty are necessary, especially in group-living animals.

The best-studied example of such a mechanism is reconciliation (see de Waal, Chapter 2, for a historical overview). The term reconciliation implies a restoration of the relationship between former opponents, and recent studies on non-human primates have shown that such post-conflict reunions function to restore tolerance around resources (Cords 1992) and reduce the likelihood of renewed attacks (Aureli & van Schaik 1991; Cords 1992; de Waal 1993; Watts 1995; Silk et al. 1996; Castles & Whiten 1998). No study has directly addressed the effects of reconciliation on the stress response and health, but there is evidence suggesting a role of reconciliation in regulating anxiety levels.

Post-conflict anxiety is an adaptive response because it increases alertness and arousal to prepare individuals for vigorous actions (Gray 1987). Since post-conflict reunions remove the causes of threat and uncertainty that may require these actions, they should reduce anxiety levels. Self-scratching rates of both former opponents are indeed lower after reconciliation than in post-conflict periods without reconciliation (Aureli & van Schaik 1991; Castles & Whiten 1998; Das et al. 1998; see Schino, Chapter 11, for evidence in non-primate species). This reduction appears to be specific to reunions between former opponents: post-conflict friendly contacts with other group members do not decrease scratching rates of the former aggressor (Das et al. 1998). As illustrated in Figure 10.3, heart rates of former opponents are reduced following post-conflict affiliation with other group members, but this reduction is faster and more pronounced following reconciliation (Smucny et al. 1997).

Another approach to the study of the function of post-conflict reunions has used playbacks of friendly grunts to wild baboons (Cheney & Seyfarth, Box 10.1). The findings suggest that post-conflict grunts serve to reduce uncertainty in the

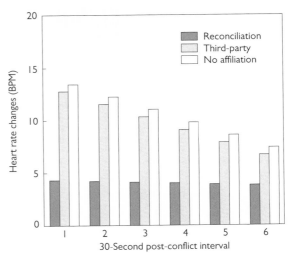

FIGURE 10.3. Heart rate (HR) changes following aggressive conflicts in captive rhesus macaques. HR changes represent the differences (in beats per minute, or BPM) from mean pre-conflict HR (i.e., during the 3 minutes preceding aggressive conflict) for six 30-second intervals following 37 high-intensity conflicts. Conflicts are distinguished into three categories, depending on whether they were followed by reconciliation, by affiliation with third parties only, or by no affiliation. HR values are derived from a two-level hierarchical model that considers post-conflict HR differences as a function of time since conflict, reconciliation, and affiliation with third parties, while controlling for movement or activity of the subject during the pre-conflict and post-conflict periods.

recipient of aggression: after the playback of grunts of the former opponent, recipients approached their former aggressor and tolerated their approaches more often than during periods without such a playback. Thus, evidence from different sources supports the view that reconciliation has an emotional impact by reducing post-conflict anxiety.

Emotion and Social Relationships

The evidence presented in the previous two sections strongly suggests that in well-established groups of nonhuman primates routine conflict and its aggressive escalation disturb the relationship between the opponents and that post-conflict

BOX 10.1

Vocal Reconciliation by Free-Ranging Baboons

Dorothy L. Cheney & Robert M. Seyfarth

Female baboons (*Papio cynocephalus ursinus*) utter low amplitude tonal grunts during many of their social interactions. The majority of grunts occur in the context of handling other females' infants, but females may also grunt as they approach one another, as they feed near or groom one another, or as they move into new areas of their range. As is true also of many other nonhuman primates' calls (e.g., cotton-top tamarins, *Saguinus oedipus:* Cleveland & Snowdon 1982; squirrel monkeys, *Saimiri sciureus:* Boinski 1992; Japanese macaques, *Macaca fuscata:* Blount 1985; stumptail macaques, *M. arctoides:* Bauers & de Waal 1991; Bauers 1993; vervet monkeys, *Cercopithecus aethiops:* Cheney & Seyfarth 1982; mountain gorillas, *Gorilla gorilla:* Harcourt et al. 1993; Seyfarth et al. 1994), baboons' grunts appear to function at least in part to facilitate social interactions. Dominant females who grunt as they approach more subordinate individuals, for example, are less likely to supplant these individuals, and more likely to handle their infants, than females who remain silent (Cheney et al. 1995). In contrast, females almost never threaten other individuals after grunting to them.

Because baboon grunts appear to be honest signals of intent that affect the behavior of lower-ranking individuals (see Silk, Box 9.1), it seems reasonable to predict that grunts might also serve to reconcile opponents following fights. In fact, baboons do occasionally grunt to each other after aggression. In a study of free-ranging baboons in the Okavango Delta, Botswana, we found that adult females grunt to subordinate recipients of aggression following approximately 13 percent of all aggressive interactions. Observations of females' behavior during the 10 minutes following an aggressive interaction showed that grunts were correlated with an increased frequency of friendly interactions between former opponents (Silk et al. 1996).

Although these observations suggest that baboons' grunts serve a reconciliatory function, their role is diffi-

cult to assess from observation alone because they often occur in conjunction with other friendly behavior such as grooming or infant handling. To test the hypothesis that grunts might act to reconcile opponents even in the absence of other friendly behavior, we carried out a playback experiment in which female baboons were played the tape recording of a dominant aggressor's scream in the minutes immediately following the dispute (Cheney et al. 1995). The scream mimicked an attack on the aggressor by an even more dominant female or a male. We hypothesized that, for subordinate listeners, the call would signal a potential retaliatory threat, because when female baboons receive aggression from higher-ranking individuals, they occasionally redirect aggression by threatening a more subordinate individual. We predicted, therefore, that subordinate subjects would orient strongly to these screams, particularly if the signaler had recently threatened them.

In conducting playback experiments, we placed a concealed loudspeaker at a distance of 5–10 meters of the subject, in the same general direction of the aggressor's last observed position. We then played a tape recording of the aggressor's scream to the subject and filmed her responses and the direction of orientation. Playback trials were conducted under three conditions: after the aggressor grunted to the subject immediately following the dispute; after the aggressor failed to grunt to her; and after a control period when the same two females had not recently interacted at all.

Results indicated that subordinate subjects oriented toward the scream of their dominant aggressor for a significantly longer duration if the aggressor had not given a "reconciliatory" grunt following the dispute than if she had (Cheney et al. 1995). Subjects oriented weakly to scream playbacks when vocal "reconciliation" had occurred; indeed, their response to scream playbacks in this context was similar to their response following a control period when the two females had not interacted.

Even in the absence of more overt friendly behavior, therefore, baboon grunts seem to act to restore the relationship of opponents to baseline tolerance levels. Moreover, in contrast to what has been reported for some species of macaques (de Waal 1989; Cords 1993),

BOX 10.1 (continued)

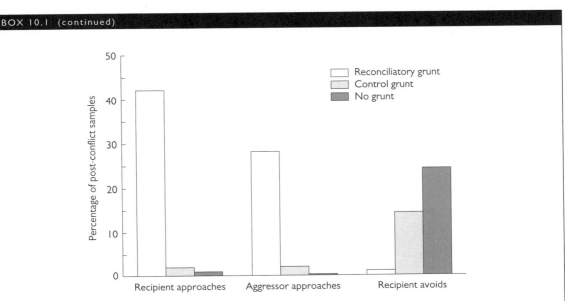

The proportion of first interactions between recipients of aggression and their aggressors following playbacks that took various forms. Histograms show means for all dyads taken together. Open histograms show recipients' behavior following playback of a "reconciliatory" grunt; gray histograms show recipients' behavior following playback of a control female's grunt; black histograms show recipients' behavior in the absence of a playback trial. Interactions were defined as follows: "Recipient approaches" = the recipient grunted to or approached her aggressor to within 2 meters; "Aggressor approaches" = the recipient allowed her aggressor to approach her without moving more than 2 meters away; "Recipient avoids" = the aggressor approached the recipient, and the recipient moved more than 2 meters away.

proximity alone is not sufficient to reconcile female baboons. Female baboons who were approached by recent aggressors in the absence of a "reconciliatory" grunt responded more strongly to their aggressors' screams than did females who were not approached or who were approached in the context of a grunt. Studies of other species have not reported whether former opponents also vocalized to each other when they came into proximity. The baboons' responses raise the strong possibility that it is not proximity but rather the presence or absence of a vocalization that serves to reconcile opponents.

Reconciliatory grunts not only reduce the anxiety of recipients of aggression; they also appear to have a strong influence on their subsequent interactions with their former aggressors. To examine the effect of apparently reconciliatory grunts on recipients' behavior, we designed another series of playback experiments in which we attempted to mimic vocal reconciliation (Cheney &

Seyfarth 1997). In these experiments, we played the grunt of dominant aggressors to subordinate subjects in the minutes immediately following the dispute and then observed their behavior for half an hour. As controls, subjects either heard no grunt at all or the grunt of another dominant female who had not been involved in the dispute.

After hearing playbacks of "reconciliatory" grunts, subjects approached their former aggressors and also tolerated their aggressors' approaches at significantly higher rates than they did under either control condition. In contrast, when subjects were played either a control vocalization from another female or no vocalization at all, they behaved as if no reconciliatory act had occurred. They did not approach their former aggressors, nor did they allow their aggressors to approach them. In fact, they were significantly more likely to be supplanted.

Playbacks of reconciliatory grunts had no effect on aggressors' tendencies to approach recipients of

BOX 10.1 (continued)

aggression or to initiate friendly interactions with them. This was no doubt due to the fact that "reconciliatory" grunts were not actually produced by the aggressors themselves but were instead mimicked through artificial playbacks. From the perspective of the recipient, reconciliation had apparently occurred, whereas from the perspective of her aggressor it had not. As a result, although former aggressors tolerated those females' more frequent approaches (Cheney & Seyfarth 1997), they were not necessarily inclined either to initiate a friendly interaction or to signal in some way their willingness to accept such an interaction.

In conclusion, there is now evidence from a number of different species that many of the most common vocalizations given by nonhuman primates function to initiate and facilitate social interactions. Playback experiments with adult female baboons have demonstrated that vocalizations also serve to diminish the strength of recipients' responses to former aggressors' potentially threatening behavior. Most important, apparently reconciliatory vocalizations influence recipients' willingness to approach and be approached by former aggressors. It is to be hoped that in future studies with other species vocalizations will be included, along with grooming and more conventionally accepted forms of reconciliation, in operational definitions of reconciliation. This, in turn, may permit more accurate estimations of reconciliation rates.

reunions restore the relationship (see also de Waal, Chapter 2; Cords & Aureli, Chapter 9). These changes in the social relationship are accompanied by emotional responses—specifically, increased anxiety after a conflict and its reduction following reconciliation. Research in other areas provides examples confirming that the disturbance and the subsequent restoration of a relationship are reflected in emotional responses.

The tendency of human beings to respond with distress to the end of a close relationship is a consistent phenomenon across different cultures and across the age span (Hazan & Shaver 1994; Fry, Chapter 16). Recent social psychology studies have concluded that threats to social bonds are a primary source of anxiety (Leary 1990) and that social exclusion is probably the most common cause of anxiety (Baumeister & Tice 1990).

Behavioral and physiological responses to the disturbance and restoration of social relationships emerge early in life. These responses have been extensively studied in nonhuman primates through experimental manipulations of the mother-infant relationship. This relationship is unique not only because it satisfies the basic needs of the infant but also because it serves as a primary source of emotional security (Harlow 1958). As expected, disturbance through experimental separation provokes strong responses, such as increased plasma cortisol and heart rate, in the infant and the mother; reunions of mothers with their infants restore their physiological responses to baseline (Levine et al. 1989; Reite & Boccia 1994; in rodents: Hofer 1996). Similar physiological changes also are found when adults of established pairs of the monogamous titi monkeys (*Callicebus moloch*) are separated from each other (Mendoza & Mason 1986), suggesting that the disruption of strong bonds has similar effects throughout the life span.

The positive emotional effect of social relationships can also be seen in situations in which the presence of group members ameliorates responses to aversive stimuli. The marked stress response normally evoked by stimuli, such as exposure to novel environments or objects, exposure to naturally occurring predators (e.g., snakes), or separation of infants from their mothers, is buffered by the presence of familiar conspecifics (Levine et al.

1989; Reite & Boccia 1994). Similarly, the role of supportive social relationships has been recently emphasized as an important factor to enhance coping with stressful situations and improve health conditions in humans (Cohen 1988).

Emotion and Relationship Quality

Within a group, social relationships vary in their quality (see Cords & Aureli, Chapter 9). The quality of relationships affects the occurrence of conflict-resolution mechanisms because individuals give higher priority to preventing disturbance of their more valuable relationships and restore them (de Waal, Chapter 2; Cords & Aureli, Chapter 9; van Schaik & Aureli, Chapter 15). We should therefore expect that emotional responses following relationship disturbance would depend greatly on relationship quality.

The effect of relationship quality on post-conflict emotional responses has been examined in longtail macaques (Aureli 1997). Self-scratching rates were higher after unreconciled conflicts between "friends" (i.e., partners with high affiliation rates) than after those between other individuals. This finding can be interpreted as evidence that higher levels of anxiety are experienced when more valuable relationships are disturbed by aggressive conflicts.

Research in areas other than conflict resolution supports the view that relationship quality influences emotional responses. For example, experimental manipulation of food access, which increases competition levels or foraging time, modifies the quality of mother-infant relationship in groups of bonnet macaques (*Macaca radiata*). Emotional responses of the infant to maternal separation are in turn affected: infants growing up in a more competitive environment or under conditions in which maternal contact is compromised experience stronger emotional responses follow-

ing separation (Boccia et al. 1991; Andrews et al. 1993). Hence, variation in emotional responses to the disruption of the bond through maternal separation reflects the quality of the mother-infant relationship (Reite & Boccia 1994).

In the wild, tree shrews usually live in heterosexual pairs. In captivity, however, putting a male and a female together does not inevitably lead to the formation of a pair bond (von Holst 1998). Although in some cases this process results in serious fights, most males and females can coexist, but in only about 20 percent of all pairings (i.e., harmonious pairs) is friendly behavior exchanged from the onset. Individuals in harmonious pairs show reduced plasma levels of glucocorticoids and epinephrine compared with levels prior to pairings, whereas the opposite is true for the same individuals in disharmonious pairs (von Holst 1998). Thus, the emotional response of tree shrews to pairing reflects the quality of their relationship.

Relationship quality also influences the effects of social buffering. In guinea pigs, males develop strong bonds with some, but not all, females in the group. When males are tested singly in an unfamiliar environment, their cortisol levels increase dramatically. The presence of an unfamiliar female or a familiar female with which no bond exists has no apparent buffering effect. A sharp reduction of the stress response occurs, however, when a female with which the male has a strong bond is present in the unfamiliar environment (Sachser et al. in press).

These observations are consistent with the difference in buffering effects of group members during maternal separation in two species of macaques (Laudenslager et al. 1990). In bonnet macaques, mothers are less protective of their infants than in pigtail macaques (*Macaca nemestrina*); this condition facilitates the development of closer relationships between bonnet macaque infants and other group members. The availability of these strong relationships in the group mitigates the distress response of bonnet macaque infants during maternal separation.

Emotional Mediation of Conflict Resolution

From our review, it appears that both the emotional responses following relationship disturbance and the occurrence of conflict resolution depend on the quality of the relationship between the opponents. Can these effects be closely linked? Can differential emotional responses mediate the effect of relationship quality on conflict resolution? And can emotional mediation be at the basis of the flexibility in conflict management due to factors other than relationship quality?

A possible mechanism for the higher frequency of post-conflict reunions between partners with valuable relationships has been recently proposed by Aureli (1997). Differential post-conflict anxiety may mediate the effects of relationship value on conciliatory tendencies (Fig. 10.4). This hypothesis implies no need of conscious knowledge of relationship value but relies on differential emotional changes consequent to potential loss of benefits due to relationship disturbance: more valuable relationships provide greater benefits, and their disturbance would produce higher levels of anxiety. Since the best way to cope with post-conflict anxiety is to repair the relationship (Aureli & van Schaik 1991), higher levels of post-conflict anxiety would lead to higher conciliatory tendencies. In fact, in longtail macaques post-conflict scratching rates are higher after conflicts with "friends" (Aureli 1997), and post-conflict reunions occur more often between these individuals (Aureli et al. 1989).

Partners in a relationship are likely to differ in the benefits they can offer and receive. Accordingly, we should expect an asymmetry in the value that two partners attach to their relationship (Cords & Aureli, Chapter 9). This asymmetry should lead to differences between the partners in their respective interests in relationship repairing. To examine asymmetries, previous research has focused on the difference between the partners in taking initiative for post-conflict reunion (reviewed in de Waal 1993). This may not be the ideal method because (1) subtle signals exchanged between partners can be missed by the human observer and (2) all unsuccessful (i.e., one-sided) attempts to reconcile are not considered. The latter are expressions of interest in restoring the relationship that are not met by the other partner. An empirical approach that uses indicators of emotional responses appears to be more promising. For instance, self-scratching can provide evidence of differential post-conflict anxiety of the two former opponents, which is likely to reflect both the level of uncertainty due to the potential loss of different benefits and the degree of interest in repairing the relationship.

The role of anxiety in mediating conflict resolution probably has limitations. We can hypothesize that if an individual is more anxious as a result of the disturbance of a given relationship, he or she will be more willing to resolve that conflict and repair the relationship. However, when anxiety reaches extremely high levels, it is likely to impair actions both to reduce the distance between the two former opponents and to facilitate relationship repair, because close proximity may be perceived as too risky (i.e., a possible "ceiling effect" of anxiety levels).

Anxiety-like emotion also may be involved in the management of conflict of interest at an earlier stage (Fig. 10.4). Several lines of evidence suggest an increase of anxiety under conditions that are strongly associated with risk of conflict (see "The Pre-Escalation Phase" section). The risk of conflict escalation can produce anxiety because of the

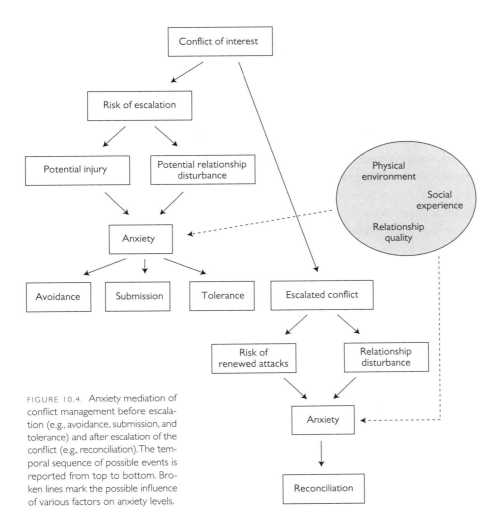

FIGURE 10.4. Anxiety mediation of conflict management before escalation (e.g., avoidance, submission, and tolerance) and after escalation of the conflict (e.g., reconciliation). The temporal sequence of possible events is reported from top to bottom. Broken lines mark the possible influence of various factors on anxiety levels.

potential for injuries and disturbance of the relationship with the opponent. As illustrated in Figure 10.4, relationship quality between opponents, social experience (e.g., developmental history, dominance rank, social support), and characteristics of the physical environment (e.g., location, population density) all likely influence anxiety levels *before* conflict escalation as well as levels *after* escalation. The internal and external stimuli before and after escalation are, however, clearly different. Pre-escalation anxiety likely leads to different responses than those mediated by post-

escalation anxiety. Whereas the latter may mediate friendly reunions between former opponents, pre-escalation anxiety may facilitate submission and avoidance or increase tolerance to limit the risk of escalated conflict (Judge, Chapter 7; Koyama, Box 7.1).

Another possible mediating role of emotion in conflict resolution is shown in the distress of young children's tantrums (Potegal, Box 12.1). Tantrums usually take place as a consequence of a conflict between child and parents. The distress associated with prolonged screaming may facilitate the child-initiated reunions with the parent by increasing the child's discomfort and thereby his or her post-tantrum need for reassurance. Prolonged screaming bouts are indeed more likely to be followed by post-tantrum child-parent affiliation.

The behavior of a distressed child seeking a positive solution shares similarities with the behavior of monkey infants during maternal separation (even though in the latter case there may not be any conflict between parent and offspring). The infant's calls and agitated activities are certainly an expression of emotional distress but are also attempts to reestablish maternal contact (Levine et al. 1989). We could hypothesize that emotional distress mediates the reunion between mother and infant, which in turn terminates the threatening situation. The high-intensity calls function to elicit retrieval by the mother, and the agitated activity reflects the infant's effort to locate the mother. In line with this view, juvenile Goeldi's monkeys (*Callimico goeldii*) that are more distressed during maternal separation seek more maternal contact upon reunion (Dettling et al. 1998).

The mediation of reunions and other forms of conflict resolution does not need to involve only "negative" emotions, such as anxiety, fear, or distress. Conflict resolution often occurs through affiliative contacts that are likely to be associated with positive sensations. Gentle touching causes relaxation and a reduction in heart rate in humans, rhesus monkeys, and horses (Drescher et al. 1980, 1982; Feh & De Mazieres 1993). Allogrooming in monkeys decreases heart rate (Boccia et al. 1989; Aureli et al. 1999) and self-directed behavior (Schino et al. 1988), indicating tension reduction. An increase of endogenous brain opioids follows the receipt of allogrooming (Keverne et al. 1989), suggesting that such affiliative behavior provides pleasure, since brain opioids are usually released during "positive" emotion created by rewarding

situations (Dum & Herz 1987). Thus, the need for relaxation and pleasant sensations, which possibly increases during conflict and its aftermath, may also facilitate affiliative reunions between former opponents.

The probable function of "positive" emotion in conflict resolution is also supported by the role of brain opioids in mother-infant separation and reunion. Although the actions of endogenous brain opioids affect and are affected by several hormones and neurotransmitters, their involvement in mother-infant attachment and separation distress is certain (Keverne 1992; Panksepp et al. 1997). The experience of separation distress likely provides the basic motivation for social reunion, and the subsequent opioid release alleviates separation distress, replacing it with positive sensations. This is probably also true for relationships between adults. For example, adult talapoin monkeys (*Miopithecus talapoin*) that have been socially isolated show a significant increase in brain opioids following affiliative behavior with their peers upon reunion (Keverne et al. 1989).

Based on the above and additional evidence, Panksepp & Bekkedal (1997) concluded that a specific form of anxiety due to social separation involves endogenous opioids, among other neuromodulators, and functions to mediate social bonding. We can then speculate that a similar neurophysiological mechanism may regulate post-conflict anxiety due to the disruption of strong bonds (i.e., psychological separation), and its mediating role in the subsequent reunion that is perceived as social reward.

Conclusion and Future Research

The use of emotion as an intervening variable during interindividual conflict (Fig. 10.1) has led us to develop a model for the possible mediating role of anxiety in conflict management (Fig. 10.4).

We have focused on anxiety because most direct and indirect evidence is available for this emotion. The model requires further tests from studies using various emotional indicators and a variety of species, since most of the relevant information thus far comes primarily from a restricted number of primate species. Although data support the mediating role of post-escalation anxiety, research should expand to test the predictions of the model for pre-escalation anxiety. Furthermore, we must explore the possible "ceiling effect" of high emotional intensity and the asymmetries in emotional responses between partners. Similar models also should be developed for other emotions and other aspects of conflict management.

The perspective of emotional mediation of social processes, including conflict resolution, does not require the identification of specific human-like emotions, especially when the conscious feeling is not considered. It is only necessary to characterize the conditions (e.g., post-escalation relationship disturbance), the various changes (e.g., increased self-scratching), and the factors affecting them (e.g., relationship quality). The emotional mediation illustrated in Figure 10.4 is an example of how this perspective can explain the effects of relationship quality on reconciliation without necessarily implying conscious knowledge of relationship value. Conflict resolution as well as other social processes involve multiple conditions, changes, and factors that can be effectively combined in the concept of emotion as a mediating variable (Fig. 10.1). This perspective can help us to understand better the mechanisms that underlie the dynamics and flexibility of social processes.

The direct correspondence between a given emotion and a specific change in a single indicator is strongly questioned (Cacioppo et al. 1993). Researchers must gather information on the multiple determinants of emotion, including the internal and external factors and conditions as well as many associated behavioral and physiological changes (Fig. 10.1). This is, of course, no easy task, but there are promising indications for the future. Research effort on the mechanisms underlying behavior has notably expanded, including the study of the neurophysiological bases of affiliative behavior and bonding (Carter et al. 1997). More widespread use of biotelemetry and behavioral indicators can provide detailed information about emotional changes during ongoing social interactions. The combination of sophisticated noninvasive techniques, such as facial electromyography (EMG: Cacioppo et al. 1990) or noninvasive brain imaging (PET: Ingvar 1997; fMRI: Kalin et al. 1997), with innovative experimental procedures (e.g., subjects viewing videotapes of social events) might offer complementary insight into the mediating role of emotion in humans as well as in other animals.

Conflict-resolution research shows obvious parallelisms with attachment theory (Bowlby 1969, 1973; Ainsworth et al. 1978). Key findings in conflict resolution include the post-conflict anxiety due to disruption of the relationship between opponents; the effect of relationship quality on the frequency of post-conflict reunions; and the mediating role of post-conflict anxiety. Similarly, the typical infant response to separation from the attachment figure is anxiety and distress; the way reunion is attempted depends strongly on the type of attachment; and separation anxiety possibly plays a critical role in the reunion process.

These similarities may provide insight for future studies of conflict management. Early attachment experiences have strong effects on the development of skills and relationships with others (Bowlby 1969; Ainsworth et al. 1978). This view has important implications for a developmental perspective on conflict resolution. For example, mother-infant attachment styles may influence conflict-resolution skills later in life (Weaver & de Waal, Box 10.2). In addition, the emotional bases underlying adult reconciliation may first be experienced by children in the aftermath of their tantrums (Potegal, Box 12.1).

Adult relationships are also interpreted from the perspective of attachment theory (Ainsworth 1989; Hazan & Shaver 1994). Research on human adults has used attachment theory to explain variation in anxious responses to separation or other disturbance of close relationships (Feeney 1998) and in conflict-resolution styles (Rholes et al. 1998). Our concept of strong, valuable relationships shares many characteristics with the concept of "affectional bonds" or attachments between human adults (cf. Ainsworth 1989). The security of a relationship, in addition to its value, has been hypothesized to affect the occurrence of reconciliation (Cords & Aureli, Chapter 9). An emotional basis for this influence is suggested by the findings that human adults with insecure attachments experience greater anxiety and distress when the relationship is threatened or lost (Feeney 1998; Mikulincer & Florian 1998). In addition, juvenile Goeldi's monkeys with insecure attachments are more distressed by separation and seek more maternal contact upon reunion (Dettling et al. 1998). Hence, attachment theory provides insight into the effects of various aspects of relationship quality (e.g., value, security) on conflict resolution.

Individual differences are present in many aspects of social processes and their underlying mechanisms (e.g., physiology: Kagan et al. 1988; Higley & Linnoila 1997; von Holst 1998; Sapolsky, Box 6.1; responses to separation: Reite & Boccia 1994; attachment: Ainsworth et al. 1978). Even though some of the mechanisms of conflict resolution (e.g., post-conflict reunions) need the participation of both opponents, individual differences are likely to play an important role in

BOX 10.2

The Development of Reconciliation in Brown Capuchins

Ann Ch. Weaver & Frans B. M. de Waal

Few studies have addressed reconciliation behavior in juvenile primates (exceptions are Cords 1988; Cords & Aureli 1993; de Waal & Johanowicz 1993), and none has thus far investigated how the mother-offspring relationship may affect the development of reconciliation. There are good reasons to expect the social environment to affect reconciliation because juvenile rhesus macaques (*Macaca mulatta*) increase their conciliatory tendencies following exposure to "tutor" stumptail macaques (*M. arctoides*) that reconcile frequently (de Waal & Johanowicz 1993).

The present study asked whether the quality of the mother-offspring relationship affects the development of reconciliation in the offspring. The answer was yes. Brown capuchin (*Cebus apella*) mother-offspring relationships were interpreted as secure or insecure in terms of attachment theory (Ainsworth et al. 1978), after which we compared secure and insecure offspring reconciliation styles with adult aggressors other than the mother. The main finding was that youngsters in secure and insecure relationships each developed a different but successful manner of reconciliation that appeared to be related to the availability of the mother for contact comfort.

The possibility that attachment is one of the biological and social bases of reconciliation is an important source of hypotheses, for example, relationship quality influences how a juvenile learns to cope with its own arousal states and uses social or asocial solutions for coping with post-conflict arousal. Based on a nonhuman primate model, these findings may also provide clinical insights into human conflict resolution.

Throughout the text, the term *juvenile* refers to 24 immature capuchins (less than five years of age) in three developmental stages: infancy (i.e., consis-

BOX 10.2 (continued)

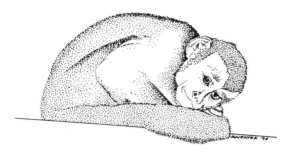

Curling by a juvenile brown capuchin. Curling is a self-protective behavior and an appeasing posture that juveniles show frequently during conflict with adults. Illustration by Ann Ch. Weaver.

tently high nursing rates), weaning (i.e., following the first significant drop in nursing rate), and juvenescence (i.e., following the last nursing episode). One subject could be observed across more than one stage, resulting in 38 mother-offspring pairs in the analysis (Weaver 1999).

We measured capuchin mother-offspring relationships qualitatively and empirically. Qualitatively, relationships were categorized as either secure or insecure based on a detailed set of criteria derived from the literature on human attachment (Ainsworth et al. 1978). Attachment is measured by the way a child acts toward his or her mother after a series of separations and reunions known as the Strange Situation. The child's Strange Situation behavior correlates with his or her behavior at home. Inspired by what is known about children's at-home behavior, we identified 18 analogous capuchin baseline behaviors to categorize mother-offspring pairs by attachment security. For example, relationships with high rates of contact and low rates of arousal fell in the secure category. There were 22 secure and 16 insecure mother-offspring pairs.

In addition, we developed a better-defined empirical measure based on the ratio of positive and negative interaction within the mother-offspring relationship. This relationship quality (RQ) index was calculated by dividing each pair's relative grooming rate by its relative agonism rate. Agonism was the hourly rate during baseline observations of low-level threats, protest, appease-

ment, and avoidance events. Relative measures were derived by dividing each pair's hourly rate of each behavior by the mean hourly rate for all pairs in the same developmental stage. Thus defined, the RQ index did not significantly vary with offspring's developmental stage or mother's dominance rank.

The qualitative and empirical measures of relationship quality appeared to measure the same relationship features: mother-offspring relationships categorized as secure also showed a high RQ index, and relationships categorized as insecure also showed a low RQ index. The mean RQ index of relationships categorized as secure was three times higher than that of relationships categorized as insecure.

We therefore consider our RQ index an empirical, continuous measure of attachment security. We examined the connection between mother-offspring relationship quality and juveniles' reconciliation behavior using juvenile conflicts with unrelated adult aggressors (hence, conflicts with the mother herself were excluded from analysis). We first demonstrated that reconciliation occurs in juvenile capuchins by showing more attraction than dispersal after conflict than in control observations (cf. Verbeek & de Waal 1997 for adults). Then we developed two measures of conciliatory tendency, that is, the percentages of juvenile-initiated and adult-initiated reconciliation of the total number of adult-juvenile conflicts followed for five minutes after the end of conflict. Conflict was defined as juvenile distress upon receipt of aggression, and the initiation of reconciliation was defined by who made the actual physical post-conflict contact (e.g., grooming, contact sitting, and playing). These measures of conciliatory tendency were then compared in a multivariate analysis with developmental stage and the RQ index as independent variables. The multiple correlation accounted for 50 percent of the variance in juvenile-initiated reconciliations and 43 percent of the variance in adult-initiated reconciliations (Weaver 1999). These results indicate that younger juveniles initiated more reconciliations than older juveniles, and insecurely attached juveniles initiated more reconciliations than securely attached ones. Further, adult aggressors initiated reconciliations with securely attached more than

BOX 10.2 (continued)

insecurely attached juveniles and were not influenced by a juvenile's age.

We also determined whether contact between a juvenile and its mother during the post-conflict period influenced reconciliation between the juvenile and its adult aggressor. An evaluation of the attraction versus dispersal hypothesis for mother-offspring contacts following juvenile conflict with other adults revealed differences in post-conflict contact frequency. Members of securely attached pairs were attracted to each other: secure juveniles treated their mothers as a secure base in the post-conflict period, making frequent contact with her. Insecurely attached pairs showed marginally significant levels of attraction and no secure base behavior.

In conclusion, why would a young monkey have contact with a large adult opponent that just frightened it? One answer is that conflict is arousing and arousal elicits the contact-seeking behavior of attachment. Juveniles with a secure base may have been able to "handle" conflict better (i.e., with less distress), as appeasing postures could imply (see the figure accompanying this box), and aggressors may have responded by approaching. Afterward, secure juveniles sought their mothers, as predicted by attachment theory. The post-

conflict behavior of capuchins with secure relationships may indicate greater social competence, as with securely attached children (Main & Weston 1982).

Juveniles without a secure base responded to conflict with an adult aggressor with greater arousal and may have reconciled with aggressors to calm down while compensating for the lack of a reliable source of comfort in their mothers. The unpredictability and higher proportion of agonism in insecure capuchin mother-offspring relationships may have prepared juveniles to deal with conflict effectively and early. By failing to provide a reliable source of comfort, insecure relationships may promote comfort seeking outside the mother-offspring bond. If so, the irony would be that insecure mother-offspring relationships enhance the development of reconciliation, at least early in development.

Attachment, that is, security in the mother, may thus influence the development of reconciliation by influencing post-conflict distress of juveniles and their tendencies to seek post-conflict contact, either with the mother or the opponent, to mitigate distress when aroused. Each type of attachment security leads to different expressions.

conflict management (cf. Cords & Aureli, Chapter 9). Through careful characterization of individual temperament and emotional responses to conflict, we could better understand the involvement of individuals in one-sided conflict management processes (e.g., avoidance or submission) and asymmetries in those processes that require mutual participation (e.g., reconciliation).

Finally, a role of emotion can also be expected in conflict management processes that involve individuals other than the opponents. Triadic processes have been studied in various species of nonhuman primates (Das, Chapter 13; Petit & Thierry,

Box 13.1; Watts et al., Chapter 14). We would expect that the level of distress or anxiety of the opponents (e.g., expressed in their screams) could trigger intervention by third parties during the conflict or post-conflict consolatory behavior by bystanders. The participation of third parties could be mediated by emotional contagion or empathy (de Waal & Aureli 1996). At this stage, these are mere speculations, but the study of differences in emotional responses of opponents and bystanders to conflict will surely help us better understand the variation of triadic forms of conflict management and their functions.

References

Ainsworth, M. D. S. 1989. Attachments beyond infancy. *American Psychologist*, 44: 709–716.

Ainsworth, M. D. S., Blehar, M., Waters, E., & Wall, S. 1978. *Patterns of Attachment*. Hillsdale, N.J.: Lawrence Erlbaum Associates.

Andrews, M. W., Sunderland, G., & Rosenblum, L. A. 1993. Impact of foraging demand on conflict within mother-infant dyads. In: *Primate Social Conflict* (W. A. Mason & S. P. Mendoza, eds.), pp. 229–252. New York: State University of New York Press.

Archer, J. 1988. *The Behavioral Biology of Aggression*. Cambridge: Cambridge University Press.

Archer, J. 1991. The influence of testosterone on human aggression. *British Journal of Psychology*, 82: 1–28.

Arsenio, W. F., & Killen, M. 1996. Conflict-related emotions during peer disputes. *Early Education and Development*, 7: 43–57.

Aureli, F. 1992. Post-conflict behaviour among wild long-tailed macaques (*Macaca fascicularis*). *Behavioral Ecology and Sociobiology*, 31: 329–337.

Aureli, F. 1997. Post-conflict anxiety in nonhuman primates: The mediating role of emotion in conflict resolution. *Aggressive Behavior*, 23: 315–328.

Aureli, F., & van Schaik, C. P. 1991. Post-conflict behaviour in long-tailed macaques (*Macaca fascicularis*): II. Coping with the uncertainty. *Ethology*, 89: 101–114.

Aureli, F., van Schaik, C. P., & van Hooff, J. A. R. A. M. 1989. Functional aspects of reconciliation among captive long-tailed macaques (*Macaca fascicularis*). *American Journal of Primatology*, 19: 39–51.

Aureli, F., Das, M., & Veenema, H. C. 1997. Differential kinship effect on reconciliation in three species of macaques (*Macaca fascicularis, M. fuscata,* and *M. sylvanus*). *Journal of Comparative Psychology*, 111: 91–99.

Aureli, F., Preston, S. D., & de Waal, F. B. M. 1999. Heart rate responses to social interactions in free-moving rhesus macaques: A pilot study. *Journal of Comparative Psychology*, 113: 59–65.

Averill, J. R. 1982. *Anger and Aggression*. New York: Springer-Verlag.

Baker, K. C., & Aureli, F. 1997. Behavioural indicators of anxiety: An empirical test in chimpanzees. *Behaviour*, 134: 1031–1050.

Bauers, K. A. 1993. A functional analysis of staccato grunt vocalizations in the stumptailed macaque (*Macaca arctoides*). *Ethology*, 94: 147–161.

Bauers, K. A., & de Waal, F. 1991. "Coo" vocalizations in stumptailed macaques: A controlled functional analysis. *Behaviour*, 119: 143–160.

Baumeister, R. F., & Tice, D. M. 1990. Anxiety and social exclusion. *Journal of Social and Clinical Psychology*, 9: 165–195.

Berkowitz, L. 1992. *Aggression: Its Causes, Consequences, and Control*. New York: McGraw-Hill.

Berman, M. E., Tracy, J. I., & Coccaro, E. F. 1997. The serotonin hypothesis of aggression revisited. *Clinical Psychology Review*, 17: 651–665.

Bernstein, I. S., & Ehardt, C. L. 1985. Intragroup agonistic behavior in rhesus monkeys *Macaca mulatta. International Journal of Primatology*, 6: 209–226.

Blanchard, D. C. 1984. Applicability of animal models to human aggression. In: *Biological Perspectives on Aggression* (K. J. Flannelly, R. J. Blanchard, & D. C. Blanchard, eds.), pp. 49–74. New York: Alan R. Liss.

Blount, B. 1985. "Girney" vocalizations among Japanese macaque females: Context and function. *Primates*, 26: 424–435

Boccia, M. L., Reite, M., & Laudenslager, M. L. 1989. On the physiology of grooming in a pigtail macaque. *Physiology and Behavior*, 45: 667–670.

Boccia, M. L., Reite, M., & Laudenslager, M. L. 1991. Early social environment may alter the development of attachment and social support: Two case reports. *Infant Behavior and Development*, 14: 252–260.

Boinski, S. 1992. Ecological and social factors affecting the vocal behavior of adult female squirrel monkeys. *Ethology*, 92: 316–330.

Bowlby, J. 1969. *Attachment and Loss: Attachment*. New York: Basic Books.

Bowlby, J. 1973. *Attachment and Loss: Separation: Anxiety and Anger*. New York: Basic Books.

Brain, P. F. 1980. Adaptive aspects of hormonal correlates of attack and defense in laboratory mice: A study in ethobiology. *Progress in Brain Research*, 53: 391–413.

Brain, P. F., & Haug, M. 1992. Hormonal and neurochemical correlates of various forms of animal "aggression." *Psychoneuroendocrinology*, 17: 537–551.

Brothers, L. 1990. The neural basis of primate social communication. *Motivation and Emotion*, 14: 81–91.

Cacioppo, J. T., Tassinary, L. G., & Fridlund, A. J. 1990. The skeletomotor system. In: *Principles of Psychophysiology* (J. T. Cacioppo & L. G. Tassinary, eds.), pp. 325–384. Cambridge: Cambridge University Press.

Cacioppo, J. T., Klein, D. J., Berntson, G. G., & Hartfield, E. 1993. The psychophysiology of emotion. In: *The Handbook of Emotions* (M. Lewis & J. M. Haviland, eds.), pp. 119–142. New York: Guilford Press.

Carter, C. S., De Vries, A. C., & Getz, L. L. 1995. Physiological substrates of mammalian monogamy: The prairie vole model. *Neuroscience and Biobehavioral Reviews*, 19: 303–314.

Carter, C. S., Lederhendler, I. I., & Kirkpatrick, B. 1997. *The Integrative Neurobiology of Affiliation*. New York: New York Academy of Sciences.

Castles, D. L., & Whiten, A. 1998. Post-conflict behaviour of wild olive baboons. II. Stress and self-directed behaviour. *Ethology*, 104: 148–160.

Cheney, D. L., & Seyfarth, R. M. 1982. How vervet monkeys perceive their grunts: Field playback experiments. *Animal Behaviour*, 30: 739–751.

Cheney, D. L., & Seyfarth, R. M. 1997. Reconciliatory grunts by dominant female baboons influence victims' behavior. *Animal Behaviour*, 54: 409–418.

Cheney, D. L., Seyfarth, R. M., & Silk, J. B. 1995. The role of grunts in reconciling opponents and facilitating interactions among adult female baboons. *Animal Behaviour*, 50: 249–257.

Cleveland, J., & Snowdon, C. T. 1982. The complex vocal repertoire of the adult cotton-top tamarin (*Saguinus oedipus oedipus*). *Zeitschrift für Tierpsychologie*, 58: 231–270.

Clutton-Brock, T. H., & Albon, S. D. 1979. The roaring of the red deer and the evolution of honest advertisement. *Behaviour*, 68: 145–170.

Cohen, S. 1988. Psychological models of the role of social support in the etiology of physical disease. *Health Psychology*, 7: 269–297.

Cords, M. 1988. Resolution of aggressive conflicts by immature long-tailed macaques, *Macaca fascicularis*. *Animal Behaviour*, 36: 1124–1135.

Cords, M. 1992. Post-conflict reunions and reconciliation in long-tailed macaques. *Animal Behaviour*, 44: 57–61.

Cords, M. 1993. On operationally defining reconciliation. *American Journal of Primatology*, 29: 255–267.

Cords, M., & Aureli, F. 1993. Coping with aggression by juvenile long-tailed macaques (*Macaca fascicularis*). In: *Juvenile Primates: Life History, Development, and Behavior* (M. E. Pereira & L. A. Fairbanks, eds.), pp. 271–284. Cambridge: Oxford University Press.

Cords, M., & Thurnheer, S. 1993. Reconciliation with valuable partners by long-tailed macaques. *Ethology*, 93: 315–325.

Crook, J. H. 1989. Introduction: Socioecological paradigms, evolution and history: Perspectives for the 1990s. In: *Comparative Socioecology: The Behavioural Ecology of Humans and Other Mammals* (V. Standen & R. Foley, eds.), pp. 1–36. Oxford: Blackwell Scientific Publications.

Damasio, A. R. 1998. Emotion in the perspective of an integrated nervous system. *Brain Research Reviews*, 26: 83–86.

Darwin, C. 1872. *The Expression of Emotions in Man and Animals*. London: John Murray.

Das, M., Penke, Z., & van Hooff, J. A. R. A. M. 1998. Post-conflict affiliation and stress-related behavior of long-tailed macaque aggressors. *International Journal of Primatology*, 19: 53–71.

Davidson, R. J. 1995. Cerebral asymmetry, emotion, and affective style. In: *Brain Asymmetry* (R. J. Davidson & K. Hugdahl, eds.), pp. 361–387. Cambridge: MIT Press.

Davidson, R. J., & Sutton, S. K. 1995. Affective neuroscience: The emergence of a discipline. *Current Opinion in Neurobiology*, 5: 217–224.

Dettling, A., Pryce, C. R., Martin, R. D., & Dobeli, M. 1998. Physiological responses to parental separation and a strange situation are related to parental care received in juvenile Goeldi's monkeys (*Callimico goeldii*). *Developmental Psychobiology*, 33: 21–31.

de Waal, F. B. M. 1989. *Peacemaking among Primates*. Cambridge: Harvard University Press.

de Waal, F. B. M. 1993. Reconciliation among primates: A review of empirical evidence and unresolved issues. In: *Primate Social Conflict* (W. A. Mason & S. P. Mendoza, eds.), pp. 111–144. Albany: State University of New York Press.

de Waal, F. B. M., & Aureli, F. 1996. Consolation, reconciliation, and a possible cognitive difference between macaques and chimpanzees. In: *Reaching into Thought: The Minds of the Great Apes* (A. E. Russon, K. A. Bard, & S. T. Parker, eds.), pp. 80–110. Cambridge: Cambridge University Press.

de Waal, F. B. M., & Johanowicz, D. L. 1993. Modification of reconciliation behavior through social experience: An experiment with two macaque species. *Child Development*, 64: 897–908.

Diezinger, F., & Anderson, J. R. 1986. Starting from scratch: A first look at a "displacement activity" in group-living rhesus monkeys. *American Journal of Primatology*, 11: 117–124.

Drescher, V. M., Gantt, W. H., & Whitehead, W. E. 1980. Heart rate response to touch. *Psychosomatic Medicine*, 42: 559–565.

Drescher, V. M., Hayhurst, V., Whitehead, W. E., & Joseph, J. A. 1982. The effects of tactile stimulation on pulse rate and blood pressure. *Biological Psychiatry*, 17: 1347–1352.

Dum, J., & Herz, A. 1987. Opioids and motivation. *Interdisciplinary Science Reviews*, 12: 180–190.

Ekman, P., & Oster, H. 1979. Facial expressions of emotion. *Annual Review of Psychology*, 30: 527–554.

Ekman, P., Levenson, R. W., & Friesen, W. V. 1983. Autonomic nervous system activity distinguishes among emotions. *Science*, 221: 1208–1210.

Feeney, J. A. 1998. Adult attachment and relationship-centered anxiety. In: *Attachment Theory and Close Relationships* (J. A. Simpson & W. S. Rholes, eds.), pp. 189–218. New York: Guilford Press.

Feh, C., & De Mazieres, J. 1993. Grooming at a preferred site reduces heart rate in horses. *Animal Behaviour*, 46: 1191–1194.

Frijda, N. H. 1994. Emotions are functional, most of the time. In: *The Nature of Emotion: Fundamental Questions* (P. Ekman & R. J. Davidson, eds.), pp. 112–122. New York: Oxford University Press.

Gouzoules, H., Gouzoules, S., & Ashley, J. 1995. Representational signaling in non-human primate vocal communication. In: *Current Topics in Primate Vocal Communication* (E. Zimmermann, ed.), pp. 235–252. New York: Plenum Press.

Gray, J. A. 1987. *The Psychology of Fear and Stress*. Cambridge: Cambridge University Press.

Haemisch, A. 1990. Coping with social conflict, and the short-term changes of plasma cortisol titers in familiar and unfamiliar environments. *Physiology and Behavior*, 47: 1265–1270.

Hannes, R. P., Franck, D., & Liemann, F. 1984. Effects of rank-order fights on whole-body and blood concentrations of androgens and corticosteroids in the male swordtail (*Xiphophorus helleri*). *Zeitschrift für Tierpsychologie*, 65: 53–65.

Harcourt, A. H., Stewart, K. J., & Hauser, M. D. 1993. Functions of wild gorilla "close" calls: I. Repertoires, context, and interspecific comparison. *Behaviour*, 124: 89–122.

Harlow, H. F. 1958. The nature of love. *American Psychologist*, 13: 673–685.

Hazan, C., & Shaver, P. R. 1994. Attachment as an organizational framework for research on close relationships. *Psychological Inquiry*, 5: 1–22.

Henry, J. P., & Stephens, P. M. 1977. *Stress, Health, and the Social Environment: A Sociobiological Approach to Medicine*. New York: Springer-Verlag.

Higley, J. D., & Linnoila, M. 1997. Low central nervous system serotonergic activity is traitlike and correlates with impulsive behavior. *Annals of the New York Academy of Sciences*, 836: 39–56.

Higley, J. D., Mehlman, P. T., Poland, R. E., Taub, D. M., Vickers, J., Suomi, S. J., & Linnoila, M. 1996. CSF testosterone and 5-HIAA correlate with different types of aggressive behaviors. *Biological Psychiatry*, 40: 1067–1082.

Hinde, R. A. 1985. Expression and Negotiation. In: *The Development of Expressive Behavior* (G. Zivin, ed.), pp. 103–116. Orlando, Fla.: Academic Press.

Hofer, M. A. 1996. On the nature and consequence of early loss. *Psychosomatic Medicine*, 58: 570–581.

Huntingford, F. A., & Turner, A. K. 1987. *Animal Conflict*. London: Champman and Hall.

Ingvar, D. H. 1997. History of brain imaging in psychiatry. *Dementia and Geriatric Cognitive Disorders*, 8: 66–72.

Kagan, J., Reznick, J. S., & Snidman, N. 1988. Biological bases of childhood shyness. *Science*, 240: 167–171.

Kalin, N. H., Davidson, R. J., Irwin, W., Warner, G., Orendi, J. L., Sutton, S. K., Mock, B. J., Sorenson, J. A., Lowe, M., & Turski, P. A. 1997. Functional magnetic resonance imaging studies of emotional processing in normal and depressed patients: Effects of venlafaxine. *Journal of Clinical Psychiatry*, 58 (suppl. 16): 32–39.

Keverne, E. B. 1992. Primate social relationships: Their determinants and consequences. *Advances in the Study of Behavior*, 21: 1–37.

Keverne, E. B., Martensz, N. D., & Tuite, B. 1989. Beta-endorphin concentrations in cerebrospinal fluid of monkeys are influenced by grooming relationships. *Psychoneuroendocrinology*, 14: 155–161.

Kiecolt-Glaser, J. K., Malarkey, W. B., Chee, M., Newton, T., Cacioppo, J. T., Mao, H., & Glaser, R. 1993. Negative behavior during marital conflict is associated with immunological down-regulation. *Psychosomatic Medicine*, 55: 395–409.

Laudenslager, M. L., Held, P. E., Boccia, M. L., Reite, M., & Cohen, J. J. 1990. Behavioral and immunological consequences of brief separation: A species comparison. *Developmental Psychobiology*, 23: 247–264.

Leary, M. R. 1990. Responses to social exclusion: Social anxiety, jealousy, loneliness, depression, and low self-esteem. *Journal of Social and Clinical Psychology*, 9: 212–229.

LeDoux, J. E. 1996. *The Emotional Brain*. New York: Simon and Schuster.

Leshner, A. I. 1980. The interaction of experience and neuroendocrine factors in determining behavioral adaptations to aggression. *Progress in Brain Research*, 53: 427–438.

Leshner, A. I. 1983. Pituitary-adrenocortical effects on intermale agonistic behavior. In: *Hormones and Aggressive Behavior* (B. S. Svare, ed.), pp. 27–38. New York: Plenum Press.

Levenson, R. W. 1992. Autonomic nervous system differences among emotions. *Psychological Science*, 3: 23–27.

Levine, S., Coe, C., & Wiener, S. G. 1989. Psychoneuroendocrinology of stress: A psychobiological perspective. In: *Psychoneuroendocrinology* (F. R. Brush & S. Levine, eds.), pp. 341–377. San Diego: Academic Press.

Lott, D. F. 1991. *Intraspecific Variation in the Social System of Wild Vertebrates*. Cambridge: Cambridge University Press.

MacLean, P. D. 1952. Some psychiatric implications of physiological studies on frontotemporal portion of limbic system (visceral brain). *Electroencephalography and Clinical Neurophysiology*, 4: 407–418.

Maestripieri, D. 1993. Maternal anxiety in rhesus macaques (*Macaca mulatta*) I. Measurement of anxiety and identification of anxiety-eliciting situations. *Ethology*, 95: 19–31.

Maestripieri, D., Schino, G., Aureli, F., & Troisi, A. 1992. A modest proposal: Displacement activities as indicators of emotions in primates. *Animal Behaviour*, 44: 967–979.

Main, M., & Weston, D. R. 1982. Avoidance of the attachment figure in infancy: Descriptions and interpretations. In: *The Place of Attachment in Human Behavior* (C. M. Parkes & J. Stevenson-Hinde, eds.), pp. 31–59. New York: Basic Books.

Malarkey, W. B., Kiecolt-Glaser, J. K., Pearl, D., & Glaser, R. 1994. Hostile behavior during marital conflict alters pituitary and adrenal hormones. *Psychosomatic Medicine*, 56: 41–51.

Mendoza, S. P., & Mason, W. A. 1986. Contrasting responses to intruders and to involuntary separation by monogamous and polygynous New World monkeys. *Physiology and Behavior*, 38: 795–801.

Mikulincer, M., & Florian, V. 1998. The relationship between adult attachment styles and emotional and cognitive reactions to stressful events. In: *Attachment Theory and Close Relationships* (J. A. Simpson & W. S. Rholes, eds.), pp. 143–165. New York: Guilford Press.

Moody, T. W., Merali, Z., & Crawley, J. N. 1988. The effects of anxiolytics and other agent on rat grooming behavior. *Annals of the New York Academy of Sciences,* 525: 281–290.

Newman, J. D. 1991. Vocal manifestations of anxiety and their pharmacological control. In: *Psychopharmacology of Anxiolytics and Antidepressants* (S. E. File, ed.), pp. 251–260. New York: Pergamon Press.

Panksepp, J. 1989. The psychobiology of emotions: The animal side of human feelings. In: *Emotions and the Dual Brain* (G. Gainotti & C. Caltagirone, eds.), pp. 31–55. Berlin: Springer-Verlag.

Panksepp, J., & Bekkedal, M. 1997. Neuropeptides and the varieties of anxiety in the brain. *Giornale Italiano di Psicopatologia,* 1: 18–27.

Panksepp, J., Nelson, E., & Bekkedal, M. 1997. Brain systems for the mediation of social separation-distress and social-reward. *Annals of the New York Academy of Sciences,* 807: 78–100.

Pavani, S., Maestripieri, D., Schino, G., Turillazzi, P. G., & Scucchi, S. 1991. Factors influencing scratching behavior in long-tailed macaques. *Folia primatologica,* 57: 34–38.

Potegal, M. 1994. Aggressive arousal: The amygdala connection. In: *The Dynamics of Aggression* (M. Potegal & J. F. Knutson, eds.), pp. 73–111. Hillsdale, N.J.: Lawrence Erlbaum Associates.

Preuschoft, S. 1992. "Laughter" and "smile" in Barbary macaques (*Macaca sylvanus*). *Ethology,* 91: 220–236.

Preuschoft, S., Paul, A., & Kuester, J. 1998. Dominance styles of female and male Barbary macaques (*Macaca sylvanus*). *Behaviour,* 135: 731–755.

Rasmussen, K. L. R., & Suomi, S. J. 1989. Heart rate and endocrine responses to stress in adolescent male rhesus monkeys on Cayo Santiago. *Puerto Rico Health Science Journal,* 8: 65–71.

Reite, M., & Boccia, M. L. 1994. Physiological aspects of adult attachment. In: *Attachment in Adults* (M. B. Sperling & W. H. Berman, eds.), pp. 98–127. New York: Guilford Press.

Rholes, W. S., Simpson, J. A., & Stevens, J. G. 1998. Attachment orientations, social support, and conflict resolution in close relationships. In: *Attachment Theory and Close Relationships* (J. A. Simpson & W. S. Rholes, eds.), pp. 166–188. New York: Guilford Press.

Rolls, E. T. 1995. A theory of emotion and consciousness, and its application to understanding the neural basis of emotion. In: *The Cognitive Neurosciences* (M. S. Gazzaniga, ed.), pp. 1091–1106. Cambridge: MIT Press.

Rowell, T. E., & Hinde, R. A. 1962. Vocal communication by the rhesus monkey (*Macaca mulatta*). *Proceedings of the Zoological Society of London,* 138: 279–294.

Rubinov, D. R., & Schmidt, P. J. 1996. Androgens, brain, and behavior. *American Journal of Psychiatry,* 153: 974–984.

Sachser, N., & Lick, C. 1991. Social experience, behavior, and stress in guinea pigs. *Physiology and Behavior,* 50: 83–90.

Sachser, N., Durschlag, M., & Hirzel, D. In press. Social relationships and the management of stress. *Psychoneuroendocrinology.*

Sapolsky, R. 1993. Endocrinology alfresco: Psychoendocrine studies of wild baboons. *Recent Progress in Hormone Research,* 48: 437–468.

Sapolsky, R. 1998. *Why Zebras Don't Get Ulcers: A Guide to Stress, Stress-Related Disease and Coping,* 2nd edn. New York: W. H. Freeman.

Schino, G., Scucchi, S., Maestripieri, D., & Turillazzi, P. G. 1988. Allogrooming as a tension-reduction mechanism: A behavioral approach. *American Journal of Primatology,* 16: 43–50.

Schino, G., Perretta, G., Taglioni, A. M., Monaco, V., & Troisi, A. 1996. Primate displacement activities as an ethopharmacological model of anxiety. *Anxiety,* 2: 186–191.

Schuurman, T. 1980. Hormonal correlates of agonistic behavior in adult male rats. *Progress in Brain Research,* 53: 415–420.

Seyfarth, R. M., Cheney, D. L., Harcourt, A. H., & Stewart, K. J. 1994. The acoustic features of gorilla double grunts and their relation to behavior. *American Journal of Primatology,* 33: 31–50.

Silk, J. B., Cheney, D. L., & Seyfarth, R. M. 1996. The form and function of reconciliation among baboons, *Papio cynocephalus ursinus. Animal Behaviour,* 52: 259–268.

Smith, O. A., Astley, C. A., Chesney, M. A., Taylor, D. J., & Spelman, F. A. 1986. Personality, stress and cardiovascular disease: Human and nonhuman primates. In: *Neural Mechanisms and Cardiovascular Disease* (B. Lown, A. Malliani, & M. Prosdomici, eds.), pp. 471–484. Padua, Italy: Liviana Press.

Smucny, D. A., Price, C. S., & Byrne, E. A. 1997. Postconflict affiliation and stress reduction in captive rhesus macaques. *Advances in Ethology*, 32: 157.

Soubriè, P. 1986. Reconciling the role of central serotonin neurons in human and animal behavior. *Behavioral and Brain Sciences*, 9: 319–364.

Spoont, M. R. 1992. Modulatory role of serotonin in neural information processing: Implications for human psychopathology. *Psychological Bulletin*, 112: 330–350.

Troisi, A., & Schino, G. 1987. Environmental and social influences on autogrooming behaviour in a captive group of Java monkeys. *Behaviour*, 100: 292–302.

Vellucci, S. V., Herbert, J., & Keverne, E. B. 1986. The effect of midazolam and beta-carboline carboxylic acid ethyl ester on behavior, steroid hormones and central monoamine metabolytes in social groups of talapoin monkeys. *Psychopharmacology*, 90: 367–372.

Verbeek, P., & de Waal, F. B. M. 1997. Postconflict behavior in captive brown capuchins in the presence and absence of attractive food. *International Journal of Primatology*, 18: 703–725.

von Holst, D. 1998. The concept of stress and its relevance for animal behavior. *Advances in the Study of Behavior*, 27: 1–131.

Watts, D. P. 1995. Post-conflict social events in wild mountain gorillas (*Mammalia, Hominoidea*) I. Social interactions between opponents. *Ethology*, 100: 139–157.

Waxer, P. H. 1977. Nonverbal cues for anxiety: An examination of emotional leakage. *Journal of Abnormal Psychology*, 86: 306–314.

Weaver, A. 1999. The role of attachment in the development of reconciliation in captive brown capuchins, *Cebus apella*. Ph.D. diss., Emory University, Atlanta.

Whiten, A. 1996. When does smart behaviour-reading become mind-reading? In: *Theories of Theories of Mind* (P. Carruthers & P. K. Smith, eds.), pp. 277–292. Cambridge: Cambridge University Press.

Zillmann, D. 1979. *Hostility and Aggression*. Hillsdale, N.J.: Lawrence Erlbaum Associates.

CHAPTER 11

Beyond the Primates

Expanding the Reconciliation Horizon

Gabriele Schino

Female squirrel Leftie spots male Corners approaching the site where she is feeding and immediately charges him. The male flees and zips into a nestbox. Ninety seconds later, Leftie climbs to sit on top of the box into which Corners disappeared and begins uttering soft, chirping calls. Corners can be heard calling antiphonally back out to Leftie, and soon his face is visible in the nestbox entrance. The two squirrels continue to exchange vocalizations for 40 seconds more. Then, Leftie bends over the top of the nestbox, "kisses" the male, and proceeds calmly away.

M. E. Pereira & J. Smith, unpublished observation

This beautiful description of an episode of possible reconciliation in eastern gray squirrels (*Sciurus carolinensis*) exemplifies the state of our knowledge on conflict resolution in non-primates: sparse (although insightful) observations, but very little quantitative data.

A Failure of Scientific Communication

The first paper that dealt explicitly with reconciliation (i.e., a friendly reunion between former opponents) was published by de Waal & van Roosmalen in 1979. It soon appeared evident that reconciliation was a widespread phenomenon. Primatologists confirmed the existence of conciliatory mechanisms in one primate species after the other, from apes to lemurs (Appendix A).

In the 20 years that have elapsed since publication of the first paper on reconciliation, several authors (always primatologists) have suggested that reconciliation should be expected to occur also in non-primate species (e.g., de Waal 1986; Cords & Thurnheer 1993). The rest of the scientific community, however, did not respond to such suggestions, and we had to wait until 1993 to see the first papers that explicitly mentioned reconciliation in non-primates (East et al. 1993; Rowell & Rowell 1993) and 1998 for the first paper that actually presented data on post-conflict affiliation in a non-primate (Schino 1998).

We may wonder about the reasons underlying this reluctance of nonprimatologists to tackle the issue of conflict resolution and relationship repair while among primatologists such research was flourishing.

A first possibility is that nonprimatologists simply did not have access to the relevant literature because most papers on reconciliation were published in specialized primatological journals. To evaluate this possibility I referred to the "Conflict Resolution Bibliography" compiled by the organizers of the Reconciliation Study Group (and available over the Internet at http://www.primate.wisc.edu/pin/rsgbib.html). This bibliography (as updated in August 1997) listed 74 papers on nonhuman primates with specific focus on reconciliation. Only 32 of these were published in specialized primatological journals or in books about primate behavior; the remaining 42 (56.8 percent of the total) were published in general ethological journals or books. Therefore, we can dismiss the possibility that nonprimatologists were not exposed to the primate research on conflict resolution.

A second possibility derives from the widespread belief that primates (all of them) are in some way "special" and that the degree of social sophistication they show is unmatched in the animal kingdom. Nonprimatologists may thus have refrained from tackling a problem they felt would be inappropriate for the species they were studying. Primatologists themselves have endorsed this view (Wrangham 1983), but I would like to argue that it is not based on any well-founded observation (see below). A consequence of this belief is that primatologists and nonprimatologists differ in the kind of research questions they pursue. For example, the study of primate social behavior has been heavily influenced by the work of Robert Hinde and Hans Kummer on interindividual relationships (Hinde 1976, 1979, 1983; Kummer 1978). Detailed investigations of the development and management of primate social relationships derived from this approach. In contrast, the study of interindividual relationships has never been a fashionable topic for nonprimatologists, and Hinde's and Kummer's theorizing has had little

impact on them (for two exceptions see Moss & Poole 1983; Rowell & Rowell 1993). Such differences between primatologists and nonprimatologists have probably been exacerbated by the different traditions of the disciplines in which primate ethologists and non-primate ethologists are usually trained (psychology/anthropology vs. ecology/zoology; see also Rowell, Box 11.1). Interestingly, two out of four of the studies that specifically addressed reconciliation in non-primates were carried out by "former primatologists" (Schino, 1998; Samuels & Flaherty, Box 11.2). It seems that research questions were not able to cross the barrier between primates and non-primates by means of standard scientific communication but had to be helped by scientists themselves who crossed the barrier and began studying different species, bringing their own background and research questions—a real failure of scientific communication.

The Existing Evidence

Anecdotal descriptions of post-conflict affiliation between former opponents have been reported for several non-primate species before the "discovery" of primate reconciliation (mouflon, *Ovis ammon musimon*: Pfeffer 1967; spotted hyena, *Crocuta crocuta*: Kruuk 1972; lion, *Panthera leo*: Schaller 1972; dwarf mongoose, *Helogale undulata rufula*: Rasa 1977). These authors seemed to appreciate the possible conciliatory function of post-conflict affiliation. For example, Schaller (1972, p. 88) wrote, "After a fight the greeting helps to reestablish amicable relations." However, no systematic observation of post-conflict behavior or test of the hypothetical conciliatory function of post-conflict affiliation was carried out.

Following the rise of primate reconciliation studies, Rowell & Rowell (1993) reported that among feral sheep (*Ovis aries*) affiliative behaviors were particularly common in aggressive contexts,

BOX 11.1

The Ethological Approach Precluded Recognition of Reconciliation

Thelma E. Rowell

Reconciliation has been one of those elegant concepts, that, once someone has "discovered" it, leaves everyone else thinking, "That is so obvious, we should have noticed it before." Logically, the process of reconciliation is essential to maintaining permanent groups of animals that, individually, have partly conflicting requirements. Prerequisites are merely that the animals must recognize one another and that group integrity is preserved through the maintenance of relationships between recognized group members. It may also be necessary to specify that conflicts are actually expressed by members of the group—there would be no reconciliation in an utterly despotic group. The function of the group, its particular value to its members, is a separate issue, not relevant to understanding the mechanism by which it is maintained.

The onus of demonstration should therefore shift to anyone who claimed that there was no reconciliation behavior in such a group, and I would then ask: What alternative mechanism is available to maintain the group cohesion that we observe? Nevertheless, reconciliation has hardly been recognized in non-primates (Schino, Chapter 11) even though there are many other mammals and birds that live in permanent groups and form long-term relationships.

The concept of reconciliation became usable when de Waal and co-workers (de Waal & van Roosmalen 1979; de Waal & Yoshihara 1983) defined a method for recognizing the pattern of interaction that could be applied consistently among primates and so lifted it beyond the level of anecdote. Whereas it is possible that the details of their method will not be appropriate even for all primates and will need further modification for more diverse species, the fundamental idea of comparing behavior immediately after a conflict with that in an equivalent period not preceded by a conflict is very

strong. It is, however, incompatible with one of the most fundamental methods in ethology. People well trained in a strong ethological tradition must have needed to make a major conceptual leap to reach the working definition of *reconciliation*.

Ethology originated in attempts to describe and interpret behavior of insects, birds, and fish whose behavior is not readily interpreted empathetically by a human observer. The method that was developed is contextual: for example, a motor pattern that occurs with a fight expresses fear or aggression, or perhaps a particular balance between the two (see Hinde 1970, p. 370). This interpretation derives from the theory of drives, which assumes that an animal is motivated by the need to fulfill immediate deficits such as food or water. In social interaction its behavior is determined by sometimes conflicting tendencies to attack and flee, for example, or in the case of courtship to attack, flee from, and mate with a member of the opposite sex, all of which are responses to the immediate situation. At any time, everything the animal does is driven by the currently overriding motivation. Using this simple system, ethologists have been highly successful in describing and predicting behavior of a wide range of animals, including social mammals. The method was applied to primates, notably in a comparative study by van Hooff (1962) but also by Hinde & Rowell (1962), Moynihan (1967), and Anthoney (1968).

The formal ethological approach to interpreting behavior has not been much used for primates after the early years. In part, this has been simply because interest in motivation as an explanatory tool has become unfashionable. In part, it is because few primatologists have a background in zoology and hence in ethology. Mainly, I think, it is because, since we are primates, we find primate signal systems fairly easy to interpret and thus anthropomorphism is reasonably justified; we have the luxury of being able to take shortcuts in interpretation with reasonable confidence.

The primatologist sees a friendly gesture in the context of a fight. It is obviously a friendly gesture because it is common to many primates, including ourselves. There seems to be no need for a strictly contextual interpretation, and reconciliation is a concept with

BOX 11.1 (continued)

which we easily empathize. The zoologist sees a novel gesture in the context of a fight. It requires interpretation, and in the ethological tradition, the gesture is attributed the same motivation as the adjacent, obviously functional behavior of fight or threat and so is described as a gesture of threat or of dominance or submission.

Zoologists used to watching monkeys looked at a ram rubbing his face on another's horn (which "makes no sense" to primates and requires interpretation) following a clash (which seems clearly to be aggression). They saw a friendly gesture, deferential to age rather than force, and wanted to interpret it, in that context, as reconciliation (Rowell & Rowell 1993)—although the formal test of reconciliation is still lacking. Geist (1971) had described the same gesture as submission, which was an ethologically correct contextual interpretation.

In recognizing the existence of social organization in permanent groups, we have also recognized that animals have long-term as well as short-term agendas. Immediate responses, such as those needed to satisfy hunger or achieve reproduction, may, however, be in conflict with the long-term maintenance of group cohesion that probably enhances reproductive effectiveness. For this reason alone, the simple motivational analysis based on immediate needs is inadequate.

After reading Schino's (1998) article on reconciliation in goats, I turned to Walther's (1984) definitive comparative description of communication among ungulates. This is a classical ethological analysis, and it leaves no doubt that these animals have a range and subtlety of expression that easily match those of primates. There is, however, simply no discussion of individual relationships (beyond the mother-infant pair), and communication among adults is described almost entirely in terms of pairwise interactions, devoid of a larger social context, even though most species live in herds that probably have permanent membership. This seems strange to a modern primatologist, but it should by no means be taken as evidence that long-term or permanent relationships between individuals do not exist among ungulates or, indeed, that Walther was not aware of them. They are simply not a topic that seemed to him to be relevant.

This difference of research focus and emphasis has led to the perception that primates are different from other mammals (Rowell in press), and that perception has in turn led to further separation, as primatologists have not bothered to read non-primate research and vice versa.

A further barrier to appreciating the mechanisms of social organization, such as reconciliation, in nonprimates is our perception of domestic animals, especially the intensely social ungulates. In modern agricultural practice, potential long-term relationships are systematically disrupted and so have been assumed not to exist. Young of all species are usually separated from their mothers as soon as they can be weaned, and nearly all males are slaughtered before adulthood. Females, and the few breeding males, are continually sorted and separated by criteria that have nothing to do with their social history and are rarely allowed to complete a normal life span. Yet, left to become feral, even the most "modern" breeds show themselves capable of forming enduring relationships.

Sheep and other ungulates are highly seasonal in their behavior, so that they show an annual cycle of two or three entirely different patterns of social organization. In particular, they are completely reorganized for the brief mating season, the rut. Patterns of (probable) reconciliation among rams are easy to see in the weeks leading up to the rut, vanish during the rut, and are extremely rare during the rest of the year (personal observation).

Reconciliation may be suppressed under some conditions: whereas Schino (1998) could clearly demonstrate it among goatlings in the absence of their mothers, I have been unable to see it in a small flock of goats with a strongly hierarchical matriarchy. As Schino (Chapter 11) says, absence of evidence is not evidence of absence. It suffices to demonstrate that a species has the capacity for reconciliation behavior under some conditions. The question of which circumstances permit and which inhibit its expression requires a great deal more research, and the answers should help us to understand mechanisms of reconciliation in both primates and non-primates.

and they hypothesized that these behaviors may serve as conciliatory gestures. Affiliative behaviors among sheep could follow as well as precede aggression and could involve third parties. Clearly, a more detailed investigation is needed in order to clarify the relationship between affiliative and agonistic (i.e., both aggressive and submissive) behaviors in sheep and to evaluate their functional significance.

I am aware of only four studies that investigated quantitatively post-conflict behavior and reconciliation in non-primates. In studying captive bottlenose dolphins (*Tursiops truncatus*), Samuels & Flaherty (Box 11.2) reported increased approaches and affiliative interactions following an aggressive conflict. Such interactions generally occurred within one minute of the conflict. Similarly, Hofer & East (Box 11.3) reported increased post-conflict friendly reunions in wild spotted hyenas meeting at the communal den. The likelihood of friendly reunion was not affected by degree of relatedness. Recipients of aggression were primarily responsible for the increased affiliation.

Following a procedure similar to that used by Cords (1988) for longtail macaques (*Macaca fascicularis*), Schino (1998) observed the behavior of female domestic goats (*Capra hircus*) after experimentally induced conflicts. Goats were tested in pairs, and the conflict was induced by offering a very small amount of a highly preferred food (Fig. 11.1). Compared with control observations, a dramatic increase in the frequency of affiliative interactions between former opponents was recorded following a conflict (Fig. 11.2). Conciliatory tendency in these experimental pair tests (calculated following Veenema et al. 1994) was 16.5 percent, within the range reported for primate species.

The post-conflict period was also characterized by frequent renewed aggression and, possibly as a

---

BOX 11.2

Peaceful Conflict Resolution in the Sea?

Amy Samuels & Cindy Flaherty

As attested by many chapters in the present volume, there are abundant examples of peaceful post-conflict behavior among monkeys and apes. The present volume also underscores the scarcity of similar observations in studies of other taxonomic groups (Schino, Chapter 11). Does this reflect a true difference among mammalian taxa? Probably not. More likely, this is evidence of our limitations as humans as we strive to make sense of the behavior of other animals. Rowell (Box 11.1) suggests that this imbalance in our knowledge stems from a greater familiarity with the gestures and social exchanges of our own relatives, the nonhuman primates, and a lesser expertise with the social behavior of other species. Careful observations over time may provide insights into the curious (to us humans) behavior of other species.

Present-day primate research may be able to point the way toward likely candidates for peaceful post-conflict behavior among other species. Among primates that exhibit peaceful conflict resolution, common elements appear to be individual recognition and good memory (de Waal & Yoshihara 1983), characteristics that often result in long-term relationships. This suggests that a logical first step in our quest would be to examine the behavior of animals that share those features.

Thus, we might expect to see peaceful post-conflict behavior among certain toothed whales because they live in social groups characterized by long-term relationships between individual animals. For example, there is a matrilineal basis to social groupings in some species such that females tend to have long-lasting, close associations with female kin (e.g., bottlenose dolphins, *Tursiops* spp.: Wells et al. 1987; Smolker et al. 1992; sperm whales, *Physeter macrocephalus*: Richard et al. 1996). In addition,

BOX 11.2 (continued)

Two adult male bottlenose dolphins engage in "gentle rubbing," an affiliative behavior sometimes seen after aggressive conflicts. Photograph by Mike Greer, courtesy of Chicago Zoological Society.

male bottlenose dolphins often form strong bonds with one or two other adult males, and these relationships may persist for years (Wells et al. 1987; Connor et al. 1992). Natal philopatry of both sexes may occur in such species as killer whales (*Orcinus orca:* Bigg et al. 1990) and pilot whales (*Globicephala melas:* Amos et al. 1993), presumably resulting in lifelong associations among male as well as female relatives.

These long-term relationships suggest that peaceful conflict resolution may be found in some whale and dolphin species. But we have a long way to go before we can understand how cetaceans resolve conflicts. We have little idea about the functions of many of their behaviors, and we know virtually nothing about their agonistic (including aggressive and submissive interactions) and affiliative relationships. Studies of cetacean social behavior have been hampered by more than our unfamiliarity with their gestures and behavioral displays. As terrestrial beings, we humans are at considerable disadvantage in our attempts to study aquatic animals at sea. As a result, our knowledge of cetacean grouping patterns far exceeds our understanding of the intricacies of their social relationships (Samuels & Tyack in press).

Sophisticated research on cetacean social relationships is becoming more widespread as an increasing number of cetacean biologists adopt the systematic sampling methods developed for behavioral studies of terrestrial animals (e.g., Smolker et al. 1993; Samuels & Gifford 1997; Mann & Smuts 1998). Recent studies of bottlenose dolphins, in particular, reveal a complex network of social relationships within a fission/fusion society (e.g., Smolker et al. 1992). Not only do female dolphins have long-lasting relationships with female kin, but they also have changeable associations with unrelated females based on their reproductive condition (Wells et al. 1987). Among independent juvenile offspring, daughters are more likely than sons to maintain an association with their mothers even though both male and female offspring continue to live in their natal home range (Samuels et al. 1996). And, despite the dolphin's mystical reputation and perpetual smile, we know that bottlenose dolphins do have agonistic conflicts. Male bottlenose dolphins that are allies appear to have cooperative relations with some male alliances and competitive relations with others (Connor et al. 1992). In addition, some sexual consortships are aggressively maintained by male alliance partners, while others may be based on affiliative relations between the female and interested males (Connor et al. 1996).

Recent research on agonistic behavior (aggression and submission) among bottlenose dolphins at the Chicago Zoological Society's Brookfield Zoo represents a first attempt to evaluate long-term dominance relationships systematically in a cetacean species (Samuels & Gifford 1997). Large underwater windows at the zoo provided a close-up view of the dolphins' social interactions. Modeled on methods used to assess dominance relations of baboons (e.g., Hausfater 1975), this study revealed behavioral patterns common among other sexually dimorphic mammals. The patterns included higher rates of agonism among males, male dominance over females, and greater stability in the dominance relationships among females. Although the two zoo males behaved like allied males, they also exhibited a changeable dominance relationship with periods of stability and low-level agonism interspersed with episodes of intense competition.

The complexity of dolphin social relationships provides a framework for evaluating post-conflict behavior in cetaceans, and the Brookfield Zoo study provides a

BOX 11.2 (continued)

starting place for looking at this phenomenon. Although Samuels & Gifford's (1997) study was designed to address basic questions about dolphin agonism, we were able to use a subset of data to examine what dolphins did after aggressive conflicts. We compared the behavior of former opponents during the minutes following an aggressive encounter with the behavior of the same pair during the 10 days that preceded and followed each conflict.

Preliminary results are intriguing, despite a small sample size. We found that during a 10-minute period following an agonistic conflict, former opponents approached each other and engaged in affiliative interactions with each other at higher rates than at other times. First approaches after a conflict were typically immediate (within one minute), brief, and without physical contact. Former antagonists sometimes engaged in prolonged affiliative interactions within five minutes after the conflict ended.

Common affiliative behaviors were seen in these post-conflict encounters. For example, former opponents engaged in "gentle rubbing" interactions in which one dolphin rubbed against the other's outstretched

flipper, and the pair took turns giving and taking rubs. Dolphins were also observed to "swim in contact" with each other, with one dolphin leaning a flipper against the other's side and being towed through the water.

We are currently conducting a study that focuses specifically on post-conflict behavior among dolphins. The jury is still out on peaceful conflict resolution among toothed whales until the present study is completed and additional investigations of other cetaceans are undertaken. However, our preliminary findings and the rich social lives of many toothed whales suggest that these studies will be fruitful. In particular, reliance of the toothed whales on acoustic communication (reviewed in Tyack 1986) suggests that looking at both social and acoustic behavior might be particularly useful in understanding peaceful conflict resolution (as demonstrated for baboons in Silk et al. 1996). A review of the social systems of toothed whales reveals both convergence and novel solutions to aquatic life when compared with social systems of terrestrial mammals (Connor et al. 1998). Thus, it is likely that future studies of conflict resolution in these marine mammals will contribute to a broader understanding of this phenomenon among all mammals.

consequence of this increased tension, by increased displacement activities (scratching and self-grooming). Displacement activities are behavior patterns (often body care activities) exhibited by an animal that are "apparently irrelevant" to its ongoing activity (Tinbergen 1952; Zeigler 1964) and are thought to occur in situations of motivational conflict. Maestripieri et al. (1992; see also Schino et al. 1996), after reviewing behavioral, physiological, and pharmacological evidence, proposed that displacement activities may be used as a behavioral measure of anxiety. As with primates (Aureli & van Schaik 1991), friendly post-conflict reunions among goats reduced displacement activities shown by the recipient of aggression (but not by the former aggressor), suggesting that recon-

ciliation may help restore the emotional status of the recipient of aggression.

Finally, van den Bos (1997, 1998) studied post-conflict behavior in confined groups of domestic cats (*Felis silvestris catus*). Like goats and primates, cats showed a temporary increase in displacement activities (self-grooming and scratching) following an aggressive conflict. However, no evidence of post-conflict affiliation between former opponents was found.

Table 11.1 presents a comparison of what is known about post-conflict behavior in nonhuman primates and in four non-primate species. In reviewing the table, one obviously has to keep in mind that, while primate data are based on a variety of studies and species, non-primate data

BOX 11.3

Conflict Management in Female-Dominated Spotted Hyenas

Heribert Hofer & Marion L. East

Among social mammals, spotted hyenas (*Crocuta crocuta*) exhibit a number of unusual traits, including high maternal investment, strong sibling rivalry that may lead to siblicide, female dominance, and the masculinization of female genitalia (Kruuk 1972; Frank et al. 1991; Hofer & East 1993c, 1997; Frank 1997; Golla et al. 1999). By emphasizing the last two traits, spotted hyenas have been portrayed as a species in which females have achieved dominance by becoming larger and more aggressive than males (Gould 1981). Selection for large size and aggressiveness is supposed to have created a "hyperaggressive" species in which evolutionary limits of aggressiveness were reached or exceeded because the evolution of these traits was assumed to be associated with hormonal and morphological changes causing masculinization (Frank 1997).

The evolutionary costs and benefits of masculinization and aggressiveness in spotted hyenas are being debated (reviewed by East et al. 1993; Hofer & East 1995; Frank 1997), yet until recently few field data were available to test predictions of some of the hypotheses put forward. We have studied the social behavior of free-ranging female spotted hyenas in the Serengeti, Tanzania, to test these predictions. Our results reveal that the prevailing view of a near pathological hyperaggressive female society is inappropriate and simplistic.

Spotted hyenas live in a fission/fusion society, so clan members are often on their own or in small groups. Female clan members encounter one another regularly when they come together at the communal den, the social center of the clan, and when they feed at carcasses of medium- to large-sized herbivores. A number of factors create the potential for numerous conflicts among females. These include large group sizes of on average 45 adults and subadults; the strong dependence of reproductive success on (1) access to food resources

mediated by social status and (2) the level of maternal care; and the relationship between social status of a mother and her ability to attend a denning cub (Frank et al. 1995; Hofer & East 1993abc, 1996). Less appreciated is the fact that female spotted hyenas frequently cooperate in the acquisition and defense of carcasses against intra- or interspecific competitors, the defense of territories and communal dens against intruders and predators, hunting, allosuckling, and the formation of coalitions to maintain or advance their social status (Kruuk 1972; Mills 1985; East & Hofer 1991; Knight et al. 1992; East et al. 1993; Hofer & East 1993b, 1998). Thus, each clan female needs to strike a balance between conflicts that arise in competitive situations and the cooperative pursuit of shared interests against other clan members, intruders, interspecific competitors, or predators.

In an earlier study (East et al. 1993) we examined greeting ceremonies in which the erect, masculinized clitoris is used to signal submission. We demonstrated that subordinates use greeting ceremonies to signal submission actively to dominants and to build relationships with other group (clan) females in a manner closely reminiscent of primate grooming. The actual distribution of greeting partners was consistent with the distribution predicted by models of primate affiliative behavior (Seyfarth 1977) that assume that participating individuals value relationships as a resource.

We used focal samples of individual females collected as part of a long-term field study to look at conflict management and test the following predictions. If spotted hyena females value relationships as a resource in a way similar to that of social primates, then a conflict that is behaviorally expressed as an aggressive encounter might damage that relationship (de Waal & van Roosmalen 1979). This hypothesis would predict that (1) opponents should take active steps to repair the damage by initiating neutral and friendly reunions, and thus the rate of these interactions should increase after an aggressive encounter; (2) both winners and losers should show such behavior.

We observed 61 focal females when they arrived at the communal den during a total of 492 hours interacting in 496 dyads with 230 other females (H. Hofer &

BOX 11.3 (continued)

A greeting ceremony between two adult female spotted hyenas in which they examine each other's anogenital region. After conflicts, greeting ceremonies may be initiated by the subordinate opponent as one form of friendly reunion and help repair the damage to the relationship caused by the conflict. Photograph by Marion L. East and Heribert Hofer.

M. L. East unpublished). In 253 out of 267 dyads with conflicts, the outcome was unambiguous and won by the initiator in 99 percent of cases. Conflicts were significantly less likely if two opponents had previously been coalition partners against a third party (39 percent of dyads had subsequent conflicts) compared with if they had not (55 percent) or if they were from the same matriline (45 percent) rather than unrelated (57 percent). The likelihood of conflict was not reduced by prior affiliative behavior, nor was it influenced by the difference in social status. In 15 percent of dyads with conflicts, a friendly reunion (licking the partner, "muzzling," sniffing with friendly vocalizations, e.g., groaning) occurred after about four minutes. Focal females were significantly more likely to increase (more than double) rather than decrease their rate of friendly interactions per dyad after a conflict compared with the period before.

Friendly reunions were initiated by only the loser in 7 percent, by only the winner in 3 percent, and by both of them in 5 percent of dyads; losers were significantly more likely to initiate a friendly reunion than winners. The likelihood of a friendly reunion was independent of

the degree of relatedness or the difference in social status. It significantly increased if after the initial conflict additional conflicts occurred ("escalation"). Established protocols to study conflict resolution (e.g., matched-control observations made the next possible day after post-conflict observations) assume that both partners are reliably present. This may not be the case in free-ranging fission/fusion societies in which weeks could pass before two individuals meet again. With this proviso, a preliminary value of conciliatory tendency using the formula of Veneema et al. (1994) for the small sample of dyads with matched controls available within 30 days is 50 percent (n = 4 pairs) and for dyads with matched controls within 60 days is 33.3 percent (n = 9 pairs).

Our results are consistent with the idea that spotted hyenas use friendly reunions to restore relationships, that is, for the purpose of reconciliation. However, at this preliminary stage we have no direct evidence that friendly reunions do in fact restore relationships. We used aggressive encounters as a behavioral indicator of conflict, but this does not imply that the expression of conflicts requires aggressive behavior. By analyzing behaviors after aggressive encounters, we focused on mechanisms of conflict management that are conspicuous. Yet many conflicts may be "managed" by one partner with tactics that minimize the possibility of an encounter, for instance, by preferring resources that are unlikely to be detected or usurped by dominants. In our population, subordinates are more likely to undertake long-distance foraging trips outside the territory than dominants, thus managing a conflict when food resources are low by avoiding the superior competitor (H. Hofer & M. L. East unpublished data). Such tactics also include avoidance of locations where encounters with dominants are more likely (communal dens, kills, communal resting sites), or visiting them when dominants are unlikely to be present (e.g., during daytime), or vacating sites when dominants arrive (H. Hofer & M. L. East unpublished data). We predict that these tactics are more common in fission/fusion societies such as spotted hyenas, chimpanzees (*Pan troglodytes*: Goodall 1986), or African elephants (*Loxodonta africana*: Moss & Poole 1983) than in many close-knit primate societies. Similarly, conflicts due

BOX 11.3 (continued)

to competition for valuable relationships with attractive (usually dominant) individuals will arise less often in fission/fusion societies than in close-knit societies because competitors are less often present and cannot interfere. We have tested this idea by comparing the distribution of greeting ceremonies in spotted hyenas with the selection of grooming partners in vervet monkeys (*Cercopithecus aethiops*) recorded by Cheney & Seyfarth (1990). As predicted, the fission/fusion society of spotted hyenas allowed greater access to attractive partners than the close-knit society of vervet monkeys (East et al. 1993).

Our findings suggest that spotted hyenas use reconciliation and a variety of other mechanisms to restore relationships and limit escalation of aggression. These and previous results support a change of view of hyena society characterized not simply by hyperbellicosity but by the integration of competition and cooperation where aggression is balanced by mechanisms to prevent it and repair its damage.

FIGURE 11.1. Goats fighting over a preferred food source. Note the small basin between the goats that contains the oats. Photograph by Gabriele Schino.

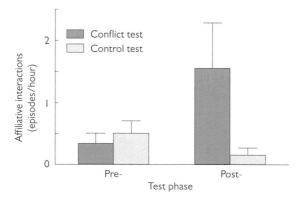

FIGURE 11.2. Rates of affiliative interactions (mean + standard error) in post-conflict and control observations in domestic goats (after Schino 1998). In conflict tests "Pre-" and "Post-" refer to observations preceding and following the induced conflict, respectively. In control tests they refer to corresponding time periods in the absence of induced conflicts.

derive from single studies of single species. They are therefore inherently less reliable and not necessarily representative of other non-primate species. Even with these limitations, the list of similarities is impressive: when information is available, most rows show homogeneous entries.

Two patterns emerge from Table 11.1. First, all group-living species show increased post-conflict affiliation between former opponents; second, all species for which information is available show increased post-conflict displacement activities.

Out of four non-primate species studied thus far—the bottlenose dolphin, spotted hyena, domestic goat, and domestic cat—only the last did not show any evidence of post-conflict affiliation

TABLE 11.1
Comparison of Post-Conflict Behavior in Primates and Non-Primates

	Primates[1]	Goats[2]	Hyenas[3]	Dolphins[4]	Cats[5]
INCREASED AFFILIATION	yes	yes	yes	yes	no
INCREASED RECIPIENT'S DISPLACEMENT ACTIVITIES	yes	yes	?	?	yes
INCREASED AGGRESSOR'S DISPLACEMENT ACTIVITIES	yes	yes	?	?	?
INCREASED RENEWED AGGRESSION	yes	yes	yes	?	no
RECONCILIATION REDUCES RECIPIENT'S DISPLACEMENT ACTIVITIES	yes	yes	?	?	—
RECONCILIATION REDUCES AGGRESSOR'S DISPLACEMENT ACTIVITIES	yes	no	?	?	—
RECONCILIATION REDUCES RENEWED AGGRESSION	yes	?	?	?	—
RECONCILIATION INCREASES SOCIAL TOLERANCE	yes	?	?	?	—
INITIATIVE OF RECONCILIATION	variable	aggressor	recipient	?	—

Sources of data used in the table: 1: Cords & Aureli, Chapter 9; Aureli & Smucny, Chapter 10; Appendix A. 2: Schino 1998. 3: Hofer & East, Box 11.3. 4: Samuels & Flaherty, Box 11.2. 5: van den Bos 1997, 1998.
Note: Throughout the table, "recipient" stands for "recipient of aggression."

between former opponents. Dolphins, hyenas, and goats are all group-living animals that form stable social groups or fission/fusion societies (e.g., Shackleton & Shank 1984; Hofer & East 1993a; Connor et al. 1998). The domestic cat, on the contrary, is an opportunistic species that may aggregate at rich food sources (as seems to be the case for urban stray cats) but probably cannot be considered as a truly social species (MacDonald et al. 1987; Natoli & De Vito 1991). Accordingly, cats lack the mechanisms of conflict regulation typical of social species (van den Bos 1997).

The observation of post-conflict affiliation in species as phylogenetically distant as primates, dolphins, hyenas, and goats implies that conflict-regulation mechanisms must have evolved independently several times in association with the rise of social life. Whether group life and the consequent inevitable frictions are invariably linked to the evolution of conflict-regulation mechanisms or whether only a fraction of group-living species

have evolved post-conflict conciliatory behaviors is a question that awaits further theoretical (van Schaik & Aureli, Chapter 15; see also below) and empirical studies.

The observation of conciliatory behaviors in species that differ widely in their cognitive capacities suggests that reconciliation does not require special abilities such as introspection and self-awareness but that good memory and individual recognition probably suffice (see de Waal & Yoshihara 1983). Obviously, stating that high cognitive abilities are not necessary does not imply that they cannot actually be involved in the process of reconciliation, at least in a few species. In other words, it is unlikely that the cognitive mechanisms underlying reconciliation are entirely the same in goats and chimpanzees (*Pan troglodytes*). Similar functions do not imply similar mechanisms.

All species for which information is available (several primates, goats, and cats) show an increase in the frequency of displacement activities after a conflict. Furthermore, in both goats and primates reconciliation caused a faster return of displacement activities to their baseline frequency. These results suggest that the post-conflict increase in displacement activities may be a general phenomenon and that one of the functions of post-conflict friendly reunions may be a reduction in the anxiety and arousal experienced by the recipient of aggression (see Aureli & Smucny, Chapter 10).

Open Questions

What (If Anything) Is a Friendly Reunion?

Cords (1993, 1994) has repeatedly stressed the importance of distinguishing between "friendly reunion" and "reconciliation," noting that although most primate studies have demonstrated the occurrence of post-conflict friendly reunions, very few have actually investigated the consequences of such reunions and thus justified the use of the more functionally loaded term *reconciliation*.

The same argument can be applied to present and future non-primate studies. These will, however, be likely to suffer from a more fundamental problem: the identification of friendly reunions. In fact, probably because of a lack of interest toward social relationships and social behavior, affiliative behavior patterns are rarely mentioned in ethological studies of non-primate species. For example, when planning my study on reconciliation in goats, I was not able to find a single reference to affiliative behaviors in goats. Such behavior patterns are simply not mentioned in goat ethograms (Hafez et al. 1969; Shank 1972; Shackleton & Shank 1984).

Thus, in many cases, studies on non-primate species will have to determine first which behavioral patterns are to be considered affiliative, a task that may appear to be deceptively simple. In fact, whereas aggressive behaviors are often unambigu-

ously identified by their immediate consequences, this is rarely so for affiliative behaviors. Moreover, whereas primatologists may be helped by empathy toward their subjects (and homology between theirs and their subjects' gestures) in recognizing (for example) allogrooming as a friendly behavior, observers of birds or non-primate mammals may find an equivalent task more problematic (see Rowell, Box 11.1). Contextual or sequential analyses may prove useful but could also be misleading when post-conflict behavior is observed. For example, the stumptail macaque (*Macaca arctoides*) hold-bottom ritual would probably be included among agonistic behaviors by a contextual analysis because of its close temporal relation with aggression (de Waal & Ren 1988).

It is likely that for any given taxon the progress in the study of conflict resolution would be facilitated if integrated with a more general understanding of the dynamics of social behavior and social relationships. For example, among goats friendly reunions were more often initiated by the former aggressor, suggesting that aggressors may be more interested in achieving reconciliation. However, a decrease in post-conflict displacement activities following reunions was observed only in the recipient of aggression. The paucity of information on goats' social dynamics and behavioral ecology makes interpretation of such results problematic.

Is Reconciliation a Homogeneous Phenomenon?

Silk (1996) has recently proposed that primate post-conflict friendly reunions should be viewed more as the signaling of the termination of hostilities than as an attempt to repair a damaged social relationship. Cords & Aureli (1996) have correctly argued that these two functions may in fact be indistinguishable, since signaling the termination of hostilities would allow resumption of friendly interactions and thus, by definition, repair the relationship (see also Silk, Box 9.1). Recent data, furthermore, suggest that not only may primates be interested in such short-term resumption of friendly interactions but they may also be concerned about disturbance of their longer-term relationships. Aureli (1997; see also Aureli & Smucny, Chapter 10) observed that behavioral measures of post-conflict anxiety were higher after conflicts with "friends" than with "nonfriends."

No comparable data exist for any non-primate species. It is possible that such effects of relationship quality on post-conflict anxiety would be limited to species able to conceptualize their social relationships (as macaques seem to do). For example, like macaques, goats show an increase in post-conflict displacement activities. Such an increase, however, may depend more on the concurrent increased probability of renewed aggression than on any concern about long-term relationships. Research both on the neurophysiological basis of post-conflict anxiety and on the cognitive mechanisms underlying reconciliation in different species is badly needed. Although reconciliation may be a phenomenon common to many social species, its proximate mechanisms and short-term consequences may differ widely, but our current knowledge of these comparative aspects of reconciliation is virtually nil.

Are Primates Different?

It is widely believed that primate social behavior is more complex than and qualitatively different from that of non-primates (e.g., Wrangham 1983). As already noted, it is likely that such belief may have hindered the progress in the study of both post-conflict behavior and general social behavior in non-primates.

Whether such differences actually exist with regard to post-conflict behavior is a question that calls for further empirical studies, although the evidence reviewed above suggests that this may not be the case. Other domains of social behavior remain similarly unexplored. For example, in a paper explicitly aimed at comparing the complexities of aggressive coalitions in primates and non-primates, Harcourt (1992) noted that the major differences seem to be in the use that primates (but not, apparently, non-primates) make of affiliative behaviors as a tool for manipulating support and competing for support. The reported evidence was mainly based on Old World monkeys' grooming behavior, and particularly on its distribution according to the social rank of the recipient (Seyfarth 1977). Recent data, however, suggest that grooming in other primate taxa follows different rules (e.g., Parr et al. 1997). Furthermore, as already noted, the attention paid to non-primate affiliative behaviors has been so scarce that the absence of comparable evidence for any non-primate species need not surprise us. In fact, when coalition formation has been looked for, it has been observed also in non-primates (plains zebras, *Equus burchelli:* Schilder 1990; bottlenose dolphins: Connor et al. 1992; spotted hyenas: Zabel et al. 1992; white-nosed coatis, *Nasua narica:* Gompper et al. 1997); in addition, the distribution of affiliation according to the social rank of the recipient seems not to be unique to primates (spotted hyenas: East et al. 1993).

Wrangham (1983) compared social relationships of primates and non-primates in terms of species constancy, individuality, openness of the social network, role of kinship, and complexity of social structure. Although his conclusion was that primates share a unique set of characteristics, most aspects of primate social relationships were found to be present in some non-primate species. Again, absence of evidence seemed much more common than evidence of absence.

This is not to say that cognitive or ecological constraints may not limit the complexities of the social life of non-primates, but we are certainly very far from having identified such limits. It is to be hoped that the recent evidence on conflict resolution among non-primates may promote further studies aimed at investigating how complex (or simple) their social life can be.

Predictions

If friendly post-conflict reunions function as a conflict-resolution mechanism, they can be expected to be especially common among group-living animals. This is more a truism than a prediction, but it is difficult to go much further given our current knowledge of the functional consequences of both relationship deterioration and reconciliation.

It is possible, however, to formulate some tentative predictions about the distribution of conciliatory behavior among group-living species on the basis of the hypothesized costs and benefits for the individuals involved.

For the recipient of aggression, short-term costs of relationship deterioration include increased renewed aggression, observed in primates and goats. Such costs are likely to be higher in those species more easily capable of inflicting physical injuries, such as primates, carnivores, and some horned ungulates. If reconciliation reduces the risk of renewed aggression, then it should be more common in those species in which the risk of physical injury is higher.

Recipients of aggression may avoid short-term costs by temporarily leaving the group. The viability of this option would depend on local ecological conditions (especially predation pressure) but could in general be predicted to be higher for animals living in fission/fusion societies than for members of more cohesive groups. The latter could therefore be predicted to reconcile their conflicts more frequently.

For the aggressor, long-term costs of relationship deterioration may include a decrease in group size below optimal values due to the recipients of aggression leaving the group either temporarily or definitively. Reconciliation may therefore be more common in those species that rely on number for predation protection (primates, social ungulates) or in social carnivores that rely on number for subduing large preys (although the role of group hunting in promoting carnivore sociality is currently much debated; Creel 1997; Packer & Caro 1997). In both cases an active role by the former aggressor in achieving reconciliation could be expected, especially when group size is below an optimum. Generally speaking, increased predation pressure should lead to an increased frequency of reconciliation and to an increased symmetry between aggressor and recipient of aggression in the initiative of reconciliation.

Van Schaik (1989; Sterck et al. 1997; see also van Schaik & Aureli, Chapter 15) modeled ecological influences on female primate social relationships and predicted that reconciliation should be particularly common in species experiencing high levels of within- or between-group contest competition. The extent to which the van Schaik model can be applied to other taxonomic groups is unclear (Sterck et al. 1997). Based on this model, however, social carnivores such as hyenas, which experience high levels of intragroup contest competition, could be tentatively predicted to reconcile their conflicts more often than social ungulates, which exploit more uniformly distributed resources. Among ungulates, browsers such as goats can be predicted to reconcile more than grazers such as cattle or sheep. Similarly, species that defend group feeding territories and therefore presumably experience strong between-group contest competition (e.g., hyenas again) may be particularly likely to have evolved conciliatory mechanisms in order to promote group cohesion.

Available data do not allow the above predictions to be tested. It should be noted, however, that at least four of the species in which reconciliation has been documented live in fission/fusion societies (chimpanzee, bonobo [*Pan paniscus*], spotted hyena, and bottlenose dolphin). Thus, at least one of the above predictions seems *not* to be confirmed.

A proper test of the above predictions should involve quantitative comparisons of conciliatory tendencies. It is hoped that future studies of reconciliation in non-primates will include comparable measures of conciliatory tendencies (Veenema et al. 1994), so as to allow the (admittedly slow)

accumulation of the data necessary to test the predictions of socioecological models.

Conclusions

The study of conflict resolution and, more generally, of relationship management among non-primate species is still in its infancy. As already noted, it is likely that the understanding of conflict-resolution strategies among non-primates would benefit from the more general study of their social relationships (see Rowell & Rowell 1993). Up to now, however, little has been done in this direction. This chapter was originally meant to be a review of existing knowledge on conflict resolution in non-primate species. Eventually, it turned out to be a cry for more data.

References

Amos, B., Schlotterer, C., & Tautz, D. 1993. Social structure of pilot whales revealed by analytical DNA profiling. *Science*, 260: 670–672.

Anthoney, T. R. 1968. The ontogeny of greeting, grooming, and sexual motor patterns in captive baboons (superspecies *Papio cynocephalus*). *Behaviour*, 31: 358–372.

Aureli, F. 1997. Post-conflict anxiety in nonhuman primates: The mediating role of emotion in conflict resolution. *Aggressive Behavior*, 23: 315–328.

Aureli, F., & van Schaik C. P. 1991. Post-conflict behavior in long-tailed macaques (*Macaca fascicularis*): II. Coping with the uncertainty. *Ethology*, 89: 101–114.

Bigg, M. A., Olesiuk, P. F., Ellis, G. M., Ford, J. K. B., & Balcomb, K. C., III. 1990. Social organization and genealogy of resident killer whales (*Orcinus orca*) in the coastal waters of British Columbia and Washington State. *Reports of the International Whaling Commission, Special Issue*, 12: 383–405.

Cheney, D. L., & Seyfarth, R. M. 1990. *How Monkeys See the World*. Chicago: University of Chicago Press.

Connor, R. C., Smolker, R. A., & Richards, A. F. 1992. Dolphin alliances and coalitions. In: *Coalitions and Alliances in Humans and Other Animals* (A. H. Harcourt & F. B. M. de Waal, eds.), pp. 415–443. Oxford: Oxford University Press.

Connor, R. C., Richards, A. F., Smolker, R. A., & Mann, J. 1996. Patterns of female attractiveness in Indian Ocean bottlenose dolphins. *Behaviour*, 133: 37–69.

Connor, R. C., Mann, J., Tyack, P. L., & Whitehead, H. 1998. Social evolution in toothed whales. *Trends in Evolutionary Ecology*, 13: 228–232.

Cords, M. 1988. Resolution of aggressive conflicts by immature long-tailed macaques, *Macaca fascicularis*. *Animal Behaviour*, 36: 1124–1135.

Cords, M. 1993. On operationally defining reconciliation. *American Journal of Primatology*, 29: 255–267.

Cords, M. 1994. Experimental approaches to the study of primate conflict resolution. In: *Current Primatology: Social Development, Learning and Behaviour*, Vol. 2 (J. J. Roeder, B. Thierry, J. R. Anderson, & N. Herrenschmidt, eds.), pp. 127–136. Strasbourg: Université Louis Pasteur.

Cords, M., & Aureli, F. 1996. Reasons for reconciling. *Evolutionary Anthropology*, 5: 42–45.

Cords, M., & Thurnheer, S. 1993. Reconciling with valuable partners in long-tailed macaques. *Ethology*, 93: 315–325.

Creel, S. 1997. Cooperative hunting and group size: Assumptions and currencies. *Animal Behaviour*, 54: 1319–1324.

de Waal, F. B. M. 1986. The integration of dominance and social bonding in primates. *Quarterly Review of Biology*, 61: 459–479.

de Waal, F. B. M., & Ren, R. M. 1988. Comparison of the reconciliation behavior of stumptail and rhesus macaques. *Ethology*, 78: 129–142.

de Waal, F. B. M., & van Roosmalen, A. 1979. Reconciliation and consolation among chimpanzees. *Behavioral Ecology and Sociobiology*, 5: 55–66.

de Waal, F. B. M., & Yoshihara, D. 1983. Reconciliation and redirected affection in rhesus monkeys. *Behaviour*, 85: 224–241.

East, M. L., & Hofer, H. 1991. Loud-calling in a female-dominated society: II. Behavioural contexts and functions of whooping of spotted hyaenas, *Crocuta crocuta*. *Animal Behaviour*, 42: 651–669.

East, M. L., Hofer, H., & Wickler, W. 1993. The erect "penis" is a flag of submission in a female-dominated society: Greetings in Serengeti spotted hyenas. *Behavioral Ecology and Sociobiology*, 33: 355–370.

Frank, L. G. 1997. Evolution of genital masculinization: Why do female hyaenas have such a large "penis"? *Trends in Ecology and Evolution*, 12: 58–62.

Frank, L. G., Glickman, S. E., & Licht, P. 1991. Fatal sibling aggression, precocial development and androgens in neonatal spotted hyaenas. *Science*, 252: 702–704

Frank, L. G., Holekamp, K. E., and Smale, L. 1995. Dominance, demography, and reproductive success of female spotted hyenas In: *Serengeti II: Dynamics, Conservation and Management of an Ecosystem* (A. R. E. Sinclair & P. Arcese, eds.), pp. 364–384. Chicago: University of Chicago Press.

Geist, V. 1971. *Mountain Sheep: A Study in Behavior and Evolution*. Chicago: University of Chicago Press.

Golla, W., Hofer, H., & East, M. L. 1999. Within-litter sibling aggression in spotted hyaenas: Effects of maternal nursing, sex and age. *Animal Behaviour*, 58: 715–726.

Gompper, M. E., Gittleman, J. L., & Wayne, R. K. 1997. Genetic relatedness, coalitions and social behavior of white-nosed coatis, *Nasua narica*. *Animal Behaviour*, 53: 781–797.

Goodall, J. 1986. *The Chimpanzees of Gombe*. Cambridge: Harvard University Press.

Gould, S. J. 1981. Hyena myths and realities. *Natural History*, 90: 16–24.

Hafez, E. S. E., Cairns, R. B., Hulet, C. V., & Scott, J. P. 1969. The behaviour of sheep and goats. In: *The Behaviour of Domestic Animals*, 2nd edn. (E. S. E. Hafez, ed.), pp. 296–348. London: Balliere Tyndall and Cox.

Harcourt, A. H. 1992. Coalitions and alliances: Are primates more complex than non-primates? In: *Coalitions and Alliances in Humans and Other Animals* (A. H. Harcourt & F. B. M. de Waal, eds.), pp. 445–471. Oxford: Oxford University Press.

Hausfater, G. 1975. Dominance and reproduction in baboons (*Papio cynocephalus*): A quantitative analysis. In: *Contributions to Primatology*, Vol. 7. Basel: S. Karger.

Hinde, R. A. 1970. *Animal Behaviour*. 2nd edn. London: McGraw-Hill.

Hinde, R. A. 1976. Interactions, relationships and social structure. *Man*, 11: 1–17.

Hinde, R. A. 1979. *Towards Understanding Relationships*. London: Academic Press.

Hinde, R. A., ed. 1983. *Primate Social Relationships*. Oxford: Blackwell Scientific Publications.

Hinde, R. A., & Rowell, T. E. 1962. Communication by postures, and facial expressions in the rhesus monkey (*Macaca mulatta*). *Proceedings of the Zoological Society of London*, 138: 1–21.

Hofer, H., & East, M. L. 1993a. The commuting system of Serengeti spotted hyaenas: How a predator copes with migratory prey. I. Social organization. *Animal Behaviour*, 46: 547–557.

Hofer, H., & East, M. L. 1993b. The commuting system of Serengeti spotted hyaenas: How a predator copes with migratory prey. II. Intrusion pressure and commuters' space use. *Animal Behaviour*, 46: 559–574.

Hofer, H., & East, M. L. 1993c. The commuting system of Serengeti spotted hyaenas: How a predator copes with migratory prey. III. Attendance and maternal care. *Animal Behaviour*, 46: 575–589.

Hofer, H., & East, M. L. 1995. Virilized sexual genitalia as adaptations of female spotted hyaenas. *Revue Suisse de Zoologie*, 102: 895–906.

Hofer, H., & East, M. L. 1996. The components of parental care and their fitness consequences: A life history perspective. *Verhandlungen der Deutschen Gesellschaft für Zoologie*, 89.2: 149–164.

Hofer, H., & East, M. L. 1997. Skewed offspring sex ratios and sex composition of twin litters in Serengeti spotted hyaenas (*Crocuta crocuta*) are a consequence of siblicide. *Applied Animal Behavior Sciences*, 51: 307–316.

Hofer, H., & East, M. L. 1998. Coalitions, conflicts, and conflict resolution among female spotted hyaenas: Mechanisms and fitness consequences. Abstracts 7th International Behavioral Ecology Congress, Asilomar Conference Grounds, Pacific Grove, Calif., Abstract Book, #183.

Knight, M. H., van Jaarsveld, A., & Mills, M. G. L. 1992. Allo-suckling in spotted hyaenas (*Crocuta crocuta*): An example of behavioural flexibility in carnivores. *African Journal of Ecology*, 30: 245–251.

Kruuk, H. 1972. *The Spotted Hyena*. Chicago: University of Chicago Press.

Kummer, H. 1978. On the value of social relationships to nonhuman primates: A heuristic scheme. *Social Science Information*, 17: 687–705.

MacDonald, D. W., Apps, P. J., Carr, G. M., & Kerby G. 1987. Social dynamics, nursing coalitions and infanticide among farm cats. *Advances in Ethology*, 28: 1–66.

Maestripieri, D., Schino, G., Aureli, F., & Troisi, A. 1992. A modest proposal: Displacement activities as an indicator of emotions in primates. *Animal Behaviour*, 44: 967–979.

Mann, J., & Smuts, B. B. 1998. Natal attraction: Allo-maternal care and mother-infant separations in wild bottlenose dolphins. *Animal Behaviour*, 55: 1–17.

Mills, M. G. L. 1985. Related spotted hyaenas forage together but do not cooperate in rearing young. *Nature*, 316: 61–62.

Moss, C. J., & Poole, J. H. 1983. Relationships and social structure of African elephants. In: *Primate Social Relationships* (R. A. Hinde, ed.), pp. 315–325. Oxford: Blackwell Scientific Publications.

Moynihan, M. 1967. Comparative aspects of communication in New World Primates. In: *Primate Ethology* (D. Morris, ed), pp. 236–266. London: Weidenfeld and Nicolson.

Natoli, E., & De Vito, E. 1991. Agonistic behaviour, dominance rank and copulatory success in a large multi-male feral cat, *Felis catus* L., colony in central Rome. *Animal Behaviour*, 42: 227–241.

Packer, C., & Caro, T. M. 1997. Foraging costs in social carnivores. *Animal Behaviour*, 54: 1317–1318.

Parr, L. A., Matheson, M. D., Bernstein, I. S., & de Waal, F. B. M. 1997. Grooming down the hierarchy: Allo-grooming in captive brown capuchin monkeys, *Cebus apella*. *Animal Behaviour*, 54: 361–367.

Pfeffer, P. 1967. Le mouflon de Corse (*Ovis ammon musimon*): Position systématique, écologie et éthologie comparées. *Mammalia* (Suppl.), 31: 1–262.

Rasa, O. A. E. 1977. The ethology and sociology of the dwarf mongoose (*Helogale undulata rufula*). *Zeitschrift für Tierpsychologie*, 43: 337–406.

Richard, K. R., Dillon, M. C., Whitehead, H., & Wright, J. M. 1996. Patterns of kinship in groups of free-living sperm whales (*Physeter macrocephalus*) revealed by multiple molecular genetic analyses. *Proceedings of the National Academy of Sciences of the USA*, 93: 8792–8795.

Rowell, T. E. In press. The myth of peculiar primates. *Symposium of the Zoological Society of London*, 72.

Rowell, T. E., & Rowell, C. A. 1993. The social organisation of feral *Ovis aries* ram groups in the pre-rut period. *Ethology*, 95: 213–232.

Samuels, A., & Gifford, T. 1997. A quantitative assessment of dominance relations among bottlenose dolphins. *Marine Mammal Science*, 13: 70–99.

Samuels, A., & Tyack, P. L. In press. Flukeprints: A history of studying cetacean societies. In: *Cetacean Societies* (J. Mann, R. C. Connor, P. L. Tyack, & H. Whitehead, eds.). Chicago: University of Chicago Press.

Samuels, A., Richards, A. F., & Mann, J. 1996. Sex difference in the association of wild juvenile bottlenose dolphins with their mothers. In: A. Samuels,

A systematic approach to measuring the social behavior of bottlenose dolphins. Ph.D. diss., Woods Hole Oceanographic Institution.

Schaller, G. B. 1972. *The Serengeti Lion*. Chicago: University of Chicago Press.

Schilder, M. B. H. 1990. Interventions in a herd of semi-captive plains zebras. *Behaviour*, 112: 53–83.

Schino, G. 1998. Reconciliation in domestic goats. *Behaviour*, 135: 343–356.

Schino, G., Perretta, G., Taglioni, A., Monaco, V., & Troisi, A. 1996. Primate displacement activities as an ethopharmacological model of anxiety. *Anxiety*, 2: 186–191.

Seyfarth, R. M. 1977. A model of social grooming among adult female monkeys. *Journal of Theoretical Biology*, 65: 671–698.

Shackleton, D. M., & Shank, C. C. 1984. A review of the social behavior of feral and wild sheep and goats. *Journal of Animal Science*, 58: 500–509.

Shank, C. C. 1972. Some aspects of social behaviour in a population of feral goats (*Capra hircus* L.). *Zeitschrift für Tierpsychologie*, 30: 488–528.

Silk, J. B. 1996. Why do primates reconcile? *Evolutionary Anthropology*, 5: 38–41.

Silk, J. B., Cheney, D. L., & Seyfarth, R. M. 1996. The form and function of postconflict interactions between female baboons. *Animal Behaviour*, 52: 259–268.

Smolker, R. A., Richards, A. F., Connor, R. C., & Pepper, J. W. 1992. Sex differences in patterns of association among Indian Ocean bottlenose dolphins. *Behaviour*, 123: 38–69.

Smolker, R. A., Mann, J., & Smuts, B. B. 1993. Use of signature whistles during separations and reunions by wild bottlenose dolphin mothers and infants. *Behavioral Ecology and Sociobiology*, 33: 393–402.

Sterck, E. H. M., Watts, D. P., & van Schaik, C. P. 1997. The evolution of female social relationships in nonhuman primates. *Behavioral Ecology and Sociobiology*, 41: 291–309.

Tinbergen, N. 1952. "Derived" activities: Their causation, biological significance, origin, and emancipation during evolution. *Quarterly Review of Biology*, 27: 1–32.

Tyack, P.L. 1986. Population biology, social behavior, and communication in whales and dolphins. *Trends in Evolutionary Ecology*, 1: 144–150.

van den Bos, R. 1997. Conflict regulation in groups of domestic cats (*Felis silvestris catus*) living in confinement. *Advances in Ethology*, 32: 149.

van den Bos, R. 1998. Post-conflict stress response in confined group-living cats (*Felis silvestris catus*). *Applied Animal Behaviour Science*, 59: 323–330.

van Hooff, J. A. R. A. M. 1962. Facial expressions in higher primates. *Symposium of the Zoological Society of London*, 8: 97–125.

van Schaik, C. P. 1989. The ecology of social relationships amongst female primates. In: *Comparative Socioecology: The Behavioural Ecology of Humans and Other Mammals* (V. Standen & R. A. Foley, eds.), pp. 195–218. Oxford: Blackwell Scientific Publications.

Veenema, H. C., Das, M., & Aureli, F. 1994. Methodological improvements for the study of reconciliation. *Behavioural Processes*, 31: 29–38.

Walther, F. R. 1984. *Communication and Expression in Hoofed Mammals*. Bloomington: Indiana University Press.

Wells, R. S., Scott, M. D., & Irvine, A. B. 1987. The social structure of free-ranging bottlenose dolphins. In: *Current Mammalogy*, Vol. 1 (H. H. Genoways, ed.), pp. 247–305. New York: Plenum Press.

Wrangham, R. W. 1983. Social relationships in comparative perspective. In: *Primate Social Relationships* (R. A. Hinde, ed.), pp. 325–334. Oxford: Blackwell Scientific Publications.

Zabel, C. J., Glickman, S. E., Frank, L. G., Woodmansee, K. B., & Keppel, G. 1992. Coalition formation in a colony of prepubertal spotted hyaenas. In: *Coalitions and Alliances in Humans and Other Animals* (A. H. Harcourt & F. B. M. de Waal, eds.), pp. 113–135. Oxford: Oxford University Press.

Zeigler, H. P. 1964. Displacement activity and motivational theory: A case study in the history of ethology. *Psychological Bulletin*, 61: 362–376.

CHAPTER 12

A Multicultural View of Peacemaking among Young Children

Marina Butovskaya, Peter Verbeek, Thomas Ljungberg, & Antonella Lunardini

Introduction

Conflict and peace are integral aspects of the human experience; until recently, however, conflict has received more scholarly attention than peace. Much of the past emphasis on conflict has been motivated by the assumption that a better understanding of the causes of conflict and aggression will help us build a peaceful world. From this perspective peace can exist only in the absence of conflict, a view of peace commonly referred to as *negative peace*. During the past decades behavioral scientists have increasingly turned their attention to *positive peace*, that is, the relational processes that actively promote peace through, for instance, avoidance of aggressive confrontation (*peacekeeping*, or *sociative peace*; Gregor 1996) or through relational repair in the aftermath of conflict, violence, and aggression (*peacemaking*, or *restorative peace*; Gregor 1996). To illustrate this trend, there is a growing interest within anthropology in determining what makes peaceful societies peaceful (Howell & Willis 1989; Sponsel & Gregor 1994), and peace psychologists are turning their attention from psychological obstacles to peace (Wessells 1993) to psychological facilitation of peace (Staub 1996).

Ethologists have been among the most active investigators of sociative and restorative peace in nonhuman animals, in particular nonhuman primates. During the past two decades ethologists have identified behavioral strategies that forestall aggression, allow avoidance of confrontation, or promote restoration of relations in the aftermath of aggression (de Waal, Chapter 2; Judge, Chapter 7).

The field of child development mirrors other behavioral disciplines in its approach to conflict and peace. Traditionally concerned with children's individual differences in aggression (e.g., Perry et al. 1992), the field has recently seen an increase in studies on normative patterns of conflict resolution (for a review, see Verbeek et al., Chapter 3).

Learning to resolve conflicts constructively is an important aspect of learning about peace. As such, the increasing body of research on conflict resolution in children has much to offer to students of peace. However, resolution studies commonly focus on the immediate resolution of conflict and tell us nothing about whether young children are likely to make peace with one another with some delay following a conflict-induced separation. In this chapter we review four studies that are among the first to demonstrate that a significant percentage of young children indeed do make peace with their opponents following a conflict-induced separation. The studies were conducted in five cultures: Italy, Kalmykia, Russia, Sweden, and the United States. These cultures present an interesting contrast, and one of the purposes of this chapter is to explore whether children's peacemaking style and tendency partly reflect their society's expectations. The studies were inspired by ethological findings of post-conflict peacemaking in nonhuman primates (*reconciliation*; for reviews, see de Waal 1993; de Waal, Chapter 2). In fact, each study reported here adopted controlled observation methods that were developed to capture post-conflict peacemaking in monkeys and apes (e.g., de Waal & Yoshihara 1983). Throughout our review we compare our findings with the primate literature on post-conflict peacemaking.

The studies reviewed in this chapter focus on peacemaking in peer groups. Peers are potentially rich social resources for the developing child; they are sources of information, assistance, comfort, and shared fun (Booth et al. 1991), and naturalistic observations have shown that most children attempt to optimize peer relations (Hartup

1983). The idea that the peer group may constitute fertile ground for learning about peacemaking has been around for decades (e.g., Dawe 1934; Katz et al. 1992); however, prior to these four studies little or no systematic research on post-conflict peacemaking among young peers has been conducted.

To understand fully the roots of post-conflict peacemaking in early development, we need to measure children's natural behavior as well as their thoughts about peace. Research on children's thoughts about peace is as scarce as systematic research on post-conflict peacemaking, but there is a small literature that provides some insight into the development of a concept of peace in children. Studies conducted in Western cultures suggest that children commonly possess a basic concept of war from about the age of 8 but do not acquire a concept of peace until around the age of 11 (Cooper 1965; Alvik 1968; Rosell 1968; Hakvoort & Oppenheimer 1993). Moreover, developmentally, peace defined by the absence of war and violence (negative peace) appears to precede the concept of peace as defined by cooperation and coexistence (positive peace), which makes its first appearance in early adolescence (Hakvoort & Oppenheimer 1993).

The fact that children in Western cultures commonly acquire a concept of war before they acquire a concept of peace is in line with contemporary theories of concept development. Siegler (1991), for instance, proposes that the number of available exemplars positively predicts the general level of understanding of a natural concept, and artifacts depicting war are considerably more pervasive in contemporary Western society than those depicting peace. From this perspective children of societies in which artifacts of peace are more prevalent than artifacts of war should develop a concept of peace prior to a concept of war. Unfortunately, little research has been done on concept development in non-Western societies that uses the same methods as those used to study this in Western societies, so this assumption remains an empirical question for now. One of the projects reviewed in this chapter sheds some light, however, on young children's perceptions of their peacemaking with peers. The Russian team interviewed both Russian and Kalmyk children about their peacemaking with peers, and we refer to these interviews throughout this chapter.

We start the main portion of this chapter with a brief introduction to the methods and sample characteristics of the four studies. We then proceed with a discussion of differences and similarities in children's peacemaking across the different cultures and age groups. We conclude the chapter with thoughts about future directions for this new line of research.

The Four Studies

The Russian Study

Three peer groups were studied by the Russian team, two consisting of Kalmyk children and one of Russian children. The Republic of Kalmykia is part of Russia and is situated in the northern Caucasian steppe west of the Caspian Sea. Kalmyks, who are descendants of the Oirats, or Western Mongols, migrated there from the lower Irtysh, in southern Siberia (Erdniev 1980). Traditionally nomadic herdsmen, Kalmyks now generally live sedentary lives in villages and towns. Kalmyks have professed Lamaism (a northern version of Buddhism) since the 1500s. Kalmyks tend to be well informed about their genealogies, and village and town people often regard themselves as relations.

The first group of Kalmyk children, consisting of 9 girls and 11 boys aged six to seven years, was observed in an elementary school in a large city, and the second group, including 8 girls and 9 boys

of the same age, was observed in a small village (Butovskaya & Kozintsev unpublished data). Children in both groups were ethnic Kalmyks from middle-class homes.

The Russian children were observed at an elementary school in a large city and included 12 girls and 13 boys aged six to seven years (Butovskaya & Kozintsev 1998, 1999). Most children in this group were ethnic Russians and came from middle-class families.

The U.S. Study

The U.S. study was conducted at a private preschool located in a suburb of a large city in the Southeast (Verbeek 1997). The 56 girls and 64 boys in the study were enrolled in six classrooms and ranged in age from three to five years. The majority of the children were European American and came from middle- and upper-income homes.

The Swedish Study

The Swedish research was carried out at three day-care centers located in and near a large city in southern Sweden (Lindqvist-Forsberg, Ljungberg, Jansson, & Westlund unpublished data; Ljungberg et al. 1999). A total of 22 girls and 34 boys, aged three to six years, were observed. Most children were from Swedish middle-class homes.

The Italian Study

The Italian research was conducted at two preschools in a medium-sized city in northern Italy (Lunardini unpublished data). This project included the youngest subjects of the four studies: 6 girls and 17 boys, all two-year-olds. The children were from middle-class Italian homes.

Procedures

In the Russian, U.S., and Italian studies, observations were conducted by trained observers during periods of free play, both in classrooms and on outside playgrounds. In the Swedish study free-play interactions were videotaped and subsequently scored for conflict and post-conflict interactions. Interobserver and intercoder agreement was determined in each of the four projects and was generally within acceptable range.

The Russian, U.S., and Italian research used the PC/MC method of observation (de Waal & Yoshihara 1983; Veenema, Box 2.1). In this method a focal observation is conducted on a child immediately following a conflict in which the child participated (*post-conflict observation*, or *PC*). The PC is matched to a control observation of the same duration on the same child, in the same context, during the next possible observation day, and starting at the same time as the PC (*matched-control observation*, or *MC*). Peacemaking, defined as a friendly post-conflict reunion between former opponents (e.g., invitation to play; apology; object offer; hug), is determined by comparing the first friendly interaction between opponents recorded in the PC and the MC. If a friendly interaction took place earlier, or only, in the PC, the opponent pair is said to be "attracted." If the friendly interaction took place earlier, or only, in the MC, the pair is said to be "dispersed." An opponent pair is considered "neutral" when no interaction occurred in either the PC or the MC, or when the interaction occurred at the same time in both observations. Attracted and dispersed pairs are then tallied across all observations, and if the ratio of attracted and dispersed pairs differs significantly from the 1:1 ratio expected by chance and favors attracted pairs, post-conflict reunion is inferred.

This method of determining post-conflict reunions also allows the determination of a *reunion tendency index* (R) for either groups or individuals. The following formula (Veenema et al. 1994) was used to determine the R for the Russian, U.S., and Italian groups:

$$\frac{(\text{Attracted pairs} - \text{dispersed pairs})}{\text{Total pairs}}$$

The U.S. research differed from the Russian and Italian studies in the way peacemaking was conceptualized. Conflict outcome in the U.S. study was recorded as either *together* or *separate* as a function of whether the opponents stayed together after the end of the conflict or went their separate ways (cf. Sackin & Thelen 1984; see also Verbeek et al., Chapter 3). In the original study together outcomes were analyzed separately, and post-conflict peacemaking was measured following separate outcomes. For the purpose of this chapter, together outcomes were treated as attracted pairs to allow comparison with the Russian and Italian studies. (Note: In the U.S. study, opponents were never together at the onset of MCs.)

Additional focal observations were conducted by the Russian and the U.S. team to determine relationships among the children. Children were classified as *friends, acquaintances,* or *nonfriends* according to how often they interacted.

The Swedish team found the PC/MC method to be inappropriate for the specific setting (see Ljungberg et al. 1999). The classrooms at the day-care center were small, and children had little or no opportunity to avoid one another or to withdraw to a secluded area. As a consequence affiliative behavior was quite intense also in "control" sessions. The Swedish team decided instead to score the first post-conflict affiliative interaction as "peacemaking," but only when the recipient of the conciliatory gesture in fact accepted it as such.

Both Russian and Kalmyk children were interviewed at the end of the study after all observations were completed. The children were interviewed individually by the first author and were asked questions such as: "If you start a conflict, should you be the one to make up?" or "How long will you wait before you make up?" The interview was open ended, and the children were free to answer in as much detail as they wanted.

Peacemaking Tendency

In each study in which the PC/MC method was feasible, it generated remarkably consistent results. In the Italian, Russian, and U.S. studies, the ratio of attracted and dispersed pairs differed significantly from the 1:1 ratio expected by chance. Post-conflict peacemaking, defined in the PC/MC method as a friendly post-conflict reunion between opponents, was thus demonstrated in a statistically significant way among Italian, Kalmyk, Russian, and U.S. children (Table 12.1). Moreover, in the Swedish study post-conflict peacemaking was evidenced by the fact that former opponents were significantly more likely to seek each other out for peaceful interaction than children who had not engaged in conflict. In each study that reported significant PC/MC findings post-conflict reunions occurred during the first four minutes after conflict, and the majority of reunions occurred during the first two minutes after conflict. A similar trend was found in the Swedish study (see respective studies for further details).

We found that older children were generally more inclined to make peace than younger children. However, considering that conditions varied significantly across the various studies, the developmental findings discussed below should be viewed as tentative at best. Perhaps the most convincing age trend comes from the Swedish study: post-conflict peacemaking was more than 60 percent in five- and six-year-olds and 42 percent

TABLE 12.1

Post-Conflict Peacemaking in Five Cultures

Subject Sample	Pairs			R^a
	Attracted	Dispersed	Neutral	
RUSSIAN CHILDREN (AGE 6–7)				
friends	23	10	9	31
acquaintances	19	0	14	58
nonfriends	33	1	50	38
girls	21	4	9	50
boys	22	6	28	29
mixed	32	1	36	46
total	75	11	73	40
U.S. CHILDREN (AGE 3–5)				
friends	17	3	2	64
acquaintances	31	7	10	50
nonfriends	105	14	84	45
girls	32	6	13	51
boys	86	14	39	52
mixed	35	4	44	37
total	153	24	96	47
KALMYK CHILDREN (AGE 6–7)				
girls	12	1	0	85
boys	68	9	12	66
mixed	36	1	12	71
total	116	11	24	70
ITALIAN CHILDREN (AGE 2)				
total	65	15	94	29
SWEDISH CHILDREN (AGE 3–6)				
girls				61[b]
boys				60[b]

[a]Post-conflict reunion tendency: $\dfrac{\text{(Attracted pairs} - \text{dispersed pairs)}}{\text{Total pairs}}$

[b]Percentage accepted conciliation in a post-conflict observation.

in four-year-olds, compared with only 17 percent in three-year-olds. The peacemaking tendency of the Italian two-year-olds (29 percent) contrasted with 40 percent in the Russian children (six- to seven-year-olds) and 70 percent in the Kalmyk children (six- to seven-year-olds). Interestingly, the overall peacemaking tendency of the U.S. preschoolers (three- to five-year-olds; 47 percent) was rather similar to that of the Russian elementary school children (six- to seven-year-olds; 40 percent). As mentioned earlier, the U.S. study distinguished between two peacemaking strategies: immediate peacemaking (staying together after conflict) and delayed peacemaking (friendly reunions following conflict-induced separation), and age was related to choice of peacemaking strategy. The older U.S. preschoolers were more likely to make peace immediately, while the younger ones were more inclined to make peace with some delay (Verbeek 1997).

When compared with their Russian age-mates, the Kalmyk elementary school children showed a significantly higher peacemaking tendency, which may, in part, reflect their society's expectations. Kalmyk society is characterized by mutual responsibility, kin preferences, and close communal relations (Erdniev 1980; Solso 1988; Ionin, 1996). The majority of Kalmyks are Buddhist, and artifacts relating to peaceful coexistence figure prominently in Kalmyk culture (Butovskaya & Guchinova 1998). Kalmyk culture may thus facilitate the development of a concept of peace in young children.

It is also likely that the adults of Kalmyk society influence children's tendency to make peace, through either direct instruction or modeling. Although it is often difficult to pinpoint the specific psychological mechanisms involved, links between adult and child conflict behavior have been well documented. Research conducted early in the previous century, for instance, showed that

whereas the Great Whale River Eskimos stressed peace and harmony in their social relationships and taught their children accordingly (Honigmann 1954), the Mundugumor of eastern New Guinea encouraged their children to be aggressive and emotionally unresponsive to others (Mead 1935). In a more recent study Fry (1988) found that children from a Mexican Zapotec community that was described as peaceful were significantly less aggressive than children from a neighboring Zapotec community that was described as competitive and violent-prone (see also Fry, Chapter 16).

To summarize this section, we did not set out to compare age or cultural influences systematically, and our developmental and cross-cultural findings should thus be viewed with caution. However, each of our studies that used the method with controlled observation (PC/MC method) consistently demonstrated a tendency in young children to make peace with their peers following a conflict-induced separation. This is an important finding when we consider that conflict-resolution studies usually tend to report destructive, rather than constructive, patterns of conflict resolution during early and middle childhood (e.g., Laursen et al. 1998; see Verbeek et al., Chapter 3). Our consistent findings of conciliatory reunions thus suggest that young children may be considerably more constructive in their approach to peer conflict than one would infer from the existing conflict-resolution literature.

How Do Children Make Peace?

Nonhuman primates show considerable variation in post-conflict peacemaking behavior. Some species use conspicuous behaviors that are rarely seen in other social contexts, while other species have a more implicit style of peacemaking (de Waal 1993). The children in our studies used both explicit and implicit strategies to make peace with

FIGURE 12.1. An example of post-conflict peacemaking in Russian elementary school children. The aggressor initiates a friendly contact by offering the object of dispute to his opponent. The aggressor maintains eye contact with his opponent and has a remorseful expression on his face. The opponent remains motionless, and his expression includes downcast eyes and pouted lips. Illustration by T. N. Shmelioffa from a photograph by M. Butovskaya.

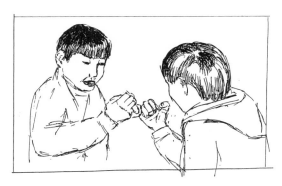

FIGURE 12.2. Ritualized peacemaking in Kalmyk elementary school boys. The boys are reciting a *mirilka* while smiling and holding hands. Illustration by T. N. Shmelioffa from a photograph by M. Butovskaya.

their peers. In the case of the U.S. children, for instance, explicit conciliatory behavior (e.g., apologies, offers to share the disputed object, hugs) strongly predicted continued peaceful interaction, in particular when the opponents were friends and when they had been playing together prior to their conflict. Explicit behavior was also effective in reuniting opponents after an initial separation.

However, post-conflict peacemaking strategies were also often more implicit, for instance, when former opponents resumed friendly play without further reference to the previous conflict.

Invitations to resume play also accounted for about half of the post-conflict contacts in the Swedish study, and they were accepted in two-thirds of the cases. Explicit strategies were also common among the Swedish children, but apologies were less effective: less than 40 percent were accepted by the recipient. Many of the conflicts of the Italian children centered around the possession of objects, a common cause of conflict for children of this young age. It is thus not surprising that object offers were effective in restoring peace among these young children, and so were invitations to resume play (Fig. 12.1).

Like the Italian, Swedish, and U.S. children, the Kalmyk and Russian children employed both explicit and implicit peacemaking behavior. An explicit strategy that was exclusively observed among Kalmyk and Russian children was the use of ritualized peacemaking rhymes (*mirilka* in Russian). Former opponents commonly held hands when they recited peacemaking rhymes to each other such as: "Make peace, make peace, don't fight, if you fight, I'll bite, and we can't bite since we're friends," or "If you bite, I'll hit you with a brick, but the brick breaks, and friendship begins again" (Fig. 12.2). In some instances other group members brought former opponents together and encouraged them to recite peacemaking rhymes to each other.

Although peacemaking rhymes were exclusively observed among Kalmyk and Russian children, the Swedish children also occasionally used rhymes as a peacemaking device. The children in the Swedish study often used counting rhymes to determine who would be the chaser in a game of chase, or how to distribute toys among themselves. On a few occasions counting rhymes were used as a peacemaking strategy by rigging the outcome so as to benefit a former opponent. Thus whereas the Kalmyk and Russian children used rhymes as an explicit peacemaking tool, the Swedish children employed rhymes as a more implicit means toward peace.

The use of peacemaking rhymes has also been observed among adults in hunter-gatherer and early agriculturist societies (Strathern 1971; Koch 1974; Eibl-Eibesfeldt 1979). However, adults in these societies use peacemaking rhymes exclusively to settle conflicts with neighboring communities and not to settle conflicts with group members.

Across the five cultures girls and boys did not differ significantly in their peacemaking tendency (Table 12.1). However, girls and boys sometimes differed in the acts and gestures they used to make peace. The Swedish study, for instance, found that while invitations to resume play were significantly more common among Swedish girls, object offers, apologies, and acting silly were more typical for Swedish boys. An example of acting silly as a peacemaking strategy is described in the U.S. study: a three-year-old girl did not allow a three-year-old boy to join her in play with dress-up clothes and aggressively chased him away. The boy retreated to another area in the playroom and played independently for a short while. He then walked over to the girl and made a funny face at her with the apparent intent to make her laugh. The girl broke into a broad smile and allowed him to join her in playing with the dress-up clothes. Acting silly to make others laugh may be a male strategy. Interviews of Russian adolescents, for instance, revealed that playing the fool enhanced the social status of boys but not girls (Butovskaya & Kozintsev unpublished data). Moreover, girls may be especially responsive to clowning by boys. Research in adults has shown that women laugh and smile more often when they are observers rather than actors, especially when the speaker is male (Provine 1993).

Third-Party Mediation

Although U.S. and Swedish children regularly intervened in the conflicts of others, they usually did not do so as mediators but rather as supporters of one of the opponents. Peer mediation was also rare among Russian and Kalmyk children, but it was effective when it occurred. Kalmyk children who acted as mediators were observed to push the opponents apart and to persuade them to stop quarreling. On a few occasions a mediator took one of the opponents by the hand and led him to the other, urging them both to make peace. Both Russian and Kalmyk children stated during the interviews that they viewed mediation as an honorable thing to do.

In all five cultures teachers intervened in children's conflict, but they were not always successful in mediating a peaceful outcome. In fact, in the Italian study teacher intervention at times resulted in a resumption of aggressive conflict rather than in peaceful interaction. Teacher mediation in the U.S. study did not affect the likelihood of peacemaking. This finding resembled a pilot study by the second author on U.S. preschoolers in which it was found that teacher mediation was highly ineffective in bringing about peaceful outcomes (Verbeek & Creveling unpublished data). In summary, our studies suggest that although adult and peer intervention may positively affect children's peacemaking, the rate of occurrence as well as the effectiveness of third-party intervention are likely to vary as a function of situational factors (cf. Verbeek et al., Chapter 3).

Who Made Peace?

Among both the U.S. and Swedish children the initiative to make peace was evenly divided between initiators and recipients of conflict. In contrast, about two-thirds of the peaceful reunions among the Russian and Kalmyk children were initiated by the child who had started the conflict. A similar pattern was found for the Italian toddlers.

In nonhuman primate species with a despotic dominance style, post-conflict peacemaking is commonly initiated by aggressors (de Waal & Yoshihara 1983). Subordinates are often fearful of dominants and are hesitant to approach them. In species with a more tolerant dominance style, it is expected that the recipient of aggression initiates post-conflict reunions; the recipient's greater need for stress reduction immediately after conflict is believed to motivate its peacemaking effort (Aureli et al. 1989). There are exceptions to this pattern, however. For instance, in bonobos (*Pan paniscus*), generally considered a tolerant species, aggressors show most of the initiative (de Waal 1993), whereas in longtail macaques (*Macaca fascicularis*), a despotic species, recipients of aggression commonly initiate post-conflict reunion (Aureli et al. 1989).

In our studies cultural expectations may have affected peacemaking initiative. For instance, both the Russian and Kalmyk children told the researchers during the interviews that they believed that reconciliation should be initiated by the individual who started the conflict. Most Russian and Kalmyk children also stated that if they were at the receiving end of a conflict their friend would intervene on their behalf and urge the aggressor to apologize. Considering that the Russian and Kalmyk children were older than the U.S. and Swedish children, a more developed sense of remorse may also have played a role in the aggressors' motivation to make peace.

Why Make Peace?

Evidence gathered in captive and natural settings in more than 20 Old and New World species strongly suggests that post-conflict peacemaking in nonhuman primates has a biological basis. But why has natural selection favored behaviors that promote peaceful coexistence? One explanation may be that checks on aggression are adaptive, because rampant aggression limits the chances of successful reproduction and jeopardizes the survival of aggressors and recipients of aggression alike (de Waal 1993). There are other reasons why peacekeeping and peacemaking may be adaptive. "Hawkish" competitive strategies are costly—in terms of both energy expenditure and risk of injury or death (Maynard Smith and Price 1973; Matsumura & Okamoto, Box 5.1). And since cooperative relationships—in theory at least—are likely to enhance the chances of survival and reproductive success of individuals, they are worth protecting from damage caused by conflicts of interests that inevitably arise in social groups (de Waal, Chapter 2).

The latter hypothesis, known as the Valuable Relationships Hypothesis, has been tested extensively by primatologists and with relative success; individuals with close bonds generally tend to make peace more often than individuals who do not maintain a close relationship (de Waal, Chapter 2; Cords & Aureli, Chapter 9). However, the long-term effect of post-conflict peacemaking on close relationships remains to be determined (for a discussion, see Cords & Aureli 1996; Silk 1996). What further complicates research on this issue is the fact that close relationships are dynamic dyadic processes. Close relationships are initiated, maintained, and terminated, and the significance of post-conflict peacemaking can be expected to vary as a function of the particular stage of development of a given relationship. These and other questions have inspired an alternative view with regard to the biological basis of post-conflict peacemaking. Silk (Box 9.1) has proposed that, given the state of the evidence, a more parsimonious explanation of post-conflict interactions is that they serve to signal that no further aggressive actions will follow.

In order to investigate the Valuable Relationships Hypothesis, both the Russian and the U.S. team determined friendship relations among the children and compared the reunion tendency of friends and nonfriends. Dyads were classified as friends, acquaintances, or nonfriends, as a function of time spent together. Children classified as acquaintances interact differently than those classified as friends. For instance, acquaintances know less about each other than friends (Ladd & Emerson 1984) and refer less frequently to common activities in describing their relationship (Hayes et al. 1980). Moreover, Hartup et al. (1988) showed that acquaintances engaged in more intense conflicts and were less likely to use conciliatory resolution strategies than friends. The distinction between friends and acquaintances is thus an important one.

As illustrated in Table 12.1, in the Russian study acquaintances were most likely to engage in post-conflict peacemaking, and friends were least likely to do so. Conversely, in the U.S. study the relationship of the opponents did not predict post-conflict reunion tendency. When friends and acquaintances are combined, the Russian and U.S. studies show comparable reunion tendencies for close associates: $R = 43$ and $R = 54$, respectively.

Taken together, these results do not provide clear evidence that post-conflict peacemaking specifically aims at restoring close relationships among young children. In both the Russian and the U.S. sample nonfriends were just as likely as close associates to make peace after conflict. By distinguishing between immediate peacemaking (together outcomes) and delayed peacemaking (post-conflict reunions), the U.S. study offered a more detailed insight into the peacemaking tendencies of friends and nonfriends (Verbeek 1997).

Friends in the U.S. study were more likely to stay together through conflict than nonfriends. Nonfriends, in contrast, were more likely to engage in post-conflict peacemaking, whereas acquaintances engaged in a more or less even measure of immediate and delayed peacemaking.

What do the U.S. and Russian findings tell us? Perhaps the simplest way of explaining these results is to suggest that both *interactions* and *relationships* matter to young children and that young children's peacemaking reflects children's motivation to repair damage to both. Friends may be motivated to protect their existing relationship from damage caused by conflict, whereas nonfriends may be motivated to restore peaceful interaction simply because playing together is more fun than playing alone. Peacemaking between acquaintances may reflect the desire of one child to protect his or her relationship with another child, or it may reflect a relationship in transition. Proximate factors such as post-conflict distress and a corresponding motivation to restore a sense of well-being may mediate post-conflict peacemaking in friends and nonfriends alike (Aureli & Smucny, Chapter 10).

Conclusions

In this chapter we reviewed evidence of post-conflict peacemaking among young children from five different cultures. This multicultural evidence is the first of its kind and complements ethological findings of post-conflict peacemaking in nonhuman primates. Differences among the studies with respect to both context and methodology precluded systematic comparisons, but these comparative data nevertheless suggest that children's tendency to make peace with peers may increase with age and may be linked to cultural expectations.

Our studies showed that reunions between former opponents generally occurred within the first two minutes after the end of the conflict. This timing is remarkably similar to the timing of post-conflict peacemaking in most nonhuman primate species (Kappeler & van Schaik 1992). There were other similarities with peacemaking in nonhuman primates. For instance, species differ in conciliatory behavior, and so did the children from the various countries. We observed culture-specific behavior such as peacemaking rhymes, as well as more universal conciliatory behaviors, such as hugs and object offers.

Nonhuman primates species vary in peacemaking tendency, and so did the children from the different cultures. When we control for age, Kalmyk children appear to be the most peace-loving, and this may reflect the emphasis placed on peaceful coexistence in their culture. Kalmyk children also successfully mediated peace among their peers and believed that a child who is a target of aggression should always be defended.

Post-conflict peacemaking in adult nonhuman primates is often the domain of individuals who share a valuable relationship; alliance partners or close kin are usually more inclined to make peace than opponents who do not share a close relationship (Cords & Aureli, Chapter 9). Moreover, certain relationships may accrue value during development. Juvenile longtail macaques, for instance, do not preferentially reconcile with kin (Cords 1988; Cords & Aureli 1993), whereas adult individuals usually do (Aureli et al. 1989; Aureli et al. 1997). Our studies showed that among Russian and U.S. children, friends and nonfriends were equally likely to engage in post-conflict peacemaking. However, the U.S. study showed that friends were more likely than nonfriends to make peace immediately by staying together after conflict. Peacemaking among young children in these

---

BOX 12.1

Post-Tantrum Affiliation with Parents

The Ontogeny of Reconciliation

Michael Potegal

Tantrums are a common, if not ubiquitous, phenomenon of early childhood. Tantrums also occur in infant and juvenile nonhuman primates (e.g., Weaver & de Waal, Box 10.2). Trivers (1985) proposes that tantrums may be a solution to a basic intergenerational conflict: the young benefit by maintaining parental nurturance as long as possible while parents' inclusive fitness is eventually better served in other ways. When a juvenile primate has a tantrum following maternal rejection (e.g., while being weaned), its mother usually accepts it (e.g., de Waal 1996). Historically, the "terrible twos" are the age of weaning in Western culture, as it remains in other cultures to this day. A disposition to have tantrums that develop at an age when weaning occurred over most of human history might well reflect an evolutionary strategy to retain parental attention.

If tantrums are a behavioral adaptation, the necessity for mother and offspring to maintain social bonds might predispose a potential for happy endings. Highlighting an ontogenetically very early form of reconciliation in human children, an initial survey carried out in England by Einon & Potegal (1994) found 35 percent of tantrums reportedly ending in a cuddle. Sometimes the cuddle was solicited or initiated by the child. However, the modal post-tantrum response was for the child to behave "as if nothing had happened." Although Trivers's (1985) hypothesis provides a framework of the evolutionary processes, proximal behavioral/motivational mechanisms affecting post-tantrum affiliation require elucidation.

The recent survey of Potegal et al. (1996) suggests that tantrums consist of at least two partially

BOX 12.1 (continued)

independent emotional/behavioral processes: anger and distress. The reconciliation literature indicates that in some species the instigator (not the recipient) of aggression typically initiates reconciliation (de Waal 1993). The supposition of a similarity between anger in humans and aggression in other animals (Blanchard 1984) implies that the likelihood or intensity of affiliation after a tantrum may be a function of anger expressed during the tantrum. An alternative hypothesis is suggested by the presence of distress in tantrums, reflected primarily in crying. Crying grows during the tantrum with a maximum closer to the tantrum's end. Post-tantrum initiation or solicitation of verbal or physical interaction, or both, may be a request for proximity, contact comfort, and reassurance to relieve the immediately preceding distress. Thus, a reasonable alternative to an anger-reconciliation linkage is that distress during the tantrum drives affiliation afterward.

As detailed by Potegal & Davidson (1997), our recent survey highlighted age-related changes in tantrums by selecting noncontiguous groups of children aged 18–24 months, 30–36 months, 42–48 months, and 54–60 months. A telephone interview with their parents determined if the selected child had tantrums, defined as episodes of dropping to the floor, shouting, screaming, crying, pushing/pulling, stamping, hitting, kicking, throwing something, or running away more frequently than once a month. Parents agreeing to provide a written description of one of the child's tantrums received a form on which to report the sequence and duration of behavior observed. They also recorded visible facial autonomic signs (flushing, sweating, drooling, and running nose) and respiratory distress (gasping, panting, and so forth). Each of 331 usable narratives, representing about a third of the largely white, college-educated, middle-class parents interviewed, was divided into consecutive 30-second intervals. Behavior in each of 13 different categories was scored as occurring or not within each interval.

The primary analysis suggested that tantrums consist of at least two emotional/behavioral processes, anger and distress, with different temporal signatures. Hitting, kicking, pulling and pushing, throwing things, and stamping are gross, vigorous behaviors indicating anger.

Their time courses all peaked at tantrum onset and then declined. The opposite pattern was shown by crying that increased progressively throughout the tantrum. A principal component analysis of behavior durations confirmed and extended this model. Angry behaviors cohered within components at three intensity levels. High Anger was defined by high loadings on kicking, stiffening the body, hitting, and screaming. Intermediate Anger was defined by shouting and throwing; Low Anger, by stamping. High Anger strongly correlated with tantrum duration, Intermediate Anger less strongly so, whereas Low Anger was uncorrelated with duration. Distress, the fourth component, was defined by loadings on whining, crying, and reassurance seeking during the tantrum.

Instances in which the child initiated or solicited physical contact (e.g., raising the arms indicating a wish to be picked up) within a few minutes after the tantrum were scored as physical post-tantrum affiliation. If the child initiated a verbal exchange with the parent (including, but not limited to, apologies), verbal post-tantrum affiliation was scored. Each child was scored as demonstrating neither, one, or both forms of post-tantrum affiliation. Ambiguous cases and cases of simultaneous parent/child affiliation were not counted. Child-initiated post-tantrum affiliation occurred following 29 percent of tantrums. Comparison with the 35 percent post-tantrum cuddle rate reported by Einon & Potegal (1994) suggests that the majority of post-tantrum reconciliations are initiated by the child.

Four factors were found to affect the probability of post-tantrum affiliation:

1. Post-tantrum affiliation increased with age. The distress characterizing the tantrum's end persists longer as children develop, so that they may be more driven to seek comfort. Other possible contributors to the age effect include the emergence of children's shame, guilt, and embarrassment in the period between 12 to 36 months and their increasing understanding of the rules of social interaction or of their parents' feelings, or both.

2. Post-tantrum affiliation probability was also increased, directly and independently, by "separation" between child and parent during the tantrum. "Separation"

BOX 12.1 (continued)

was a post hoc variable composed of the parental behaviors of leaving the room or imposing a time-out. There was no association between the child's running away during the tantrum and post-tantrum affiliation, suggesting that it is not physical distance between child and parent but parent-enforced separation that affects post-tantrum affiliation.

3. A composite measure of visible and audible markers of "physiological stress," that is, autonomic activity and respiratory distress, also predicted increased post-tantrum affiliation.

4. Post-tantrum affiliation probability increased with screaming, especially if the screaming persisted for six or more minutes. Screaming, which occurred in 47 percent of tantrums, had the largest effect size of the four factors. It may increase post-tantrum affiliation through several routes. Separation was associated with screaming and may partially mediate its effects. Separation was found to be a parental response to, rather than a stimulus for, the child's screaming. Prolonged screaming appeared to be aversive enough that parents distanced themselves from the child. Physiological stress was also significantly associated with screaming. It made little independent contribution to post-tantrum affiliation but appeared to mediate the effects of screaming. We received reports of children whose intense and prolonged screaming caused them to vomit and others whose eyes became bloodshot and in whose cheeks capillaries burst. Perhaps prolonged screaming, like simi-

larly protracted exercise, reduces blood oxygenation. Compensatory hyperventilation is then followed by bronchospasm, contributing to respiratory distress. Screaming also causes acute pathology of the vocal folds (e.g., Case 1991). Chronic vocal fold lesions ("screamer's nodules"; Lancer et al. 1988) are found at higher rates in tantrum-prone children (Green 1989). Thus, screaming may increase children's physical discomfort, exacerbating post-tantrum need for reassurance. If physiological stress is a mediator of screaming effect on post-tantrum affiliation, rather than merely its by-product, there may be a parallel between physiological stress and post-tantrum affiliation in children and the effects of arousal or anxiety on reconciliation in monkeys (Aureli 1997; Aureli & Smucny, Chapter 10). Finally, identification of screaming as a vocal component of High Anger suggests, in addition, a possible coupling between post-tantrum affiliation and anger.

The finding that forced separation prompts post-tantrum affiliation suggests that one function of such affiliation is to restore the child's connection to the resource controller. These analyses are exploratory, not confirmatory; identified variables are potential, not proven, generators of post-tantrum affiliation. This work must be replicated and extended with direct observation of tantrums and their consequences. However, these caveats do not affect the conclusion that the ontogeny of post-conflict reconciliation in humans may involve the aftermath of children's tantrums.

two cultures thus seemed motivated by a desire both to restore peaceful interaction and to repair damaged relationships.

In conclusion, we hope that our initial findings will inspire others to investigate peacemaking among young children. Throughout this chapter we have offered suggestions for future research, and we would like to end with a few additional

suggestions. First, we need to know more about to what extent the intensity of conflict may affect the likelihood of peacemaking (cf. Verbeek 1997). Recent work on temper tantrums and reconciliation in early parent-child conflict (Potegal, Box 12.1) can be helpful in defining testable hypotheses in this area. Future research should also focus on possible relationships among structural dimensions of peer groups, such as dominance relations and peer status, and peacemaking. We also need to gain a better understanding of the cognitive correlates of peacemaking, especially with regard to issues such as concept development and perspective taking. Finally, recent research on juvenile

nonhuman primates suggests that peacemaking tendency may be modified by observational learning (de Waal & Johanowicz 1993). If the same can be demonstrated with young children, this will open up promising opportunities for peer-based conflict intervention programs.

References

Alvik, T. 1968. The development of views on conflict, war, and peace among school children. *Journal of Peace Research*, 5: 171–198.

Aureli, F. 1997. Post conflict anxiety in nonhuman primates: The mediating role of emotion in conflict resolution. *Aggressive Behavior*, 23: 315–328.

Aureli, F., van Schaik, C., & van Hooff, J. A. R. A. M. 1989. Functional aspects of reconciliation among captive longtailed macaques (*Macaca fascicularis*). *American Journal of Primatology*, 19: 39–52.

Aureli, F., Das, M., & Veenema, H. C. 1997. Differential kinship effect on reconciliation in three species of macaques (*Macaca fascicularis*, *M. fuscata*, and *M. sylvanus*). *Journal of Comparative Psychology*, 111: 91–99.

Blanchard, D. C. 1984. Applicability of animal models to human aggression. In: *Biological Perspectives on Aggression* (K. J. Flannelly, R. J. Blanchard, & D. C. Blanchard, eds.), pp. 49–74. New York: Alan Liss.

Booth, C. L., Rose-Krasnor, L., & Rubin, K. H. 1991. Relating preschoolers' social competence and their mothers' parenting behaviors to early attachment security and high-risk status. *Journal of Social and Personal Relationships*, 8: 363–382.

Butovskaya, M. L., & Guchinova, E. M. 1998. Men and women of Kalmykia: Intergeneration differences and gender stereotypes. In: *Gender Problems in Ethnography* (I. M. Semashko & A. N. Sedlovskaya, eds.), pp. 60–76. Moscow: Institute of Ethnology and Anthropology.

Butovskaya, M. L., & Kozintsev, A. G. 1998. Aggression and reconciliation in primary school children (Ethological analyses of social tension control mechanisms in human groups). *Ethnographic Review*, 4: 122–139 (in Russian).

Butovskaya, M. L., & Kozintsev, A. G. 1999. Aggression, friendship, and reconciliation in Russian primary school children. *Aggressive Behavior*, 25: 125–139.

Case, J. L. 1991. *Clinical Management of Voice Disorders*. Austin, Tex.: Pro-Ed.

Cooper, P. 1965. The development of the concept of war. *Journal of Peace Research*, 2: 1–17.

Cords, M. 1988. Resolution of aggressive conflicts by immature long-tailed macaques (*Macaca fascicularis*). *Animal Behaviour*, 36: 1124–1135.

Cords, M., & Aureli, F. 1993. Patterns of reconciliation among juvenile long-tailed macaques. In: *Juvenile Primates* (M. E. Pereira & L. A. Fairbanks, eds.), pp. 271–283. New York: Oxford University Press.

Cords, M., & Aureli, F. 1996. Reasons for reconciling. *Evolutionary Anthropology*, 5: 42–45.

Dawe, H. C. 1934. An analysis of two hundred quarrels of preschool children. *Child Development*, 5: 139–157.

de Waal, F. B. M. 1993. Reconciliation among primates: A review of empirical evidence and unresolved issues. In: *Primate Social Conflict* (W. A. Mason & S. P. Mendoza, eds.), pp. 111–144. Albany: State University of New York Press.

de Waal, F. B. M. 1996. Conflict as negotiation. In: *Great Ape Societies* (W. C. McGrew, L. F. Marchant, & T. Nishida, eds.), pp. 159–172. Cambridge: Cambridge University Press.

de Waal, F. B. M., & Johanowicz, D. 1993. Modification of reconciliation behavior through social experience: An experiment with two macaque species. *Child Development*, 64: 897–905.

de Waal, F. B. M., & Yoshihara, D. 1983. Reconciliation and re-directed affection in rhesus monkeys. *Behaviour*, 85: 224–241.

Eibl-Eibesfeldt, I. 1979. *The Biology of Peace and War.* London: Thames and Hudson.

Einon, D. F., & Potegal, M. 1994. Temper tantrums in young children. In: *The Dynamics of Aggression: Biological and Social Processes in Dyads and Groups* (M. Potegal & J. Knutson, eds.), pp. 157–194. Hillsdale, N.J.: Lawrence Erlbaum Associates.

Erdniev, E. 1980. *Kalmyks.* Elista, Kalmykia: Kalmykia Publishers.

Fry, D. P. 1988. Intercommunity differences in aggression among Zapotec children. *Child Development*, 59: 1008–1019.

Green, G. 1989. Psycho-behavioral characteristics of children with vocal nodules: WPBIC ratings. *Journal of Speech and Hearing Disorders*, 54: 306–312.

Gregor, T. 1996. *A Natural History of Peace.* Nashville, Tenn.: Vanderbilt University Press.

Hakvoort, I., & Oppenheimer, L. 1993. Children and adolescents' conceptions of peace, war, and strategies to attain peace: A Dutch case study. *Journal of Peace Research*, 30: 65–77.

Hartup, W. W. 1983. Peer relations. In: *Socialization, Personality, and Social Development* (P. H. Mussen, gen. ed.), pp. 103–196. New York: Wiley.

Hartup, W. W., Laursen, B., Stewart, M. I., & Eastenson, A. 1988. Conflict and the friendship relations of young children. *Child Development*, 59: 1590–1600.

Hayes, D. S., Gershmann, E., & Bolin, L. J. 1980. Friends and enemies: Cognitive bases for preschool children's unilateral and reciprocal relationships. *Child Development*, 51: 1276–1279.

Honigmann, J. J. 1954. *Culture and Personality.* New York: Harper.

Howell, S., & Willis, R. 1989. *Societies at Peace: Anthropological Perspectives.* London: Routledge.

Ionin, I. N. 1996. Historical unconscious and the political myth. In: *Modern Political Mythology: Context and Mechanisms of Functioning* (A. P. Logunov & T. V. Evgenieva, eds.), pp. 5–21. Moscow: Russian State University for Humanities Press.

Kappeler, P. M., & van Schaik, C. P. 1992. Methodological and evolutionary aspects of reconciliation among primates. *Ethology*, 92: 51–69.

Katz, L. F., Kramer, L., & Gottman, J. M. 1992. Conflict and emotions in marital, sibling, and peer relationships. In: *Conflict in Child and Adolescent Development* (C. U. Shantz & W. W. Hartup, eds.), pp. 122–149. Cambridge: Cambridge University Press.

Koch, K. F. 1974. *War and Peace in Jalemo: The Management of Conflict in Highland New Guinea.* Cambridge: Harvard University Press.

Ladd, G. W., & Emerson, E. S. 1984. Shared knowledge in children's friendships. *Developmental Psychology*, 20: 932–940.

Lancer, J. M., Syder, O., Jones, A. S., & LeBoutillier, A. 1988. Vocal cord nodules: A review. *Clinical Otolaryngology*, 13: 43–51.

Laursen, B., Betts, N. T., & Finkelstein, B. D. 1998. The resolution of conflict with peers: A developmental meta-analysis. Manuscript, Florida Atlantic University.

Ljungberg, T., Westlund, K., & Lindqvist Foresberg, A. J. 1999. Conflict resolution in 5 year old boys: Does post-conflict affiliative behaviour have a reconciliatory role? *Animal Behaviour*, 58: 1007–1016.

Lorenz, K. 1966. *On Aggression.* San Diego: Harcourt Brace Jovanovich.

Maynard Smith, J., & Price, G. 1973. The logic of animal conflict. *Nature*, 246: 15–18.

Mead, M. 1935. *Sex and Temperament in Three Primitive Societies.* New York: Morrow.

Perry, D. G., Perry, L. C., & Kennedy, E. 1992. Conflict and the development of antisocial behavior. In: *Conflict in Child and Adolescent Development* (C. U. Shantz & W. W. Hartup, eds.), pp. 301–329. Cambridge: Cambridge University Press.

Potegal, M., & Davidson, R. J. 1997. Young children's post tantrum affiliation with their parents. *Aggressive Behavior*, 23: 329–342.

Potegal, M., Kosorok, M. R., & Davidson, R. J. 1996. The time course of angry behavior in the temper tantrums of young children. *Annals of the New York Academy of Sciences*, 749: 31–45.

Provine, R. R. 1993. Laughter punctuates speech: Linguistic, social and gender contexts of laughter. *Ethology*, 95: 291–298.

Rosell, L. 1968. Children's view of war and peace. *Journal of Peace Research*, 5: 268–276.

Sackin, S., & Thelen, E. 1984. An ethological study of peaceful associative outcomes to conflict in preschool children. *Child Development,* 55: 1098–1102.

Siegler, R. S. 1991. *Children's Thinking.* Englewood Cliffs, N.J.: Prentice-Hall.

Silk, J. B. 1996. Why do primates reconcile? *Evolutionary Anthropology,* 5: 39–42.

Silk, J. B. 1997. The function of peaceful post-conflict contacts among primates. *Primates,* 38: 265–279.

Solso, R. L. 1988. *Cognitive Psychology.* Newton, Mass.: Allyn and Bacon.

Sponsel, L. E., & Gregor, T. 1994. *The Anthropology of Peace and Nonviolence.* Boulder, Colo.: Lynne Rienner Publishers.

Staub, E. 1996. The psychological and cultural roots of group violence and the creation of caring societies and peaceful group relations. In: *A Natural History of Peace* (T. Gregor, ed.), pp. 129–155. Nashville, Tenn.: Vanderbilt University Press.

Strathern, A. 1971. *The Rope of Moka: Big-Men and Ceremonial Exchange in Mount Hagen New Guinea.* London: Cambridge University Press.

Trivers, R. 1985. *Social Evolution.* Cummings, Calif.: Benjamin.

Veenema, H. C., Das, M., & Aureli, F. 1994. Methodological improvements for the study of reconciliation. *Behavioural Processes,* 31: 29–38.

Verbeek, P. 1997. Peacemaking of young children (Ph.D. diss., Emory University 1996). *Dissertation Abstracts International:* Section B: Sciences & Engineering. Vol. 57, 7253.

Wessells, M. G. 1993. Psychological obstacles to peace. In: *Nonviolence: Social and Psychological Issues* (V. K. Kool, ed.), pp. 26–35. Lanham, Md.: University Press of America.

PART IV

Triadic Affairs

Introduction

The previous two sections have mainly focused on conflict management mechanisms involving only the individuals in conflict. Most of these conflicts, however, do not occur in a social vacuum, but they are likely to be influenced by other group members. Interventions by third parties during conflict and post-conflict interactions with bystanders may mitigate and mediate conflict outcome. This section aims to present the current knowledge and highlight new avenues of research on the role of third parties in conflict management.

In Chapter 13, **Das** focuses on the post-conflict friendly contacts of third parties with the former aggressor. Most of the systematic research on this topic has been carried out on macaques. The chapter necessarily has limitations in the selection of species treated; however, special effort is made to extrapolate general principles and expectations from the available knowledge. Macaque aggressors tend to increase contacts with various classes of individuals following a conflict. Post-conflict affiliation with own kin and allies may lower the aggressor's tension and reinforce the existing dominance hierarchy by strengthening the social bond between the aggressor and the third party. Post-conflict affiliation with kin of the target of aggression, on the other hand, could help to mitigate antagonism and avoid the spread of hostility to other family members of the former opponents. Das emphasizes that the identity of third parties and the functions of post-conflict friendly contacts of the former aggressor are likely to depend on the degree of tolerance within the group, and she concludes her chapter with predictions on the distributions of these interactions across species.

Interactions of third parties that serve conflict management goals may occur during aggressive confrontations as well. In Box 13.1, **Petit and Thierry** present various categories of interventions according to their aggressive or peaceful nature and the degree of directness toward one of the opponents. Often such interventions result in the end of the confrontation, which can be beneficial for both opponents. Petit and Thierry question specifically the level of impartiality of the various types of interventions. Relatively impartial interventions may maintain relationships between group members under control and avoid the excess use of force by some individuals, while at the same time avoiding the disturbance of the relationships with such individuals. In turn, the tolerance level in a group probably affects the degree of partiality in interventions. The study of impartial interventions as devices for conflict management needs therefore to consider the general features of social organization as much as the acts of the single individuals.

Third-party interactions involving the target of aggression can also play an important role in conflict management. In Chapter 14, **Watts, Colmenares, and Arnold** review the evidence for interactions between targets and bystanders. These interactions can be positive or negative from the perspective of the target; targets initiate some, and third parties initiate others. Variation within and across species in the value of particular classes of social relationships and in the risks and benefits associated with further interaction with aggressors presumably explains much of the variation in interactions with third parties. Watts, Colmenares, and Arnold focus on male interventions in female conflicts especially in harem groups, and they interpret such interventions as male policing with beneficial reproductive consequences for

the males. Male policing, as well as consolatory contacts with bystanders or redirected aggression against third parties, may decrease stress of targets and protect them against further aggression; they may thereby serve as alternatives to reconciliation in contexts in which conciliatory attempts are too risky. Male policing and consolation may provide benefit to the third party as well in the form of reinforcing the relationship with the target. This self-interest, however, does not necessarily exclude the potential role of empathy in such interactions.

More in-depth discussion of the role of empathy and other cognitive abilities in third-party interactions can be found in Box 14.1. Here, **Castles** examines whether triadic conflict management may require more complex abilities than dyadic interactions between the opponents (e.g., reconciliation). Consolatory contacts initiated by third parties may require empathy with the distressed targets of aggression. This consideration implies that species showing consolatory behavior may possess the ability to take others' perspectives, that is, they may have a "theory of mind." For simple quantitative reasons (i.e., because they involve more individuals) triadic interactions may be more demanding. Additionally, they are likely to require knowledge of the relationships of others. Castles points out that the lack of quantitative demonstrations of the presumed functions of triadic interactions limits our ability to make clear distinctions of cognitive challenge. He concludes that distinctions of the cognitive abilities between dyadic and triadic resolutions may be further complicated by the fact that individuals capable of dealing with the cognitive load of triadic interactions will employ such cognitive skills to better execute dyadic resolutions.

CHAPTER 13

Conflict Management via Third Parties

Post-Conflict Affiliation of the Aggressor

Marjolijn Das

Introduction

Reconciliation (i.e., affiliation between opponents shortly after an aggressive conflict) has been extensively studied in primates for nearly two decades (de Waal, Chapter 2). The phenomenon of post-conflict third-party affiliation (i.e., affiliation between opponents and other group members) has received far less attention. Although de Waal & van Roosmalen described "consolation" (i.e., post-conflict affiliation between targets of aggression and third parties) in chimpanzees (*Pan troglodytes*) in 1979, it was not until the 1990s that research explicitly focused on post-conflict third-party affiliation. Most of these studies did not investigate the function of post-conflict third-party affiliation but merely demonstrated its occurrence or absence. In this chapter, I discuss post-conflict third-party affiliation involving the aggressor (see Watts et al., Chapter 14, for a discussion on post-conflict third-party affiliation involving targets of aggression). The chapter focuses primarily on macaques because most studies of this affiliative behavior have been conducted on these species (Table 13.1). This is a relatively new field of study, and we are still in the phase of formulating working hypotheses for the function of this behavior. The chapter presents an overview of different types of post-conflict third-party affiliation and discusses hypotheses regarding their function with the intention to stimulate thinking and generate directions for future research. Throughout the chapter, the term *conflict* is used in the meaning of aggressive conflict.

TABLE 13.1

Overview of Different Types of Post-Conflict Third-Party Affiliation
of Aggressors in Primate Species

Class of Third Party		Aggressor	Unspecified Opponent
OWN KIN	YES	pigtail macaques[1]	patas monkeys[4]
		longtail macaques[2]	
	NOT	stumptail macaques[3]	olive baboons[5]
OPPONENT'S KIN	YES	pigtail macaques[1]	patas monkeys[4]
		longtail macaques[2]	olive baboons[5]
		stumptail macaques[3]	vervet monkeys[7]
		Barbary macaques[6]	
NONKIN	YES	longtail macaques[2]	olive baboons[5]
		Barbary macaques[6]	
		rhesus macaques[8]	
	NOT	pigtail macaques[1]	patas monkeys[4]
		stumptail macaques[3]	
		capuchin monkeys[9]	

Sources: 1: Judge 1991; 2: Das et al. 1997; 3: Call, Aureli, & de Waal unpublished data; 4: York & Rowell 1988 (data on own kin and nonkin deduced from their Table 3); 5: Castles & Whiten 1998; 6: Aureli & Das unpublished data; 7: Cheney & Seyfarth 1989; 8: de Waal & Yoshihara 1983; 9: Verbeek & de Waal 1997.

Note: The first column classifies the third parties. Studies of species in which post-conflict affiliation with that class of third party was found are listed after YES. Studies in which the occurrence of this post-conflict affiliation was investigated but could not be demonstrated are listed after NOT. First section (own kin): the third party is related to the contestant with which it has contact. Second section (opponent's kin): the third party is related to the opponent of the contacted individual. Third section (nonkin): the third party is unrelated to both opponents. Studies listed in the second column focused specifically on third-party affiliation of the aggressor, whereas those in the third column did not distinguish between opponents. Third-party affiliation with nonkin of olive baboons took place with individuals that had provided aggressive support to the contestant during the conflict. Only studies that used the PC/MC method (de Waal & Yoshihara 1983; Veenema, Box 2.1) are reported in this table.

Group Influence and the
Consequences of Aggression

Within primate groups, interactions between two individuals are influenced by other group members. There are many examples of this influence. For instance, young female macaques show more aggressive behavior toward adult females subordinate to their mothers in the presence rather than in the absence of their mothers. Furthermore, juveniles are only able to elicit submissive responses from those lower-ranking females while in the presence of their mothers (Walters & Seyfarth 1987). Thus, simple maternal presence influences interactions and relationships of juvenile macaques. Among captive chimpanzees, the alpha male may, at times, be completely dependent on an alliance with another male for maintaining his top position (de Waal 1982). Cheney & Seyfarth (1986) showed that vervet monkeys (*Cercopithecus aethiops*) recognize relationships between others and that social actions are often based on that knowledge. For instance, two vervet monkeys are more likely to attack each other if their respective kin have recently been involved in an aggressive conflict.

Oftentimes, the influence of dominants is much more obvious than that of subordinates. In many primate species, for example, the alpha male plays a prominent role during aggressive conflicts, halting aggression by intervening on behalf of the target of aggression (Watanabe 1979; de Waal 1987; Watts 1997). In a number of studies of cercopithecines, it was found that females selectively direct grooming to higher-ranking females, presumably in return for future benefits such as support during conflicts (Gouzoules & Gouzoules 1987). In an experiment using Japanese macaques (*Macaca fuscata*), Chapais et al. (1995) found that subordinates who maintained strong grooming relations with dominant females were in turn supported by them in aggressive conflicts. Thus, affiliation with dominants is advantageous, and subordinate females compete with one another for access to dominant females (cf. Seyfarth 1977).

These examples show that the social structure of a group is more than the sum of all its dyadic relationships. Moreover, especially dominant individuals exert a large influence on interactions between others. Consequently, it is vital to study conflict resolution and conflict management not only at the dyadic level but also at the group level, taking into account the role of the other group members as well.

In this chapter, I address conflict resolution and management using a functional approach. This approach focuses on the ultimate advantages of a behavior in terms of its effects on fitness. This approach does not exclude proximate explanations

that focus on the immediate causation of behavior or on cognitive preconditions. The two latter approaches merely emphasize different types of questions. Proximate issues are treated here to highlight underlying functional issues. For example, stress is an adaptive mechanism to cope with disadvantageous situations, and if a certain friendly interaction reduces an individual's stress, this is taken as an indication that the interaction is advantageous for the individual. Such an interaction might, for instance, reduce the risk of being attacked by the interaction partner. For a discussion of cognitive issues related to third-party affiliation, see Castles, Box 14.1.

Both parties involved in post-conflict third-party affiliation are expected to derive benefits. It is easy to think of advantages of post-conflict third-party affiliation for the contestants, such as protection, support, or alleviation of distress (see below). For the third party, affiliation with the aggressor may be advantageous because aggression can spread far beyond the two original opponents. In cercopithecines, in which kin usually maintain strong bonds, conflicts can be extended to family members; kin tend to support their relatives (de Waal 1977; Bernstein & Ehardt 1985; Netto & van Hooff 1986; see Silk 1987 for a review), and individuals may be attacked by their relatives' opponent or even by his or her relatives (Judge 1982; Cheney & Seyfarth 1986, 1989; Aureli & van Schaik 1991a; Aureli et al. 1992). Additionally, in cercopithecines unrelated individuals aggressively support contestants, usually choosing to aid the aggressor (de Waal 1977; Watanabe 1979; Prud'homme & Chapais 1996). Group members even attack the target of aggression at increased rates after the original aggression has ended (Aureli & van Schaik 1991b). Lastly, the target may redirect aggression to other group members (Scucchi et al. 1988; Aureli & van Schaik 1991a; Watts et al., Chapter 14).

Involvement of third parties may also precipitate aggressive outcomes. De Waal (1989b) describes a particularly dramatic escalation of aggression in chimpanzees in which an adult male was killed by two male rivals. In the captive groups of longtail macaques (*Macaca fascicularis*) at Utrecht University, we have witnessed several hierarchy overthrows through the years, in which subordinate females attempted to kill dominant females and sometimes succeeded. Escalation of everyday aggression was usually the direct trigger for such overthrows, though not the underlying cause (personal observation). Ehardt & Bernstein (1986) report similar matrilineal occurrences in rhesus macaques (*Macaca mulatta*).

Even when escalation of aggression has less severe consequences, it can still have costs for all group members. It may damage relationships between the various combatants, and it may increase social instability in the group. As the probability for individuals of becoming targets of aggression increases, the stress levels of these individuals may also increase, thereby diverting energy and attention to social events and away from vital functions, such as foraging (cf. Aureli 1992) or antipredatory vigilance. On the other hand, some individuals may benefit from escalation. For instance, in cercopithecines individuals often support the aggressor for opportunistic reasons, such as the reinforcement of their own dominance position with respect to the target or the improvement of their relationship with the original aggressor (de Waal 1977; Bernstein & Ehardt 1985; Chapais et al. 1995; Prud'homme & Chapais 1996). In the case of hierarchical overthrows, the newly dominant females obviously benefit. Either way, aggression has a significant impact on group members, and post-conflict third-party affiliation may be a way to exert influence on post-conflict events.

Types of Post-Conflict Third-Party Affiliation

Affiliation between opponents and third parties can take place either during or after the aggressive conflict. Affiliative interactions of opponents with third parties during the aggressive interaction have been carefully studied by de Waal & van Hooff (1981) for chimpanzees and by Petit & Thierry (1994a) for Tonkean macaques (*Macaca tonkeana*). In chimpanzees, different types of third-party interactions with different functions can be distinguished. One type involves body contact, such as contact sitting, and often marks the end of the conflict; it then functions as appeasement. Two other types of third-party interaction involve vocalizations, invitations to the third party with, for instance, begging gestures, and displays involving additional aggressive behavior directed at the opponent. These types of interactions were interpreted as support-recruiting and as encouragement. For instance, after mounting or being mounted by another individual, the contestant often returns to his or her adversary to continue the fight. In Tonkean macaques, affiliative interventions are often effective in stopping aggression. Mounting, lip smacking, clasping, and social play are the most frequent patterns of nonagonistic intervention in this species.

Some of the patterns of peaceful interventions of Tonkean macaques resemble those found for post-conflict third-party affiliation between longtail macaque aggressors and nonkin (Das 1998; see below): Males are more likely to intervene than females, and interveners are most often dominant to the opponent with whom they affiliate. The behavioral exchanges, however, differ between the species. In Tonkean macaques, brief explicit contacts, mounts, or clasps seem to be the rule during interventions. In longtail macaques, on the other hand, post-conflict third-party affiliation may consist of a prolonged contact, such as contact sitting or grooming, or of an inconspicuous brief touch when one individual passes the other. As in nonconflict situations, genital investigations and mounts are often seen when males are involved. Thus, post-conflict third-party affiliation in longtail macaques superficially resembles "normal" contacts outside conflict situations. In this respect, post-conflict third-party affiliation of longtail macaques is similar to reconciliation because in this species there is no typical reconciliation gesture as there is in other primates (Aureli et al. 1989; de Waal 1993).

Petit & Thierry (Box 13.1) discuss peaceful interventions in more depth. In this chapter I concentrate on different types and functions of post-conflict third-party affiliation of the aggressor. To give an impression of the actual events that take place during post-conflict third-party affiliation, I shall provide an example taken from my own data on longtail macaques for each of the types of post-conflict third-party affiliation.

Post-Conflict Third-Party Affiliation with Own Kin

Opa, an adult middle-ranking female, chases Bilboa, a juvenile female of the lowest rank. Bilboa flees. Then, Opa walks around for a while and sits down. After a few moments, Sjattoa, Opa's juvenile daughter, approaches her and sits in contact with her.

In species with female philopatry (i.e., females remain in their natal group for the entire life), we expect that adult females and their immature offspring are mostly involved in these interactions. Post-conflict third-party affiliation with own kin was found in the two studies on macaque aggressors that investigated it (Judge 1991; Das et al. 1997; Table 13.1), whereas none of the studies of macaques that focused on the targets of aggression found evidence for post-conflict third-party

---

BOX 13.1

Do Impartial Interventions in Conflicts Occur in Monkeys and Apes?

Odile Petit & Bernard Thierry

Principles of justice and processes of mediation are typically developed within human institutions (Fry, Chapter 16; Yarn, Chapter 4). Interventions aiming to settle disputes are common within the human family (Ross et al. 1990; Cords & Killen 1998). The characteristic of such interventions is that they aim to stop a dispute and to prevent escalation, and they appear to be impartial. Do such interventions occur in monkeys and apes? Do these animals act evenhandedly toward both opponents? For impartial interventions to occur, the intervener should not favor one opponent more than the other (Boehm 1994). In order to assess the degree of impartiality of individuals, we need to distinguish instances in which the intervener's behavior is directed toward a single opponent from those in which it is displayed toward both opponents.

Entering an ongoing conflict on behalf of one of the two opponents is the most common form of intervention in nonhuman primates. Aggressive interventions may lead to coalitions in which two individuals threaten or attack a third one. The occurrence of such interventions may be a means by which support is given to a preferred partner, the relationship with the beneficiary is reinforced, or other goals, such as increasing social dominance over the targeted individual, are attained (Datta 1983; Harcourt & de Waal 1992; Chapais 1995). In these situations, there is no doubt about the partiality of the intervener's behavior: one opponent is aided against the other. High-ranking males who play control roles, however, often intervene against aggressors and protect weaker individuals. It is assumed that males break up fights in order to eliminate social disruption, thus promoting group stability (Boehm 1981; Ehardt & Bernstein 1992).

BOX 13.1 (continued)

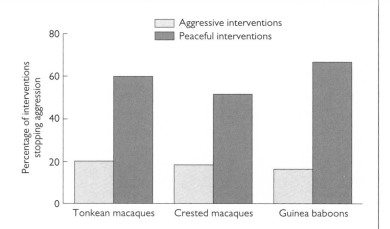

Aggressive and peaceful interventions in three cercopithecine species. The criterion of effectiveness (i.e., stopping aggression) was the end of conflict within five seconds of the onset of intervention (for more information, see Petit & Thierry 1994a; Petit et al. 1997).

In chimpanzees (*Pan troglodytes*), some high-ranking males intervene in a manner disconnected from their affiliative preferences (de Waal 1984). It is difficult, however, to determine the proximate cues that trigger an intervention: Is it the occurrence of the disturbance itself or is it the opportunity to pursue selfish goals (e.g., increasing their dominance over the target) that stimulates the intervention?

In addition to aggressive interventions, third parties sometimes direct nonaggressive behaviors toward an aggressor. In macaques, mothers give silent bared-teeth displays or scream toward individuals threatening their offspring (*Macaca nemestrina*: Massey (1977); *M. mulatta* and *M. fascicularis*: Thierry personal observation). Such interventions are submissive and may be qualified as partial. Peaceful unilateral interventions occur when the intervener displays appeasement behaviors (e.g., lip smacking, clasping, mounting, social play) toward one of the opponents. Peaceful interventions have been reported for chimpanzees (de Waal 1982, 1992), gorillas (*Gorilla gorilla*: Yamagiwa 1987; Sicotte 1995), white-throated capuchins (*Cebus capucinus*: Leca personal communication), golden monkeys (*Rhinopithecus roxellanae*: Ren et al. 1991), Guinea baboons (*Papio papio*), and several macaque species (*Macaca* spp.: Massey 1977; Petit & Thierry 1994a; Petit et al. 1997). Such interventions were thoroughly studied in Tonkean macaques (*M. tonkeana*), in which they regularly occur

(Petit & Thierry 1994a). In this species, interveners preferentially target the aggressor, who is usually the highest-ranking opponent. Males tend to intervene more often than females and use peaceful interventions more frequently. The intervener is typically dominant over both parties. Peaceful interventions halt aggression more effectively than aggressive ones. This applies to Guinea baboons and crested macaques (*M. nigra*) as well.

When peacefully intervening, the third party does not increase its social rank and does not need to make a pro/contra choice as in aggressive interventions. In Tonkean macaques, peaceful interventions are frequently followed by an affiliative interaction, such as grooming, between target and intervener. The latter takes sides in the sense of targeting one opponent on behalf of another, but its behavior appears to protect the beneficiary while preserving its social relationship with the target (Petit & Thierry 1994a). The degree of partiality in peaceful interventions is therefore weaker than in aggressive ones.

Interventions in conflicts may be bilateral in that interventions may not be clearly directed to either opponent. For instance, a female gorilla emits aggressive cough grunting toward both opponents (Watts 1997), or a dominant female chimpanzee barks and makes arm gestures in the direction of two juveniles fighting, who then stop their quarrel (de Waal 1982). Male rhesus macaques (*M. mulatta*) may intervene without taking

BOX 13.1 (continued)

sides (Lindburg 1971, p. 68), and in Japanese macaques (*M. fuscata*), the highest-ranking male looks or moves toward juveniles fighting, who then disperse (Kurland 1977, p. 92). Instances of interpositions have also been observed in juvenile Tonkean macaques (Petit personal observation) and in gorillas, whereby individuals may move between opponents and vocalize (Sicotte 1995).

Whereas the above interventions are undirected, others are specifically performed toward each opponent. Bilateral aggressive interventions, however, have been rarely reported in primates. In chimpanzees, a dominant male was observed breaking up fights by pulling or beating the two adversaries apart, after which it stood between them to prevent further aggression (de Waal 1982, 1992). Aggression by the highest-ranking male toward both opponents was also observed in gorillas (Watts 1997) and Japanese macaques (Chaffin personal communication).

Bilateral peaceful interventions were described in chimpanzees: a male put its hands between two adult females and forced them apart, standing between them until they had stopped screaming (de Waal 1982, p. 124). In bonobos (*Pan paniscus*), a mother retrieved its offspring and calmed the offspring's opponent by putting an arm around its shoulder (de Waal 1989b, p. 218). Several cases were observed in Tonkean macaques; for example, a female lip smacked toward each of the opponents alternately (Petit personal observation). Similarly, in crested macaques, a subadult female was observed touching one adversary with one hand while directing an affiliative bared-teeth display toward the other; then the conflict stopped (Petit personal observation). In golden monkeys, a male interposed itself between female opponents, contacting them while displaying appeasement behaviors (Ren et al. 1991).

In bilateral interventions, whether aggressive or peaceful, the same kind of behavior is directed toward each opponent. Generally, the effect of bilateral interventions tends to be termination of the aggressive conflict. If impartiality is defined from observed acts, then we may say that impartial interventions occur in nonhuman primates, even though such interventions are rarely reported. Alternatively, impartiality may be defined based on the intervener's intentions, in that the motives for such interventions might arise out of concern for the preservation of social relationships, or from a desire to protect one partner without harming the other, or from concern over counterbalancing the respective strengths of the opponents. It is difficult, however, to operationalize such intentions and motives, especially since the goals of these intentions and motives may not always be achieved.

By distinguishing degrees of partiality, we can recognize that some unilateral interventions involve a low degree of partiality. Control and peaceful interventions in which high-ranking individuals terminate aggressive interactions by aiding the recipient of aggression or appeasing the aggressor are examples of such cases.

Yet, if (relatively) impartial interventions can successfully end conflicts, why do they not occur more frequently?

In control interventions, interveners should have little concern for kinship bonds if they are to avoid partiality. Additionally, there should be a strong power imbalance between interveners and targets; otherwise conflicts would escalate rather than end. The highest-ranking individuals are the best candidates for the control role because the costs for their unilateral and bilateral interventions are negligible.

In peaceful interventions, the intervener appears able to protect one opponent while preserving its relationship with the other. Neither high power asymmetry between intervener and target nor suspension of concerns for relationships is needed. The ability to appease partners is likely the relevant factor (Petit & Thierry 1994a); accordingly, peaceful interventions are more common in species characterized by highly developed appeasement behavior.

By improving interindividual tolerance, relatively impartial interventions may keep the dominance gradient among group members low. In turn, more relaxed dominance relationships would affect the degree of partiality in interventions. Hence, when studying conflict management mechanisms such as impartial interventions, we should examine the general features of social organization as much as the acts and intentions of single individuals.

affiliation with own kin (reviewed in Watts et al., Chapter 14). Thus, it seems that at least in macaques only aggressors, and not targets, are involved in this type of affiliation. The study on longtail macaques distinguished between male and female aggressors and found that, as expected, only female aggressors affiliated with their own kin (Das et al. 1997). It must be noted, however, that although in longtail macaques affiliation between the aggressor and her own kin increased following a conflict, this increase was not selective with regard to the general increase in affiliation with other group members.

Post-Conflict Third-Party Affiliation with the Opponent's Kin

Ikea, an adult female of the highest-ranking matriline, directs an "open-mouth threat" to Moa, an adult female of the lowest matriline. Moa moves away from Ikea, whereupon Ikea lunges toward her. After a few seconds Orka, Moa's mother, approaches Ikea. Ikea and Orka first exchange friendly facial expressions. Then, Ikea presents to Orka for grooming, and Orka grooms her. All this happens within 30 seconds after the lunge against Moa.

This behavior is expected only when kin maintain strong bonds. Again, in species with female philopatry, it is expected that only adult females and their immature offspring engage in post-conflict affiliation with the opponent's kin. Unlike most other types of third-party affiliation, this behavior has received a fair amount of attention by researchers. It has been described in four species of macaques and probably occurs in patas monkeys (*Erythrocebus patas*), olive baboons (*Papio anubis*), and vervet monkeys (Table 13.1). Studies on the last three species did not, however, make the distinction between aggressor and target. In the macaque studies, no clear evidence was found for the occurrence of post-conflict third-party affiliation between the target of aggression and the aggres-

sor's kin, but affiliation between the aggressor and the target's kin was found.

Post-Conflict Third-Party Affiliation with Nonkin

Opa bites Rosso, a low-ranking juvenile male, and Antonello, a juvenile male from the highest-ranking matriline, joins Opa in the attack. After the conflict has ended, Xiano, the beta male, approaches Opa and performs a genital inspection. They then sit in proximity for about 30 seconds until Opa leaves.

Three studies that investigated post-conflict behavior of the aggressor in rhesus macaques, Barbary macaques (*Macaca sylvanus*), and longtail macaques found an elevation of friendly interactions between the aggressor and third parties. Two other studies on pigtail macaques (*Macaca nemestrina*) and stumptail macaques (*Macaca arctoides*) did not find affiliation of the aggressor with nonkin (Table 13.1).

Our study on longtail macaques, like those on pigtail macaques and stumptail macaques, made an unambiguous classification of the interaction partner of the aggressor: these interaction partners consisted of individuals that were matrilineally unrelated to both the aggressor and the target. These contacts are referred to as post-conflict third-party affiliation with nonkin (Fig. 13.1). According to the definition to which we adhered, the affiliation itself could consist of any physical contact as long as it was not aggressive or semi-aggressive, or it could consist of friendly gestures or facial expressions over a distance if performed by both individuals. In practice, we found that the vast majority of the affiliative interactions consisted of body contact. As was mentioned earlier, the type of behavior performed during a "typical" post-conflict interaction did not differ in appearance from affiliative interactions in nonconflict situations. The main difference was in the increased frequency with which the behaviors were performed.

FIGURE 13.1. A typical post-conflict third-party affiliation in captive longtail macaques at Utrecht University. The former aggressor, a high-ranking adult male, is groomed by an unrelated low-ranking adult female. The yawning is a sign of post-conflict tension. Photograph by Hans C. Veenema.

Another factor to be considered in the analysis of post-conflict third-party affiliation with nonkin is whether the identity of the third party differs systematically. In support of the idea that the identity of the third party does differ systematically, we found that the aggressor's interaction partner after a conflict was more often higher ranking than the aggressor's interaction partner in a nonconflict situation (Das 1998).

Sex Differences in Post-Conflict Third-Party Affiliation

In captive longtail macaques, male aggressors, unlike female aggressors, do not affiliate at increased rates with their own kin or with kin of the target after a conflict (Das et al. 1997). This is the pattern that indeed would be expected on the basis of known dispersal patterns in macaques, with males migrating from their natal group around adolescence. Male aggressors, however, tend to be involved in post-conflict third-party affiliation with nonkin more than female aggressors do (Das et al. 1997). Furthermore, in this type of affiliation third parties are more often males than females, and males affiliate with the aggressor regardless of the sex of the aggressor or the sex of the target (Das 1998). Thus, whereas females use strategies related to the matrilineal network, males seem to be more active than females in post-conflict strategies outside the matrilineal network, interacting both with other males and with females. In our study, many of the males were young and still lived in their natal group. We need studies of groups living in the wild in order to determine whether these findings present a true picture of male post-conflict behavior in macaques and other primate species.

Function of Post-Conflict Affiliation of the Aggressor

Most evidence of post-conflict third-party affiliation of the aggressor comes from studies on macaques. Macaques have a relatively despotic "dominance style" (de Waal 1989a), although there is some variation between macaque species in degree of despotism. In general, dominance style ranges from despotic to tolerant. Compared with tolerant species, despotic species have more rigid dominance hierarchies, aggression is generally of higher intensity, and counteraggression or bidirectional aggression is rare (de Waal & Luttrell 1989; van Schaik 1989; Thierry 1990). In addition, in despotic macaque species levels of affiliation and reconciliation are generally lower and more kin-biased than in less despotic macaques (Thierry 1986, 1990; Veenema et al. 1994; Aureli et al. 1997).

Consequently, in despotic macaques the aggressor has more influence over the situation during and after an aggressive interaction than does the target. The aggressor is the individual that initiates the aggressive interaction, is generally also the winner (de Waal et al. 1976), and, in the majority of the cases, is the higher ranking of the two opponents (Netto & van Hooff 1986). As a result of this power asymmetry, group members who seek to influence events following aggression would be more successful if they contact the aggressor than if they contact the target of aggression. This may explain why in (despotic) macaque species the target of aggression does not show an increase in post-conflict affiliation with group members other than the aggressor (Aureli & van Schaik 1991a; Judge 1991; Aureli et al. 1993; Aureli et al. 1994), whereas affiliation of the aggressor with own kin, target's kin, and/or others has been described in five species (Table 13.1). Below, I examine various potential functions of post-conflict third-party affiliation of aggressors using data of my recent study on longtail macaques (Das 1998), which are the only available data at the moment.

The Function of Post-Conflict
Third-Party Affiliation with Own Kin

No studies have yet investigated the function of affiliation with kin. The two studies that found post-conflict affiliation between the aggressor and his or her kin were carried out on pigtail (Judge 1991) and longtail macaques (Das et al. 1997). Longtail macaques are characterized by a fairly despotic dominance style (Preuschoft 1995; Aureli et al. 1997), and there are indications that pigtail macaques are also relatively despotic (Castles et al. 1996). In despotic species, the risks for a dominant individual that attacks a subordinate are low because there is hardly any counteraggression or aggression directed up the hierarchy. Therefore, the aggressor probably does not need protection by his or her kin. Whether distressed aggressors can, or need to, be calmed by affiliation with kin is unclear. Although in longtail macaques aggressors show signs of stress after conflicts, this stress

is directly related to relationships with the targets and can probably only be reduced by restoring these relationships (Das et al. 1998).

Affiliation with kin after aggression may reconfirm bonds between the participants and serve as a signal to subordinates that aggressors maintain strong bonds with stable alliance partners. This idea is discussed further in the section on post-conflict third-party affiliation with nonkin.

The Function of Post-Conflict
Third-Party Affiliation with the Opponent's Kin

It has been proposed that this interaction functions as a substitute for reconciliation or is a "next best option" when reconciliation with the opponent is impossible (Aureli & van Schaik 1991a; Judge 1991). If this is true, then this interaction should ultimately help to restore the relationship between the opponents and should produce effects similar to those of reconciliation, such as reduction of aggression directed to the target or alleviation of distress of contestants. In longtail macaques, however, post-conflict third-party affiliation between the aggressor and the target's kin does not reduce stress-related behavior (Maestripieri et al. 1992) of the aggressor, whereas reconciliation does (Das et al. 1998). Thus, post-conflict third-party affiliation with the opponent's kin is obviously not interpreted by the aggressor as equivalent to, or as a substitute for, reconciliation. It is therefore unlikely that this behavior directly restores the relationship between the opponents.

It could be that post-conflict third-party affiliation with the opponent's kin is a conflict management strategy on a more complex level, concerned with relationships between whole matrilines instead of the dyadic relationship between the opponents. It could, for instance, serve to reduce the likelihood that aggression will spread to other members of the opponents' matrilines (Judge 1991). In this way, it may somehow help to restore the relationship between entire matrilines that could have been disturbed by the conflict between two of their members. To test this hypothesis, studies are needed that investigate the effects of these post-conflict third-party affiliations on stress responses of opponents' kin, on aggression rates between the involved matrilines, and on social tolerance between matrilines.

The Function of Post-Conflict Third-Party Affiliation with Nonkin

Several hypotheses can be formulated with regard to the function of post-conflict third-party affiliation with nonkin. As it does during aggressive conflicts (de Waal & van Hooff 1981), this affiliation might serve to recruit supporters for the aggressor or to encourage the aggressor. Alternatively, it could function as appeasement if it reduces subsequent aggression involving the aggressor. It could also function as consolation, alleviating distress of the aggressor, but this function should be less important for aggressors than for targets (de Waal & Aureli 1996). We first tested these hypotheses in our study on longtail macaque aggressors, and we then proposed one additional hypothesis, that is, this type of affiliation may serve to strengthen alliances between dominant individuals and reinforce the existing dominance hierarchy (Das 1998).

Recruitment of Support and Encouragement

In longtail macaques, post-conflict third-party affiliation with nonkin was not effective in recruiting supporters for aggressors. The third party was never observed to aid the aggressor aggressively following the affiliation. There is some indication, however, that this type affiliation might encourage the aggressor. In fact, the level of aggression performed by the aggressor tended to be higher in post-conflict observations after third-party affiliation with non-

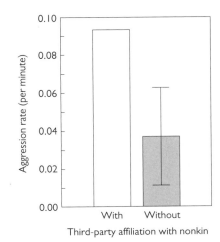

FIGURE 13.2. The frequency of aggression by the former aggressor directed to any group member in the first five minutes following a conflict depending on whether third-party affiliation with nonkin had taken place. The 95 percent confidence limits are shown for post-conflict situations without third-party affiliation with nonkin.

kin had taken place than in post-conflict observations without such affiliation (Fig. 13.2).

Appeasement and Consolation Post-conflict affiliation is often implicitly assumed to be "appeasing," that is, to reduce the risk of aggression. It is also assumed to alleviate distress of individuals ("consolation"). Our results on longtail macaques show that these functions are not necessarily present. We found no indication that post-conflict third-party affiliation with nonkin reduces aggression performed by the aggressor (Fig. 13.2) or that it influences aggression directed at the aggressor. Moreover, post-conflict third-party affiliation with nonkin did not reduce stress-related behavior of aggressors (Das et al. 1998).

If post-conflict third-party affiliation with nonkin serves as appeasement, we would expect the subordinate member of the dyad—that is, the individual with the greater interest in reducing the risk of aggression—to take the initiative to establish such an interaction, but this was not the case. The aggressor initiated post-conflict third-party affiliation toward both dominant and subordinate nonkin. When an unrelated individual initiated post-conflict third-party affiliation, he or she was usually higher ranking than the aggressor. Apparently, in longtail macaques post-conflict third-party affiliation with nonkin occurs for reasons other than appeasement and distress alleviation.

Reinforcement of the Hierarchy

After the overall failure of our study to support the previous functions, we proposed that in longtail macaques third-party affiliation with the aggressor is a nonaggressive means to signal "approval" to the aggressor. It may at the same time reconfirm and improve the relationship between the interaction partners and, in the case of high-ranking partners, give a signal to subordinates that the alliance between the dominant partners remains strong. Post-conflict third-party affiliation with nonkin can therefore reinforce the existing hierarchy (Das 1998). This mechanism might be comparable to that of aggressor support, which is thought to function to establish and reinforce the dominance hierarchy (Chapais et al. 1995; Prud'homme & Chapais 1996) and which can reinforce alliances with the aggressor (de Waal 1977; Bernstein & Ehardt 1985). Moreover, even aggression itself often has no apparent cause and seems to be performed solely in order to reinforce the hierarchy (Walters & Seyfarth 1987).

The hypothesis that post-conflict third-party affiliation with nonkin functions to reconfirm bonds between the partners and to reinforce the hierarchy is supported by the tendency for increased attack rates of the aggressor following such affiliation (Fig. 13.2). Aggressors might indeed be encouraged in their aggressive behavior if another individual signals approval of their actions, especially when this individual is dominant; in fact, nonkin that initiate post-conflict third-party affiliation are usually higher ranking than the aggressor. Affiliation between a female aggressor and her own kin could be interpreted in the same way. In macaques, kin are important alliance partners for adult females, and it might be advantageous for the aggressor and her kin to communicate to other group members, and to each other, that these bonds are still strong.

No study has provided direct evidence to support this tentative hypothesis. A number of working hypotheses can be derived. For instance, it is expected that individuals that need each other for alliances are mainly interested in this type of affiliation; hence, individuals with weak relationships should not engage in this affiliation. Furthermore, if it is especially important for dominants to reinforce their alliances against subordinates, we expect that the third party, if not higher in rank than the aggressor, would often be higher ranking than the target of aggression. With regard to interspecific differences, it is expected that despotic species, or, in general, species in which dominants derive benefits from a strict hierarchy, should have high levels of post-conflict third-party affiliation with nonkin (see below). It should be noted that this type of affiliation could well have different functions in different primate species and that it could even serve two different functions within one species, depending, for example, on the identity of the third party (see below).

Conclusions

Is Post-Conflict Third-Party Affiliation a Conflict-Resolution Mechanism?

In some cases, third parties may contribute toward resolution of the conflict by engaging in post-conflict third-party affiliation. For instance, in chimpanzees the situation has been described in which an outsider actively leads former opponents toward each other—that is, mediated reconciliation (de Waal & van Roosmalen 1979). Resolution of the conflict may also be facilitated by post-conflict third-party affiliation with kin of the opponent, if this functions to restore the relationship between matrilines that was disturbed by the conflict. In other cases, however, post-conflict third-party affiliation may not be aimed at resolution of the conflict at all.

In longtail macaques, this might be the case for post-conflict third-party affiliation between the aggressor and nonkin and also for post-conflict third-party affiliation between a female aggressor and her own kin. If these types of post-conflict third-party affiliation indeed function to strengthen bonds between the two individuals and reinforce the existing hierarchy, they do not restore the relationship between the opponents. The fact that aggressive behavior of the former aggressor seems to increase after third-party affiliation with nonkin supports the idea that this type of affiliation does not function as conflict resolution. Individuals using these types of post-conflict third-party affiliation then merely make opportunistic use of the aggressive interaction. Individuals certainly derive benefits, but these are not directly related to the conflict. If this affiliation increases the advantage gained from the conflict by the aggressor, it may be considered a conflict management strategy. Although conflict management in the strict sense is defined as serving to reduce the costs of the conflict (Cords & Killen 1998), the net result of this affiliation is then the same, that is, a shift in the ratio of costs and benefits of the conflict in favor of the benefits.

Post-Conflict Third-Party Affiliation and Relationship Value

The distinctions of post-conflict third-party affiliation have so far been based mainly on genetic relatedness, dominance position in the group, or role in the conflict. Some of the proposed functions for post-conflict third-party affiliation with nonkin imply that these interactions should occur specifically among individuals with valuable relationships, for example, when it reconfirms the strength of bonds between alliance partners. In other cases, such as when affiliation appeases the aggressor, we could expect individuals with a valuable relationship with the target to engage in affiliation with the aggressor.

In the case of reconciliation, it has become clear that bond strength determines much of the variation in reconciliation rates (de Waal & Yoshihara 1983; Aureli et al. 1989; Castles et al. 1996; de Waal, Chapter 2; Cords & Aureli, Chapter 9; van Schaik & Aureli, Chapter 15). We expect this to be the case for third-party affiliation as well; there are some indications for the importance of bond strength in third-party affiliation. The third-party affiliation with own kin (Judge 1991; Petit & Thierry 1994b; Das et al. 1997) is certainly one example. In addition, in olive baboons, opponents affiliate preferentially with supporters in the previous conflict (Castles & Whiten 1998). In gorillas (*Gorilla gorilla beringei*), immature targets affiliate with their mothers, and adult female targets affiliate with the alpha male, with whom they maintain strong bonds (Watts 1995). In Hamadryas baboons (*Papio hamadryas*), third-party affiliation occurs between adult males and their harem females (Watts et al., Chapter 14).

This range of findings suggests that the critical factor for third-party affiliation is relationship value rather than matrilineal relatedness. Kin can be considered merely as a "special case" of a valuable partner (in that there is the additional advantage of accruing benefits to inclusive fitness). In sum, relationship value might prove a fundamental underlying factor in many conflict management strategies, and the routine assessment of relationship quality between each opponent and third parties in post-conflict affiliation might bring useful insight.

Interspecific Differences: Predictions for Post-Conflict Third-Party Affiliation

Obviously there are numerous important differences in the social lives of different primate species, not to mention non-primates (van Schaik & Aureli, Chapter 15), which may influence the occurrence and patterns of post-conflict third-party affiliation. The cognitive capacities of a species are one such difference; this factor may, for instance, determine the occurrence or absence of consolation if the capacity for empathy is a precondition for this behavior (de Waal & Aureli 1996; see also Castles, Box 14.1). Furthermore, aspects of group life, such as dispersal patterns and the relative number of males and females, may influence the quality of the relationships between group members and thus third-party affiliation.

Here, I focus on one social factor that may have a large influence on third-party affiliation, namely, dominance style (de Waal 1989a), which varies across species and ranges from despotic to tolerant. Despotic societies are thought to have developed under conditions of fierce intragroup contest competition, whereas more tolerant societies are thought to have developed under conditions in which cohesiveness was favored, such as severe intergroup competition or the presence of significant external threat (van Schaik 1989). We can derive predictions for the occurrence of post-conflict third-party affiliation of aggressors in despotic versus more tolerant species, based on the variation in relationship value and the need for a strict hierarchy. Let us examine the predictions for each type of post-conflict third-party affiliation.

Post-Conflict Third-Party Affiliation with Own Kin In both despotic and tolerant species, kin are important alliance partners. Thus, the value of

kin relationships should not vary much with dominance style of the species. This claim is supported by the finding that conciliatory tendencies between close kin in two despotic species of macaques did not differ from the conciliatory tendency between close kin in a less despotic macaque species (Aureli et al. 1997). It is therefore expected that post-conflict affiliation with kin will occur with approximately equal frequencies in both despotic and tolerant societies if this type of post-conflict third-party affiliation indeed reconfirms alliances between kin.

Post-Conflict Third-Party Affiliation with the Opponent's Kin The use of this strategy is probably correlated with the quality of relationships between families, for example, with the need for stable alliances between matrilines. In tolerant species, individuals are thought to benefit from the development of strong bonds with all group members, both kin and nonkin, since high levels of cooperation between group members are needed against external threats. Therefore, relationships with individuals from other matrilines should be relatively more valuable in tolerant species than in despotic species; post-conflict third-party affiliation with kin of the opponent is expected to occur at higher rates in tolerant than in despotic species (similar to differences in reconciliation rates with nonkin between tolerant and despotic societies: Aureli et al. 1997).

Post-Conflict Third-Party Affiliation with Nonkin Predictions for interspecific differences related to dominance style in this type of affiliation are dependent on the function of this affiliation. In cases in which post-conflict third-party affiliation with nonkin serves to strengthen bonds between high-ranking individuals, thus maintaining the hierarchy, it could be expected to occur more frequently in despotic species. In cases in which post-conflict third-party affiliation with nonkin functions as appeasement or distress alleviation, as may be the case for peaceful interventions during the conflict (Petit & Thierry, Box 13.1) and for consolation (de Waal & van Roosmalen 1979), it is expected to occur more frequently in tolerant species than in despotic, since they have a greater need for a stable, close-knit group. In general, we would expect affiliation with an appeasing function to be initiated by subordinates if they run a risk of becoming the target of further aggression. We could even imagine this type of behavior to serve two different functions in the same species, depending on whether it takes place with dominants or with subordinates.

Directions for Future Research

How can the work discussed in this chapter, which was conducted mainly on macaques, guide us in future research on other animal species? Obviously, in group-living species dyadic relationships cannot be seen in isolation from the relationships with and between other group members. It has also been shown here that the consequences of aggressive conflicts between two individuals concern and involve other group members. Thus, conflicts, conflict resolution, and conflict management should be studied taking the social environment made up of other group members into account.

Furthermore, affiliation after a conflict might have numerous different functions depending on, among other things, the participating individuals and the species. Reconciliation restores social relationships in many primate species, consolation in

chimpanzees might alleviate distress of the target, whereas affiliation between longtail macaque aggressors and third parties could reconfirm alliances. Many of these strategies probably help to resolve conflicts or reduce their adverse effects, but not all. For instance, affiliation between longtail macaque aggressors and nonkin reduces neither stress of the aggressor nor post-conflict aggression. We are still left with many unanswered questions. The wide variety of possibilities regarding the function of these interactions calls for studies that combine descriptions of occurrence of the various types of third-party affiliation with investigations of their function.

The function may depend on a number of factors. Specifically, future studies on primates and non-primates should consider third-party affiliation in the context of the social structure and socioecology of the species under study. Aspects such as dispersal patterns, degree of despotism, and male/female ratio are expected to be important because each of these aspects plays a role in determining the value of the relationships between group members (cf. van Schaik & Aureli, Chapter 15). Differences in relationship value probably explain much of the variation in third-party affiliation.

There is still much to be discovered in the study of third-party affiliation in primates, especially non-cercopithecines. An even greater challenge is to combine this effort with the study of conflict management in non-primates. Knowledge of third-party affiliation may also be an important source of information on the nature and complexity of social relationships in less known primate and non-primate species.

References

Aureli, F. 1992. Post-conflict behaviour among wild long-tailed macaques (*Macaca fascicularis*). *Behavioural Ecology and Sociobiology*, 31: 329–337.

Aureli, F., & van Schaik, C. P. 1991a. Postconflict behaviour in long-tailed macaques (*Macaca fascicularis*): I. The social events. *Ethology*, 89: 89–100.

Aureli, F., & van Schaik, C. P. 1991b. Post-conflict behaviour in long-tailed macaques (*Macaca fascicularis*): II. Coping with the uncertainty. *Ethology*, 89: 101–114.

Aureli, F., van Schaik, C. P., & van Hooff, J. A. R. A. M. 1989. Functional aspects of reconciliation among captive long-tailed macaques (*Macaca fascicularis*). *American Journal of Primatology*, 19: 39–51.

Aureli, F., Cozzolino, R., Cordischi, C., & Scucchi, S. 1992. Kin-oriented redirection among Japanese macaques: An expression of a revenge system? *Animal Behaviour*, 44: 283–291.

Aureli, F., Veenema, H. C., van Panthaleon van Eck, J. C., & van Hooff, J. A. R. A. M. 1993. Reconciliation, consolation, and redirection in Japanese macaques (*Macaca fuscata*). *Behaviour*, 124: 1–21.

Aureli, F., Das, M., Verleur, D., & van Hooff, J. A. R. A. M. 1994. Postconflict social interactions among Barbary macaques (*Macaca sylvanus*). *International Journal of Primatology*, 15: 471–485.

Aureli, F., Das, M., & Veenema, H. C. 1997. Differential kinship effect on reconciliation in three species of macaques (*Macaca fascicularis*, *M. fuscata*, and *M. sylvanus*). *Journal of Comparative Psychology*, 111: 91–99.

Bernstein, I. S., & Ehardt, C. L. 1985. Agonistic aiding: Kinship, rank, age, and sex influences. *American Journal of Primatology*, 8: 37–52.

Boehm, C. 1981. Parasitic selection and group selection: A study of conflict interference in rhesus and Japanese macaque monkeys. In: *Primate Behavior and Sociobiology* (A. B. Chiarelli & R. S. Corrucini, eds.), pp. 161–182. Berlin: Springer-Verlag.

Boehm, C. 1994. Pacifying interventions at Arnhem Zoo and Gombe. In: *Chimpanzee Cultures* (R. W. Wrangham, W. C. McGrew, F. B. M. de Waal, & P. G. Heltne, eds.), pp. 211–226. Cambridge: Harvard University Press.

Castles, D. L., & Whiten, A. 1998. Post-conflict behaviour of wild olive baboons. 1. Reconciliation, redirection and consolation. *Ethology*, 104: 126–147.

Castles, D. L., Aureli, F., & de Waal, F. B. M. 1996. Variation in conciliatory tendency and relationship quality across groups of pigtail macaques. *Animal Behaviour*, 52: 289–403.

Chapais, B. 1995. Alliances as a means of competition in primates: Evolutionary, developmental, and cognitive aspects. *Yearbook of Physical Anthropology*, 38: 115–136.

Chapais, B., Gauthier, C., & Prud'homme, J. 1995. Dominance competition through affiliation and support in Japanese macaques: An experimental study. *International Journal of Primatology*, 16: 521–536.

Cheney, D. L., & Seyfarth, R. M. 1986. The recognition of social alliances by vervet monkeys. *Animal Behaviour*, 34: 1722–1731.

Cheney, D. L., & Seyfarth, R. M. 1989. Redirected aggression and reconciliation among vervet monkeys, *Cercopithecus aethiops*. *Behaviour*, 110: 258–275.

Cords, M., & Killen, M. 1998. Conflict resolution in human and non-human primates. In: *Piaget, Evolution, and Development* (J. Langer & M. Killen, eds.), pp. 193–218. Mahwah, N.J.: Lawrence Erlbaum Associates.

Das, M. 1998. Conflict management and social stress in long-tailed macaques. Ph.D. diss., Utrecht University, the Netherlands.

Das, M., Penke, Zs., & van Hooff, J. A. R. A. M. 1997. Affiliation between aggressors and third parties following conflicts in long-tailed macaques (*Macaca fascicularis*). *International Journal of Primatology*, 18: 157–179.

Das, M., Penke, Zs., & van Hooff, J. A. R. A. M. 1998. Postconflict affiliation and stress-related behavior of long-tailed macaque aggressors. *International Journal of Primatology*, 19: 53–71.

Datta, S. B. 1983. Patterns of aggressive interference. In: *Primate Social Relationships* (R. A. Hinde, ed.), pp. 289–297. Oxford: Blackwell Scientific Publications.

de Waal, F. B. M. 1977. The organization of agonistic relations within two captive groups of Java-monkeys (*Macaca fascicularis*). *Zeitschrift für Tierpsychologie*, 44: 225–282.

de Waal, F. B. M. 1982. *Chimpanzee Politics: Power and Sex among Apes*. London: Jonathan Cape.

de Waal, F. B. M. 1984. Sex differences in the formation of coalitions among chimpanzees. *Ethology and Sociobiology*, 5: 239–255.

de Waal, F. B. M. 1987. Dynamics of social relationships. In: *Primate Societies* (B. B. Smuts, D. L. Cheney, R. M. Seyfarth, R. W. Wrangham, & T. T. Struhsacker, eds.), pp. 121–134. Chicago: University of Chicago Press.

de Waal, F. B. M. 1989a. Dominance "style" and primate social organization. In: *Comparative Socioecology: The Behavioural Ecology of Humans and Other Mammals* (V. Standen & R. Foley, eds.), pp. 195–218. Oxford: Blackwell.

de Waal, F. B. M. 1989b. *Peacemaking among Primates.* Cambridge: Harvard University Press.

de Waal, F. B. M. 1992. Coalitions as part of reciprocal relations in the Arnhem chimpanzee colony. In: *Coalitions and Alliances in Human and Other Animals* (A. H. Harcourt & F. B. M. de Waal, eds.), pp. 233–258. Oxford: Oxford University Press.

de Waal, F. B. M. 1993. Reconciliation among primates: A review of empirical evidence and unresolved issues. In: *Primate Social Conflict* (W. A. Mason & S. P. Mendoza, eds.), pp. 111–144. Albany: State University of New York Press.

de Waal, F. B. M., & Aureli, F. 1996. Consolation, reconciliation, and a possible cognitive difference between macaques and chimpanzees. In: *Reaching into Thought: The Minds of the Great Apes* (A. E. Russon, K. A. Bard, & S. T. Parker, eds.), pp. 80–110. Cambridge: Cambridge University Press.

de Waal, F. B. M., & Luttrell, L. M. 1989. Towards a comparative socioecology of the genus *Macaca*: Different dominance styles in rhesus and stumptail macaques. *American Journal of Primatology*, 19: 83–109.

de Waal, F. B. M., & van Hooff, J. A. R. A. M. 1981. Side-directed communication and agonistic interactions in chimpanzees. *Behaviour*, 77: 164–198.

de Waal, F. B. M., & van Roosmalen, A. 1979. Reconciliation and consolation among chimpanzees. *Behavioural Ecology and Sociobiology*, 5: 55–66.

de Waal, F. B. M., & Yoshihara, D. 1983. Reconciliation and redirected affection in rhesus monkeys. *Behaviour*, 85: 223–241.

de Waal, F. B. M., van Hooff, J. A. R. A. M., & Netto, W. J. 1976. An ethological analysis of types of agonistic interactions in a captive group of Java-monkeys (*Macaca fascicularis*). *Primates*, 17: 257–290.

Ehardt, C. L., & Bernstein, I. S. 1986. Matrilineal overthrows in rhesus monkey groups. *International Journal of Primatology*, 7: 157–179.

Ehardt, C. L., & Bernstein, I. S. 1992. Conflict intervention behaviour by adult male macaques: Structural and functional aspects. In: *Coalitions and Alliances in Human and Other Animals* (A. H. Harcourt & F. B. M. de Waal, eds.), pp. 83–111. Oxford: Oxford University Press.

Gouzoules, S., & Gouzoules, H. 1987. Kinship. In: *Primate Societies* (B. B. Smuts, D. L. Cheney, R. M. Seyfarth, R. W. Wrangham, & T. T. Struhsacker, eds.), pp. 299–305. Chicago: University of Chicago Press.

Harcourt, A. H., & de Waal, F. B. M. 1992. *Coalitions and Alliances in Human and Other Animals.* Oxford: Oxford University Press.

Judge, P. G. 1982. Redirection of aggression based on kinship in a captive group of pigtail macaques. *International Journal of Primatology*, 3: 301.

Judge, P. G. 1991. Dyadic and triadic reconciliation in pigtail macaques (*Macaca nemestrina*). *American Journal of Primatology*, 23: 225–237.

Kurland, J. A. 1977. *Kin Selection in the Japanese Monkey.* Basel: Karger.

Lindburg, D. G. 1971. The rhesus monkey in North India: An ecological and behavioral study. In: *Primate Behavior*, Vol. 2 (L. A. Rosenblum, ed.), pp. 2–106. New York: Academic Press.

Maestripieri, D., Schino, G., Aureli, F., & Troisi, A. 1992. A modest proposal: Displacement activities as an indicator of emotions in primates. *Animal Behaviour*, 44: 967–979.

Massey, A. 1977. Aggressive aids and kinship in a group of pigtail macaques. *Behavioral Ecology and Sociobiology*, 2: 31–40.

Netto, W. J., & van Hooff, J. A. R. A. M. 1986. Conflict interference and the development of dominance relationships in immature *Macaca fascicularis*. In: *Primate Ontogeny, Cognition and Social Behaviour* (J. G. Else & P. C. Lee, eds.), pp. 291–313. Cambridge: Cambridge University Press.

Petit, O., & Thierry, B. 1994a. Aggressive and peaceful interventions in conflicts in Tonkean macaques. *Animal Behaviour*, 48: 1427–1436.

Petit, O., & Thierry, B. 1994b. Reconciliation in a group of Guinea baboons. In: *Current Primatology*, Vol. 2 (J. J. Roeder, B. Thierry, J. R. Anderson, & N. Herrenschmidt, eds.), pp. 137–145. Strasbourg: Université Louis Pasteur.

Petit, O., Thierry, B., & Abegg, C. 1997. A comparative study of aggression and conciliation in three cercopithecine monkeys (*Macaca fuscata, Macaca nigra, Papio papio*). *Behaviour*, 134: 415–432.

Preuschoft, S. 1995. "Laughter" and "smiling" in macaques—an evolutionary perspective. Ph.D. diss., Utrecht University, the Netherlands.

Prud'homme, J., & Chapais, B. 1996. Development of intervention behaviour in Japanese macaques: Testing the targeting hypothesis. *International Journal of Primatology*, 17: 429–443.

Ren, R., Yan, K., Su, Y., Qi, H., Liang, B., Bao, W., & de Waal, F. B. M. 1991. The reconciliation behavior of golden monkeys (*Rhinopithecus roxellanae roxellanae*) in small breeding groups. *Primates*, 32: 321–327.

Ross, H., Tesla, C., Kenyon, B., & Lollis, S. 1990. Maternal intervention in toddler peer conflict: The socialization of principles of justice. *Developmental Psychology*, 26: 994–1003.

Scucchi, S., Cordischi, C., Aureli, F., & Cozzolino, R. 1988. The use of redirection in a captive group of Japanese monkeys. *Primates*, 29: 229–236.

Seyfarth, R. M. 1977. A model of social grooming among adult female monkeys. *Journal of Theoretical Biology*, 65: 671–698.

Sicotte, P. 1995. Interpositions in conflicts between males in bimale groups of mountain gorillas. *Folia primatologica*, 65: 14–24.

Silk, J. B. 1987. Social behavior in evolutionary perspective. In: *Primate Societies* (B. B. Smuts, D. L. Cheney, R. M. Seyfarth, R. W. Wrangham, & T. T. Struhsacker, eds.), pp. 318–329. Chicago: University of Chicago Press.

Thierry, B. 1986. A comparative study of aggression and response to aggression in three species of macaques. In: *Primate Ontogeny, Cognition and Social Behaviour* (J. G. Else & P. C. Lee, eds.), pp. 307–313. Cambridge: Cambridge University Press.

Thierry, B. 1990. Feedback loop between kinship and dominance: The macaque model. *Journal of Theoretical Biology*, 145: 511–521.

van Schaik, C. P. 1989. The ecology of social relationships among female primates. In: *Comparative Socioecology: The Behavioural Ecology of Humans and Other Mammals* (V. Standen & R. Foley, eds.), pp. 195–218. Oxford: Blackwell.

Veenema, H. C., Das, M., & Aureli, F. 1994. Methodological improvements for the study of reconciliation. *Behavioural Processes*, 31: 29–38.

Verbeek, P., & de Waal, F. B. M. 1997. Postconflict behavior of captive brown capuchins in the presence and absence of attractive food. *International Journal of Primatology*, 18: 703–725.

Walters, J. M., & Seyfarth, R. M. 1987. Conflict and cooperation. In: *Primate Societies* (B. B. Smuts, D. L. Cheney, R. M. Seyfarth, R. W. Wrangham, & T. T. Struhsacker, eds.), pp. 306–317. Chicago: University of Chicago Press.

Watanabe, K. 1979. Alliance formation in a free-ranging troop of Japanese macaques. *Primates*, 20: 459–474.

Watts, D. P. 1995. Post-conflict social events in wild mountain gorillas (Mammalia, Hominoidea). II. Redirection, side direction, and consolation. *Ethology*, 100: 158–174.

Watts, D. P. 1997. Agonistic interventions in wild mountain gorilla groups. *Behaviour*, 134: 23–57.

Yamagiwa, J. 1987. Intra- and inter-group interactions of an all-male group of Virunga mountain gorillas (*Gorilla gorilla beringei*). *Primates*, 28: 1–30.

York, A. D., & Rowell, T. E. 1988. Reconciliation following aggression in patas monkeys, *Erythrocebus patas. Animal Behaviour*, 36: 502–509.

CHAPTER 14

Redirection, Consolation, and Male Policing

How Targets of Aggression Interact with Bystanders

David P. Watts, Fernando Colmenares, & Kate Arnold

Pandora is an immigrant adult female mountain gorilla (Gorilla gorilla beringei) whose two immature offspring are her only relatives in Group 5. Puck is a natal female with several adult relatives in the group. Puck and Pandora rarely have friendly interactions, and aggression between them is common. The two females are feeding near each other when Ziz, the dominant male, starts a procession to a new feeding area. They follow, and their paths intersect; they jostle each other repeatedly and exchange "cough grunts." Suddenly Puck attacks, and the two scream as they grapple and bite each other. Puck's mother and sister also scream at Pandora. Within seconds, Ziz runs over and uses his tremendous size and strength advantage to stop the fight: he grabs Puck's shoulder with one hand and Pandora's with the other, pulls them apart, and restrains them until their struggles cease. Both females, still agitated, "grumble" to Ziz, apparently to appease him. Pandora puts her hand on his back as she grumbles and maintains physical contact with him as he

walks off and then stops to feed. Puck glares at her as she leaves and then goes elsewhere to feed.

The literature on primate behavior is replete with descriptions and analyses of such interactions. Compared with most other mammals, many gregarious primate species are distinguished by high frequencies of aggressive interactions involving more than two individuals and by complex tactics to gain, provide, and manipulate support (Harcourt 1992). Opponents also commonly interact with third parties after aggressive conflicts. Such interactions occur in a rich social milieu and take many forms. Their frequency varies considerably across species and across classes of individuals within species, as does responsibility for their initiation.

Consider the options available to the recipient of aggression in the example above. Avoiding an opponent is a temporarily expedient but not a long-term option for group-living animals. Reconciliation with the former aggressor could minimize the negative effects of the escalated conflict and make further attack unlikely (i.e., it allows conflict management: Cords 1997; Cords & Killen 1998). Reconciliation is also probably the best way (Kappeler & van Schaik 1992) to minimize incompatibilities between the goals of the two opponents and to repair any damage to their social relationship (i.e., to resolve conflicts: Cords 1997; Cords & Killen 1998). But approaching an opponent can be dangerous, especially when the relationship may already be a bad one (as in Puck and Pandora's case). Redirecting aggression at a bystander may defuse tension and allow conflict management (Kappeler & van Schaik 1992). Friendly contact with an aggressor's relative may do the same and perhaps could indirectly resolve the conflict (Cheney & Seyfarth 1989; Aureli & van Schaik 1991a; Judge 1991), but aggressors' kin may initiate further aggression. Approaching one's own kin, if some are available, may have a calming effect but may expose those kin to attack by aggressors and their kin. Friendly contact with a physically and socially available third party, who does not face such risks, may allow conflict management. Ziz was not at risk, and he offered immediate protection; his actions may have also calmed Pandora and perhaps lowered the amount of aggression that she would have received from Puck and her relatives over the long term.

In this chapter, we review information on the forms, contexts, and consequences of interactions that recipients of dyadic aggression have with third parties, and we indicate major unresolved questions about their functions (see Das, Chapter 13, for interactions of third parties with aggressors). Conflict does not always involve aggression (Mason 1993; Cords & Killen 1998), but we focus on aggressive conflicts because most research has

been carried out on third-party interactions around such conflicts. Third-party interventions in ongoing conflicts are treated extensively elsewhere (e.g., Chapais, 1992; Noë 1992; Petit & Thierry, Box 13.1); except for male interventions in conflicts between females, that is, "male policing," we consider here only post-conflict interactions. We particularly focus on redirection of aggression and consolation and on male policing and provision of post-conflict reassurance to females, especially in species with harem-polygynous mating systems.

Costs and Benefits of Third-Party Interactions

Post-conflict interactions between bystanders and targets can be affiliative or aggressive, and either party can initiate them. Possible benefits to targets are easily imagined, although we should stress that none has been adequately investigated (cf. Kappeler & van Schaik 1992; Aureli & Smucny 1998, for reconciliation). They include reduced risk of receiving further aggression, tension reduction, and reinforcement of the relationships with bystanders or indirectly with aggressors themselves.

Targets, for example, often have the opportunity to initiate friendly or appeasing contact with aggressors' kin, which might remove any threat of attack by those kin (or of further attack by aggressors) and alleviate physiological stress. This interaction also could substitute for reconciliation if it restores tolerance between targets and aggressors (Cheney & Seyfarth 1989; Judge 1991), and it could be a safe alternative to approaching the aggressor when this carries a high risk of further aggression. Relationship quality, however, is likely to influence the probability of such interactions: a bad relationship with the opponent can mean bad relationships with its relatives and generate strong constraints on affiliative post-conflict contact with them (Aureli & van Schaik 1991a). Still, aggressors' kin may sometimes benefit by accepting affiliative contact; the target may be an ally against other group members. In addition, bystanders may benefit from a general decrease in social tensions (Das et al. 1997, 1998).

Interests of targets in initiating or receiving post-conflict behavioral acts do not always coincide with those of bystanders. For example, bystanders can sometimes reinforce or help achieve dominance over targets, or can improve their access to resources, by initiating further aggression. Conversely, bystanders who interact affiliatively with targets or simply stay near them may risk also becoming victims of the aggressor or its allies (Aureli et al. 1994). For third parties to initiate interactions with targets or to accept affiliative overtures from them demands consideration of whether they thereby take risks. Several factors could explain risk taking. Like aggressive support for kin, post-conflict friendly contact with related targets could have indirect fitness payoffs if it decreases their exposure to further attack and quickly alleviates stress that receipt of aggression causes. Reconciliation can apparently have these effects (Aureli & van Schaik 1991b). Third parties should only initiate, or accept, post-conflict affiliation with unrelated targets if they receive some compensation for associated risks or if no risks exist (Das 1998). Male mountain gorillas are not at risk from females, for example, although subordinate males in multimale groups risk aggression from dominant males if they intervene in female conflicts (Watts 1997). Unrelated third parties should intervene aggressively or peacefully in conflicts, or provide post-conflict reassurance to targets, only when their social relationships with the targets are valuable (cf. de Waal & Aureli 1996). Allies and present or prospective mates are two classes of valuable social partners. For example, male baboons often support their female

friends in conflicts between females or make reassuring contact with their friends after such conflicts (Smuts 1985; Castles and Whiten 1998b). Preferential post-conflict affiliation between allies and between mates is typical of reconciliation ("valuable relationships hypothesis": Cords & Aureli, Chapter 9; de Waal, Chapter 2). It may also apply to third-party interactions if bystanders tend to decrease damage to the well-being of preferred targets.

Socioecology and Variation in Third-Party Involvement

The quality of relationship between former opponents and between targets and bystanders may strongly affect the occurrence of third-party interactions during and after conflicts because it determines the costs and benefits of the individuals involved. Variation in relationship quality must be examined within the socioecological context, as is true for variation in conciliatory tendency across species and across classes of social partners within species (Kappeler & van Schaik 1992; de Waal 1993; Aureli et al. 1997; Watts 1995a; Thierry, Chapter 6; van Schaik & Aureli, Chapter 15).

In gregarious primates, food distribution is the main factor affecting social relationships between females because of its influence on the nature and intensity of feeding competition and, consequently, on dispersal costs; male mating strategies also influence female social relationships (Wrangham 1980; van Schaik 1989; Sterck et al. 1997; van Schaik & Aureli, Chapter 15). Ecological situations in which within-group contest competition for food can strongly influence female reproductive success favor female philopatry, strong or "despotic" female dominance hierarchies, and cooperation among high-ranking related and unrelated females in rank acquisition and maintenance ("nepotistic" dominance systems). Female kin, besides sharing genes identical by descent, are therefore highly valuable social partners for one another, as are some unrelated females. Variation in tolerance for subordinates presumably relates to variation in between-group contest competition for food (van Schaik 1989; Sterck et al. 1997). As for reconciliation (e.g., de Waal & Ren 1988; Aureli et al. 1997), this variation should influence third-party interactions; relatively high tolerance should weaken constraints on post-conflict affiliation between targets and third parties.

Ecological situations in which contest feeding competition is unimportant and scramble competition predominates favor more individualistic dominance relationships and reduce the importance of cooperation between females, kin and nonkin. Ecological and social costs to female dispersal can be low and female transfer common (e.g., mountain gorillas: Watts 1994, 1996). Reconciliation may then be uncommon, even between female relatives who reside together (Watts 1995a); the general unimportance of relationships between females may also mean that female third parties give little or no conflict management aid to female targets.

Male distribution depends mostly on that of fertile females, although predation pressure may also influence the number of males per group (van Schaik & Aureli, Chapter 15). Male dispersal is the rule in most species. Males are often important social partners for females, and female need for male protection against conspecifics, particularly infanticidal males, can fundamentally influence male-female social relationships (Sterck et al. 1997). Males can be expected to improve their chances of mating with particular females by helping those females to manage their conflicts. This should especially be the case when female transfer between single-male breeding groups is common (e.g., mountain gorillas: Watts 1996), but it may also occur in some cases in which females are philopatric and groups typically contain multiple males.

Males in multimale groups are usually mating competitors with little to gain by helping one another other to manage conflicts. However, conflict management services given to allies who have received aggression (e.g., consolation between allied male chimpanzees: de Waal & van Roosmalen 1979; de Waal 1982) could help to maintain alliances in those cases in which they occur and can influence reproductive success.

The Occurrence of Third-Party Interactions

Data on at least some types of interactions between targets and third parties exist for various species (Table 14.1). Evidence for redirection, consolation, and male policing is reviewed in separate sections below. Here we focus on the other interactions.

Targets of aggression may receive further aggression from initially uninvolved individuals shortly after the end of dyadic conflicts (Table 14.1). A tendency for further aggression occurs in species with clear female dominance hierarchies, either despotic (e.g., longtail macaques [*Macaca fascicularis*]: Aureli & van Schaik 1991a) or more tolerant (e.g., vervet monkeys [*Cercopithecus aethiops*]: Cheney & Seyfarth 1989), and in at least some with more individualistic female dominance relationships (mountain gorillas: Watts 1995a). Such aggression may reinforce dominance relationships or allow bystanders to score psychological victories. It could also be a form of reproductive competition that induces stress and thereby impairs ovarian functioning or that makes targets more likely to emigrate or to transfer. Reconciliation between opponents may forestall such sequences of aggression. If so, targets should receive less aggression from third parties in post-conflict periods during which reconciliation occurs than during those

TABLE 14.1
Classification of Interactions between Opponents and Third Parties after Conflicts

Type	Initiator	Recipient	Present
FURTHER AGGRESSION	third party	target	YES: *Macaca fascicularis,*[1, 2] *M. mulatta,*[3] *Erythrocebus patas,*[4] *Cercopithecus aethiops,*[5] *Papio anubis,*[6] *Gorilla g. beringei*[9]
REDIRECTION	target	third party	YES: *M. fascicularis,*[1, 2] *M. fuscata,*[8] *Papio hamadryas,*[18] *G. g. beringei* (males, juveniles)[9]
			NO: *M. sylvanus,*[7] *Trachypithecus obscurus,*[10] *P. anubis,*[11] *G. g. beringei* (females)[9]
"KIN-ORIENTED REDIRECTION"[13]	target	aggressor's kin	YES: *M. fascicularis,*[12] *M. fuscata*[13]
	unspecified	opponent's kin	YES: *C. aethiops*[5]
"COMPLEX REDIRECTION"[5]	opponent's kin	opponent's kin	YES: *C. aethiops,*[5] *M. fuscata*[13]
THIRD-PARTY KIN AFFILIATION[a]	target	aggressor's kin	NO: *M. fascicularis,*[12] *M. fuscata,*[8] *M. nemestrina,*[14] *M. sylvanus,*[7] *P. papio,*[15] *M. nigra,*[16] *G. g. beringei* (females),[9] *T. obscurus*[10]
	target	own kin	YES: *P. papio*[15]
			NO: *M. fascicularis,*[12] *M. fuscata,*[8] *M. nemestrina,*[14] *M. nigra,*[16] *M. sylvanus*[7]
"SOLICITED CONSOLATION"[17]	target	third party	YES: *Cebus apella,*[17] *P. hamadryas,*[18] *G. g. beringei* (juveniles, females),[9] *Pan troglodytes*[19, 20]
CONSOLATION	third party	target	YES: *P. troglodytes*[19, 20]
			NO: *M. fascicularis,*[12] *M. mulatta,*[3] *M. arctoides,*[21] *M. fuscata,*[8] *M. sylvanus,*[7] *C. aethiops,*[5] *E. patas,*[4] *P. anubis,*[6] *P. hamadryas,*[18] *Eulemur fulvus,*[22] *Lemur catta,*[22] *G. g. beringei,*[9] *Pan paniscus*[23]

Sources: 1: Aureli & van Schaik 1991b; 2: Aureli 1992; 3: de Waal & Yoshihara 1983; 4: York & Rowell 1988; 5: Cheney & Seyfarth 1989; 6: Castles & Whiten 1998b; 7: Aureli et al. 1994; 8: Aureli et al. 1993; 9: Watts 1995b; 10: Arnold & Barton unpublished; 11: Castles & Whiten 1998a; 12: Aureli & van Schaik 1991a; 13: Aureli et al. 1992; 14: Judge 1991; 15: Petit & Thierry 1994c; 16: Petit & Thierry 1994b; 17: Verbeek & de Waal 1997; 18: Zaragoza & Colmenares unpublished; 19: de Waal & van Roosmalen 1979; 20: de Waal & Aureli 1996; 21: de Waal & Ren 1988; 22: Kappeler 1993; 23: de Waal 1987.

Note: No standard nomenclature exists for interaction types; we report some of the labels used. "Unspecified" indicates that authors did not distinguish whether aggressors or targets initiated interactions. Entries under "Present" refer to comparisons between post-conflict samples and either matched-control samples or baseline samples (see Veenema, Box 2.1, for details). "YES" indicates that rates were higher in post-conflict samples than in matched-control or baseline samples or than baseline rates; "NO" indicates that no difference between the two samples was found.

[a]Authors do not always distinguish whether third parties or targets initiate interactions; if third parties are initiators, these are special cases of "solicited consolation."

without reconciliation (cf. Aureli & van Schaik 1991b).

There is not much evidence for post-conflict affiliation between targets and their own kin or the aggressor's kin. Predictions about the variation in these interactions are not straightforward. Despotic dominance may make it difficult for female targets to approach their aggressor or its kin soon after a conflict: most aggression goes down the hierarchy, so that targets usually approach an individual who outranks them and who may be inclined to continue the hostility. Approaches to the aggressor's kin may then be more likely in more tolerant species, in which such approaches are less dangerous. In fact, no clear tolerance-related distinctions are known for various cercopithecines (Table 14.1). Post-conflict friendly interaction between targets and aggressors' kin was not elevated in four species with relatively despotic female dominance hierarchies (pigtail macaque [*Macaca nemestrina*]: Judge 1991; longtail macaques: Aureli & van Schaik 1991a; Japanese macaques [*M. fuscata*]: Aureli et al. 1993; guinea baboons [*Papio papio*]: Petit & Thierry 1994c). However, neither did it increase in two more tolerant macaque species (black macaques [*M. nigra*]: Petit & Thierry 1994b; Barbary macaques [*M. sylvanus*]: Aureli et al. 1994).

Similar arguments about variation in tolerance can apply to post-conflict interactions with targets' own kin. In most studies, however, targets did not show elevated post-conflict affiliation with their own kin (Table 14.1). In despotic species, the benefits of approaching them may be counterbalanced by the risks of exposing them to aggression. Post-conflict increase was reported for despotic guinea baboons (Petit & Thierry 1994c), but no evidence was found for either despotic (e.g., Aureli & van Schaik 1991a) or tolerant macaques (e.g., Petit & Thierry 1994b).

Post-conflict levels of friendly contact between targets and aggressors' kin also did not differ significantly from control or baseline levels in two non-cercopithecine species with more individualistic female dominance relationships (mountain gorillas: Watts 1995a; spectacled langurs [*Trachypithecus obscurus*]: Arnold & Barton unpublished). Small family size in these species acts against such differences, but their absence may also reflect the general unimportance and high predictability of relationships between females. The "good" relationships that characterize a few dyads, for example, those between maternal relatives in mountain gorillas (Watts 1995a), may be resilient in the face of conflict-induced social tension. Conflict may have little effect on tension in other relationships, or females may only be able to get conflict management aid from the few partners with whom they have good relationships, notably adult males (Fig. 14.1). In fact, in mountain gorillas female targets increased their rates of post-conflict affiliative or appeasing contact with third parties unrelated to either opponent, such as adult males (Watts 1995b; see below).

Redirection

Papoose and Tuck, two female mountain gorillas who often interact aggressively, engage in a screaming match while they feed close together. Neither abandons her spot, and although both fall silent, they seem tense. When Tuck moves off three minutes later, she sees Shangaza, Papoose's adult daughter. Tuck cough grunts harshly at Shangaza and makes a sudden movement in her direction; Shangaza avoids her.

An individual redirects aggression when it threatens or attacks a third party shortly after being the target of aggression. Redirection is common in primates and other mammals (e.g., rats: Levine et al. 1989). However, not all studies have shown significant increases in targets' aggression toward

FIGURE 14.1. Adult male (above) and adult female mountain gorillas. Males often intervene in conflicts between females and end many of these without either female winning. Photograph by David P. Watts.

third parties in post-conflict situations relative to control situations or to baseline rates (Table 14.1); thus redirection is not always a typical sequel to receipt of aggression.

Presumably redirection is partly an alternative to counteraggression toward opponents (Thierry 1985; Aureli et al. 1992; Castles & Whitten 1998a), and its occurrence should vary with risks attached to counteraggression and with risks that targets of redirection will themselves retaliate. For example, redirection at opponents' relatives by female targets (i.e., kin-oriented redirection) may limit immediate

and future aggression from those opponents because it imposes indirect fitness costs on them (Cheney & Seyfarth 1989; Aureli & van Schaik 1991a; Aureli et al. 1992). It should be rare, however, in species with "nepotistic" female dominance hierarchies because it could easily lead to further attacks from aggressors and their kin. At least, kin-oriented redirection should usually be directed at vulnerable kin in low-risk contexts (Aureli et al. 1992). This was the case in a group of Japanese macaques in which targets preferentially redirected against aggressors' kin that were younger than and still subordinate to them, and usually joined aggressive interactions in which risks of retaliation were presumably low (Aureli et al. 1992).

Regular, effective support from nonrelatives also constrains redirection. A female in a species with "nepotistic" female dominance hierarchies risks retaliation from females of all matrilines that outrank her own if she redirects aggression at a member of any one of them. This may help to explain why little redirection by females against other females occurred in wild olive baboons (*Papio anubis;* Castles & Whiten 1998a). Another reason for the rarity of redirection in this case was that adult males often intervened in female-female conflicts. Redirecting aggression at another female might have caused retaliation from that female's male friend (Castles & Whiten 1998a).

High rates of bidirectional aggression can also constrain redirection by forcing targets to choose their own victims carefully so as to minimize costly counteraggression (Castles & Whiten 1998b). For example, bidirectional aggression was common in the same group of olive baboons with little redirection. Juvenile and subordinate male mountain gorillas commonly redirected aggression, but adult females did not (Watts 1995a). Common bidirectional aggression, including damaging fights, between females (Watts 1994) may have made most other females, and even their immature offspring (at least when in sight of their mothers), unsafe targets for redirection by a given female. Similar explanations may hold for the absence of redirection among female spectacled langurs (Arnold & Barton unpublished).

Functions of Redirection

One important potential function of redirection is to reduce aggression-induced stress. In rats, directing aggression at conspecifics reduces the secretion of glucocorticoids (i.e., stress-related hormones) after they have received unpredictable electric shocks (Levine et al. 1989). Receipt of aggression may often entail a loss of social control that parallels this unpredictability, and redirection could have similar effects in this context. The finding that male wild baboons who redirected aggression had lower basal glucocorticoid levels than those who did not is consistent with the stress-reduction hypothesis (Sapolsky & Ray 1989; Virgin & Sapolsky 1997; Sapolsky, Box 6.1), although the immediate neuroendocrine effects of redirection in this case were unknown. Redirection may also decrease physiological or behavioral indicators of stress (e.g., scratching rates: Maestripieri et al. 1992) compared with values during post-conflict periods with neither redirection nor reconciliation (reconciliation can have this effect: Aureli & van Schaik 1991b; Castles & Whiten 1998b; Das et al. 1998). Aureli & van Schaik (1991b) found support for this hypothesis in the only study that reports relevant data: scratching rates were lower during post-conflict periods with redirection than during those with neither redirection nor reconciliation.

Redirection might also have other functions mediated by social effects. First, it may modify the immediate attitudes of aggressors or third parties so that they are less likely to make further attacks on targets, perhaps because their attention is deflected to third parties (de Waal & van Hooff 1981; Aureli & van Schaik 1991a). Rates of aggression against targets during post-conflict periods should then be lower when targets redirect aggression than when they do not. Data on longtail macaques support this prediction (Aureli & van Schaik 1991b). Kin-oriented redirection can be considered an interesting, special case: it could induce aggressors to become more tolerant of subordinate opponents over the long term by threatening aggressors with indirect fitness costs (Aureli et al. 1992).

Another possible function of redirection is to facilitate reconciliation. If targets manage to induce the former aggressor to join them against third parties, redirection could directly function in restoring the relationship between target and aggressor (Aureli & van Schaik 1991b). There is some evidence for a facilitating effect of redirection on the occurrence of reconciliation. In a group of longtail macaques, reconciliation was more likely to occur after redirection took place than vice versa (Aureli & van Schaik 1991a).

Consolation

Git, an adult female hamadryas baboon, is quietly grooming her harem male Abo. Chi, a higher-ranked female of the harem, approaches the grooming dyad and stares at Git. She stops grooming, scratches her chest, and looks away. Chi keeps staring at Git, who suddenly pushes off, screams against Chi, and then walks away. Chi agitatedly starts to grooms Abo while raising eyebrows at Git. The harem male walks off. Chi and Git nervously sweep the ground for a while. Then, Chi approaches her mother, Fav, the harem's most dominant female, and grooms her. Neg, Git's adult sister, approaches and grooms Git.

De Waal & van Roosmalen (1979) defined *consolation* as any increase in affiliative interaction between targets and third parties specific to post-conflict contexts. Following this definition both the grooming of Chi on Fav and the grooming of Neg on Git would qualify as instances of consolation. The term *consolation*, however, is best reserved for interactions that third parties initiate (de Waal 1993, 1996; Cords & Aureli 1996; de Waal & Aureli 1996), because it raises the issue of whether these interactions depend on empathy, that is, the ability to recognize the situation of another, to distinguish it from one's own, and to act out of concern for the other's well-being (de Waal 1993, 1996; de Waal & Aureli 1996). Calling post-conflict target-initiated affiliation *solicited consolation* (Verbeek & de Waal 1997) sidesteps this issue.

Overall, there is little evidence for bystander-initiated consolation, whereas solicited consolation has been reported in a few species (Table 14.1). De Waal & Aureli (1996) proposed two alternative, but not necessarily mutually exclusive, explanations for why consolation occurs in chimpanzees but apparently not in macaques:

1. The *Social Cognition Hypothesis* (see also Castles, Box 14.1): Consolation depends on empathy, which may be beyond the cognitive abilities of most nonhuman primates. The apparent inability of macaques to attribute knowledge states to others (Povinelli et al. 1991) buttresses this claim. Strong evidence that chimpanzees can attribute knowledge, or at least emotions, would support this hypothesis, although their abilities in this regard are unclear (Povinelli & Eddy 1996).

2. The *Social Constraints Hypothesis*: Third parties initiate affiliation only when their own risks of becoming targets of further aggression are low or when their potential gains are sufficiently high. Risks depend on power asymmetries: when these are pronounced and strongly enforced and tolerance for unrelated

---

BOX 14.1

Triadic versus Dyadic Resolutions

Cognitive Implications

Duncan L. Castles

How intelligent do primates have to be to resolve conflicts? Is the cognition involved in triadic conflict resolution special?

In principle, when just two parties are involved (*dyadic* conflicts), the cognitive demands of conflict resolution are low: if a disposition toward friendly contact following a conflict exists, individual recognition and the ability to remember the identity of the former opponent are the minimum requirements (de Waal & Yoshihara 1983)—that is to say, basic abilities that neither are unique to nor differentiate cognitive function among primates (e.g., Manning & Dawkins 1992; Byrne 1995). However, in reality conflict resolution appears considerably more complex: it involves skills of behavior reading while avoiding, for example, becoming embroiled in further aggression that is unlikely to be trivial. Furthermore, if reconciliation is selectively employed to manipulate relationship quality, a "Machiavellian" dimension is added to the cognitive challenge (Cords 1997).

Conflict resolution does not consist exclusively of affiliation between former combatants. *Triadic* resolutions comprise various forms of (1) post-conflict affiliation between aggressors and third parties uninvolved in the original aggressive episode (Das, Chapter 13), (2) affiliation between targets of aggression and third parties (Watts et al., Chapter 14), and (3) mediating interventions by third parties in conflicts (Petit & Thierry, Box 13.1). Here, I examine the cognitive demands of triadic resolutions.

For simple quantitative reasons we might expect triadic resolutions, because they involve more individuals, to be more computationally complex and thus more cognitively taxing; but will there also be

BOX 14.1 (continued)

differences in the kind of cognition required? In assessing the cognitive load of triadic resolutions, several factors are important: (1) the nature and supposed function of the post-conflict interaction, (2) evidence that the supposed function is actual, and (3) analysis of the cognitive ability necessary to achieve the function. In these regards it is important to note that empirical demonstrations that triadic interactions function to resolve conflicts are as yet rare (Das, Chapter 13).

However, it may be possible to use variation in the post-conflict options of aggressors, targets of aggression, and third parties to deduce differences in the cognitive demands placed on each class of participant. For example, during one type of triadic resolution, the aggressor increases post-conflict affiliation rate with the kin of its former target ("triadic reconciliation," Judge 1991; Das, Chapter 13). In comparison with dyadic resolutions, such interactions may demand more sophisticated observational skills and more often require careful reading of dangerous situations. Certainly, when aggressors are the initiators, knowledge of the target's social relations is a prerequisite of this style of third-party resolution.

Similar considerations apply to post-conflict affiliation with aggressors initiated by third parties. Knowledge of the aggressor's relationships will be of value in minimizing chances of receiving aggression from close affiliates of the aggressor. Risk and the importance of behavior reading will vary with the danger presented by the aggressor, or its affiliate, which will commonly be a function of relative rank. In these respects triadic post-conflict affiliation seems to share cognitive aspects with coalitions and alliances (Harcourt 1988).

Tonkean macaque (*Macaca tonkeana*) third parties peacefully intervene in conflicts to effect triadic resolutions (Petit & Thierry 1994a). Again, social knowledge will be necessary if third parties selectively intervene according to the nature of the opponents' relationships. Social knowledge may also be used when aggression spreads between matrilines, particularly during kin-oriented redirection in which targets of aggression (or their kin) attack the kin of targets' opponents (Aureli & van Schaik 1991a; Aureli et al. 1992).

Knowledge of the relationships of others is a form of information qualitatively distinct from knowledge of an individual's own relationships (de Waal 1982; Cheney & Seyfarth 1990). Yet, though obtaining social knowledge may demand more time and cognitive effort, the cognition employed need not be of a different kind than that used elsewhere: although the complexity of managing triadic social relationships has been implicated in the evolution of intentional reasoning (Humphrey 1976) and theory of mind, associative learning is considered sufficient to cope with the triadic interactions of monkeys (Whiten & Byrne 1988).

Both targets and initiators of aggression show increased rates of post-conflict anxiety (Aureli & Smucny, Chapter 10). Thus both need anxiety reduction, which, when reconciliation does not occur, could be derived from consolatory affiliation with third parties. Considerable discussion has centered around the proposition that the absence of elevated rates of post-conflict affiliation between third parties and targets of aggression in monkeys is the product of an inability to empathize with the distress of combatants (de Waal & Aureli 1996). This hypothesis is grounded on a failure to demonstrate increased rates of apparently consolatory behavior in several cercopithecine monkeys coupled with its presence in chimpanzees (*Pan trogoldytes*) (see Watts et al., Chapter 14) and assumptions that (1) such behavior reduces stress, (2) chimpanzees have a theory of mind, absent in monkeys, enabling empathy with distressed recipients of aggression and leading them, at times, to console them; and (3) that it is *not* the case that the social environment of monkeys is too dangerous to allow frequent consolatory behavior (i.e., the Social Constraints Hypothesis: de Waal & Aureli 1996).

Debate continues over the idea that apes possess a theory of mind that monkeys do not (e.g., Povinelli & Eddy 1996; Whiten & Byrne 1997). Moreover, restrictions on cercopithecine consolation seem marked. Since recipients of aggression are (1) liable to be attacked again and (2) inclined to redirect aggression toward lower-rankers, third parties run the risk of being drawn into conflict themselves (Aureli & van Schaik 1991b). Therefore it remains unclear whether empathy is a necessary condition for consolation. Theoretically, selection for a behavioral tendency to affiliate with and calm conflict participants might be sufficient.

BOX 14.1 (continued)

A further consideration, sometimes overlooked in conflict-resolution research, is that *absence of an effect* should not invariably be regarded as evidence that there is *no effect*. For example, if approaching a target of aggression is dangerous yet individuals nevertheless interact with targets at equal rates in post-conflict and control situations, one interpretation is that bystanders are electing to suffer an increased risk of received aggression to preserve their rate of interaction with targets. Fortunately, this hypothesis is testable. How third parties perceive the risks of post-conflict proximity with the target of aggression and how they respond to the target's distress can be measured by paying attention to post-conflict interactions as well as signs of anxiety, such as self-scratching, by third parties during post-conflict and control conditions (Castles et al. in press).

Overall, as the functions of most post-conflict triadic interactions are undemonstrated, it is possible to make only limited inferences about the cognitive implications of such interactions. Furthermore, though as observers we can attempt to use variation in interaction patterns to make some a posteriori inferences about difference in cognitive demands, we should ask if ours is a realistic account of the challenge faced by animals engaged in aggression and resolution. To take an alternative perspective, a dyadic resolution attempt remains such only while just two parties are involved. As animals involved in any kind of resolution attempt are required both to read the behavior of others and to time their actions skillfully to avoid becoming embroiled in further aggression, it seems likely that individuals capable of dealing with the cognitive load of triadic interactions will employ any distinct cognitive skills involved to execute dyadic resolutions better. As a result, there may be little difference in the cognition employed in dyadic and triadic resolutions by species capable of both. However, this does suggest that such species' dyadic resolutions will be more cognitively sophisticated than those of species incapable of triadic resolution.

subordinates is low, third parties often cannot support or affiliate with targets without high risk. This characterization applies to most of the macaques that have been the main focus of research, whereas relationships in chimpanzees involve considerably more tolerance.

How could the Social Constraints Hypothesis be tested? One experimental test that has not yet been done (de Waal & Aureli 1996) would be to allow third parties to interact with targets in situations in which they faced no immediate threat of reprisal from higher-ranking aggressors. The prediction is that consolation would then occur in monkeys with despotic dominance.

Comparative data on species that lack despotic dominance could also serve as a test (de Waal & Aureli 1996). The prediction is that consolation would occur in species (or situations within a given species) in which third parties do not risk retaliation by aggressors, either because they are dominant over them or because tolerance is a key component of social relationships. For example, the presence of consolation in stumptail macaques, in which female relationships involve considerable tolerance (de Waal & Ren 1988), would support this prediction. However, "tolerant" macaques still have formal dominance hierarchies, and species that do not may be better test cases. Here we report data from three studies on species in which female social relationships are individualistic and harem-holding males are clearly dominant over females.

In one study of mountain gorillas (Watts 1995b), rates of affiliative interaction between female targets and males were higher after conflicts between females than they were during matched-control periods (and were higher than baseline rates). Targets, however, initiated most interactions, many of which followed male interventions in the conflicts and may have served to appease the males (see below).

In a second study, Arnold & Barton (unpublished) did not find elevated rates of post-conflict affiliative interactions between targets and bystanders in spectacled langurs (Fig. 14.2). They found, however, that third parties and targets embraced each other at higher rates in post-conflict samples. Embraces occurred almost exclusively in this context, and third parties did not face reprisals from aggressors when they interacted with targets.

In the third study, Zaragoza & Colmenares (unpublished) adapted the index of conciliatory tendency of Veenema et al. (1994; see Veenema, Box 2.1) to calculate "consolatory tendencies" for the Madrid Zoo hamadryas baboon (*Papio hamadryas*) colony. Although affiliative contacts between adult targets and third parties were higher after conflicts than during control periods, the initiative for these post-conflict affiliative contacts was mostly taken by targets. No evidence for bystander-initiated consolation was found.

In hamadryas baboons, males intervene aggressively or peacefully in many of their females' conflicts, and females often seek out males during conflicts (see below). Male protection and post-conflict affiliation may stem from self-interest, if they help to retain female loyalty. In fact, female loyalty to males could influence male success at preventing aggressive harem takeovers and female resistance to attempts by extragroup males to appropriate them from their units (Colmenares unpublished). This condition could also apply to other species in which female transfer is common (e.g., mountain gorillas, many langurs).

Affiliative contacts between targets and individuals who had just intervened are problematic cases of consolation. In hamadryads baboons, for

FIGURE 14.2. A spectacled langur. This species lives in single-male groups in which females have individualistic dominance relationships. Photograph by Michael Seres.

example, males typically intervene during a conflict between their females and then groom or, more often, are groomed by the original target of aggression. These males no longer qualify as uninvolved third parties. This temporal pattern of intervention in conflicts and post-conflict affiliation might also render problematic the interpretation of consolatory behavior of males in mountain gorillas and in some macaques. The possibility that this and other factors, including present and potential relationship value, may motivate male and female behavior during or following dyadic aggressive interactions should also caution against easy acceptance of pure, empathy-based explanations (cf. Castles & Whiten 1998a).

Functions of Consolation

The major proposed functions of consolation, which could apply to solicited consolation as well, echo some proposed for reconciliation, such as stress alleviation and protection against further aggression. Consolation may even substitute for reconciliation in stress reduction and protection. Data on hamadryas baboons indirectly support the "substitution" hypothesis: hamadryas males typically groom with one of their harem females after conflicts with other males, and females groom with their harem males after conflicts with other harem females (Zaragoza & Colmenares unpublished). For accurate testing, we need direct comparison of stress indicators during post-conflict samples with consolation, but not reconciliation, with those during samples with neither reconciliation nor consolation and those during control samples to test the stress-reduction hypothesis. We then need to compare these rates with those during samples with reconciliation only to test the substitution hypothesis. Similar comparisons using data on aggression rates could test the protective function.

If consolation provides these benefits, it can be expected to occur between kin or individuals with strong, mutual attachments (de Waal & Aureli 1996). High frequencies of post-conflict third-party affiliation in male-female dyads of hamadryas baboons (Zaragoza & Colmenares unpublished) and male-female "friends" of olive baboons (Castles & Whiten 1998a) offer circumstantial support (see also below). An adequate test, however, requires data on the long-term effects of such interactions and on the effects of experimental manipulations that prevent them (cf. Cords & Aureli 1996; Silk 1996 for reconciliation).

Third parties who are potential recipients of further aggression might also benefit from initiating affiliative post-conflict contact with targets if this reduces general social tension, although contact with aggressors is probably more effective (Das et al. 1997, 1998). This interaction would fit the definition of third-party consolation, but it may not depend on empathy if the goal is not to alleviate the stress of the target.

Male Policing

Why should males intervene in conflicts that involve females, as in our opening example? Smuts's (1985) proposal that male olive baboons develop and maintain preferential mating relationships with their female "friends" partly by giving them support in aggressive interactions implies that conflict management help given by males to females is mating effort. Castles & Whiten (1998b) extended this argument to post-conflict affiliation that male baboons initiate with females: calming females and protecting them against further aggression may also help to develop and maintain friendships and thereby have reproductive payoffs. The same logic applies forcefully to species in which breeding groups typically have single males and in which female loyalty to males influences male reproductive success. It also applies to "control interventions," that is, those that stop conflicts impartially (de Waal 1977; Petit & Thierry, Box 13.1), like Ziz's intervention in our opening example.

Control interventions should mainly occur when males have equally valuable relationships with the two contestants. Conversely, support for targets should mostly take place when males have more valuable relationships with them than with aggressors or are trying to improve their relationships with targets, when males' relationships with targets are less resilient than those with aggressors, or when the targets are close kin of the males.

Males should intervene at particularly high rates when females can easily transfer between groups (or at least when female dispersal is common) if male protection discourages females from leaving the group.

Data on mountain gorillas support these predictions (Watts 1991, 1997). Males intervened in many conflicts between females and mostly performed control interventions. Most exceptions involved either support for close kin by an old, apparently nonbreeding male or support by a breeding male for recently immigrated females, with whom he had not yet established stable mating relationships. Data on captive hamadryas baboons also support the predictions (Zaragoza & Colmenares unpublished) in that males commonly intervened in conflicts between females (28 percent); most interventions (79 percent) were aggressive, particularly when conflicts occurred between females of different one-male units; and males supported their own females in almost all of their conflicts with extragroup males (96 percent). The frequent male policing of conflicts between females, and the observation that females often sought males for help or reassurance, are consistent with data from studies of wild hamadryas baboons (Kummer 1967, 1968) and earlier findings in captivity (Colmenares & Lázaro-Perea 1994).

Male golden monkeys (*Rhinopithecus roxellanae*) in two small captive groups also often intervened in within-group conflicts between females (Ren et al. 1991). Most interventions (74 percent) were peaceful; aggressive intervention was mild and directed mostly at aggressors. In the wild, this species and some other Asian colobines typically form very large groups that may consist of multiple one-male units between which females transfer, and female transfer is also common in Asian colobine species that typically form one-male groups (Ren et al. 1991; Kirkpatrick 1998; Yeager et al. 1998). Maintaining harmonious relationships

among females may have reproductive payoffs for the male (Ren et al. 1991), and peaceful interventions could achieve this goal without the risk for the male of damaging his relationships with either opponent (Petit & Thierry 1994a, Box 13.1). Few data on interventions by males in wild groups are available. In Thomas's langurs (*Presbytis thomasi*), a male "almost always intervenes and tries to stop aggression between" females (Steenbeek 1997, p. 174). In this species, however, female transfer occurs, but males seem unable to control female residency.

Gelada baboons (*Theropithecus gelada*) provide an important comparative case because females of this species remain in their natal one-male units as adults, form strict, "nepotistic" dominance hierarchies, and maintain alliances with their close female kin (Dunbar 1980, 1983b, 1984, 1988). This contrasts with the small one-male unit size and female dispersal of hamadryas baboons that results in the usual absence of closely related females in the same group. Closely related female mountain gorillas sometimes reside together, but male control of female aggression limits the effectiveness of kin alliances (Watts 1997). This is also the case in hamadryas baboons, when female kin happen to reside in the same one-male unit (Colmenares unpublished). In gelada baboons, on the other hand, because of their social structure, most interventions in conflicts between females are by female allies of one or both opponents; male interventions are mostly restricted to support for any females who lack female allies (Dunbar 1980, 1983a, 1984). Still, support for females without allies and control interventions supplement grooming (Dunbar 1984) as means for males to retain female loyalty in the face of harem fissions and takeover attempts by extragroup males. In addition, in gelada baboons frequent male defense of females against aggression from extragroup females (Dunbar 1983b) could reduce adverse effects of female-female harassment (Dunbar 1984) while demonstrating male protective ability.

Males are socially peripheral in some species in which females are philopatric and single males compete to gain access to female groups (e.g., redtail monkeys [*Cercopithecus ascanius*]: Struhsaker & Leland 1979). Male conflict management help would not be expected in such cases. No relevant data, however, are available.

Functions of Male Policing

In the short term, male control of aggression between females could make targets less likely to lose contested resources, less likely to experience further attacks and suffer any negative effects associated with stress, and less likely to receive potentially debilitating wounds. No study has tested the stress-reduction hypothesis. Data on mountain gorillas are consistent with the other functions (Watts 1994, 1995b, 1997). Male interventions were associated with targets being less likely to lose contests, even when they faced female coalitions rather than single opponents. Opponents commonly engaged in further aggression during post-conflict periods, but aggression rates were lower after male interventions. Females commonly inflict wounds during fights, and males intervened more often in escalated fights than in conflicts that did not escalate.

Several medium- or long-term functions are also possible. Effective male interventions may lower rates of dyadic aggression if they prevent those females who would otherwise win dyadic contests (because they are larger or can get more effective support from female allies) from gaining any associated benefits. Initiating aggression is not risk-free; any factor that lowers its potential payoffs should lower its frequency. Female immigration provides one context in which to look for this effect: when immigrants meet resistance from resident females, male control of aggression should

diminish this resistance. Mountain gorilla data are consistent with this prediction: high rates of aggression toward newcomers after a wave of immigration into one group declined in association with high rates of male interventions in contests between females (Watts 1991). Dyadic aggression rates should also increase with female group size, because male ability to police female conflicts declines with female group size (Sterck et al. 1997). Dyadic aggression rates and female group size, however, are not clearly associated in mountain gorillas, and data are not available from other species (Watts 1996; Sterck et al. 1997).

If effective policing of female contests is a case of reproductive effort, it should increase male reproductive success by making females less likely to emigrate, less likely to mate with other sexually mature males in their groups, and more resistant to the mating tactics of extragroup males. Support for recent immigrants with whom males have not yet established stable mating relationships, as in mountain gorillas (Watts 1991), could have these effects. Females who fare poorly in dyadic contests with other females (e.g., because they are relatively small) may be persuaded to stay with males who prevent them from losing contests (Sterck et al. 1997; Watts 1997). As this implies, control interventions are not necessarily neutral: a male may thereby prevent one opponent from losing (cf. Petit & Thierry, Box 13.1). If so, males should intervene relatively often in conflicts that involve females who often lose dyadic contests, and female emigration should be independent of female success in contests in which males do not intervene. These predictions have not been systematically investigated. Even in harem-polygynous species, females often have opportunities to mate with follower males (e.g., hamadryas baboons: Kummer 1968; mountain gorillas: Sicotte 1993; Watts 2000) or with extragroup males (e.g., guenons: Cords 1987). Whether the effectiveness of male policing influences female mate choice in these contexts is unknown, although the presence of conflicts between males in multimale gorilla groups over which male can intervene in female conflicts (Watts 1997) may be interpreted as indirect evidence for this function. Male and female interests do not always coincide, and factors such as infanticide avoidance can lead females to seek mating with multiple males.

General Discussion

The material above, heterogeneous in some respects, shares several main themes. Targets of aggression have several social tactics for conflict management, some of which may be as effective as reconciliation. Socioecological variation influences both the ability of targets to use these tactics and the ability and willingness of third parties to provide conflict management services, either as recipients or as initiators of post-conflict interactions. As a corollary, third parties should be more inclined to engage in post-conflict interactions that benefit targets when their own relationships with those targets are more valuable and hence they are more likely to receive direct or indirect benefits from these relationships. High direct benefits are an especially likely explanatory factor when third parties are not closely related to the targets. In particular, the conflict management role of male third parties may be an important aspect of male-female relationships when it can influence female mate choice.

We need more detailed investigation of particular cases and more systematic comparative analysis across species to test the proposed short-term conflict management functions of interactions between targets and third parties (e.g., stress reduction) and the hypothesis that relationship quality influences the involvement of third parties in dyadic conflicts. The possibility that male interventions in female conflicts, and male initiation of

and acceptance of post-conflict affiliative contact with females, represent male mating effort also requires systematic testing. Such male behavior, however, raises several more general issues, and we conclude by considering some.

One is both conceptual and methodological. Suppose a male hamadryas baboon or mountain gorilla intervenes aggressively in a conflict between females, and the target then initiates nonaggressive contact with him (male interventions often involve rapid interaction sequences like this). Is she trying to manage a conflict with him, with her opponent, or with both? Female mountain gorillas often give formal signals of submission to males, and seem to be appeasing them, in this context (Watts 1995b). Perhaps the same tactic can serve both the female-male and female-female conflict management goals. A hypothetical sequence of events could be the following: a female attacks another female, the male intervenes, the target appeases the male, and finally the target redirects aggression at a fourth individual. If this was the typical sequence, we might conclude that male intervention does not facilitate management of the initial conflict because redirection usually follows. At least in mountain gorillas, however, this is not the case, and male interventions may both help targets to deal with the immediate effects of aggression from other females (e.g., stress reduction) and reinforce the relationship between the target female and the male.

The influence of relationship quality raises the issue of reproductive success with regard to male interventions, just as it raises the issue of inclusive fitness for reconciliation (Cords & Aureli 1996; Silk 1996, Box 9.1). We suggest that the high tendency for males to make impartial interventions in within-group contests in hamadryas baboons, mountain gorillas, langurs, and probably other species (perhaps including some non-primates, such as gregarious equids) is an important mating tactic, favored by female choice (cf. Watts 1996, 1997). This interpretation is in line with recent theoretical developments on male-female conflict of reproductive interest, male sexual coercion of females, and other social influences on reproductive success. Long-term social bonds and exclusive mating associations between males and females are viewed as favoring male involvement in the protection of females and infants against social harassment from within and outside their groups (Smuts 1985; Whitten 1987; van Schaik & Dunbar 1990; Smuts & Smuts 1993; Brereton 1995; van Schaik 1996; Sterck et al. 1997). In line with this reasoning, we predict that males who protect their females (and infants) effectively in contests within groups, as well as against outside threats, are preferred as mating partners and consequently have larger, more cohesive, and more stable female groups. If female group size indicates the owner male's capacity to defend group members and hence improve female reproductive success, then (within the constraints set by competition among females) females should prefer to enter large, rather than small, groups and to mate with group-holding males, not bachelors. Long-term data on hamadryas baboons (Colmenares unpublished) and mountain gorillas (Watts 1997, 2000) support this argument.

A final issue is the possible complementarity of self-interested male behavior and consolation. Since Bernstein (1976) described the "control role" in capuchins and rhesus macaques, many investigators have investigated this role, usually by the group dominant male (reviews: Boehm 1981, 1994; Ehardt & Bernstein 1992; see also de Waal

1982). Many interventions in the Madrid hamadryas colony were peaceful ("greeting and grooming interventions": Colmenares & Lázaro-Perea 1994). Like peaceful interventions in other species (Ren et al. 1991; Petit & Thierry 1994a, Box 13.1), these constitute bystander-initiated "consolation" *sensu* de Waal & Aureli (1996) if they occurred during post-conflict periods. Furthermore, male agonistic interventions often pacify opponents (e.g., chimpanzees: Boehm 1981, 1994; de Waal & van Hooff 1981; de Waal 1982; gorillas: Watts 1991, 1997). Consolation, impartial interventions, and interventions that are partial but that pacify opponents all presumably help to reestablish social order and may make social relationships among group members more relaxed. They could therefore be precursors of morality (de Waal 1991, 1996; Killen & de Waal, Chapter 17).

We also agree with Castles & Whiten (1998b) that pacifying and protective behavior by males in response to within-group aggression could be primarily selfish, with side effects that look moralistic, especially when males face no serious immediate physical risks. By protecting targets and stopping fights between females, males are defending their social and sexual interests. Effective maintenance of social order and reduction of the amount of stress that females experience could increase female loyalty and attraction and thereby increase male reproductive success. Still, such "selfish" behavior may serve to "console" and presumably stems from a tendency to be bothered by the conflicts of others. This may be the same tendency that underlies true consolation based on empathy and sympathy (de Waal 1996; de Waal & Aureli 1996).

References

Aureli, F. 1992. Post-conflict behaviour among wild long-tailed macaques (*Macaca fascicularis*). *Behavioral Ecology and Sociobiology*, 31: 329–337.

Aureli, F., & Smucny, D. 1998. New directions in conflict resolution research. *Evolutionary Anthropology*, 6: 115–119.

Aureli, F., & van Schaik, C. P. 1991a. Post-conflict behaviour in long-tailed macaques (*Macaca fascicularis*). I: The social events. *Ethology*, 89: 89–100.

Aureli, F., & van Schaik, C. P. 1991b. Post-conflict behaviour in long-tailed macaques (*Macaca fascicularis*). II: Coping with the uncertainty. *Ethology*, 89: 101–114.

Aureli, F., Cozzolino, R., Cordischi, C., & Scucchi, S. 1992. Kin-oriented redirection among Japanese macaques: An expression of a revenge system? *Animal Behaviour*, 44: 283–291.

Aureli, F., Veenema, H. C., van Panthaleon van Eck, J. C., & van Hooff, J. A. R. A. M. 1993. Reconciliation, redirection, and consolation in Japanese macaques (*Macaca fuscata*). *Behaviour*, 124: 1–21.

Aureli, F., Das, M., Verleur, D., & van Hooff, J. A. R. A. M. 1994. Post-conflict social interactions among Barbary macaques (*Macaca sylvanus*). *International Journal of Primatology*, 15: 471–485.

Aureli, F., Das, M., & Veenema, H. C. 1997. Differential kinship effect on reconciliation in three species of macaque (*Macaca fascicularis, M. fuscata,* and *M. sylvanus*). *Journal of Comparative Psychology*, 111: 91–99.

Bernstein, I. S. 1976. Dominance, aggression, and reproduction in primate societies. *Journal of Theoretical Biology*, 60: 459–472.

Boehm, C. 1981. Parasitic selection and group selection: A study of conflict interference in rhesus and Japanese macaques. In: *Primate Behavior and Sociobiology* (A. B. Chiarelli & R. S. Corrucini, eds.), pp. 161–182. New York: Springer-Verlag.

Boehm, C. 1994. Pacifying interventions at Arnhem Zoo and Gombe. In: *Chimpanzee Cultures* (R. W. Wrangham, W. C. McGrew, F. B. M. de Waal, & P. G. Heltne, eds.), pp. 211–226. Cambridge: Harvard University Press.

Brereton, A. R. 1995. Coercion-defense hypothesis: The evolution of primate sociality. *Folia primatologica*, 64: 207–214.

Byrne, R. W. 1995. *The Thinking Ape: Evolutionary Origins of Intelligence.* Oxford: Oxford University Press.

Castles D. L., & Whiten, A. 1998a. Post-conflict behaviour of wild female olive baboons. I. Reconciliation, redirection, and consolation. *Ethology*, 104: 126–147.

Castles D. L., & Whiten, A. 1998b. Post-conflict behaviour of wild olive baboons. II. Stress and self-directed behaviour. *Ethology*, 104: 148–160.

Castles, D. L., Aureli, F., & Whiten, A. In press. Social anxiety, relationships, and self-directed behaviour among wild olive baboons. *Animal Behaviour.*

Chapais B. 1992. The role of alliances in the social inheritance of rank among female primates. In: *Coalitions and Alliances in Humans and Other Animals* (A. H. Harcourt & F. B. M. de Waal, eds.), pp. 29–60. Cambridge: Cambridge University Press.

Cheney, D. L., & Seyfarth, R. M. 1989. Redirected aggression and reconciliation among vervet monkeys, *Cercopithecus aethiops. Behaviour*, 110: 258–275.

Cheney, D. L., & Seyfarth, R. M. 1990. *How Monkeys See the World.* Chicago: University of Chicago Press.

Colmenares, F., & Lázaro-Perea, C. 1994. Greeting and grooming during social conflicts in baboons: Strategic uses and social functions. In *Current Primatology,* Vol. 2: *Social Development, Learning, and Behaviour* (J. J. Roeder, B. Thierry, J. R. Anderson, & N. Herrenschmidt, eds.), pp. 165–174. Strasbourg: Université Louis Pasteur.

Cords, M. 1987. Forest guenons and patas monkeys: Male-male competition in one-male groups. In: *Primate Societies* (B. B. Smuts, D. L. Cheney, R. M. Seyfarth, R. W. Wrangham, & T. T. Struhsaker, eds.), pp. 98–111. Chicago: University of Chicago Press.

Cords, M. 1997. Friendship, alliances, reciprocity, and repair. In: *Machiavellian Intelligence II: Evaluations and Extensions* (A. Whiten & R. W. Byrne, eds.), pp. 24–49. Cambridge: Cambridge University Press.

Cords, M., & Aureli, F. 1996. Reasons for reconciling. *Evolutionary Anthropology*, 5: 42–45.

Cords M., & Killen, M. 1998. Conflict resolution in human and nonhuman primates. In: *Piaget, Evolution, and Development* (J. Langer & M. Killen, eds.), pp. 193–219. Mahwah, N.J.: Lawrence Erlbaum Associates.

Das, M. 1998. Conflict management and social stress in long-tailed macaques. Ph.D. diss., University of Utrecht, the Netherlands.

Das, M., Penke, Z., & van Hooff, J. A. R. A. M. 1997. Affiliation between aggressors and third parties following conflicts in long-tailed macaques (*Macaca fascicularis*). *International Journal of Primatology*, 18: 157–179.

Das, M., Penke, Z., & van Hooff, J. A. R. A. M. 1998. Post-conflict affiliation and stress-related behavior of long-tailed macaque aggressors. *International Journal of Primatology*, 19: 53–71.

de Waal, F. B. M. 1977. The organization of agonistic relations in two captive groups of Java monkeys (*Macaca fascicularis*). *Zeitschrift für Tierpsychologie*, 44: 225–282.

de Waal, F. B. M. 1982. *Chimpanzee Politics: Power and Sex among Apes*. New York: Harper & Row.

de Waal, F. B. M. 1987. Tension regulation and non-reproductive functions of sex among captive bonobos (*Pan paniscus*). *National Geographic Research*, 3: 318–335.

de Waal, F. B. M. 1991. The chimpanzee's sense of social regularity and its relation to the human sense of justice. *American Behavioral Scientist*, 34: 335–349.

de Waal, F. B. M. 1993. Reconciliation among primates: A review of empirical evidence and unresolved issues. In: *Primate Social Conflict* (W. A. Mason & S. P. Mendoza, eds.), pp. 111–144. Albany: State University of New York Press.

de Waal, F. B. M. 1996. *Good Natured*. Cambridge: Harvard University Press.

de Waal, F. B. M., & Aureli, F. 1996. Consolation, reconciliation, and a possible cognitive difference between macaques and chimpanzees. In: *Reaching into Thought: The Minds of the Great Apes* (A. E. Russon, K. A. Bard, & S. T. Parker, eds.), pp. 80–110. Cambridge: Cambridge University Press.

de Waal, F. B. M., & Ren, R. M. 1988. Comparison of the reconciliation behavior of stumptail and rhesus macaques. *Ethology*, 78: 129–142.

de Waal, F. B. M., & van Hooff, J. A. R. A. M. 1981. Side-directed communication and agonistic interactions in chimpanzees. *Behaviour*, 77: 164–198.

de Waal, F. B. M., & van Roosmalen, A. 1979. Reconciliation and consolation among chimpanzees. *Behavioral Ecology and Sociobiology*, 5: 55–66.

de Waal, F. B. M., & Yoshihara, D. 1983. Reconciliation and redirected affection in rhesus monkeys. *Behaviour*, 85: 225–241.

Dunbar, R. I. M. 1980. Determinants and evolutionary consequences of dominance among female gelada baboons. *Behavioral Ecology and Sociobiology*, 7: 253–265.

Dunbar, R. I. M. 1983a. Structure of gelada baboon reproductive units. III. The male's relationship with his females. *Animal Behaviour*, 31: 565–575.

Dunbar, R. I. M. 1983b. Structure of gelada baboon reproductive units. IV. Integration at group level. *Zeitschrift für Tierpsychologie*, 63: 265–282.

Dunbar, R. I. M. 1984. *Reproductive Decisions: An Economic Analysis of Gelada Baboon Social Strategies*. Princeton: Princeton University Press.

Dunbar, R. I. M. 1988. *Primate Social Systems*. Ithaca, N.Y.: Cornell University Press.

Ehardt, C. L., & Bernstein, I. S. 1992. Intervention behavior by adult male macaques: Structural and functional aspects. In: *Coalitions and Alliances in Humans and Other Animals* (A. H. Harcourt & F. B. M. de Waal, eds.), pp. 83–111. Cambridge: Cambridge University Press.

Harcourt, A. H. 1988. Alliances in contests and social intelligence. In: *Machiavellian Intelligence: Social Expertise and the Evolution of Intellect in Monkeys, Apes, and Humans* (R. W. Byrne & A. Whiten, eds.), pp. 132–152. Oxford: Oxford University Press.

Harcourt, A. H. 1992. Coalitions and alliances: Are primates more complex than non-primates? In: *Coalitions and Alliances in Humans and Other Animals* (A. H. Harcourt & F. B. M. de Waal, eds.), pp. 29–60. Cambridge: Cambridge University Press.

Humphrey, N. K. 1976. The social function of intellect. In: *Growing Points in Ethology* (P. P. G. Bateson & R. A. Hinde, eds.), pp. 303–317. Cambridge: Cambridge University Press.

Judge, P. 1991. Dyadic and triadic reconciliation in pigtail macaques (*Macaca nemestrina*). *American Journal of Primatology*, 23: 225–237.

Kappeler, P. M. 1993. Reconciliation and post-conflict behaviour in ringtailed lemurs, *Lemur catta*, and redfronted lemurs, *Eulemur fulvus rufus*. *Animal Behaviour*, 45: 901–915.

Kappeler, P. M., & van Schaik, C. P. 1992. Methodological and evolutionary aspects of reconciliation among primates. *Ethology*, 92: 51–69.

Kirkpatrick, R. 1998. Ecology and behavior of the snub-nosed and douc langurs. In: *The Natural History of the Doucs and Snub-Nosed Monkeys* (N. Jablonski, ed.), pp. 155–190. Singapore: World Scientific Press.

Kummer, H. 1967. Tripartite relations in hamadryas baboons. In: *Social Communication among Primates* (S. A. Altmann, ed.), pp. 63–71. Chicago: University of Chicago Press.

Kummer, H. 1968. *Social Organization of Hamadryas Baboons.* Chicago: University of Chicago Press.

Levine, S., Coe, C., & Wiener, S. 1989. The psycho-neuroendocrinology of stress: A psychobiological perspective. In: *Psychoendocrinology* (S. Levine & R. Bursh, eds.), pp. 181–207. New York: Academic Press.

Maestripieri, D., Schino, G., Aureli, F., & Troisi, A. 1992. A modest proposal: Displacement activities as an indicator of emotions in primates. *Animal Behavior,* 141: 744–753.

Manning, A., & Dawkins, M. S. 1992. *An Introduction to Animal Behaviour,* 4th edn. New York: Cambridge University Press.

Mason, W. A. 1993. The nature of social conflict: A psycho-ethological perspective. In: *Primate Social Conflict* (W. A. Mason & S. P. Mendoza, eds.), pp. 13–48. Albany: State University of New York Press.

Noë, R. 1992. Alliance formation among male baboons: Shopping for profitable partners. In: *Coalitions and Alliances in Humans and Other Mammals* (A. H. Harcourt & F. B. M. de Waal, eds.), pp. 285–321. Cambridge: Cambridge University Press.

Petit, O., & Thierry, B. 1994a. Aggressive and peaceful interventions in conflicts in Tonkean macaques. *Animal Behaviour,* 48: 1427–1436.

Petit, O., & Thierry, B. 1994b. Reconciliation in a group of black macaques. *Dodo,* 30: 89–95.

Petit, O., & Thierry, B. 1994c. Reconciliation in a group of guinea baboons. In: *Current Primatology,* Vol. 2: Social Development, Learning, and Behavior (J. J. Roeder, B. Thierry, J. R. Anderson, & N. Herrenschmidt, eds.), pp. 137–145. Strasbourg: Université Louis Pasteur.

Povinelli, D. J., & Eddy, T. J. 1996. What chimpanzees know about seeing. *Monographs of the Society for Research in Child Development,* 61, no. 3: 1–189.

Povinelli, D. J., Parks, K. A., & Novak, M. A. 1991. Do rhesus monkeys (*Macaca mulatta*) attribute knowledge and ignorance to others? *Journal of Comparative Psychology,* 105: 318–325.

Ren, R., Yan, K., Su, Y., Qi, H., Liang, B., Bao, W., & De Waal, F. B. M. 1991. The reconciliation behavior of golden monkeys (*Rhinopithecus roxellanae roxellanae*) in small breeding groups. *Primates,* 32: 321–327.

Sapolsky, R., & Ray, J. 1989. Styles of dominance and their endocrine correlates among wild olive baboons (*Papio anubis*). *American Journal of Primatology,* 18: 1–13.

Sicotte, P. 1993. Inter-group encounters and female transfer in mountain gorillas: Influence of group composition on male behavior. *American Journal of Primatology,* 36: 21–36.

Silk, J. 1996. Why do primates reconcile? *Evolutionary Anthropology,* 5: 39–42.

Smuts, B. B. 1985. *Sex and Friendship in Baboons.* Chicago: Aldine.

Smuts, B. B., & Smuts, R. W. 1993. Male aggression and sexual coercion of females in nonhuman primates and other mammals: Evidence and theoretical implication. *Advances in the Study of Behavior,* 22: 1–63.

Steenbeek, R. 1997. What a maleless group can tell us about the constraints on female transfer in Thomas's langurs. *Folia primatologica,* 67: 169–181.

Sterck, E. H. M., Watts, D. P., & van Schaik, C. P. 1997. The evolution of social relationships in female primates. *Behavioral Ecology and Sociobiology,* 41: 291–309.

Struhsaker, T. T., & Leland, L. 1979. Socioecology of five sympatric monkey species in the Kibale Forest, Uganda. *Advances in the Study of Behavior,* 9: 159–228.

Thierry, B. 1985. Patterns of agonistic interactions in three species of macaque (*Macaca mulatta, M. fascicularis, M. tonkeana*). *Aggressive Behavior,* 11: 223–233.

van Schaik, C. P. 1989. The ecology of social relationships amongst female primates. In: *Comparative Socioecology* (V. Standen & R. Foley, eds.), pp. 195–218. London: Blackwell.

van Schaik, C. P. 1996. Social evolution in primates: The role of ecological factors and male behaviour. *Proceedings of the British Academy,* 88: 9–31.

van Schaik, C. P., & Dunbar, R. I. M. 1990. The evolution of monogamy in large primates: A new hypothesis and some crucial tests. *Behaviour,* 115: 30–62.

Veenema, H. C., Das, M., & Aureli, F. 1994. Methodological improvements for the study of reconciliation. *Behavioural Processes,* 31: 29–38.

Verbeek, P., & de Waal, F. B. M. 1997. Postconflict behavior of captive brown capuchins in the presence and absence of attractive food. *International Journal of Primatology,* 18: 703–726.

Virgin, C. E., & Sapolsky, R. 1997. Styles of male social behavior and their endocrine correlates among low-ranking baboons. *American Journal of Primatology,* 42: 25–39.

Watts, D. P. 1991. Harassment of immigrant female mountain gorillas by resident females. *Ethology,* 89: 135–153.

Watts, D. P. 1994. Agonistic relationships of female mountain gorillas. *Behavioral Ecology and Sociobiology,* 34: 347–358.

Watts, D. P. 1995a. Post-conflict social events in wild mountain gorillas (Mammalia, Hominoidea). I. Social interactions between opponents. *Ethology,* 100: 139–157.

Watts, D. P. 1995b. Post-conflict social events in wild mountain gorillas. II. Redirection, side direction, and consolation. *Ethology,* 100: 158–171.

Watts, D. P. 1996. Comparative socioecology of mountain gorillas. In: *Great Ape Societies* (W. C. McGrew, L. F. Marchant, & T. Nishida, eds.), pp. 16–28. Cambridge: Cambridge University Press.

Watts, D. P. 1997. Agonistic interventions in wild mountain gorilla groups. *Behaviour,* 134: 23–57.

Watts, D. P. 2000. Variation in male mountain gorilla life histories. In: *The Socioecology of Primate Males* (P. Kappeler, ed.), pp. 169–179. Cambridge: Cambridge University Press.

Whiten, A., & Byrne, R. W. 1988. Tactical deception in primates. *Behavioral and Brain Sciences,* 11: 233–244.

Whiten, A., & Byrne, R. W. 1997. *Machiavellian Intelligence II: Evaluations and Extensions.* Cambridge: Cambridge University Press.

Whitten, P. 1987. Infants and adult males. In: *Primate Societies* (B. B. Smuts, D. L. Cheney, R. M. Seyfarth, R. W. Wrangham, & T. T. Struhsaker, eds.), pp. 343–357. Chicago: University of Chicago Press.

Wrangham, R. W. 1980. An ecological model of female-bonded primate groups. *Behaviour,* 75: 262–300.

Yeager, C. P., Kirkpatrick, R., & Craig, R. 1998. Asian colobine social structure: Ecological and evolutionary constraints. *Primates,* 39: 147–156.

York, A. D., & Rowell, T. E. 1988. Reconciliation following aggression in patas monkeys (*Erythrocebus patas*). *Animal Behaviour,* 36: 502–509.

PART V

Ecological and Cultural Contexts

Introduction

Conflict resolution takes place in the larger context of ecology and culture. To understand how and why conflict resolution techniques evolved and how these techniques reflect the demands of society in a given context, we need to investigate the environmental pressures. In this section, several authors attempt to outline this larger framework for both humans and other animals.

Van Schaik and Aureli start out, in Chapter 15, explaining why animals live in groups and how, within these groups, enduring alliances are formed. Like team sports, alliances bridge the gap between cooperation and competition in that they are cooperative contracts that individuals enter into for the sake of competition: alliances provide a competitive edge. Because these contracts are so valuable, the relationships are serviced by means

of grooming and reconciliation after fights. The concept of social relationship, so central in primate research, is highlighted by these authors by reviewing various forms of valuable relationships that may differ between the sexes and between adults and immatures. Some of their points are illustrated by the case of the muriqui, an endangered neotropical primate studied in the field by **Strier, Carvalho, and Bejar,** who in Box 15.1 describe their remarkably peaceful society. Females keep their distance from one another, thus avoiding competition, and males establish close bonds in which conflicts are minimized for the sake of group solidarity. They need to maintain a united front in the face of competition with neighbors.

Pereira and Kappeler, in Box 15.2, compare two member species of an early branch on the primate tree and examine the complementarity of mechanisms of conflict management. One species, the ringtailed lemur, has a well-developed dominance hierarchy, whereas the other, the redfronted lemur, does not. The mechanism of dominance-submission makes group life predictable and helps reduce conflict (Preuschoft & van Schaik, Chapter 5). The authors suggest that reconciliations are all the more needed in redfronted lemurs because they have no dominance hierarchy to help modulate aggression.

After this, we move to the specific human situation with its incredible variability in cultural practices and moral systems. In Chapter 16, **Fry** explores the cross-cultural variation in the self-control of anger, concern for others, and procedures of mediation, peacemaking, and adjudication. He compares three cultures in detail, showing how each culture ends up developing its own specific style of conflict resolution. Theoretically interesting parallels probably exist between the interspecific variation in conflict resolution in the primate order and the cultural variation in our own species: that is, in both cases conflict resolution can be expected to vary as a result of environmental pressures. Fry makes clear, however, that processes of reconciliation are a totally neglected topic in cultural anthropology.

In the final chapter of this section, **Killen and de Waal** compare human morality with the tendencies for sympathy, care, and conflict resolution in nonhuman primates. Instead of looking at human morality as coming from the outside—imposed by adults on the child or imposed by culture on a fundamentally nasty human nature—the authors see it as generated from the inside and as an integrated part of humanity's evolutionary background. They consider conflict resolution a key element in understanding the origins of morality. Conflict and its resolution provide the experiential basis by which moral judgments are formed in childhood.

Cultural variation with regard to a moral tendency, such as forgiveness, is indicated in Box 17.1 by **Park and Enright**, who compare the tendency to forgive in three human cultures. In some cultures, people forgive by replacing the negative feelings toward an offender with positive feelings, whereas in other cultures forgiveness involves merely a reduction in negative feelings. Forgiveness is a rapidly growing area of research, and even though reconciliation and forgiveness are not identical (the first occurs between parties, the second within the mind of one party), obvious connections exist.

CHAPTER 15

The Natural History of Valuable Relationships in Primates

Carel P. van Schaik & Filippo Aureli

Introduction

The most important generalization to emerge from two decades of work on reconciliation (i.e., post-conflict friendly reunion between opponents) in primates is that individuals that reconcile are likely to have a strong social bond (de Waal, Chapter 2; Cords & Aureli, Chapter 9). Depending on the species, strong bonds are characterized by more time in close proximity, more friendly behavior such as grooming, lower rates of agonistic conflict (i.e., conflict including aggressive and submissive patterns), and more mutual agonistic support than the average dyad in the group (review: Cords 1997). Animals with a strong bond are likely to derive considerable value from their relationship (Kummer 1978).

Many examples of reconciliation fit this pattern. In chimpanzees (*Pan troglodytes*), in which males but not females form strong intrasexual

bonds, reconciliation after agonistic conflicts is far more common between males than between females (de Waal 1986; also muriquis [*Brachyteles arachnoides*]: Strier et al., Box 15.1). In contrast, among gorillas (*Gorilla gorilla*), strong bonds are seen only between the group's silverback male and the adult females, and these are the only types of dyads in which post-conflict reconciliation occurs (Watts 1995). Similarly, the dominant adult male in a longtail macaque (*Macaca fascicularis*) group frequently reconciles with the adult females, but not with other males in the group (Aureli & van Schaik unpublished). Among these same macaques, juvenile females, who unlike juvenile males tend to remain in their natal groups for life, form bonds with unrelated adult females and thus reconcile more with them. Juvenile males, in contrast, reconcile more with other juveniles, with whom they may develop bonds in other groups (Cords & Aureli 1993). Finally, in despotic societies (see van Schaik 1989 for definition), relatives are more likely to form strong bonds and hence reconcile conflicts than nonrelatives (Thierry 1990, Chapter 6; Aureli et al. 1997).

Far less is known about other mechanisms of conflict resolution, but we expect that relationship value will explain much of the variation in all aspects of conflict resolution, not just reconciliation, and will probably do so in both primates and non-primates.

Given the pivotal role of valuable relationships in reconciliation and probably general conflict resolution, it is natural to ask where such relationships occur in nature. The aim of this chapter is therefore to describe the natural history of valuable social relationships in primates. We first set the scene with a few comments about social relationships in general and about the role of communication, including reconciliation, in maintaining them. We then discuss the two major kinds of benefits arising from strong bonds or valuable social relationships: alliance formation within sexes and protective services, mainly between sexes. This exercise prepares the ground for an overview of the natural history of valuable relationships in primates. Although the focus is on primates, we hope to derive principles that can be useful for other animals as well.

Social Relationships

The social interactions between two animals depend not only on their individual characteristics (e.g., age, sex, dominance rank, temperament) but also on the history of interactions between them, provided that they possess the capacity for individual recognition, have sufficient memory to remember the outcome of social interactions, and repeatedly meet each other. The two animals can thus be said to have developed a social relation-

ship. This relationship is not directly visible. Observers decide it exists because the history of previous interactions allows them to predict the outcome of subsequent interactions between the same animals (Hinde 1976). Of course, it also allows the animals to predict the actions and responses of the partner with reasonable accuracy. This makes social intercourse much more efficient, obviating the need for thorough reassessment of the partner's strength and possible other qualities every time they come close, which is why the establishment of relationships is adaptive for the participants. The relationship concept also helps to explain why the same animal can be hostile to one conspecific and yet be friendly to another.

Social behavior is about conflict and cooperation, and relationships characterized by only one of these two probably do not exist. Even the most collaborative relationships tend to contain an element of competition—for instance, because collaborators need to decide on how to divide the benefits of their cooperative effort. Likewise, in competitive relationships, it may be in both contestants' interest not to let every encounter escalate into a potentially damaging fight and instead settle through some signals with mutually agreed-on meanings (Maynard Smith & Price 1973; Preuschoft & van Schaik, Chapter 5). Thus, social relationships always contain elements of both cooperation and competition.

Because valuable relationships require frequent social interactions, frequent spatial association is a precondition for their occurrence. Hence, they are certainly found among animals living in groups with stable membership. Our focus is group living primates, in which this subject is well studied.

Some valuable relationships involve interactions in which one partner incurs costs that will be recouped much later in time. Hence, especially relationships with a cooperative element are not established overnight but instead require some investment in which mutual trust is gradually built up (Kummer 1978). Kummer (1975, 1995) showed that in several cercopithecine monkeys, two strange conspecifics go through a fixed sequence of behavioral exchanges while they establish their relationship. Aggression usually characterizes the first exchange, after which presenting, mounting, and grooming predictably follow.

Little is known, however, about the process of building up trust in more cooperative relationships. Roberts & Sherrat (1998) suggested that stable cooperative relationships can be established by gradually raising the investment in subsequent interactions and exchanging less costly behaviors first (e.g., grooming) before investing in more costly behaviors (e.g., defense). Intuitively, one expects that relationship establishment will be easier in dyads with overlapping interests and hence among relatives. Not only will relatives be more familiar with each other, but they also incur greater costs from harming each other and more benefits from helping each other, through inclusive fitness effects.

Communication about the Relationship

The partners in an established relationship need to communicate about their relative balance of power (i.e., physical strength and social influence) and about the exchange rate of services. The need for some form of communication is clear in even the simplest relationship. For example, individuals can alternate grooming each other and thus provide benefits in terms of hygiene (Hutchins & Barash 1976), tension reduction (Schino et al. 1988; Aureli et al. 1999), and endorphin release (Keverne et al. 1989). By varying the duration and frequency of their own grooming bouts, the partners can express their perceived relative power in the relationship and, through refusals of grooming invitations or variations in length, negotiate the actual power balance.

Communication about the relationship is all the more important because the partners' values change all the time. Individual qualities, such as strength or experience, change over time, as do their needs for services. External factors—for instance, group composition—may change the value of a partner because they affect the number of other group members who can offer the same service (market effects: Noë et al. 1991; Noë & Hammerstein 1994). Intuitively, we expect that animals have ways to reassess continually their own value relative to that of their partner.

How do partners communicate about these assessments? In most cases, this communication takes place *during* or *after* interactions. This is expected when the exchanges do not involve much risk or when partners have strong relationships and usually can rely on each other. Changing assessments can be communicated by ceasing to support or share, by taking larger shares than before, or even by punishing partners after they failed to reciprocate (Trivers 1971; Clutton-Brock & Parker 1995). Post-interaction communication will still work when partners exchange different kinds of behavior. The best-studied examples involve grooming for support (Seyfarth 1980; de Waal & Luttrell 1988; Hemelrijk & Ek 1991; Silk 1992; Hemelrijk 1994), for reduced aggression (Fairbanks 1980; Silk 1982), for tolerance around resources (Weisbard & Goy 1976; de Waal 1989; Kapsalis & Berman 1996), and directly for food (de Waal 1997). The exchange rates of the different forms of behavior can be negotiated by refusing aid, tolerance, or invitations for grooming.

Communication *before* engaging in the interaction, on the other hand, is critical when interactions involve high risks, especially when the partners do not have reliable relationships. Agonistic support between males is a prime example because it involves high risks and because male alliances are fickle. In chimpanzees, for instance, an ally can be dropped overnight in favor of another (cf. de Waal 1982; Nishida 1983). It is therefore understandable that a male starting an attack must ensure that the partner will provide support in the high-risk fight. A well-known example is that of male baboons involved in an alliance. They exchange specialized signals before taking part in an aggressive coalition to confirm their willingness to support each other (Noë 1990; Smuts & Watanabe 1990; cf. Colmenares et al., Box 5.3).

We noted that even among collaborators interests never coincide completely and that most relationships must maintain a balance between competition and cooperation. Hence, aggressive conflicts among valuable partners, although they may be rare, are unavoidable. Given the dynamic nature of relationships, it is critical for one or both partners to establish whether the conflict signifies a growing mismatch in the assessments of each other's value or whether it is a mere hiccup in an otherwise unchanged relationship. We believe this is why reconciliation evolved (Aureli et al. unpublished). Reconciliation can be viewed as communication about the value of the relationship: it shows how much each partner is interested in the relationship and thereby is willing to repair it after the disturbance due to the conflict.

According to this view, reconciliation is not only an effective way to end the conflict but is also a primary tool for relationship management. This managing function becomes most apparent when a relationship undergoes a major reassessment such as during dominance challenges. For example, during periods of dominance struggle between male chimpanzees, conflict frequency increases, and reconciliation ceases to occur. Reconciliation is not granted by the future dominant male until the partner shows the first signs of submission (de Waal 1982, 1986).

Asymmetry in Value

The relationship may not be equally valuable to both partners (cf. Cords & Aureli, Chapter 9). In stable groups, decided dominance relationships usually produce tolerance around resources, or at least proximity to the dominant. The value of this general tolerance is greatest to the subordinate. Asymmetries in value are also found in more collaborative relationships. For instance, when two individuals have an agonistic alliance, the stronger partner is of greater value to the weaker than vice versa. Indeed, the stronger partner may not even need the weaker one all the time because there are alternative partners available (Noë 1990).

Such asymmetries in value create an asymmetry in power, which is translated into payoff imbalances among the partners. However, they should also create an asymmetry in the motivation to resolve conflict. There has been little systematic study of this question. In a preliminary compilation of the available work, de Waal (1993) did not find that subordinates consistently took the initiative toward reconciliation more than expected by chance. However, motivation may not necessarily be expressed directly in initiative. Subordinates may, for instance, refrain from reconciliation attempts for fear of renewed attacks unless the dominant has signaled its willingness to reconcile by moving toward the subordinate in a nonhostile manner.

Sources of Relationship Value

The nature and range of possible social benefits depend largely on a taxon's natural history and on its cognitive capabilities. Hence, in each particular taxon, we tend to find only a small number of realized social benefits.

We already noted that social benefits can arise from all relationships, be they mainly competitive or mainly cooperative. Some benefits arise from mere gregariousness (e.g., the reduced per capita risk of being captured by a predator, the dilution effect: Bertram 1978). Others, however, require stable association in which familiar individuals exchange low-risk behaviors and mutual benefits. The exchange of signals for conflict regulation is more efficient under such conditions (Preuschoft & van Schaik, Chapter 5). Group members tend to approach food calls made by any individuals who discover new sources of food (e.g., Hauser & Marler 1993). Likewise, the exchange of alarm calls does not require strong bonds (cf. Cheney & Seyfarth 1990). The meaning of these calls is learned by juveniles, and again no strong bonds with the tutors are required (Cook et al. 1985). By associating with males, females may derive enhanced protection from predation in species in which males tend to be more vigilant than females and more likely to counterattack predators (e.g., van Schaik & van Noordwijk 1989). Finally, animals can exchange grooming simply to improve hygiene. All these benefits merely require stable association and some level of mutual tolerance and exchange.

Associates tend to be animals with similar ecological interests. Thus, females with infants in a group often form a distinct cluster (e.g., Wasser 1982). They follow similar feeding and resting schedules (e.g., van Schaik & van Noordwijk 1986) and spend more time grooming and socializing with the infants than other group members. Similarly, juveniles and subadult or low-ranking adult males tend to form distinct clusters within a group.

Relationship value may vary greatly even within the same group. Some social benefits require cooperative bonds. Indeed, reaping them is the very reason for the existence of valuable relationships (Kummer 1978). In primates, the benefits include selective tolerance around resources (provided gregariousness is generally advantageous; e.g., Janson 1985), cooperative hunting (e.g., Boesch & Boesch 1989), food sharing (e.g., de Waal 1997), services for mating privileges (e.g., Stanford 1998; but see Hemelrijk et al. 1992), agonistic support, and protection against harassment. The last two are probably the most widespread and critical benefits related to relationship-dependent cooperation in primates. In both, the interests of the cooperating partners are served at the expense of the interests of a third party. If these interactions occur repeatedly between two partners, special relationships will develop. We examine these relationships in detail next.

Alliances

A coalition takes place when individuals support one another in an agonistic conflict (Fig. 15.1). We are mainly concerned here with coalitions of two individuals, although larger coalitions occur occasionally. An alliance is a type of relationship in which the two partners repeatedly form coalitions. Thus, coalitions are interactions that can be formed on a case-by-case basis, whereas alliances are enduring cooperative relationships (de Waal & Harcourt 1992).

The basic rule for coalitions and alliances is simple: they should be formed when they improve access to limiting resources for both partners (or at least for the weaker one when the partners are close relatives). This rule therefore assumes that (1) there is competition for limiting resources; (2) this competition has an important contest component; and (3) the coalition or alliance improves the fitness return for both partners. These three components need some illustration.

The limiting resources are usually different for the two sexes because the fitness of males and females is limited by very different factors. Female

FIGURE 15.1. Three related rhesus females (top) form a coalition against a member of another family. Photograph by Frans B. M. de Waal.

reproductive success in species with internal fertilization, gestation, and lactation (i.e., mammals) depends on the resources that the mother can garner to invest in the offspring and the safety she can provide to them. Male mammals, in contrast, usually contribute little to the development of the young. Unlike the females, they are ready to breed at virtually all times. Male reproductive success is therefore limited largely by the number of fertilizations they can achieve. Thus females usually compete for food or shelter, whereas males compete for mating opportunities (Emlen & Oring 1977).

The occurrence of alliances depends on the competitive regime. Competition for resources can be broken down into two components: contest and scramble (van Schaik 1989). In contest, access depends on agonistic strength. This occurs when the resource is patchy, distributed in defensible clusters. For instance, in a rain forest, most trees do not bear fruit, and individual trees of one species tend to be scattered, so a fruit-bearing tree is a discrete patch. Often such trees are not large enough to satiate all available individuals, and access to it can be contested. In scramble, animals are forced to share equally because monopolization is impossible. In our rain forest example, scramble would ensue when trees are very large relative to group size or when fruits are small, inconspicuous, and highly dispersed (e.g., Mitchell et al. 1991). It is only in contest competition that we expect animals to show aggression to improve access to resources and to develop decided dominance relations.

Decided dominance relationships may lead to alliances by two pathways (cf. Wrangham 1980, 1982). First, some animals may improve their access to the limiting resources by teaming up with other individuals, including nonrelatives. Obviously, dominants usually do not need allies, and the lowest-ranking individuals may not be able to bring enough strength into the alliance to make much difference, but mid-rankers might be able to improve access this way. Such an alliance will only succeed, however, if both partners gain, that is, if sharing access to the contested resource does not reduce the absolute return for the dominant of the two partners. When the resource is a patch of food, this condition is easily met, but when it is an estrous female, this sharing may be less straightforward, and more complicated negotiation may be necessary. Thus, alliances among nonrelatives occur but are not universal.

A second pathway to alliances relies on kin selection. If dominant animals derive important fitness benefits from unconstrained access to resources, they may also support weaker relatives in their struggle against others so that the relatives get to reap these benefits as well. The supporter gains through increased inclusive fitness, even when sharing access to the resources leads to a slight reduction in direct fitness returns.

In practice, the two pathways interact. Over time, such improved access to limiting resources will also improve the growth and physical strength of the relatives and thus increase their value as allies. In primate species in which females compete through contest for access to food and are the philopatric sex, all females are involved in lifelong alliances with their relatives (Sterck et al. 1997). These alliances also affect rank acquisition and result in matrilineal social systems in which the adult females belonging to a family outrank all the adult females of other families (Kawai 1965 [1958]; Chapais 1995).

Protective Bonds

Some of the most significant threats faced by female primates are social ones, in particular, sexual harassment and infanticide by males. Protective bonds between females, and especially those between males and females, may serve to reduce these social threats (Fig. 15.2).

For a variety of reasons, females may prefer to mate with some males and thus avoid mating with others (Andersson 1994). However, inevitably, the nonpreferred males are still interested in mating. Several factors affect the likelihood that these males will succeed in securing copulations (Smuts & Smuts 1993). First, the pressure from nonpreferred males depends on the operational sex ratio (i.e., males ready to mate/females ready to mate), which is generally male-biased. The male bias is

FIGURE 15.2. A resident male baboon (middle) defends a female against an immigrant male (left). Photograph by Frans B. M. de Waal.

stronger when life history factors cause slow female reproduction. As mammals go, primate females have unusually slow reproduction. Second, males can more easily force matings when they are larger and stronger than females and can hold them. Finally, they are more likely to be successful when females are not permanently associated with protectors, be they females or males (Smuts & Smuts 1993). Strong bonds involving females may therefore function to reduce the risk of harassment by males (Smuts & Smuts 1993; van Schaik 1996; Mesnick 1997).

Protective bonds are also likely to depend on the risk of male infanticide. Species with lactational amenorrhea are vulnerable to infanticide by males unlikely to have fathered the infant, because this will speed up the female's next conception. Many primates have lactational amenorrhea; male infanticide is reported for a remarkably high proportion of primate species (Hausfater & Hrdy 1984). Yet male infanticide is rare in most of the species in which it occurs. Perhaps this is because of effective social counterstrategies: infants and their mothers are usually protected by the likely sire, and perhaps by others. Primates are the mammalian order with by far the highest percentage of species in which males and females form stable associations, and the likely sire is often most actively involved in these protective associations (van Schaik & Kappeler 1997). In other words, prevention of male infanticide is thought to be ultimately responsible for the predominance of permanently bisexual groups in primates and thus of the rich variety of male-female social relationships in this order (see below).

Valuable Relationship in Nature

Readers not familiar with primates may conclude from such general accounts that different primate species have basically the same social system. Instead, there is rich interspecific variation. Primate socioecology aims to explain this variability using a limited set of principles that specify the interplay between social behavior on the one hand and ecological factors and life history variables on the other hand. Here, we first provide a sketch of primate socioecology in order to place valuable relationships into the broader context of social systems (based largely on Wrangham 1980; van Schaik 1989, 1996; Sterck et al. 1997; C. Nunn et al. unpublished) and then discuss the socioecology or natural history of valuable relationships in primates by dyad type (age-sex class combination).

Primate Socioecology

Associations and social relationships can often be seen as strategies to reduce the negative impact of some challenge. Paradoxically, the selective forces that produced particular social behavior are often hard to discern because the behavior it has produced is effective in eliminating the fitness impact of these ultimate causes. Hence, their action is rarely apparent. For instance, grouping behavior in mobile animals is often regarded as a strategy

against predation risk, but actual cases of predation may be so rare that we can only deduce the selective impact of predation risk indirectly. In addition, experimental elimination of the threat of predation will not always produce the expected result because animals may be selected not to respond to apparent disappearance of predators. Hence, it may be very difficult to identify the myriad of factors affecting social relationships and to disentangle the important ones.

Fortunately, this Herculean task has been made easier by the application of a successful deductive rule, based on evolutionary biology. It assumes that the fitness of males and females is limited by different factors (see above), and their social strategies are thought to differ accordingly (Emlen & Oring 1977). Theories of social evolution generally start by considering females and add males later (Fig. 15.3, page 321).

Group living primarily depends on whether females are associating with one another. Among the various possible selective factors favoring female gregariousness, predation avoidance may be the most universal (van Schaik 1983; Dunbar 1988; Janson 1992; cf. Isbell 1994). Diurnal primates actively avoid predation by detecting approaching predators before they can strike and taking evasive action or, more rarely, organizing collective counterdefense. This early detection is achieved through vigilance and rapid communication (e.g., alarm calls) of the threat throughout the group.

If predation avoidance favors gregariousness, competition for access to vital resources limits it. Female social relationships in their groups depend primarily on the intensity and nature of the competition for food, water, and shelter (Sterck et al. 1997).

Males compete for mating access to females. When females live in groups, males compete for membership in these groups. If groups are small, one male may be able to exclude all other males, especially if the periods of sexual activity of the females are not synchronized. When males cannot exclude their rivals, competition focuses mainly on dominance rank among the male group members. When both exclusion and dominance do not produce differential sexual access to fertilizable females, a mating scramble ensues, in which sperm competition is likely to be a major theme (cf. Strier et al., Box 15.1).

These ideas may account for much of the variation in intrasexual relationships in primate societies, but they do not explain male-female relationships. As we noted above, the latter are likely to depend mostly on social factors, especially protection against the risk of male infanticide (Fig. 15.3). Stable male-female association produces stable bisexual groups when females are gregarious. It produces pair-living animals when females are solitary; this is perhaps why pair living is surprisingly common among primates (van Schaik & Kappeler 1997). However, male infanticide may affect not merely male-female association and relationships but associations among females as well. Another social factor, protection against sexual harassment, may also have affected the nature of male-female relationships (Mesnick 1997) and even female-female relationships (Brereton 1995; Treves 1998).

In conclusion, the main factors thought to have affected associations and relationships among primates are both ecological and social. These various factors interact in numerous ways, depending on the environment and on the lifestyles and life histories of the species involved, producing a rich variety of social systems. Species therefore differ dramatically in the extent to which strong bonds, or valuable relationships, are formed and in the types of dyads that form them. Pereira & Kappeler (Box 15.2) discuss variation in bonding among closely related species. Let us now examine these

---

BOX 15.1

Prescription for Peacefulness

Karen B. Strier, Dennison S. Carvalho, & Nilcemar O. Bejar

The traditional prescription for evolutionary success involves competition over resources that improve fitness. These resources differ for males, who typically compete over mates, and females, who typically compete over food (see van Schaik & Aureli, Chapter 15). The distribution of potential mates and preferred food resources in time and space determines the optimal dosages of competition for each sex. Thus, whether competition includes direct aggression or indirect maneuvering, and whether it leads to chronic hierarchical relationships or acute interactions among otherwise peaceful group members, vary with resource distribution in predictable ways across primates.

Muriqui monkeys (*Brachyteles arachnoides*) appear to subscribe to a different evolutionary practice than that of other primates living in similar multimale, multifemale societies. Early investigations in 1983–1984 of one muriqui group inhabiting a forest of more than 800 hectares at the Estação Biologica de Caratinga, on Fazenda Montes Claros in Minas Gerais, Brazil, revealed a remarkably peaceful society. Females avoided overt contests over food by avoiding close proximity with other group members while feeding (Strier 1990). Philopatric males avoided contests over access to sexually receptive females because of their dependence on one another as allies in aggressive contests with neighboring groups of related males seeking access to the same mates. In addition, group males never tried to coerce or threaten females, who freely solicited and avoided sexual advances from preferred partners. Instead of competing aggressively with one another, males maintained close, affiliative relationships with one another to reinforce solidarity in between-group competition and tolerated one another's sexual interactions with assertive females (Strier 1992b, 1997b).

BOX 15.1 (continued)

Muriqui males engage in a group embrace.
Photograph by Paulo Coutinho.

The muriquis' prescription for peace was attributed to a unique suite of locomotor and energetic constraints that led, over the course of their evolutionary divergence from their closest atelin relatives (spider monkeys and woolly monkeys), to reduced sexual dimorphism in both body and canine size (Milton 1985a; Rosenberger & Strier 1989). Weak social relationships and a generalist feeding strategy, which can include high quantities of leaves, permit the formation of facultatively fluid patterns of association among females (Milton 1984; Strier 1990, 1991, 1992a). Female choice and the benefits of cooperating with male kin in between-group contests over access to females favor more subtle forms of competition instead of overt conflicts (Milton 1985a, 1985b; Strier 1994).

Muriqui males have large testes relative to their body size, a trait that has been associated with sperm competition in other primates. Although we know nothing about the number or quality of sperm muriquis produce, we can see visible ejaculate plugs that form right after they mate. These plugs create a physical barrier that may give one male's sperm a head start toward fertilizing an egg before it is manually removed by the female or the next male in line to mate (Strier 1999a).

By competing for fertilizations instead of copulations, male muriquis can minimize the overt aggression for access to mates that would be disruptive to their otherwise strong affiliative relationships.

In the original study, only 31 aggressive interactions were observed during more than 1,200 contact hours with the 23–26 individuals in the group at that time (Strier 1992b). Of the 30 dyadic conflicts observed, 30 percent occurred between the 8 males (6 adults and 2 subadults), and 40 percent occurred between the 10 females (8 adult and 2 adolescent immigrants).

Much has changed in the size and composition of the group over the 17-plus years that its members have been monitored. Yet the muriquis' pattern of peacefulness has proved to be uncommonly resistant to change. With the increase of group size, currently up to more than 60 members, the original tendency of females to avoid close proximity while feeding has led them to fission into smaller, temporary subgroups (Strier et al. 1993; Strier 1999b). Their greater tendency to fission may explain why female-female conflicts accounted for only 8 percent of the 48 aggressive interactions observed during a recent study conducted from August 1997 to April 1998.

BOX 15.1 (continued)

Increases in the ratio of sexually active females to males, from 1.1 to 1.4, coupled with more fluid female associations, may have also reduced the ability of group males to rebuff incursions from extragroup males, resulting in an increase from 0 to 12 percent in the proportion of copulations involving male outsiders (Strier 1994). Nonetheless, cooperative affiliations among male kin are still effective antidotes to between-group competitive pressures, provided that male kin groups are large enough to deter outsiders and that females continue to prefer males in their own group as mates (Strier 1997a).

The persistent strength of affiliations among male kin is evident from their still infrequent aggression and from the distribution of embraces following conflicts between males compared with other age-sex classes. During our recent study, male-male conflicts accounted for only 17 percent of the total 48 aggressive interactions observed among group members. All three of the polyadic interactions (i.e., interactions involving more than two individuals) and two of the five dyadic interactions among males were followed by embraces among opponents (i.e., 63 percent of the eight interactions). In a third dyadic interaction, opponents were subsequently observed in close proximity (< 5 m radius). No instances of post-conflict reconciliatory embraces or other friendly physical contact were recorded for any other age-sex classes, although occasionally opponents were subsequently observed in proximity to one another.

It is difficult to evaluate the significance of these recent findings without reference to baseline data on rates of aggression, proximity, or affiliative interactions among group members (Carvalho unpublished data). However, these recent findings raise tantalizing questions about how male muriquis may increase their probability of fertilizing females while still avoiding overt competition with one another.

Today, as in the past, muriqui copulations occur in full view of other group members without any visible tension or interference attempts by other males (Strier 1997a). Although most females mate with multiple partners both during and across multiple ovulatory cycles, no copulations were observed for three out of five conception cycles identified from noninvasive fecal steroid assays (Strier & Ziegler 1997). Thus, it is possible that male muriquis may sometimes employ evasive tactics in their efforts to avoid overt competition and achieve fertilizations.

There is no reason to suspect observer biases that would have increased the probability of missing conceptive matings for these females. Rather, what appears to be a strategic disappearance of some mating consorts during times when conception probabilities are high suggests that female choice may give preferred mates reproductive advantages without necessitating that males resort to competitive aggression with one another. The close associations maintained among males may be a strategy to make it more difficult for individuals to elude one another in the presence of a sexually receptive female (Strier 1997b). Thus, the low incidence of aggressive conflicts and the high incidence of post-conflict reconciliation may be among the active ingredients that make it possible for muriqui males to monitor—and gain a share in—one another's reproductive opportunities.

Whether male solidarity is effective at intercepting and altering female mate choices is one of the many outstanding questions about muriquis now under investigation (Strier 1997b). The evidence for indirect forms of competition operating among same-sexed muriquis, including the avoidance of proximity (Strier 1990) and subgrouping by females and the maintenance of affiliative proximity among males, implies that conflicts between male and female strategies should be similarly subtle.

As life history data on individual females accumulate during this study, it will be possible to evaluate the reproductive consequences of male and female reproductive strategies. Comparable analyses of variance in male reproductive success are constrained, however, by the still nascent state of noninvasive methods for detecting genetic paternity in muriquis. The revelations from such analyses will provide further clues into the muriqui's prescription for peacefulness.

BOX 15.2

Divergent Social Patterns in Two Primitive Primates

Michael E. Pereira & Peter M. Kappeler

Lemur research is important for a variety of reasons, but two are most relevant here. First, along with bush-babies and lorises, lemurs are prosimians, or "primitive" primates, having diverged least from the forms taken by the first primates more than 70 million years ago. Thus, lemurs offer some perspective on the foundation for everything primate that has evolved since. In addition, only lemurs among prosimians evolved a range of gregarious lifestyles. Indeed, we will show how even some closely related lemurs manifest remarkably different social systems. Thus, lemur work can help to evaluate not only phylogenetic constraints on primate social behavior but also hypotheses on socioecology formulated during work on anthropoid taxa (van Schaik & Kappeler 1996).

We study ringtailed (*Lemur catta*) and redfronted lemurs (*Eulemur fulvus rufus*), two Malagasy primates sharing many important features. First, they are the same size and shape and closely related, having evolutionarily diverged only recently (Martin 1990). In addition, they co-occur in some forests where they use many of the same resources. These primates show identical seasonality of reproduction, growth, and fattening (Pereira 1993b; Pereira et al. in press), and both species characteristically form multimale-multifemale social groups. As in monkeys and apes, females carry their infants on their bodies for months of lactational dependency (Jolly 1966; Sussman 1977). Given this context of pervasive similarity, we were intrigued that the social systems of ringtailed and redfronted lemurs are distinctly different. Early fieldwork suggested that female ringtailed lemurs remained in their natal groups while males dispersed soon after maturing, for example, whereas it was unclear that this pattern held for redfronted lemurs (Richard 1987). Moreover, ringtailed lemurs manifested dominance relations within their groups, whereas social dominance appeared relatively unimportant to redfronted lemurs (Sussman & Richard 1974).

Because lemur social behavior had yet to be dissected thoroughly when we started our work, our first objective was to illuminate the agonistic systems of both species, beginning with unequivocal determination of gestures used to signal aggressive and submissive motivation (*agonistic* refers to both aggressive and submissive elements). Basic discoveries followed our unscrambling of the two agonistic repertoires (Pereira & Kappeler 1997). We found, for example, that roughly half of agonistic interactions between ringtailed lemurs involved no aggression whatsoever, only submissive signaling by remarkably respectful subordinates. Our work also corroborated that ringtailed lemurs differ from virtually all other mammals in exhibiting unconditional adult female social dominance over males (Kappeler 1993a).

But, whereas all lemurs had earlier been assumed to show some form of female dominance (Jolly 1984; Richard 1987), our data extended the earliest reports by showing that redfronted lemurs do not manifest formal dominance behavior at all. Redfronted lemurs shared ringtails' repertoires of aggressive and submissive acts (e.g., "bite," "jump away") but diverged radically in having few agonistic signals (e.g., "huvv" vocalization) and none that communicated subordination. In many redfronted relationships, winners of particular conflicts often lost succeeding ones with the same social partners, and most interactions were not cleanly won or lost—typically, neither party expressed submission, or both combined aggressive and submissive acts.

Lacking reliable subordination, redfronted lemurs modulate the potential for severe within-group aggression using the common primate behavior, "look away." As in many monkeys, subordinate ringtailed lemurs look away during their interactions, often just before departing a dominant social partner. By contrast, all redfronted lemurs commonly look away when approaching or being approached by others, without exhibiting other agonistic behavior, and social departure rarely follows. Moreover, both opponents look away in most redfronted lemur conflicts. Often an individual will lunge and strike an opponent and immediately look away. More than half the time, such aggression invokes no response other than looking away. Other times, recipients of aggression return aggression before looking away, sometimes turn-

BOX 15.2 (continued)

Ringtailed lemur social groups revolve around a matriarchy. During rest, the oldest adult female in one study group is surrounded by her adult daughters and their juvenile offspring. Each pair of individuals maintains a salient dominance relationship. Photograph by Michael E. Pereira.

ing 180 degrees away from their antagonist. Extended squabbles see initiators striking back again and opponents alternating rapidly between striking and looking away from one another.

Why such different agonistic systems in such closely related and similar primates? Lemurs may have only recently begun their transition into diurnal living (van Schaik & Kappeler 1996), and ringtailed lemurs appear to have marched a bit further down that evolutionary road than have redfronted lemurs (Overdorff & Rasmussen 1995). Formalized dominance (de Waal & Luttrell 1985) is a diurnal social behavior, relying on signals unambiguously originating with and directed toward particular individuals and communicated over safe distances. Interestingly, even ringtailed lemurs retain a "nocturnal eye" (Pereira 1995) and, unlike many monkeys with female philopatry (Chapais 1992), matrilineal rank inheritance may yet be impossible for maturing ringtailed females because their mothers lack the visual acuity needed to provide agonistic support safely (Pereira 1993a, 1995; also Nakamichi & Koyama 1997). Alternatively, or in addition, the social systems of ringtailed and redfronted lemurs may represent divergent evolutionary equilibria, each comparably well promoting reproductive success for individual lemurid primates (Pereira & Kappeler 1997; Pereira & McGlynn 1997; Pereira et al. in press).

In any case, these lemurs' divergent social systems include multiple contrasts that may help to elucidate why social dominance is important for ringtailed lemurs and absent between redfronted lemurs. It is important to begin by considering where strong dyadic bonds form (de Waal 1986).

Female ringtailed lemurs form their strongest bonds with their mothers and daughters; thus, the core of a ringtailed group is a matriarchy. Under favorable conditions, groups grow, and each matriarchy collaboratively defends the boundary of its core range as a territory, seeking to exclude females easily identified to be from neighboring ranges (Jolly et al. 1993). Along with augmented predator pressure for these relatively diurnal lemurs, territoriality increases costs of female dispersal, and, consequently, ringtailed groups grow larger than other lemur foraging parties, exacerbating contest competition. Dominance relations mediate that competition. In dense populations, ringtailed females seem especially obsessed with dominance (Pereira 1993a; Nakamichi & Koyama 1997; Jolly 1998), dealing with certain peers as friends but others, including some sisters, as perennial adversaries (Kappeler 1993c; Pereira 1993a; Pereira & Kappeler 1997). Friends rarely conflict and use relaxed displays of subordination when they do. Most agonistic interaction occurs between adversaries, typically entailing moderate to high intensity aggression

BOX 15.2 (continued)

Redfronted lemur social groups revolve around several bonded male-female pairs. During rest, individual males and females sharing special, affinitive partnerships huddle closely together. The members of few, if any, adult dyads, regardless of individuals' sexes, maintain discernible dominance relationships. Photograph by Peter M. Kappeler.

(Vick & Pereira 1989; Pereira & Kappeler 1997). Finally, and of particular interest here, agonistic interaction did not elevate the probability of friendly behavior between the ringtailed lemurs that we studied, in contrast to all other primates yet investigated (Kappeler 1993b). Seemingly, reconciliation is unnecessary between friends and undesirable between adversaries, who actively strive to expel one another seasonally (Vick & Pereira 1989; Pereira 1993a).

By contrast, redfronted lemurs are not territorial, and closely related females do not bond strongly. Rather, each female forms an attachment with a particular adult male (Kappeler 1993c; Pereira & McGlynn 1997) with whom she travels and, presumably, reproduces (Overdorff 1998) over the course of one or more

years. Redfronted lemurs may generally succeed or fail primarily in relation to the efficacy of their partnerships, making dyadic dominance behavior unimportant. Pivotal competition may occur primarily between pairs whose members move about so closely that they can safely and reliably support one another (Pereira & Kappeler 1997; Overdorff 1998). Dominance within male-female pairs is similarly a nonissue for the world's monogamous primates, for example, the Hylobatidae (Leighton 1987) and Callitrichidae (Wright 1993). Friendly reunion following conflict was important to the redfronted lemurs that we studied, perhaps partly because these primates have no subordination to help modulate aggression. Recent work corroborated that most such friendly reunions occurred between partners in special relationships (H. Mennenga unpublished data; also Pereira & Kappeler 1997; Pereira & McGlynn 1997).

In sum, even closely related primates can have comprehensively different social systems, including divergent patterns of agonistic interaction and post-conflict behavior. Group-living primates manage social conflict in a variety of ways, and routine daily aggression may not need to be followed by friendly interaction to maintain social bonds in some species. Insofar as reconciliation in most primates, including redfronted lemurs, is observed most often in valuable relationships (Kummer 1978; also de Waal, Chapter 2; Cords & Aureli, Chapter 9), however, future research on ringtailed lemurs may show that mothers and daughters and other friendly adults do reconcile some of their rare and relatively mild conflicts. More research on a variety of mammals, primate and non-primate (Schino, Chapter 11), is needed to determine the taxonomic distributions of post-conflict reconciliation and the pattern of friendly pairs reconciling most frequently.

valuable relationships in some more detail (see Table 15.1 and Fig. 15.3).

Female-Female Relationships

Strong female-female bonds are especially likely in the face of contest competition for access to

resources in their groups, brought about or made worse by the physical clumping of females. These groups are characterized by the matrilineal structure, and bonds between related adult females are at the core of group life (resident-nepotistic female social systems: Sterck et al. 1997). Thus, in many

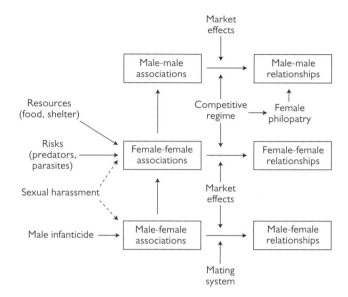

FIGURE 15.3. Factors affecting the intra- and intersexual relationships among adult primates.

groups of macaques and baboons, adult female kin have the most valuable relationships and support one another in contest competition against unrelated females through selective tolerance around resources, direct support, and the indirect effect of matrilineal dominance rank. Strong supportive bonds therefore do not characterize all female-female dyads in the group.

In some of the species with resident-nepotistic female relationships, females are relatively tolerant toward nonkin (de Waal & Luttrell 1989; Thierry 1990; Thierry, Chapter 6). It has been speculated that the tolerance is a result of the higher value of these females for collective defense against a common enemy, either neighboring groups or predators (van Schaik 1989; cf. Wrangham 1980). There is, however, little evidence for an effect of collective female resource defense on female social relationships (Cheney 1992; Matsumura & Kobayashi 1997), prompting some scholars to invoke nonadaptive explanations for the variation in tolerance (Thierry 1990; Matsumura & Kobayashi 1997). However, we can also speculate that the

females' collective defense concerns harassing males. As with other ultimate causes, this may be difficult to demonstrate: the need for this defense need not arise very often when males may be discouraged to harass because of the risks of counterattacks by female coalitions.

Females in some group-living species tend not to form strong relationships with any of the other females. They do not experience strong contest competition and live together simply because they share protection by the group male against infanticide and perhaps harassment or are better off against predators. These female groups dissolve again after the male's protective ability wanes. It is likely, therefore, that female groups serve as much to reduce the risk of infanticide as to reduce the risk of predation. All known examples of this type of female grouping occur in species with high-fiber diets (e.g., gorillas, leaf monkeys) for which feeding competition is largely through scramble (Sterck et al. 1997).

In other species, females are semisolitary, spending much time alone, and therefore have little

TABLE 15.1
Overview of Social Benefits Exchanged in Valuable Relationships among Adult Primates

	F-F	M-F	M-M
ALLOGROOMING	x	x	x
ALLIANCES FOR INTRAGROUP CONTEST FOR RESOURCES	x (food)	x	x (mates)
TOLERANCE AT RESOURCES	x	x	x?
INFANT CARE AND PROTECTION AGAINST INFANTICIDE	x	x	
ALLIANCES/PROTECTION AGAINST HARASSING MALES	x	x	
ALLIANCES/PROTECTION AGAINST PREDATORS/OTHER GROUPS	x	x	
MATING PRIVILEGES		x?	
ALLIANCES AGAINST EXTRA-GROUP MALES			x

Note: F-F = female-female relationships; M-F = male-female relationships; M-M = male-male relationships.

opportunity for social interactions and alliance formation. Examples include chimpanzees, which live in a male-bonded fission/fusion system, and orangutans (*Pongo pygmaeus*), in which males are solitary. These semisolitary females can certainly exchange some services (e.g., allogrooming, infant care), but overall the value of their relationships is lower than that of females in resident-nepotistic social systems.

Finally, in some species, only a single female breeds in each group, even though more than one female may be present. In such a setting, value in female relationships is highly asymmetrical, since females compete for nonshareable male services (e.g., infant carrying) but also cooperate in rearing offspring of the only breeding female. In these species, there is female reproductive inhibition (Schaffner & Caine, Chapter 8) and, if the

inhibition fails, even female infanticide (Digby 1995).

Male-Male Relationships

Among primate males, strong bonds to maintain alliances are expected to serve to improve mating access to females, and indeed they do. However, they are less common and shorter-lived than those among females (van Hooff & van Schaik 1994). In various primate species, only a single male stays in a group of females, preventing the establishment of male bonds. But even in the all-male bands commonly found in these species (e.g., Rajpurohit et al. 1995) or in many species with multimale groups, male bonds are not pervasive. The reasons for this are largely speculative at this stage. First, males are less likely to be philopatric and thus surrounded by relatives. However, even in groups with male philopatry, paternally based relatedness is not recognized, and thus kinship does not affect male alliances to the same extent as female alliances (Chapais 1995). Second, alliances tend not to increase male mating success because the number of conceptions in a group is a fixed quantity and dominants may lose by sharing matings with others (van Schaik 1996). This quantity is no longer constant when male alliances serve to defend access to a group of females. Male alliances may well monopolize female groups longer or defend a bigger group of females, and such between-group male alliances are thus expected to be more common.

The four contexts in which male alliances are found conform to these expectations. First, in some large multimale groups, males often form alliances (savanna baboons [*Papio cynocephalus*]: Noë 1990; Smuts & Watanabe 1990; Barbary macaques [*Macaca sylvanus*]: Kuester & Paul 1992; bonnet macaques [*M. radiata*]: Silk 1992), in many cases especially older males. Whereas in some cases these alliances serve to maintain or enhance rank, in others (e.g., baboons) their aim is to gain access to estrous females by displacing individually stronger males. The allies tend to take turns consorting with a female once this access has been acquired, even though one of the partners may end up with the female much more often than the other (Noë 1990). The male allies' bonds are expressed in mutual support, tolerance, and greetings, although grooming among them is not very common.

Second, in a few species, males are philopatric and form strong relationships characterized by a delicate balance between intracommunity competition for dominance and cooperation in intercommunity hostility. Chimpanzees are the best studied of these male-bonded species (Nishida 1979; de Waal 1982; Goodall 1986; Takahata 1990). Male chimpanzees fiercely compete with

one another for dominance rank and access to estrus females within the same community, and coalitions with other males are often critical in determining the outcome of these struggles. At the same time, it is of great importance that male chimpanzees of the same community maintain unity among themselves because of the highly hostile nature of intercommunity interactions. Thus, even though male-male relationships vary in value depending on support in intracommunity competition, they are on average highly valuable because of the male common effort during intercommunity interactions.

Third, in some species in which both single-male and multimale groups are found, the males in multimale groups may collaborate in defending the group against other males. In red howlers (*Alouatta seniculus*), such alliances are more stable and effective when formed by male relatives (Pope 1990). Likewise, in gelada (*Theropithecus gelada*) and hamadryas (*Papio hamadryas*) baboons, the one-male units may contain an additional male, the follower (Kummer 1968; Dunbar 1984). At least in gelada baboons, the function may similarly be reduced risk of takeover, but an additional benefit may be reduced risk of infanticide. In both geladas and mountain gorillas, the follower is an older male, who is probably the deposed leader staying on to protect his infants (Dunbar 1984; Robbins 1995).

Finally, in hamadryas baboons, in which one-male units are embedded in larger bands and troops, males of different one-male units within the same band show remarkable inhibitions to compete openly, as long as the units do not mingle (Kummer 1968). This tolerance may reach the point of close coordination of movements in two-male teams (Kummer 1968). The members of a team show frequent proximity and affiliation.

Male-Female Relationships

We already noted that the common occurrence of stable male-female association in primates should lead us to expect a rich variety of male-female interactions and relationships that go well beyond the brief sexual encounters characterizing most other mammals. We expect the most varied male-female relationships in situations in which both sexes can choose between multiple potential partners within the same group (market effect: Noë & Hammerstein 1994) and in which sexual dimorphism is limited, that is, the smaller sex (usually females) can be an effective ally for the larger one. We therefore discuss male-female bonds by mating system.

In most pair-living species, the pair mates are the only adults in the group, and they are expected to have rather coincident interests. Nonetheless, it is not always appropriate to characterize their bonds as strong. The bond between pair mates in white-handed gibbons (*Hylobates lar*), for instance, hardly goes beyond general travel coordination and duetting; allogrooming and other forms of affiliation are rather rare, and males are largely responsible for relationship maintenance (Palombit 1996). In the closely related and sympatric siamang (*Hylobates syndactylus*), by contrast, the relationship is much stronger and more symmetric. One possible reason for this difference is that siamang males are more involved in raising the infant (Palombit 1996). This suggestion is consistent with our impression that male-female bonds are strong in those pair-living species in which males contribute strongly to infant care by carrying and occasionally provisioning (e.g., titi monkeys [*Callicebus moloch*]: Mendoza & Mason 1986).

Infant-carrying lemurs form groups of roughly equal numbers of adult males and females (e.g., van Schaik & Kappeler 1993). Particular male-female dyads form strong bonds in some *Eulemur* species (van Schaik & Kappeler 1993; Pereira & Kappeler 1997; Pereira & McGlynn 1997), although the function of these bonds is not yet clear (cf. Overdorff 1998).

In several marmosets and tamarins with single female breeders, facultative polyandry occurs (Caine 1993; Garber 1997). Because multiple males live together with a single breeding female and provide extensive infant care, we expect male-female bonds to be strong. In addition, female influence in the relationship must be fairly strong because she can bias matings, and, consequently, male-female relationships in the same group may vary in value. Unfortunately, we are unaware of quantitative comparisons of male-female bonds in different demographic settings in these species.

In some single-male groups, the male has strong affiliative bonds with all the group's females, expressed mainly as affiliation, grooming, and agonistic protection. He leads the group in between-group interactions and defends the group against predators and male intruders (and hence infanticidal attacks). However, there are pronounced and unexplained interspecific differences in the degree to which the male is affiliating with his females. For instance, male colobus monkeys spend more time with their females than male langurs (cf. van Schaik & Hoerstermann 1994) or male patas monkeys (*Erythrocebus patas*: Cords 1987). In some of the species with single-male groups (e.g., gorillas, hamadryas), males often intervene in conflicts between females (see Petit & Thierry, Box 13.1; Watts et al., Chapter 14). Such policing makes them valuable partners for females that may compete for the male services.

As noted above, the scope for strong male-female bonds would seem to be greatest in multimale, multifemale groups. In their relationships, the sexes could exchange agonistic support for matings or reciprocal agonistic support. However, both kinds of transactions are made more difficult when males are much larger and stronger than females. Although females can potentially elicit services from males by selectively favoring them as mates, male harassment (Smuts & Smuts 1993; Manson 1994) may stymie the potential for expression of female preferences and hence for male-female bonds based on these preferences. Although this situation may lead to interesting social dynamics, including sperm competition and cryptic female choice (Eberhard 1996), it is hardly one in which one expects strong male-female bonds, apart from those involving protection.

The same problem is found for male-female alliances serving to improve or maintain the dominance position of either the male or the female. Most examples of male-female alliances are from species with limited sexual dimorphism, in which males and females can potentially provide mutual agonistic support. In those species, alliances with one or more females have been shown to play a pivotal role in males maintaining top dominance (chimpanzees: de Waal 1982; Japanese macaques [*Macaca fuscata*]: Gouzoules 1980). Conversely, females in some nepotistic societies can rise in rank by allying with a powerful male (e.g., Japanese macaques: Gouzoules 1980; Takahata 1991).

We suspect that most strong male-female bonds in multimale groups are between the likely sire of the infants and their mothers (cf. Smuts & Smuts 1993). In longtail macaques and white-faced capuchins (*Cebus capucinus*), for instance, this male is most commonly affiliating with these females, is often groomed by them, and (in longtail macaques) is supported agonistically by them when he is challenged by other males (van Noordwijk & van Schaik 1988; Perry 1997). In some baboons and macaques, particular male-female dyads form selective associations in which disproportionate amounts of grooming are exchanged, mainly from female to male, and the male occasionally supports the female in agonistic conflicts and often has friendly interactions with the female's infant. In short, they develop friendships that transgress brief mating associations (Smuts 1985). A recent study by Palombit et al. (1997) showed that in chacma baboons (*Papio ursinus*) these friendships serve to protect the infants probably sired by the male friends, often before they were friends (cf. Pereira 1988b). These male-female relationships are therefore highly valuable because infant survival strongly depends on the presence of protectors.

Relationships of Immatures

Social relationships between immatures and between adults and immatures have received much less attention, with the exception of mother-infant relationships (Pereira & Fairbanks 1993). None of the formal socioecological models has so far incorporated immature relationships. We can, however, hypothesize that valuable partners for juveniles are those that provide immediate benefits as well as those that can play a critical role in their future integration into the adult social structure.

In species with intense contest competition and steep dominance hierarchies, juveniles, being smaller and less experienced, are often the target of aggression (Dittus 1977; Silk et al. 1981; Bernstein & Ehardt 1985; Pereira 1988a). Under these circumstances, protectors are certainly valuable partners for juveniles. In addition, juveniles would benefit from adult tolerance to increase the time spent in the central part of the group and reduce predation risk to which they are particularly vulnerable (Janson & van Schaik 1993). Protective bonds with unrelated adults are, however, difficult to achieve because juveniles do not usually have much to offer to adults in terms of services. Thus, in most cases in which relationships between adults and juveniles are strong, this is likely to reflect inclusive fitness returns. Indeed, most agonistic support in favor of juveniles by females comes from mothers and other close relatives (e.g., Kurland 1977; Netto & van Hooff 1986). Likewise, male supporters tend to be likely fathers (e.g., Pereira 1989).

Juveniles could also establish bonds as investments in future relationships. In many species, males and females have predictably different social trajectories, and the relationships they develop as juveniles reflect these different futures. In catarrhine monkeys, female juveniles are attracted to adult females, especially to high-ranking ones, whereas male juveniles prefer other immature or older males (Cheney 1978; Pereira 1988a; Fairbanks 1993; van Noordwijk et al. 1993; Nikolei & Borries 1997). These partner preferences are in agreement with the expectation of future availability. In these species, juvenile females spend their entire life in their natal group, and establishing relationships with adult females is critical for their integration in the matrilineal system and for the acquisition of dominance rank (Pereira 1992; Chapais & Gauthier 1993; Fairbanks 1993). Males, on the other hand, migrate into other groups around puberty. They often migrate with several other peers and are likely to migrate into groups in which there are older males from their natal groups (Cheney & Seyfarth 1983; van Noordwijk & van Schaik 1985; Fairbanks 1993). Not only will this facilitate immigration because these relatives are less likely to attack them, but it may also provide them with a source of allies.

Conclusion

Social relationships are critically important in the life of many primates. The many examples provided above show that social relationships within a group can be highly differentiated. This fact has a major implication for the study of reconciliation and other mechanisms of conflict resolution. Group-level analyses for reconciliation may come up empty handed even if reconciliation is very common in the few valuable relationships present in the group. Hence, future attempts to demonstrate reconciliation should focus on (classes of) dyads rather than groups (cf. Watts 1995; Pereira & Kappeler, Box 15.2).

The survey of the socioecology of primate social relationships shows that the natural history (its lifestyle and its life history) of the taxon being studied is highly relevant to the nature of the benefits derived from social relationships. These benefits and hence the strength of the bonds vary among species and even among types of dyads within species. We believe that the most valuable relationships in primates are mainly those that provide mutual agonistic support or agonistic protection against male sexual coercion.

Valuable relationships are likely to provide different benefits among communally hunting carnivores, burrowing social rodents, or marine mammals. In some taxa, agonistic alliances or protective bonds may be as important as in primates, but in others, different benefits may arise, because relationships develop in different contexts, such as communal hunting and food sharing, turn taking in vigilance, or communal breeding. However, even though the nature of the benefits varies, we should see similar processes at work related to the establishment and maintenance of social relationships whenever they are critical to fitness, provided they are not always formed with close relatives. Reconciliation can be used as a gauge for the presence of those relationship-servicing mechanisms, and the recent demonstration of its existence in non-primates (Schino, Chapter 11) underscores the widespread distribution of valuable relationships in the animal kingdom.

References

Andersson, M. 1994. *Sexual Selection*. Princeton: Princeton University Press.

Aureli, F., Das, M., & Veenema, H. C. 1997. Differential kinship effect on reconciliation in three species of macaques (*Macaca fascicularis*, *M. fuscata*, and *M. sylvanus*). *Journal of Comparative Psychology*, 111: 91–99.

Aureli, F., Preston, S. D., & de Waal, F. B. M. 1999. Heart rate responses to social interactions in free-moving rhesus macaques: A pilot study. *Journal of Comparative Psychology*, 113: 59–65.

Bernstein, I. S., & Ehardt, C. L. 1985. Intragroup agonistic behavior in rhesus monkeys *Macaca mulatta*. *International Journal of Primatology*, 6: 209–226.

Bertram, B. C. R. 1978. Living in groups: Predators and prey. In: *Behavioural Ecology: An Evolutionary Approach* (J. R. Krebs & N. B. Davies, eds.), pp. 64–96. Oxford: Blackwell.

Boesch, C., & Boesch, H. 1989. Hunting behavior of wild chimpanzees in the Tai National Park. *American Journal of Physical Anthropology*, 78: 547–573.

Brereton, A. R. 1995. Coercion-defence hypothesis: The evolution of primate sociality. *Folia primatologica*, 64: 207–214.

Caine, N. G. 1993. Flexibility and co-operation as unifying themes in *Saguinus* social organization and behaviour: The role of predation pressures. In: *Marmosets and Tamarins: Systematics, Behaviour, and Ecology* (A. B. Rylands, eds.), pp. 200–219. Oxford: Oxford University Press.

Chapais, B. 1992. Role of alliances in the social inheritance of rank among female primates. In: *Coalitions and Alliances in Humans and Other Animals* (A. Harcourt & F. B. M. de Waal, eds.), pp. 29–58. New York: Oxford University Press.

Chapais, B. 1995. Alliances as a means of competition in primates: Evolutionary, developmental, and cognitive aspects. *Yearbook of Physical Anthropology*, 38: 115–136.

Chapais, B., & Gauthier, C. 1993. Early agonistic experience and the onset of matrilineal rank acquisition in Japanese macaques. In: *Juvenile Primates: Life History, Development, and Behavior* (M. E. Pereira & L. A. Fair-

banks, eds.), pp. 245–258. Oxford: Oxford University Press.

Cheney, D. L. 1978. Interactions of immature male and female baboons with adult females. *Animal Behaviour*, 26: 389–408.

Cheney, D. L. 1992. Intragroup cohesion and intergroup hostility: The relation between grooming distributions and intergroup competition among female primates. *Behavioral Ecology*, 3: 334–345.

Cheney, D. L., & Seyfarth, R. M. 1983. Nonrandom dispersal in free-ranging vervet monkeys: Social and genetic consequences. *American Naturalist*, 122: 392–412.

Cheney, D. L., & Seyfarth, R. M. 1990. *How Monkeys See the World: Inside the Mind of Another Species*. Chicago: University of Chicago Press.

Clutton-Brock, T. H., & Parker, G. A. 1995. Punishment in animal societies. *Nature*, 373: 209–216.

Cook, M., Mineka, S., Wolkenstein, B., & Laitsch, K. 1985. Observational conditioning of snake fear in unrelated rhesus monkeys. *Journal of Abnormal Psychology*, 94: 591–610.

Cords, M. 1987. Forest guenons and patas monkeys: Male-male competition in one-male groups. In: *Primate Societies* (B. B. Smuts, D. L. Cheney, R. M. Seyfarth, R. W. Wrangham, & T. T. Struhsaker, eds.), pp. 98–111. Chicago: University of Chicago Press.

Cords, M. 1997. Friendships, alliances, reciprocity and repair. In: *Machiavellian Intelligence II: Evaluations and Extensions* (A. Whiten & R. W. Byrne, eds.), pp. 24–49. Cambridge: Cambridge University Press.

Cords, M., & Aureli, F. 1993. Patterns of reconciliation among juvenile long-tailed macaques. In: *Juvenile Primates: Life History, Development, and Behavior* (M. E. Pereira & L. A. Fairbanks, eds.), pp. 271–284. Oxford: Oxford University Press.

de Waal, F. B. M. 1982. *Chimpanzee Politics: Power and Sex among Apes*. New York: Harper & Row.

de Waal, F. B. M. 1986. The integration of dominance and social bonding in primates. *Quarterly Review of Biology*, 61: 459–479.

de Waal, F. B. M. 1989. Food sharing and reciprocal obligations among chimpanzees. *Journal of Human Evolution*, 18: 433–459.

de Waal, F. B. M. 1993. Reconciliation among primates: A review of the empirical evidence and theoretical issues. In *Primate Social Conflict* (W. A. Mason & S. P. Mendoza, eds.), pp. 111–144. New York: State University of New York Press.

de Waal, F. B. M. 1997. The chimpanzee's service economy: Food for grooming. *Evolution and Human Behavior*, 18: 375–386.

de Waal, F. B. M., & Harcourt, A. 1992. Coalitions and alliances: A history of ethological research. In: *Coalitions and Alliances in Humans and Other Animals* (A. Harcourt & F. B. M. de Waal, eds.), pp. 1–19. Oxford: Oxford University Press.

de Waal, F. B. M., & Luttrell, L. M. 1985. The formal hierarchy of rhesus monkeys: An investigation of the bared-teeth display. *American Journal of Primatology*, 9: 73–85.

de Waal, F. B. M., & Luttrell, L. M. 1988. Mechanisms of social reciprocity in three primate species: Symmetrical relationship characteristics or cognition? *Ethology and Sociobiology*, 9: 101–118.

de Waal, F. B. M., & Luttrell, L. M. 1989. Toward a comparative socioecology of the genus *Macaca*: Different dominance styles in rhesus and stumptail macaques. *American Journal of Primatology*, 19: 83–109.

Digby, L. 1995. Infant care, infanticide, and female reproductive strategies in polygynous groups of common marmosets (*Callithrix jacchus*). *Behavioral Ecology Sociobiology*, 37: 51–61.

Dittus, W. 1977. The social regulation of population density and age-sex distribution in the toque monkey. *Behaviour*, 63: 281–322.

Dunbar, R. I. M. 1984. *Reproductive Decisions: An Economic Analysis of Gelada Baboon Social Strategies.* Princeton: Princeton University Press.

Dunbar, R. I. M. 1988. *Primate Social Systems.* London: Croom Helm.

Eberhard, W. G. 1996. *Female Control: Sexual Selection by Cryptic Female Choice.* Princeton: Princeton University Press.

Emlen, S. T., & Oring, L. W. 1977. Ecology, sexual selection, and the evolution of mating systems. *Science*, 197: 215–223.

Fairbanks, L. A. 1980. Relationships among adult females in captive vervet monkeys: Testing a model of rank-related attractiveness. *Animal Behaviour*, 28: 853–859.

Fairbanks, L. A. 1993. Juvenile vervet monkeys: Establishing relationships and practicing skills for the future. In: *Juvenile Primates: Life History, Development, and Behavior* (M. E. Pereira & L. A. Fairbanks, eds.), pp. 211–227. Oxford: Oxford University Press.

Garber, P. A. 1997. One for all and breeding for one: Cooperation and competition as a tamarin reproductive strategy. *Evolutionary Anthropology*, 5: 187–199.

Goodall, J. 1986. *The Chimpanzees of Gombe: Patterns of Behavior.* Cambridge: Harvard University Press.

Gouzoules, H. 1980. A description of genealogical rank changes in a troop of Japanese monkeys (*Macaca fuscata*). *Primates*, 21: 262–267.

Hauser, M. D., & Marler, P. 1993. Food-associated calls in rhesus macaques (*Macaca mulatta*): II. Costs and benefits of call production and suppression. *Behavioral Ecology*, 4: 206–212.

Hausfater, G., & Hrdy, S. B. 1984. *Infanticide: Comparative and Evolutionary Perspectives.* New York: Aldine.

Hemelrijk, C. K. 1994. Support for being groomed in long-tailed macaques, *Macaca fascicularis. Animal Behaviour*, 48: 479–481.

Hemelrijk, C. K., & Ek, A. 1991. Reciprocity and interchange of grooming and "support" in captive chimpanzees. *Animal Behaviour*, 41: 923–935.

Hemelrijk, C. K., van Laere, G. J., & van Hooff, J. A. R. A. M. 1992. Sexual exchange relationships in captive chimpanzees? *Behavioral Ecology and Sociobiology*, 30: 269–275.

Hinde, R. A. 1976. Interactions, relationships and social structure. *Man*, 11: 1–17.

Hutchins, M., & Barash, D. P. 1976. Grooming in primates: Implications for its utilitarian functions. *Primates*, 17: 145–150.

Isbell, L. A. 1994. Predation on primates: Ecological patterns and evolutionary consequences. *Evolutionary Anthropology*, 3: 61–71.

Janson, C. H. 1985. Aggressive competition and individual food consumption in wild brown capuchin monkeys (*Cebus apella*). *Behavioral Ecology and Sociobiology*, 18: 125–138.

Janson, C. H. 1992. Evolutionary ecology of primate social structure. In: *Evolutionary Ecology and Human Behavior* (E. A. Smith & B. Winterhalder, eds.), pp. 95–130. New York: Aldine de Gruyter.

Janson, C. H., & van Schaik, C. P. 1993. Ecological risk aversion in juvenile primates: Slow and steady wins the race. In: *Juvenile Primates: Life History, Development, and Behavior* (M. E. Pereira & L. A. Fairbanks, eds.), pp. 57–76. Oxford: Oxford University Press.

Jolly, A. 1966. *Lemur Behavior: A Madagascar Field Study*. Chicago: University of Chicago Press.

Jolly, A. 1984. The puzzle of female feeding priority. In: *Female Primates: Studies by Woman Primatologists* (M. F. Small, ed.), pp. 197–215. New York: Alan R. Liss.

Jolly, A. 1998. Pair-bonding, female aggression, and the evolution of lemur societies. *Folia primatologica*, 69 (suppl. 1): 1–13.

Jolly, A., Rasamimanana, H., Kinnaird, M. F., O'Brien, T. G., Crowley, H. M., Harcourt, C. S., Garnder, S., & Davidson, J. M. 1993. Territoriality in *Lemur catta* groups during the birth season at Berenty, Madagascar. In: *Lemur Social Systems and Their Ecological Basis* (P. M. Kappeler & J. U. Ganzhorn, eds.), pp. 85–109. New York: Plenum Press.

Kappeler, P. M. 1993a. Female dominance in primates and other mammals. In: *Perspectives in Ethology*, Vol. 10 (P. P. G. Bateson, P. H. Klopfer, & N. S. Thompson, eds.), pp. 143–158. New York: Plenum Press.

Kappeler, P. M. 1993b. Reconciliation and post-conflict behaviour in ringtailed lemurs, *Lemur catta*, and red-fronted lemurs, *Eulemur fulvus rufus*. *Animal Behaviour*, 45: 901–915.

Kappeler, P. M. 1993c. Variation in social structure: The effects of sex and kinship on social interaction in three lemur species. *Ethology*, 93: 125–145.

Kapsalis, E., & Berman, C. M. 1996. Models of affiliative relationships among free-ranging rhesus monkeys (*Macaca mulatta*): II. Testing predictions for three hypothesized organizing principles. *Behaviour*, 133: 1235–1263.

Kawai, M. 1965 [1958]. On the system of social ranks in a natural troop of Japanese monkeys: I. Basic rank and dependent rank. In: *Japanese Monkeys* (K. Imanishi & S. A. Altmann, eds.), pp. 66–86. Atlanta: Emory University Press.

Keverne, E. B., Martensz, N. D., & Tuite, B. 1989. Beta-endorphin concentrations in cerebrospinal fluid of monkeys are influenced by grooming relationships. *Psychoneuroendocrinology*, 14: 155–161.

Kuester, J., & Paul, A. 1992. Influence of male competition and female mate choice on male mating success in barbary macaques (*Macaca sylvanus*). *Behaviour*, 120: 192–217.

Kummer, H. 1968. *Social Organization of Hamadryas Baboons*. Chicago: University of Chicago Press.

Kummer, H. 1975. Rules of dyad and group formation among captive gelada baboons (*Theropithecus gelada*). In: *Proceedings of the Symposium of the Fifth Congress of the International Primatology Society* (S. Kondo, M. Kawai, A. Ehara, & S. Kawamura, eds.), pp. 129–159. Tokyo: Japan Science Press.

Kummer, H. 1978. On the value of social relationships to nonhuman primates: A heuristic scheme. *Social Science Information*, 17: 687–705.

Kummer, H. 1995. *In Quest of the Sacred Baboon*. Princeton: Princeton University Press.

Kurland, J. A. 1977. Kin selection in the Japanese monkey. *Contributions to Primatology*, 12.

Leighton, D. R. 1987. Gibbons: Territoriality and monogamy. In: *Primate Societies* (B. B. Smuts, D. L. Cheney, R. M. Seyfarth, R. W. Wrangham, & T. T. Struhsaker, eds.), pp. 135–145. Chicago: University of Chicago Press.

Manson, J. H. 1994. Male aggression: A cost of female mate choice in Cayo Santiago rhesus macaques. *Animal Behaviour*, 48: 473–475.

Martin, R. D. 1990. *Primate Origins and Evolution*. London: Chapman and Hall.

Matsumura, S., & Kobayashi, T. 1997. A game model for dominance relations among group-living animals. *Behavioral Ecology and Sociobiology*, 42: 77–84.

Maynard Smith, J., & Price, G. R. 1973. The logic of animal conflict. *Nature*, 246: 15–18.

Mendoza, S. P., & Mason, W. A. 1986. Parental division of labour and differentiation of attachments in a monogamous primate (*Callicebus moloch*). *Animal Behaviour*, 34: 1336–1347.

Mesnick, S. L. 1997. Sexual alliances: Evidence and evolutionary implications. In: *Feminism and Evolutionary Biology: Boundaries, Intersections, and Frontiers* (P. A. Gowaty, ed.), pp. 207–257. New York: Chapman and Hall.

Milton, K. 1984. Habitat, diet, and activity patterns of free-ranging woolly spider monkeys (*Brachyteles arachnoides* E. Geoffroy 1806). *International Journal of Primatology*, 5: 491–514.

Milton, K. 1985a. Mating patterns of woolly spider monkeys, *Brachyteles arachnoides*: Implications for female choice. *Behavioral Ecology and Sociobiology*, 17: 53–59.

Milton, K. 1985b. Multimale mating and absence of canine tooth dimorphism in woolly spider monkeys (*Brachyteles arachnoides*). *American Journal of Physical Anthropology*, 68: 519–523.

Mitchell, C. L., Boinski, S., & van Schaik, C. P. 1991. Competitive regimes and female bonding in two species of squirrel monkeys (*Saimiri oerstedi* and *S. sciureus*). *Behavioral Ecology and Sociobiology*, 28: 55–60.

Nakamichi, M., & Koyama, N. 1997. Social relationships among ring-tailed lemurs (*Lemur catta*) in two free-ranging troops at Berenty Reserve, Madagascar. *International Journal of Primatology*, 18: 73–93.

Netto, W. J., & van Hooff, J. A. R. A. M. 1986. Conflict interference and the development of dominance relationships in immature *Macaca fascicularis*. In: *Primate Ontogeny, Cognition and Social Behavior* (J. G. Else & P. C. Lee, eds.), pp. 291–300. Cambridge: Cambridge University Press.

Nikolei, J., & Borries, C. 1997. Sex differential behavior of immature hanuman langurs (*Presbytis entellus*) in Ramnagar, South Nepal. *International Journal of Primatology*, 18: 415–437.

Nishida, T. 1979. The social structure of chimpanzees of the Mahale Mountains. In: *The Great Apes* (D. A. Hamburg & E. R. McCown, eds.), pp. 73–121. Menlo Park, Calif.: Benjamin/Cummings.

Nishida, T. 1983. Alpha status and agonistic alliance in wild chimpanzees. *Primates*, 24: 318–336.

Noë, R. 1990. A veto game played by baboons: A challenge to the use of the prisoner's dilemma as a paradigm for reciprocity and cooperation. *Animal Behaviour*, 39: 78–90.

Noë, R., & Hammerstein, P. 1994. Biological markets: Supply and demand determine the effect of partner choice in cooperation, mutualism and mating. *Behavioral Ecology Sociobiology*, 35: 1–11.

Noë, R., van Schaik, C. P., & van Hooff, J. A. R. A. M. 1991. The market effect: An explanation for pay-off asymmetries among collaborating animals. *Ethology*, 87: 97–118.

Overdorff, D. J. 1998. Are Eulemur species pair-bonded? Social organization and mating strategies in *Eulemur fulvus rufus* from 1988–1995 in Southeast Madagascar. *American Journal of Physical Anthropology*, 105: 153–166.

Overdorff, D. J., & Rasmussen, M. 1995. Determinants of nighttime activity in "diurnal" lemurid primates. In: *Creatures of the Dark: The Nocturnal Prosimians* (L. Alterman, G. A. Doyle, & M. K. Izard, eds.), pp. 61–74. New York: Plenum Press.

Palombit, R. A. 1996. Pair bonds in monogamous apes: A comparison of the siamang, *Hylobates syndactylus*, and the white-handed gibbon, *Hylobates lar*. *Behaviour*, 133: 321–356.

Palombit, R. A., Seyfarth, R. M., & Cheney, D. L. 1997. The adaptive value of "friendships" to female baboons: Experimental and observational evidence. *Animal Behaviour*, 54: 599–614.

Pereira, M. E. 1988a. Agonistic interactions of juvenile savannah baboons. I. Fundamental features. *Ethology*, 79: 195–217.

Pereira, M. E. 1988b. Effects of age and sex on intra-group spacing behaviour in juvenile savannah baboons, *Papio cynocephalus cynocephalus*. *Animal Behaviour*, 36: 184–204.

Pereira, M. E. 1989. Agonistic interactions of juvenile savannah baboons. II. Agonistic support and rank acquisition. *Ethology*, 80: 152–171.

Pereira, M. E. 1992. The development of dominance relations before puberty in cercopithecine societies. In: *Aggression and Peacefulness in Humans and Other Primates* (J. Silverberg & J. P. Gray, eds.), pp. 117–149. New York: Oxford University Press.

Pereira, M. E. 1993a. Agonistic interaction, dominance relation, and ontogenetic trajectories in ringtailed lemurs. In: *Juvenile Primates: Life History, Development, and Behavior* (M. E. Pereira & L. A. Fairbanks, eds.), pp. 285–305. New York: Oxford University Press.

Pereira, M. E. 1993b. Seasonal adjustment of growth rate and adult body weight in ringtailed lemurs. In: *Lemur Social Systems and Their Ecological Basis* (P. M. Kappeler & J. U. Ganzhorn, eds.), pp. 205–221. New York: Plenum Press.

Pereira, M. E. 1995. Development and social dominance among group-living primates. *American Journal of Primatology*, 37: 143–175.

Pereira, M. E., & Fairbanks, L. A. 1993. *Juvenile Primates: Life History, Development, and Behavior.* Oxford: Oxford University Press.

Pereira, M. E., & Kappeler, P. M. 1997. Divergent systems of agonistic behaviour in lemurid primates. *Behaviour*, 134: 225–274.

Pereira, M. E., & McGlynn, C. A. 1997. Special relationships instead of female dominance for redfronted lemurs, *Eulemur fulvus rufus. American Journal of Primatology*, 43: 239–258.

Pereira, M. E., Strohecker, R., Cavigelli, S. A., Hughes, C., & Pearson, D. In press. Metabolic strategy and social behavior in Lemuridae. In: *New Directions in Lemur Research* (H. Rasamimanana, B. Rakotosamimanana, J. Ganzhorn, & S. Goodman, eds.). New York: Plenum Press.

Perry, S. 1997. Male-female social relationships in wild white-faced capuchins (*Cebus capucinus*). *Behaviour*, 134: 477–510.

Pope, T. R. 1990. The reproductive consequences of male cooperation in the red howler monkey: Paternity exclusion in multi-male and single-male troops using genetic markers. *Behavioral Ecology Sociobiology*, 27: 439–446.

Rajpurohit, L. S., Sommer, V., & Mohnot, S. M. 1995. Wanderers between harems and bachelor bands: Male hanuman langurs (*Presbytis entellus*) at Jodhpur in Rajasthan. *Behaviour*, 132: 255–299.

Richard, A. F. 1987. Malagasy prosimians: Female dominance. In: *Primate Societies* (B. B. Smuts, D. L. Cheney, R. M. Seyfarth, R. W. Wrangham, & T. T.

Struhsaker, eds.), pp. 25–33. Chicago: University of Chicago Press.

Robbins, M. M. 1995. A demographic analysis of male life history and social structure of mountain gorillas. *Behaviour*, 132: 21–47.

Roberts, G., & Sherratt, T. N. 1998. Development of cooperative relationships through increasing investment. *Nature*, 394: 175–179.

Rosenberger, A. L., & Strier, K. B. 1989. Adaptive radiation in the ateline primates. *Journal of Human Evolution*, 17: 479–488.

Schino, G., Scucchi, S., Maestripieri, D., & Turillazzi, P. G. 1988. Allogrooming as a tension-reduction mechanism: A behavioral approach. *American Journal of Primatology*, 16: 43–50.

Seyfarth, R. M. 1980. The distribution of grooming and related behaviours among adult female vervet monkeys. *Animal Behaviour*, 28: 798–813.

Silk, J. B. 1982. Altruism among female *Macaca radiata*: Explanations and analysis of patterns of grooming and coalition formation. *Behaviour*, 79: 162–188.

Silk, J. B. 1992. The patterning of intervention among male bonnet macaques: Reciprocity, revenge, and loyalty. *Current Anthropology*, 33: 318–325.

Silk, J. B., Samuels, A., & Rodman, P. S. 1981. The influence of kinship, rank, and sex on affiliation and aggression between adult female and immature bonnet macaques (*Macaca radiata*). *Behaviour*, 78: 111–177.

Smuts, B. B. 1985. *Sex and Friendship in Baboons.* New York: Aldine.

Smuts, B. B., & Smuts, R. W. 1993. Male aggression and sexual coercion of females in nonhuman primates and other mammals: Evidence and theoretical implications. *Advances in the Study of Behavior*, 22: 1–63.

Smuts, B. B., & Watanabe, J. M. 1990. Social relationships and ritualized greetings in adult male baboons (*Papio cynocephalus anubis*). *International Journal of Primatology*, 11: 147–172.

Stanford, C. B. 1998. *Chimpanzee and Red Colobus: The Ecology of Predator and Prey.* Cambridge: Harvard University Press.

Sterck, E. H. M., Watts, D. P., & van Schaik, C. P. 1997. The evolution of female social relationships in nonhuman primates. *Behavioral Ecology and Socio-biology*, 41: 291–309.

Strier, K. B. 1990. New World primates, new frontiers: Insights from the woolly spider monkey, or muriqui (*Brachyteles arachnoides*). *International Journal of Primatology*, 11: 7–19.

Strier, K. B. 1991. Diet in one group of woolly spider monkeys, or muriquis (*Brachyteles arachnoides*). *American Journal of Primatology*, 23: 113–126.

Strier, K. B. 1992a. Atelinae adaptations: Behavioral strategies and ecological constraints. *American Journal of Physical Anthropology*, 88: 515–524.

Strier, K. B. 1992b. Causes and consequences of non-aggression in the woolly spider monkey, or muriqui (*Brachyteles arachnoides*). In: *Aggression and Peacefulness in Humans and Other Primates* (J. Silverberg & J. Patrick Gray, eds.), pp. 100–116. New York: Oxford University Press.

Strier, K. B. 1993. Growing up in a patrifocal society: Sex differences in the spatial relations of immature muriquis (*Brachyteles arachnoides*). In: *Juveniles: Comparative Socioecology* (M. E. Pereira & L. A. Fairbanks, eds.), pp. 138–147. New York: Oxford University Press.

Strier, K. B. 1994. Brotherhoods among atelins: Kinship, affiliation, and competition. *Behaviour*, 130: 151–167.

Strier, K. B. 1997a. Mate preferences in wild muriqui monkeys (*Brachyteles arachnoides*): Reproductive and social correlates. *Folia primatologica*, 68: 120–133.

Strier, K. B. 1997b. Subtle cues of social relations in male muriqui monkeys (*Brachyteles arachnoides*). In: *New World Primates: Ecology, Evolution, and Behavior* (W. G. Kinzey, ed.), pp. 109–118. New York: Aldine de Gruyter.

Strier, K. B. 1999a. *Faces in the Forest: The Endangered Muriqui Monkey of Brazil*. Cambridge: Harvard University Press.

Strier, K. B. 1999b. Predicting primate responses to "stochastic" events. *Primates*, 40: 131–142.

Strier, K. B., & Ziegler, T. E. 1997. Behavioral and endocrine characteristics of the reproductive cycle in wild muriqui monkeys, *Brachyteles arachnoides*. *American Journal of Primatology*, 42: 299–310.

Strier, K. B., Mendes, F. D. C., Rímoli, J., & Rímoli, A. O. 1993. Demography and social structure in one group of muriquis (*Brachyteles arachnoides*). *International Journal of Primatology*, 14: 513–526.

Sussman, R. W. 1977. Socialization, social structure, and ecology of two sympatric species of *Lemur*. In: *Primate Bio-social Development: Biological, Social, and Ecological Determinants* (S. Chevalier-Skolnikoff & F. E. Poirier, eds.), pp. 515–528. New York: Garland Publishing.

Sussman, R. W., & Richard, A. F. 1974. The role of aggression among diurnal prosimians. In: *Primate Aggression, Territoriality, and Xenophobia: A Comparative Perspective* (R. L. Holloway, ed.), pp. 49–76. New York: Academic Press.

Takahata, Y. 1990. Social relationships among adult males. In: *The Chimpanzees of the Mahale Mountains: Sexual and Life History Strategies* (T. Nishida, ed.), pp. 149–170. Tokyo: University of Tokyo Press.

Takahata, Y. 1991. Diachronic changes in the dominance relations of adult female Japanese monkeys of the Arashiyama B group. In: *The Monkeys of Arashiyama: Thirty-Five Years of Research in Japan and the West* (L. M. Fedigan & P. J. Asquith, eds.), pp. 123–139. Albany: State University of New York Press.

Thierry, B. 1990. Feedback loop between kinship and dominance: The macaque model. *Journal of Theoretical Biology*, 145: 511–521.

Treves, A. 1998. Primate social systems: Conspecific threat and coercion-defense hypotheses. *Folia primatologica*, 69: 81–88.

Treves, A., & Chapman, C. A. 1996. Conspecific threat, predation avoidance, and resource defense: Implications for grouping in langurs. *Behavioral Ecology and Sociobiology*, 39: 43–53.

Trivers, R. L. 1971. The evolution of reciprocal altruism. *Quarterly Review of Biology*, 46: 35–57.

van Hooff, J. A. R. A. M., & van Schaik, C. P. 1994. Male bonds: Affiliative relationships among nonhuman primate males. *Behaviour*, 130: 309–337.

van Noordwijk, M. A., & van Schaik, C. P. 1985. Male migration and rank acquisition in wild long-tailed macaques (*Macaca fascicularis*). *Animal Behaviour*, 33: 849–861.

van Noordwijk, M. A., & van Schaik, C. P. 1988. Male careers in Sumatran long-tailed macaques (*Macaca fascicularis*). *Behaviour*, 107: 24–43.

van Noordwijk, M. A., Hemelrijk, C. K., Herremans, L. A. M., & Sterck, E. H. M. 1993. Spatial position and behavioral sex differences in juvenile long-tailed macaques. In: *Juvenile Primates: Life History, Development, and Behavior* (M. E. Pereira & L. A. Fairbanks, eds.), pp. 77–85. New York: Oxford University Press.

van Schaik, C. P. 1983. Why are diurnal primates living in groups? *Behaviour*, 87: 120–144.

van Schaik, C. P. 1989. The ecology of social relationships amongst female primates. In: *Comparative Socioecology: The Behavioural Ecology of Humans and Other Mammals* (V. Standen & R. Foley, eds.), pp. 195–218. Oxford: Blackwell.

van Schaik, C. P. 1996. Social evolution in primates: The role of ecological factors and male behaviour. *Proceedings of the British Academy*, 88: 9–31.

van Schaik, C. P., & Hoerstermann, M. 1994. Predation risk and the number of adult males in a primate group: A comparative test. *Behavioral Ecology and Sociobiology*, 35: 261–272.

van Schaik, C. P., & Kappeler, P. M. 1993. Life history, activity period and lemur social systems. In: *Lemur Social Systems and Their Ecological Basis* (P. M. Kappeler & J. U. Ganzhorn, eds.), pp. 241–260. New York: Plenum Press.

van Schaik, C. P., & Kappeler, P. M. 1996. The social systems of gregarious lemurs: Lack of convergence with Anthropoids due to evolutionary disequilibrium? *Ethology*, 102: 915–941.

van Schaik, C. P., & Kappeler, P. M. 1997. Infanticide risk and the evolution of male-female association in primates. *Proceedings of the Royal Society of London*, Series B 264: 1687–1694.

van Schaik, C. P., & van Noordwijk, M. A. 1986. The hidden costs of sociality: Intra-group variation in feeding strategies in Sumatran long-tailed macaques (*Macaca fascicularis*). *Behaviour*, 99: 296–315.

van Schaik, C. P., & van Noordwijk, M. A. 1989. The special role of male *Cebus* monkeys in predation avoidance and its effect on group composition. *Behavioral Ecology and Sociobiology*, 24: 265–276.

Vick, L. G., & Pereira, M. E. 1989. Episodic targeting aggression and the histories of *Lemur* social groups. *Behavioral Ecology and Sociobiology*, 25: 3–12.

Watts, D. P. 1995. Post-conflict social events in wild mountain gorillas (*Mammalia, Hominoidea*) I. Social interactions between opponents. *Ethology*, 100: 139–157.

Weisbard, C., & Goy, R. W. 1976. Effect of parturition and group composition on competitive drinking order in stumptail macaques (*Macaca arctoides*). *Folia primatologica*, 25: 95–121.

Wrangham, R. W. 1980. An ecological model of female-bonded primate groups. *Behaviour*, 75: 262–299.

Wrangham, R. W. 1982. Mutualism, kinship and social evolution. In: *Current Problems in Sociobiology* (King's College Sociobiology Group, ed.), pp. 269–289. Cambridge: Cambridge University Press.

Wright, P. C. 1993. Variations in male-female dominance and offspring care in non-human primates. In: *Sex and Gender Hierarchies* (B. D. Miller, ed.), pp. 127–145. New York: Cambridge University Press.

CHAPTER 16

Conflict Management in Cross-Cultural Perspective

Douglas P. Fry

Any other adult in camp is related to any would-be aggressor by dozens of overlapping ties of kinship and marriage. Once a person attacks his victim he is like a fly that attacks an insect already caught in a spider's web. Immediately both are caught. If the combatants forget the sticky web in the heat of their anger, the onlookers do not. Real anger frightens and sickens the !Kung, for it is so destructive of their web of relationships.

Draper 1978, pp. 43–44

In each society, a variety of culturally patterned choices exist regarding how to deal with conflict (e.g., Hickson 1986; Nader 1990). Among the !Kung of the African Kalahari, for instance, conflicts find expression and resolution through gossip, ridicule, shunning, public discussions, and occasionally violence (Draper 1978). Among the Dou Donggo of Indonesia, in addition to aggressive self-help, choices for dealing with conflict include requesting one or more elders to mediate a dispute, appealing to the village headman for a decision, or pursuing a grievance in court, an option rarely used (Just 1991). Even in cultures that lack a central authority, acephalous societies, norms are enforced and disputes resolved through a variety of informal processes (Turnbull 1961; Merry 1982; Thomas 1994).

Although considerable cross-cultural variation exists regarding how conflicts are dealt with, simultaneously it is possible to note recurring patterns. On the one hand, conflict management mechanisms can be viewed as highly specific to the particular cultural system within which they operate (Avruch 1991). On the other hand, cross-cultural comparisons reveal general mechanisms of conflict management. For example, a comparative perspective shows that *mediation* (i.e., third-party assisted negotiation of a settlement) occurs in many cultures, whereas at a more specific level, the subtleties of mediation processes vary with the plethora of cultural values, beliefs, institutions, social roles, and so on found in different societies.

The central goal of this chapter is to illustrate the cultural variation in conflict management, while simultaneously noting recurring themes and features. Anthropological discussions of conflict management tend to focus on such procedures as avoidance, coercion, negotiation, mediation, arbitration, and adjudication but rarely consider *conflict prevention* explicitly. In this chapter, the treatment of conflict management is expanded to encompass conflict prevention and *reconciliation processes*. Some techniques for managing conflicts are tightly embedded within particular cultural meaning systems, whereas other measures are more general in nature and recur across cultural settings.

The first section of this chapter contains a brief overview of existing conflict management typologies, including third-party roles. The second section expands this perspective by explicitly focusing on prevention measures. The third section examines three case studies—the Semai *becharaa'*, the Bedouin guarantee, and the Zapotec court—in order to sample the cultural variation in conflict management processes and to illustrate further certain points raised in the two previous sections. The fourth section considers reconciliation, another largely neglected topic within cultural anthropology. Reconciliations are seen as most likely to occur when relationships are important in terms of alliances, kinship, economics, and so forth.

Traditional Conflict Management Typologies

Although at times employing different terms or subclassifications, the conflict management typologies presented by Koch (1974, 1979), Nader & Todd (1978), and Black (1993) are in general agreement that disputes can be handled through (1) negotiation, (2) self-help or coercion, (3) avoidance, (4) toleration, and (5) third-party assisted settlements such as mediation, arbitration, and adjudication (Fig. 16.1). These procedures are now discussed briefly.

Negotiation involves the handling of a dispute through joint decisions made by the disputants themselves and generally results in compromises or mutually agreeable solutions (Gulliver 1979). For example, feuding Jívaro of Ecuador end hostilities by negotiating a payment, normally a pig or a shotgun, to be given in restitution for a past killing. Other examples of negotiation and a thorough analytical treatment of the topic appear in Gulliver (1979).

Self-help/coercion entails the use of unilateral action as a way of handling a grievance and may vary in severity from theft to murder. Self-help can lead to cycles of violence manifested in feuds or blood revenge. Ethnographic illustrations of violent self-help include vendettas of Italian peasants (Brögger 1968), homicides among the Nuer of Africa (Greuel 1971), blood feuds among Finnish gypsies (Grönfors 1977), revenge killings among the Waorani of Ecuador (Robarchek & Robarchek 1996), and similar phenomena (Hoebel 1967; Black 1993).

Avoidance entails ceasing or limiting interaction with a disputant, either temporarily or permanently (Koch 1974; Black 1993). Across different types of social organization, people use avoidance as a response to conflict, for example, in Finnish culture generally (Fry 1999) and among Finnish gypsies (Grönfors 1977) or among Fijians (Arno 1979; Hickson 1986) and East Indians settled in Fiji (Brenneis 1990). The Mexican "La Paz" Zapotec will walk away from a dispute (O'Nell 1989; Fry 1994), and the Toraja of Indonesia avoid contact with angry persons (Hollan 1988, 1997). In some societies, avoidance is the culturally favored way of dealing with conflict. Among the Buid of the Philippines, "the socially approved response to aggression is avoidance or even flight" (Gibson 1989, p. 66).

In *toleration*, the issue causing the conflict "is simply ignored, and the relationship with the offending party is continued" (Nader & Todd 1978, p. 9). Toleration is illustrated among the "La Paz" Zapotec when a man decided not to pursue a grievance because, he explained, "it was not good to prolong a dispute and make someone angry at you for a long time" (Fry 1994, p. 141).

Settlement is the dealing with a grievance through a nonpartisan third party and includes such roles as (here listed in order of increasing authoritativeness) (1) friendly peacemaker, (2) mediator, (3) arbitrator, (4) adjudicator, and (5) repressive peacemaker (Black 1993; cf. Yarn, Chapter 4). Instances of *friendly peacemaking* probably occur in all cultures, since this third-party role simply entails separating or distracting adversaries (Black 1993; cf. Petit & Thierry, Box 13.1, for examples from nonhuman primates). Friendly peacemakers do not delve into the issues of the conflict.

In *mediation*, neutral third parties attempt to assist antagonists in arriving at agreements. Gulliver (1979) views mediation as the intervention of a third party in the negotiation process. Mediators do not make rulings or judgments, and they lack the authority to impose an agreement on the disputants (Koch 1974; Black 1993). However, sometimes mediators use coercion to push for an agreement between parties (Merry 1982; Podolefsky 1990).

Mediation occurs in many societies, but it is relied on in some societies much more than in others. "If 'mediation' represents a wide range of procedures and styles around the world, it is because societies themselves differ, both in their own cultural preoccupations and emphases, and in the nature of their ties to external sources of authority" (Greenhouse 1985, p. 97). Hunter-gatherer societies have friendly peacemakers, but owing to their largely egalitarian social organization, they tend not to rely significantly on mediators and lack positions of higher authority prerequisite to arbitration or adjudication (Black

1993). People in many horticultural and pastoral societies do rely on mediation, usually as an alternative to aggressive self-help initiatives, and peasant agriculturalists, nowadays typically living within the jurisdiction of national governments, sometimes opt for mediation in lieu of pursuing a grievance through a governmental court system (Merry 1982).

In some societies, multiple mediators assist in settlements. For instance, several elders typically mediate disputes among East Indians of Fiji (Brenneis 1990), the Dou Donggo (Just 1991), the Limbus of Nepal (Caplan 1995), and the Abkhazians of the Caucasus (Garb 1996). Additional examples of mediation that span the cross-cultural spectrum include the Liberian Kpelle moot (Gibbs 1963), the Hawaiian family meeting called *ho'oponopono* (Shook 1985; Shook & Kwan 1991), and court-ordered mediation sessions lead by trained volunteers in Finland (Takala 1998). The Semai and Zapotec case studies presented below also pertain to mediation.

In *arbitration*, a third party renders a decision but cannot enforce it; in *adjudication*, a judge offers a verdict and can enforce it. Disputants abide by an arbitrator's decision for any number of reasons: because of the pressure of public opinion, pressure from relatives, or both; because they believe the arbitrator's ruling to be fair; to maintain a good reputation; to avoid having to continue the dispute in court; and so on. Among the case studies presented below, the Bedouin and Semai cases pertain to arbitration (cf. Black 1993), whereas the Zapotec case pertains to adjudication. Other examples of adjudication include when Fiji Islanders file suit in the government court (Arno 1979) or when the Tarahumara of Mexico appear before the local judge (Pastron 1974).

Repressive peacemaking is the most authoritative third-party settlement role and treats fighting itself "as offensive and punishable, regardless of why it

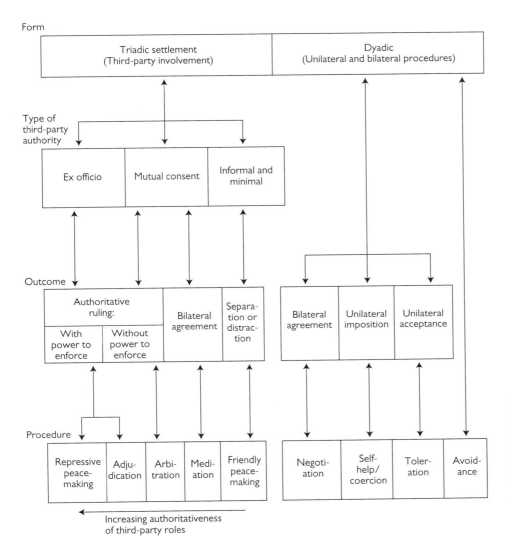

FIGURE 16.1. Nine conflict management procedures are classified according to their basic form (either triadic or dyadic) and types of outcome. For triadic procedures, the types of authority that are vested in the third parties and the relative authoritativeness of the five third-party roles also are depicted. After Koch (1974); Black (1993).

occurs" (Black 1993, p. 117). Repressive peacemaking occurs when colonial powers or national governments impose peace on warring indigenous peoples (Fry & Fry 1997). "The Jalé of the Jaxólé Valley realized that a new kind of stranger [the New Guinea government police] who neither spoke nor understood their language would punish any form of violent behavior" (Koch 1974, p. 223).

Certain features of the nine conflict management procedures discussed above are summarized in Figure 16.1. The figure shows, for instance, how arbitration is triadic in form and occupies an intermediate position on the continuum of third-party authoritativeness. The figure further shows that the mutual consent of the disputants is a fundamental feature of the arbitration process and that arbitrators issue authoritative decisions without having the power to enforce them.

With the exception of self-help (which can lead to escalating cycles of violence), the approaches for conflict management just discussed often can be seen to have preventive aspects. For instance, successful mediation agreements, arbitration rulings, or negotiated settlements not only resolve the disputes in question but also preempt the escalation or proliferation of the conflict (or both) to other parties. The next section focuses on conflict prevention in further detail.

Conflict Prevention

De Waal (1989) points out that prevention of conflict is difficult to study, because *potential* conflict situations are harder to define than post-conflict situations. Within cultural anthropology, relatively few researchers have focused on conflict prevention (exceptions are Marshall 1961; Hollan 1988, 1997; Briggs 1994; Fry 1994; Fry & Fry 1997). A survey of the cross-cultural literature on conflict management, however, suggests that individuals and societies utilize a host of conflict-prevention procedures. Some mechanisms can be seen as preventing conflict per se. Other mechanisms seem more appropriate for curtailing the *spread* of conflict to other persons, its *escalation*, or both.

Conflict prevention can involve individual decisions as well as group-level phenomena. In actuality, individual and group prevention measures are intertwined, as reflected, for instance, in the fact that individuals internalize the values and beliefs of their cultures and then act accordingly (Fry 1994). Certain psychocultural mechanisms for conflict prevention that develop within individuals during socialization are subsequently reinforced as the individuals continue to interact with other members of the culture who share the same worldviews, values, beliefs, and meanings. The following listing of conflict-prevention mechanisms begins with those that are primarily individual initiatives and moves toward more group phenomena, rec-

ognizing that sharp dichotomizing between the individual and the group does not reflect reality (Table 16.1).

First, self-restraint in the expression of anger, a mechanism that can be culturally promoted and individually manifested, can be seen as preventing conflict. Self-control regarding the expression of anger occurs among cultures differing in social complexity. Among Ju/wasi hunter-gatherers, for example, "phenomenal self-control was practiced by everyone but the smallest child. . . . Self-discipline pervaded everyday life, so the people virtually never showed hunger or pain, let alone anger" (Thomas 1994, p. 75). Hollan (1988, 1997) lists several psychocultural mechanisms that facilitate self-control of emotions and prevent the spread and escalation of conflict among the rice-farming Toraja, including a cultural belief that people eventually get what they deserve and a practice of simply not thinking about something or someone that evokes anger. "One prevalent technique . . . is to remind oneself of the dangers of strong, negative emotions: that by expressing such feelings one may upset others and suffer public censure or provoke magical retaliation; . . . that one may experience bad fortune in life as a consequence of getting angry and quarreling with others; and that by even experiencing negative emotions, one leaves oneself vulnerable to serious physical or mental illness" (Hollan 1997, p. 65).

Second, self-control also can relate to aggression when individuals avoid getting into a fight or keep a fight from becoming serious. For example, various instances of self-control have been observed among the Zapotec (Fry 1994), Finns (Fry unpublished), and the Mbuti, as "the men battled with burning logs, three or four feet long, swinging them from the cool end, but always just missing each other" (Turnbull 1961, p. 123).

Third, certain preventive techniques can be classified together under the label *attention to the*

TABLE 16.1
Conflict-Prevention Measures

Prevention Measures	Cultural Examples
SELF-RESTRAINT IN THE EXPRESSION OF ANGER	Inuit,[1] Ju/wasi,[2] !Kung,[3] Semai,[4] Toraja,[5] Tarahumara,[6] La Paz Zapotec,[7] Hawaiians,[8] Finns,[9] Norwegians[10]
SELF-CONTROL RELATED TO AGGRESSION	Inuit,[1] Ju/wasi,[2] !Kung,[3] La Paz Zapotec,[7] Tory Island Irish[11]
ATTENTION TO THE SATISFACTION AND NEEDS OF OTHERS	Inuit,[1] Chewong,[12] Semai,[4] Toraja,[5] La Paz Zapotec,[7] San Andrés Zapotec,[13] Zinacantecos[14]
SHOWING REMORSE AND APOLOGIZING	Mbuti,[15] Kpelle,[16] Talean Zapotec,[17] San Andrés Zapotec,[13] Zinacantecos,[14] Hawaiians,[8] Fijians[18]
INTERVENTION BY FRIENDLY PEACEMAKERS	Ju/wasi,[2] !Kung,[3] Mbuti,[15] Jívaro,[19] Cheyenne,[20] San Andrés Zapotec,[13] Tarahumara,[6] Tory Island Irish,[11] Finns,[9] Japanese[21]
OTHER SOCIAL MECHANISMS (e.g., institutionalized sharing and gift exchange, joking, ridicule, song duels or condemning songs, and gossip)	Baffin Islanders,[20] Inuit,[1] !Kung,[3] Mbuti,[15] Fijians,[18] Tonga,[22] La Paz Zapotec,[7] San Andrés Zapotec[13]

Sources: 1: Briggs 1994; 2: Thomas 1994; 3: Marshall 1961, Draper 1978; 4: Dentan 1968, Robarchek & Robarchek 1996, Robarchek 1997; 5: Hollan 1988, 1997; 6: Pastron 1974; 7: O'Nell 1989, Fry 1992, 1994; 8: Shook 1985, Shook & Kwan 1991; 9: Edelsward 1991, Tarasti 1991, Fry 1999; 10: Ross 1993; 11: Fox 1989; 12: Howell 1989; 13: Fry 1990, 1994; 14: Greenhouse 1979; 15: Turnbull 1961; 16: Gibbs 1963; 17: Nader 1969, 1990; 18: Arno 1979, Hickson 1979; 19: Harner 1972; 20: Hoebel 1967; 21: Black 1993; 22: Olson 1997.

satisfaction and needs of others. Briggs (1994, p. 161) catches the heart of this idea: one means of prevention for the Inuit is "keeping relations smooth; that is, keeping people happy, satisfied, unafraid, so that they will have no reason to be aggressive." The Toraja also attempt to "keep people happy" (Hollan 1988, p. 56). Similarly, the Semai pattern of nurturance, sharing, and support can be seen as preventive of dissatisfaction and conflict (Dentan 1968; Robarchek & Robarchek 1996; Robarchek 1997). Persons in various cultures attempt to avoid making or keeping others angry with them (e.g.,

Greenhouse 1979; Merry 1982; Fry 1994). Conflict escalation also can be prevented by simply ignoring another person's anger, a technique noted for the Toraja (Hollan 1988), the Chewong (Howell 1989), and the Inuit (Briggs 1994).

Preventive measures geared toward keeping other persons happy or satisfied are illustrated by the manner in which the "San Andrés" Zapotec (Fry 1992, 1994) treat intoxicated persons. Zapotecs follow these guidelines: if the drunk is not angry, prevent him from becoming so by being extra polite, showing exaggerated respect, and being very tolerant, for instance, by not getting upset when he screams in your face or vomits on your hat. If the drunk is angry, prevent him from becoming aggressive (or more aggressive) through verbal reassurances (e.g., "he is your compadre," "he was drunk," "it's not serious"), distracting him, and, if possible, spatially separating him from the sources of his anger.

Fourth, *showing remorse and apologizing* are initiatives that not only end a dispute or an avoidance condition but also simultaneously can be seen as preventing its spread or escalation (Hoebel 1967). As an example, a Talean Zapotec son's expression of regret—"I admit I was at fault, and now he can say how to punish me" (Nader 1969, p. 78)—kept a dispute from escalating.

Fifth, the *intervention of friendly peacemakers* to separate combatants prevents the escalation and spread of conflict in many cultural settings. In some cases, it is clear that the antagonists know that other persons will intervene before anybody gets seriously injured, as when fighting Tory Islanders shout "Hold me back or I'll kill him for sure!" (Fox 1989, p. 161), or when angry Finnish men exchange insults with great bravado while nonetheless allowing a couple of their peacekeeping companions to keep them from exchanging serious blows (Fry unpublished; cf. Hoebel 1967; Black 1993).

Among the "San Andrés" Zapotec (Fig. 16.2), typically several friendly peacemakers cooperatively prevent violence from escalating. "San Andrés" Zapotec know that disturbances should be taken seriously: someone is murdered every three to five years in their small community (Fry 1992). The cooperation and the use of multiple strategies by friendly peacemakers can be noted in the following episode. Two intoxicated men were attempting to cross the fast-flowing river when one of them, Antonio, fell into the ice-cold water. He became enraged at his companion and

angrily shouted that he was going home to get his pistol. When a group of four or five men coming along the road heard his tirade, they immediately intercepted him. Estanislau firmly grabbed Antonio's arm and started talking rapidly, pointing out, "You both are very drunk—he didn't know what he was doing." Estanislau continued to hold the angry Antonio by his arm, smiling and grinning constantly and nervously in what seemed to be an exaggerated display of nonaggressive intent. A second man also tried verbally to calm Antonio, who was still cussing and complaining. In a herding maneuver, the entire group of men surrounded Antonio, thereby assisting Estanislau in turning Antonio around and leading him in the opposite direction from his house—and his gun. In what seemed to be intended as a distraction, Estanislau helped Antonio to ring out his dripping sweater. Meanwhile, someone went on ahead and arranged for Antonio to receive "an invitation" to visit. As the group walked past a house, Tomas came to the front gate and invited Antonio to his house. Another man said to Antonio, "Look, you are invited to visit." In local etiquette, it would be considered extremely rude not to accept such an invitation. Antonio had calmed down somewhat as he was led into the household. It was very clear that this drunk, furious man stood no chance of getting his gun as long as other people were around. Like !Kung antagonists caught in the social web (Draper 1978), Antonio found his intentions to do violence thwarted by the network of cooperating Zapotec peacekeepers. At another level, in all likelihood, Antonio was engaging in this dramatic display of anger with some awareness that he would not be allowed to cause real violence.

Sixth, a variety of other social mechanisms and institutions may contribute to conflict prevention. For example, Marshall (1961) suggests that the highly socially prescribed patterns of meat sharing and gift exchange prevent conflicts among the

FIGURE 16.2. "San Andrés" Zapotec gather in the village churchyard during a saint's day celebration. Photograph by Douglas P. Fry.

!Kung. Whenever an animal is killed, the institutionalized patterns of meat distribution result in everyone in the band receiving some of the meat, thereby squelching dissatisfaction that might otherwise develop. Another social device that appears to prevent conflict and its escalation is joking. Joking can provide indirect feedback to individuals that other group members are dissatisfied with their behavior, ideally allowing them to "get the message" and alter their ways before conflicts are exacerbated. For example, during a social event in a Tongan village, a series of jokes were made about the inconveniences caused by one man's marauding pigs; a couple days later the man was seen mending his pig pen (Olson 1997). In the same vein as joking, ridicule (e.g., Turnbull 1961), the singing of condemning songs (e.g., Marshall 1961), and gossip (Merry 1984; Fry 1994) exert pressure on people to abide by group standards, thus preventing conflict.

Three Case Studies

The following case studies illustrate the cross-cultural variation in conflict management processes and also reinforce certain elements from the above discussions.

The Semai Becharaa'

The Semai of Malaysia live in small bands, usually consisting of less than 100 people. The Semai practice a combination of swidden gardening, fishing, hunting, and gathering. Semai bands are extremely egalitarian. Each band has a "headman" (occasionally a woman), but this leader, although respected, lacks political power.

The Semai value nonviolence and "go to great lengths to avoid conflict and will usually tolerate annoyances and sacrifice personal interests rather than precipitate an open confrontation" (Robarchek 1997, p. 54). In fact, the Semai fear a dispute more than they fear a tiger (Robarchek 1979, 1997). It is the responsibility of any member of the band who becomes aware of an interpersonal conflict to bring it to the attention of the headman for immediate resolution (Robarchek 1979). On the occasions when a dispute simply cannot be avoided, the headman oversees a dispute resolution assembly called becharaa'. The disputants and their relatives, as well as any other members of the community, convene at the headman's house.

Several of the elders present lengthy monologues in which they emphasize the mutual dependency of the band community and the need to maintain group harmony. The discussion then moves to the dispute at hand. In turn, each disputant presents his or her perspective. Other band members offer opinions, make observations, and ask questions. The disputants do not confront one another, nor do they argue with one another (Robarchek 1997). In fact, the verbal etiquette is for disputants to speak to the whole group. The *becharaa'* continues around the clock, typically lasting a day or two. During the *becharaa'*, the headman's household provides food and drink for those present. Throughout the *becharaa'*, anger and other heated emotions typically are not displayed. In a broader context, the suppression and denial of anger are typical of Semai culture (Dentan 1968).

During a *becharaa'*, the events related to the dispute are considered from "every conceivable perspective in a kind of marathon encounter group. Every possible explanation is offered, every imaginable motive introduced, every conceivable mitigating circumstance examined . . . until finally a point is reached where there is simply nothing left to say" (Robarchek 1997, p. 55). At that point, the headman takes an active role in lecturing one or both of the disputants, instructing them as to how they should have acted differently and admonishing them not to repeat their transgressions. Basically the headman voices the consensus of the group. Various elders in attendance also offer speeches that emphasize the necessity of maintaining band harmony and unity.

The Semai use the *becharaa'* to deal with serious conflicts—disputes over infidelity, divorce, property ownership, and so on—nonviolently and in such a way so as to (1) dissipate angry emotions

by allowing the dispute to be "talked to death," (2) deal with the substantive issues of the dispute, (3) promote the reconciliation of the antagonists—a primary goal of the *becharaa'*—and (4) reconfirm the interdependence of all members of the band and reiterate the value of social harmony (Robarchek 1979).

In terms of typological classification (e.g., Koch 1974; Black 1993), the *becharaa'* has elements of both mediation and arbitration. The headman, assisted by the other members of the assembly, mediates the dispute so that the disputants can resume a normal relationship. Eventually the headman issues a closing statement in which he may fine one or both of the disputants and/or specify that compensation be paid by one to the other. But the headman is basically voicing the consensus of the assembled community as to how the dispute should be settled, and since he lacks any real power to enforce his decision, this aspect of the process fits the definition of arbitration. It is significant that conflict resolution through the *becharaa'* is very much a group process, involving not only the disputants but also their relatives, other community members (sometimes the entire band), and the headman. Public opinion in general as well as specific statements by the elders, relatives, and the headman serve to apply strong social pressure on the disputants to accept the outcome favored by the group. Paralleling the social situation described for the !Kung in the opening quote of this chapter, Semai disputants are very much caught in a social web of interdependent relationships. The *becharaa'* simultaneously accomplishes conflict resolution and conflict prevention. For instance, the *becharaa'* prohibits direct confrontations between antagonists and also facilitates the self-control of anger. More generally, this mediation-arbitration process prevents the dispute from persisting, spreading, and escalating to the point of causing rifts and tensions within the entire band. Overall, Semai nonviolent values and other cultural ideals find expression within the *becharaa'* process.

The Bedouin Guarantee

Each Bedouin tribe has several respected men, whom Stewart (1990) refers to as "judges." By mutual consent, disputants may agree to appear before one of these "judges," who, on hearing the details of a particular case, renders a decision. Critically important to Bedouin social organization are *blood money groups*. Blood money groups consist of approximately 10 men who most likely are related to one another through descent from a common male ancestor (Stewart 1990). Members of a blood money group both support one another's interests and share one another's liabilities.

To consider a dispute, Frayj and another man contest the ownership of a camel and agree to take their case before a "judge" (Stewart 1990). Both men want assurances that the other will appear before the "judge" and abide by his decision. Among the Bedouin, one's good name is held to be very important. Because of the importance of one's reputation, a strong sanction that can be applied against a person who does not live up to an agreement is to proclaim publicly that he is dishonorable, in Bedouin parlance, that "his face is black" (Stewart 1990, p. 396). Additionally, to reassure one's opponent further, each disputant asks another member of his own blood money group to *guarantee* that he will appear before the "judge" and adhere to the decision. In this example, Silmiy agrees to act as Frayj's guarantor. Since normally guarantors are elder members of a disputant's own blood money group, they have the power to exert pressure on the persons whom they agree to guarantee. If Frayj reneges on his agreement to appear, he thus incites the wrath of his own guarantor, Silmiy, and he may soon find himself without the support of his own blood money group. In the

event that Frayj fails to appear, then Silmiy must appear in his place, and if the "judge" rules that the camel in question belongs to the other disputant, then Silmiy must provide compensation for it on Frayj's behalf. Should Frayj not appear, Silmiy has the right to receive from Frayj twice the compensation that Frayj cost him by not showing up.

In this Bedouin example, the "judge" is actually an arbitrator, since, like the Semai headman, he makes a decision but lacks the power to enforce it. The enforcement, if necessary, stems from pressure from within a disputant's own support group, especially from his guarantor. Since a disputant is dependent on his guarantor for his continued social, political, and economic well-being within his blood money group, the guarantor can effectively exert pressure on the disputant to comply with an arbitrator's ruling. Although typologically classifiable as arbitration, the specifics of this procedure are inseparable from Bedouin cultural meaning as reflected, for instance, in the Bedouin use of guarantors, the blood money group structure of social organization, and the cultural importance of reputation as expressed in "face blackening." The importance of reputation is certainly not unique to the Bedouin (cf. Merry 1984), but the concept here has specific cultural meanings. This Bedouin example again shows how disputes involve persons beyond the participants, such as a "judge" (arbitrator), guarantors, and other members of each disputant's blood money group. And here again, as in the Semai *becharaa'*, a disputant's relatives simultaneously support and pressure him to comply with the norms of the Bedouin legal system. This arbitration process also has preventive elements in that it provides an alternative to acts of self-help that can result in conflict escalation.

Talean Zapotec Court

Nader (1964, 1969, 1990) analyzes the disputing process in the Zapotec village of Talea (also called

Ralu'a) in rural Mexico. The peasant farmers of Talea grow maize, beans, sugarcane, and, as a cash crop, coffee. Three classes of remedy agents are used by Taleans to pursue their grievances: the family, the court, and the supernatural. Regarding conflicts among kin, older relatives may attempt to solve problems by mediating or arbitrating. Taking a family dispute to the local court authorities is also a possibility. Disputes between non-relatives, if not ignored (i.e., tolerated), are dealt with via self-help or the courts. Self-help may involve, for instance, the use of violence, supernatural channels, or malicious gossip. Nader (1990, p. 54) explains that supernatural remedy agents are "ambiguous and multifaceted"—God, saints, little gremlinlike men called *duendes*, witches, and the "man of the mountain" have supernatural powers and can be called on to punish other persons.

Taleans have access to three local courts wherein elected officials—the *presidente, alcalde,* and *sindico*—adjudicate (Nader 1990). Each court has the power to prosecute cases, make judgments, and enforce its verdicts by using the police. Sanctions invoked by court officers include levying fines to be paid in labor or cash, public reprimands, and short periods of jail time. Whereas witchcraft is for "getting even," the court is for "making the balance." In other words, the central precept of the Talean Zapotec court is to restore the relationship between disputants to equilibrium (Nader 1969, 1990). "Decision-making in a single case consists of a series of successive choices influenced by the overall pull toward harmony or 'agreement'" (Nader 1990, p. 92).

The operation of the Talean court differs in many ways from courts in large-scale, industrialized societies. First and foremost, the emphasis is on restoring a balanced, harmonious relationship between the antagonists, rather than on blaming or punishing. Second, the Talean court is not always a win-lose proposition; in a balanced way,

fault is often attributed to both parties. Third, justice is swift, with cases generally being heard on the day that they are brought to the court. Fourth, court officers allow great latitude regarding what can be discussed during a hearing, allowing hearsay, gossip, and past behavior to be brought up. Relatedly, the court officers take into account their knowledge of the disputants' reputations when making their rulings. Fifth, the court officials do not always adhere to the role of adjudicator but instead shift among roles such as mediator, judge, and even therapist (Nader 1969). Sixth, the proceedings are focused toward achieving compromises and agreements between the disputants—to "make the balance"—as the court officers strive to resolve conflict "by minimizing the sense of injustice and outrage felt by the parties to the case" (Nader 1969, p. 84).

Excerpts from the following case involving an estranged husband and wife, reported by Nader (1969, pp. 76–78), illustrate many of the above points.

"Mr. President, I am here to complain . . . my wife, had my coffee cut. . . ." Carmen Ibarra, wife of the plaintiff, said excitedly: ". . . Why shouldn't I cut the coffee as it belongs to both of us, and besides we have children to support and feed." [The husband and wife take turns presenting their complaints and perspectives at some length to the presidente, who does more listening than talking. Eventually, Mrs. Ibarra says to her husband:] ". . . Yes, I will deliver the coffee, which you said I took, but you have to pay in front of the president the bills of the treatment that our daughter had to have—poor little girl—who wants so much to be cured as she says in her postcard, 'Mama, do sell some interests of my part for the treatment.'. . ." "Look, Mr. President," explained the plaintiff, "I planted the coffee seeds for my wife and she has sufficient, and now she tries to take mine. . . ." [Having listened to both parties for some while, the presidente begins to lecture both of them:] "You should now think like mature people about what you are doing, and the only thing you should do is to get together again, forget the troubles of the past. . . . You should both be home worrying about how your daughter can be cured." [To the husband, the presidente continues:] . . . "Return to your home and think about how to resolve the problems of the home. . . . As to the woman, the obligation of the wife is to be there where the husband orders, as long as it is in agreement and to the benefit of the home." Having listened to what the presidente had to say and after a long silence and reflection, the couple . . . said that they would unite again and that they would follow the advice of the presidente.

In this example, clearly the central focus is on restoring a balance in the wife-husband relationship, thus attempting simultaneously conflict resolution, reconciliation, and conflict prevention. At least in this case, the *presidente* acts more like a mediator and a marriage counselor than a judge, imposing no fines or monetary restitution in this case but instead listening to both parties at great length and lecturing *both* of them as to the culturally appropriate roles of a married man and woman. Given the *presidente's* approach to this dispute, it is apparent why Taleans sometimes refer to their *presidente* as "the father of the town" (Nader 1990, p. 64). Reflecting Zapotec cultural values and beliefs, the Talean court delivers much more than pure adjudication and differs markedly from courts in large-scale, industrial societies.

Reconciliation

De Waal's (1989) suggestion that reconciliation has received scant attention as a topic of research seems apropos to cultural anthropology. If mentioned at all in anthropological writings, reconciliations tend to receive only a sentence or two. However, while usually neglecting the details of

reconciliation, time and again anthropological discussions emphasize that the goal of conflict resolution is to reestablish normal, harmonious relationships among the disputants (e.g., Turnbull 1961; Gibbs 1963; Brögger 1968; Nader 1969, 1990; Koch 1974, 1979; Hickson 1979, 1986; Noland 1981; Shook 1985; Hollan 1988; Caplan 1995; Robarchek 1997). The emphasis on restoring amicable relations also is apparent in both the Semai and Zapotec case studies. Just's (1991, p. 117) assessment for the Dou Donggo is rather typical: "Conflict resolution may be seen as the restoration of damaged status relationships rather than the pursuit of equity or the imposition of retributive punishment." These anthropological sources support de Waal's (1996) suggestion that it is appropriate to focus attention on relationships when studying conflict and reconciliation (cf. de Waal, Chapter 2; Cords & Aureli, Chapter 9; van Schaik & Aureli, Chapter 15).

Just as disputes often involve additional persons as supporters or peacemakers, similarly, reconciliations tend to incorporate the additional parties whose interpersonal relationships also have been affected by the conflict (cf. de Waal 1996). Among Arabian women, when close friends are engaged in a dispute, a third friend often facilitates a reconciliation by arranging a meeting at her house in which simple reconciliation rituals occur (Koch et al. 1977). Reconciliation rituals among Hawaiians involve the entire family (Shook 1985). Following a Kpelle moot, the entire assembly drinks together (Gibbs 1963). During Jalé reconciliations, the disputants, their allies, and perhaps other people share a ritually consecrated meal (Koch et al. 1977). The Ju/wasi hunter-gatherers may perform a "trance dance" during reconciliations that necessitates the cooperation and participation of the entire band (Thomas 1994). Thus, although conflict-resolution processes leading to reconciliations may be dyadic, more typically partisan and nonpartisan third parties also are involved. Group involvement in reconciliation is yet another manifestation of the interconnected and interdependent nature of most human social life.

As noted for nonhuman primates (de Waal, Chapter 2), the cross-cultural literature on disputing and conflict management suggests that reconciliation is most likely to occur when (1) relationships are important (e.g., emotionally, socially, economically, politically, and/or strategically)—in other words, when social distance is short and mutual dependency is high—and (2) in social situations in which new, replacement relationships are difficult or impossible to establish (cf. Hoebel 1967; Koch et al. 1977; Noland 1981; Hickson 1986). Conversely, reconciliation is relatively less likely among socially distant and independent parties (Koch 1974; Merry 1982).

When a group is faced with external threats, relationships within the group and with allies increase in importance. Group pressure is exerted on disputants to reconcile conflicts that threaten valued defensive alliances or solidarity. For example, among the Comanche, "fighting within the tribe was not to be countenanced when there were always outside enemies to be confronted" (Hoebel 1967, p. 139).

Furthermore, being in conflict with "close persons" appears to be an uncomfortable, emotionally upsetting experience in many, if not all, cultures. Psychological distress over disrupted important social relationships appears to motivate persons to seek reconciliations in many contexts (cf. Aureli & Smucny, Chapter 10). For instance, a Fijian man expressed the great relief that he felt following an apology ritual that he had just attended, for he had felt choked by the conflict. Similarly, the Ju/wasi perform a "trance dance" to rid the band of ill will—"star sickness"—which infects the group when persons quarrel: "After trance dancing, people feel emotionally cleansed"

(Thomas 1994, p. 77). And Turnbull (1961) contrasts the talking, laughing, and joking on a typical day within a Mbuti band with the tense, silent, heavy atmosphere when a serious conflict is brewing.

Conflict-resolution events often appear to incorporate reconciliation rituals, although as previously mentioned, the accounts typically provide few actual details. Hawaiian ho'oponopono closing rituals included, in pre-Christian times, an offering of meat to the gods, sometimes followed by a ceremonial ocean bath, and, in any case, a feast (Shook 1985). Based on her cross-cultural study of mediation, Merry (1982, p. 30) notes that "the last step in the mediation process is typically a ritual of reconciliation, whether drinking coffee together in a Lebanese village or a massive village feast financed by the loser as a public apology, as in prerevolutionary China." Koch et al. (1977, pp. 272–273) provide an unusually detailed description of the reconciliation process among the Jalé of New Guinea:

The party who desires a reconciliation enlists a curer whose job it is to supervise the slaughter of the animal and to perform the ritual. The ceremony itself is a rather simple affair, but with deep symbolic significance. The Jalé term for the condition described here as avoidance is *héléroxo*. The expression derives from *hélé*, denoting the ditch that separates two adjacent beds in a garden; the suffix—*roxo* corresponds to the English "-wise." In performing the ritual, the curer, uttering esoteric formulae, smears a mixture of soil and blood drawn from the slaughtered pig on the hams of the antagonists. That act is called *kénangenep-tuk* ("soil them up"). Soil, or *kénan*, being the substance of that which *hélé*, divides, the metaphorical aspect of the rite becomes apparent, and the expression of "seal them up" aptly to connote the nature of the event. . . .

Following the rite, the parties sit down to a communal meal to which relatives and friends may have been invited. Any person who was drawn into the avoidance relationship on the prompting of either of the disputants dissolves his condition by eating from the pork consecrated by the curer. . . . While at ordinary meals it is customary to tear off a portion of one's piece of food and hand it to a kinsman or neighbor present, on this occasion all participants, especially the reconciled parties, exaggerate these mutual exchanges. As the suspension of food-sharing has signaled the inception of the *héléroxo* relationship, so is its termination affirmed by the ostentatious resumption of commensal practice.

As is readily apparent in these descriptions, ritual aspects of reconciliation reflect unique cultural meaning systems. At the same time, however, an examination of different ethnographic accounts suggests certain recurring elements during reconciliation, most notably (1) gift giving or gift exchange, (2) payment of restitution, (3) sharing food or drink, (4) physical contact such as kissing and shaking hands, (5) appeasement postures and gestures, (6) verbal expressions of apology, remorse, or contrition, and (7) the participation of other persons besides the disputants in reconciliation rituals. Cultural examples for each point are listed in Table 16.2.

Conclusion

A culturally comparative approach suggests that the conflict management procedures utilized in human societies can be classified typologically. Across cultural settings, persons engage in *avoidance* and *toleration*, attempt *negotiations*, exercise aggressive *self-help*, and rely on the assistance of various kinds of third-party *settlement* agents to resolve conflicts. Particular cultures use some procedures more than others. The extensive use of violent self-help by the head-hunting Jívaro is reflected in their saying "I was born to die fight-

ing," but Jívaro also negotiate solutions to their differences (Harner 1972). On the other extreme, the conflict-fearing Semai employ avoidance, toleration, and mediation-arbitration sessions (i.e., the *becharaa'*) as procedures that are in accordance with their nonviolent cultural values and beliefs.

The use of particular third-party settlement agents relates in part to the degree of authoritativeness inherent in specific types of social organization (Black 1993). The friendly peacemaker role that involves separating or distracting disputants requires little or no authority and is found across societies, whereas the roles of mediator, arbitrator, adjudicator, and repressive peacemaker require increasing authority over disputants. Except within egalitarian bands, mediation is widespread across societies in which some individuals (e.g., elders) hold at least slightly more authority than others. Within hierarchical societies (e.g., chiefdoms, states), arbitration, adjudication, and/or repressive peacemaking are likely to occur as well. Conflict management typologies help to illuminate such patterns in the cross-cultural data. At the same time, it is clear that conflict management procedures are part and parcel of particular cultural meaning systems with their associated worldviews, values, and beliefs, a point illustrated by the Semai, Bedouin, and Zapotec case studies. Understanding of conflict management can best be advanced through both culturally comparative studies *and* culturally specific research.

The relatively neglected topics within cultural anthropology of reconciliation and conflict prevention have received some preliminary coverage in this chapter. One clear conclusion is that conflict resolution often focuses on restoring relationships and involves interested parties beyond the disputants themselves. A second conclusion is that when disputants have important attachments and are interdependent on one another—for example, for protection or social, economic, and political

TABLE 16.2
Common Features of Reconciliation Rituals

Features	Cultural Examples
GIFT GIVING OR GIFT EXCHANGE	Zinacantecos[1] exchange local liquor; apologizing Fijians[2] make symbolic gifts of *kava* brew, a whale's tooth, etc.
PAYMENT OF RESTITUTION	Jívaro[3] use pigs and shotguns; Cheyenne[4] used horses and other goods; Jalé[5] use pigs
SHARING FOOD OR DRINK	Hawaiians[6] and Jalé[5] feast together; Kpelle,[7] Torajans,[8] Fijians,[2] and Italian peasants[9] drink together
PHYSICAL CONTACT	East Indians of Fiji[10] and Torajans[8] shake hands; Arabian women[11] and Iranians[12] kiss
APPEASEMENT POSTURES AND GESTURES	Contrite Tarahumara[13] bow their heads; apologizing Fijians[2] bow, look down, tremble, and/or prostrate themselves
EXPRESSIONS OF APOLOGY, REMORSE, OR CONTRITION	Apologetic words are used among the Mbuti,[14] Kpelle,[7] Talean Zapotec,[15] Zinacantecos,[1] Fijians,[2] and Hawaiians[6]
PARTICIPATION OF OTHER PERSONS BESIDES THE DISPUTANTS IN RECONCILIATION RITUALS	Entire Mbuti,[14] !Kung,[16] and Ju/wasi[17] bands; Jalé[5] kin, neighbors, and allies; Semai[18] *becharaa'* attendees; Kpelle[7] moot attendees

Sources: 1: Greenhouse 1979; 2: Koch et al. 1977; 3: Harner 1972; 4: Hoebel 1967; 5: Koch 1974, Arno 1979, Hickson 1979; 6: Shook 1985, Shook & Kwan 1991; 7: Gibbs 1963; 8: Hollan 1988; 9: Brögger 1968; 10: Brenneis 1990; 11: Koch et al. 1977; 12: Noland 1981; 13: Pastron 1974; 14: Turnbull 1961; 15: Nader 1969, 1990; 16: Marshall 1961, Draper 1978; 17: Thomas 1994; 18: Robarchek 1979, 1997.

reasons—antagonists, often with the encouragement and aid of third parties, tend to mend the relationships that have become strained by conflict. As in other animals (de Waal, Chapter 2), this emphasis on restoring relationships and promoting group harmony is noteworthy for its pervasiveness from one cultural setting to the next. At the same time, it should be borne in mind that some cultures are highly conflictual or aggressive (e.g., the Nuer and the Waigali: Merry 1982) and may lack well-developed conflict-

prevention and -resolution mechanisms (e.g., the Jívaro: Harner 1972; the Waorani: Robarchek & Robarchek 1996).

It is clear that written legal codes and court systems are not the only paths to justice and social order (Hoebel 1967; Draper 1978; Yarn, Chapter 4). A great number of ethnographically diverse processes, from negotiation to arbitration, can be seen as preventing, managing, and resolving conflict. Approaches to social control and the administration of justice also include diverse *informal* mechanisms such as joking, ridicule, reprimands, gossip, discussions, debates, shunning, sorcery, threats of violence, violence, ostracism, and, in extreme cases, group-condoned executions. It was suggested in this chapter that the *prevention* of conflict and its spread may be facilitated by some such informal mechanisms as well as by systems of sharing and reciprocal cooperation, internalization of self-restraint toward expressing anger and aggression, socialized sensitivity toward the emotional state and needs of others, apologizing or showing remorse, and the activities of friendly peacemakers.

References

Arno, A. 1979. A grammar of conflict: Informal procedure on an Island in Lau, Fiji. In: *Access to Justice*, Vol. 2: *The Anthropological Perspective: Patterns of Conflict Management: Essays in the Ethnography of Law* (K.-F. Koch, ed.), pp. 41–68. Alphen aan den Rijn: Sijthoff & Noordhoff.

Avruch, K. 1991. Introduction: Culture and conflict resolution. In: *Conflict Resolution: Cross-Cultural Perspectives* (K. Avruch, P. W. Black, & J. A. Scimecca, eds.), pp. 1–17. New York: Greenwood Press.

Black, D. 1993. *The Social Structure of Right and Wrong.* San Diego: Academic Press.

Brenneis, D. 1990. Dramatic gestures: The Fiji Indian *pancayat* as therapeutic event. In: *Disentangling: Conflict Discourse in Pacific Societies* (K. A. Watson-Gegeo & G. M. White, eds.), pp. 214–238. Stanford: Stanford University Press.

Briggs, J. L. 1994. "Why don't you kill your baby brother?": The dynamics of peace in Canadian Inuit camps. In: *The Anthropology of Peace and Nonviolence* (L. E. Sponsel & T. Gregor, eds.), pp. 155–181. Boulder, Colo.: Lynne Reinner Publishers.

Brögger, J. 1968. Conflict resolution and the role of the bandit in peasant society. *Anthropological Quarterly,* 41: 228–240.

Caplan, L. 1995. The milieu of disputation: Managing quarrels in East Nepal. In: *Understanding Disputes: The Politics of Argument* (P. Caplan, ed.), pp. 137–159. Oxford: Berg.

Dentan, R. K. 1968. *The Semai: A Nonviolent People of Malaya.* New York: Holt, Rinehart & Winston.

de Waal, F. 1989. *Peacemaking among Primates.* Cambridge: Harvard University Press.

de Waal, F. 1996. *Good Natured: The Origins of Right and Wrong in Humans and Other Animals.* Cambridge: Harvard University Press.

Draper, P. 1978. The learning environment for aggression and anti-social behavior among the !Kung. In: *Learning Non-Aggression: The Experience of Non-Literate Societies* (A. Montagu, ed.), pp. 31–53. Oxford: Oxford University Press.

Edelsward, L. M. 1991. *Sauna as Symbol: Society and Culture in Finland.* New York: Peter Lang.

Fox, R. 1989. *The Search for Society: Quest for a Biosocial Science and Morality.* New Brunswick, N.J.: Rutgers University Press.

Fry, D. P. 1990. Play aggression among Zapotec children: Implications for the practice hypothesis. *Aggressive Behavior,* 16: 321–340.

Fry, D. P. 1992. "Respect for the rights of others is peace": Learning aggression versus non-aggression among the Zapotec. *American Anthropologist,* 94: 621–639.

Fry, D. P. 1994. Maintaining social tranquility: Internal and external loci of aggression control. In: *The Anthropology of Peace and Nonviolence* (L. E. Sponsel & T. Gregor, eds.), pp. 133–154. Boulder, Colo.: Lynne Reinner Publishers.

Fry, D. P. 1999. Altruism and Aggression. In: *Encyclopedia of Violence, Peace, and Conflict,* Vol. 1 (L. R. Kurtz, ed.), pp. 17–33. San Diego: Academic Press.

Fry, D. P., & Fry, C. B. 1997. Culture and conflict resolution models: Exploring alternatives to violence. In: *Cultural Variation in Conflict Resolution: Alternatives to Violence* (D. P. Fry & K. Björkqvist, eds.), pp. 9–23. Mahwah, N.J.: Lawrence Erlbaum Associates.

Garb, P. 1996. Mediation in the Caucasus. In: *Anthropological Contributions to Conflict Resolution* (A. W. Wolfe & H. Yang, eds.), pp. 31–46. Athens: University of Georgia Press.

Gibbs, J. L., Jr. 1963. The Kpelle moot: A therapeutic model for the informal settlement of disputes. *Africa*, 33: 1–11.

Gibson, T. 1989. Symbolic representations of tranquility and aggression among the Buid. In: *Societies at Peace: Anthropological Perspectives* (S. Howell & R. Willis, eds.), pp. 60–78. London: Routledge.

Greenhouse, C. J. 1979. Avoidance as a strategy for resolving conflict in Zinacantan. In: *Access to Justice*, Vol. 4: *The Anthropological Perspective: Patterns of Conflict Management: Essays in the Ethnography of Law* (K.-F. Koch, ed.), pp. 105–123. Alphen aan den Rijn: Sijthoff & Noordhoff.

Greenhouse, C. J. 1985. Mediation: A comparative approach. *Man*, 20: 90–114.

Greuel, P. J. 1971. The leopard-skin chief: An examination of political power among the Nuer. *American Anthropologist*, 73: 1115–1120.

Grönfors, M. 1977. *Blood Feuding among Finnish Gypsies: Sociology Research Report No. 213*. Helsinki: University of Helsinki.

Gulliver, P. H. 1979. *Disputes and Negotiations: A Cross-Cultural Perspective*. New York: Academic Press.

Harner, M. J. 1972. *The Jívaro: People of the Sacred Waterfall*. Garden City, N.Y.: Anchor Books/Doubleday.

Hickson, L. 1979. Hierarchy, conflict, and apology in Fiji. In: *Access to Justice*, Vol. 4: *The Anthropological Perspective: Patterns of Conflict Management: Essays in the Ethnography of Law* (K.-F. Koch, ed.), pp. 17–39. Alphen aan den Rijn: Sijthoff & Noordhoff.

Hickson, L. 1986. The social contexts of apology in dispute settlement: A cross-cultural study. *Ethnology*, 25: 283–294.

Hoebel, E. A. 1967. *The Law of Primitive Man*. Cambridge: Harvard University Press.

Hollan, D. 1988. Staying "cool" in Toraja: Informal strategies for the management of anger and hostility in a nonviolent society. *Ethos*, 16: 52–72.

Hollan, D. 1997. Conflict avoidance and resolution among the Toraja of South Sulawesi, Indonesia. In: *Cultural Variation in Conflict Resolution: Alternatives to Violence* (D. P. Fry & K. Björkqvist, eds.), pp. 59–68. Mahwah, N.J.: Lawrence Erlbaum Associates.

Howell, S. 1989. "To be angry is not to be human, but to be fearful is": Chewong concepts of human nature. In: *Societies at Peace: Anthropological Perspectives* (S. Howell & R. Willis, eds.), pp. 45–59. London: Routledge.

Just, P. 1991. Conflict resolution and moral community among the Dou Donggo. In: *Conflict Resolution: Cross-Cultural Perspectives* (K. Avruch, P. W. Black, & J. A. Scimecca, eds.), pp. 109–143. New York: Greenwood Press.

Koch, K.-F. 1974. *War and Peace in JalÇm¢: The Management of Conflict in Highland New Guinea*. Cambridge: Harvard University Press.

Koch, K.-F. 1979. Introduction: Access to justice: An anthropological perspective. In: *Access to Justice*, Vol. 4: *The Anthropological Perspective: Patterns of Conflict Management: Essays in the Ethnography of Law* (K.-F. Koch, ed.), pp. 1–16. Alphen aan den Rijn: Sijthoff & Noordhoff.

Koch, K.-F., Altorki, S., Arno, A., & Hickson, L. 1977. Ritual reconciliation and the obviation of grievances: A comparative study in the ethnography of law. *Ethnology*, 16: 270–283.

Marshall, L. 1961. Sharing, talking, and giving: Relief of social tensions among !Kung Bushmen. *Africa*, 31: 231–249.

Merry, S. E. 1982. The social organization of mediation in nonindustrial societies: Implications for informal community justice in America. In: *The Politics of Informal Justice*, Vol. 2: *Comparative Studies* (R. L. Abel, ed.), pp. 17–45. New York: Academic Press.

Merry, S. E. 1984. Rethinking gossip and scandal. In: *Toward a General Theory of Social Control*, Vol. 1: *Fundamentals* (D. Black, ed.), pp. 271–302. Orlando, Fla.: Academic Press.

Nader, L. 1964. An analysis of Zapotec law cases. *Ethnology*, 3: 404–419.

Nader, L. 1969. Styles of court procedure: To make the balance. In: *Law and Culture in Society* (L. Nader, ed.), pp. 69–91. Chicago: Aldine.

Nader, L. 1990. *Harmony Ideology: Justice and Control in a Zapotec Mountain Village*. Stanford: Stanford University Press.

Nader, L., & Todd, H. F., Jr. 1978. Introduction: The disputing process. In: *The Disputing Process: Law in Ten Societies* (L. Nader & H. F. Todd, Jr., eds.), pp. 1–40. New York: Columbia University Press.

Noland, S. 1981. Dispute settlement and social organization in two Iranian rural communities. *Anthropological Quarterly*, 54: 190–202.

Olson, E. G. 1997. Leaving anger outside the kava circle: A setting for conflict resolution in Tonga. In: *Cultural Variation in Conflict Resolution: Alternatives to Violence* (D. P. Fry & K. Björkqvist, eds.), pp. 79–87. Mahwah, N.J.: Lawrence Erlbaum Associates.

O'Nell, C. W. 1989. The non-violent Zapotec. In: *Societies at Peace: Anthropological Perspectives* (S. Howell & R. Willis, eds.), pp. 117–132. London: Routledge.

Pastron, A. G. 1974. Collective defenses of repression and denial: Their relationship to violence among the Tarahumara Indians of northern Mexico. *Ethos*, 2: 387–404.

Podolefsky, A. 1990. Mediator roles in Simbu conflict management. *Ethnology*, 29: 67–81.

Robarchek, C. A. 1979. Conflict, emotion, and abreaction: Resolution of conflict among the Semai Senoi. *Ethos*, 7: 104–123.

Robarchek, C. A. 1997. A community of interests: Semai conflict resolution. In: *Cultural Variation in Conflict Resolution: Alternatives to Violence* (D. P. Fry & K. Björkqvist, eds.), pp. 51–58. Mahwah, N.J.: Lawrence Erlbaum Associates.

Robarchek, C. A., & Robarchek, C. J. 1996. Waging peace: The psychological and sociocultural dynamics of positive peace. In: *Anthropological Contributions to Conflict Resolution* (A. W. Wolfe & H. Yang, eds.), pp. 64–80. Athens: University of Georgia Press.

Ross, M. H. 1993. *The Management of Conflict*. New Haven: Yale University Press.

Shook, E. V. 1985. *Ho'oponopono: Contemporary Uses of a Hawaiian Problem-Solving Process*. Honolulu: East-West Center/University of Hawaii.

Shook, E. V., & Kwan, L. K. 1991. *Ho'oponopono*: Straightening family relationships in Hawaii. In: *Conflict Resolution: Cross-Cultural Perspectives* (K. Avruch, P. W. Black, & J. A. Scimecca, eds.), pp. 213–229. New York: Greenwood Press.

Stewart, F. H. 1990. Schuld and Haftung in Bedouin law. In: *Zeitschrift der Savigny-Stiftung für Rechtsgeschichte, Hundertsiebenter Band, CXX* (Th. Mayer-Maly, D. Nörr, W. Waldstein, A. Laufs, W. Ogris, M. Heckel, P. Mikat, & K. W. Nörr, eds.), pp. 393–407. Vienna: Hermann Böhlaus Nachf.

Takala, J.-P. 1998. *Moraalitunteet Rikosten Sovittelussa, Oikeuspoliittisen Tutkimuslaitoksen Julkaisuja 151* [Moral emotions in victim-offender mediation—with English summary]. Helsinki: Oikeuspoliittinen Tutkimuslaitos.

Tarasti, E. 1991. Finland in the eyes of a semiotician. *Semiotica*, 87: 203–216.

Thomas, E. M. 1994. Management of violence among the Ju/wasi of Nyae Nyae: The old way and the new way. In: *Studying War: Anthropological Perspectives* (S. P. Reyna & R. E. Downs, eds.), pp. 69–84. New York: Gordon & Breach.

Turnbull, C. M. 1961. *The Forest People: A Study of the Pygmies of the Congo*. New York: Simon & Schuster.

CHAPTER 17

The Evolution and Development of Morality

Melanie Killen & Frans B. M. de Waal

What are the origins of morality, both phylogenetically and ontogenetically? Phylogeny deals with possible homologies between species and with stages of evolution over millions of years. Naturally, evolution also deals with the issue of adaptive function. After all, how could we have evolved moral tendencies if such tendencies lacked positive effects on survival and reproduction? The ontogenetic question, however, focuses on the development over a much more limited time, between infancy and adulthood, within a single species. In this chapter, we make two fundamental assertions regarding the evolution of morality: (1) there are specific types of behavior demonstrated by both human and nonhuman primates that hint at a shared evolutionary background to morality; and (2) there are theoretical and actual connections between morality and conflict reso-

lution in both nonhuman primates and human development.

In our analysis of the literature we make several points that apply to both human and nonhuman primates. First, the transition from nonmoral or premoral to moral is more gradual than commonly assumed. No magic point appears in either evolutionary history or human development at which morality suddenly comes into existence. In both early childhood and in animals closely related to us, we can recognize behaviors (and, in the case of children, judgments) that are essential building blocks of the morality of the human adult. Second, we theorize that the phylogenetic origins of human morality can be detected in the social interactions of nonhuman primates, our closest biological relatives, and the ontogenetic origins of morality can be observed in the early social interactions of children. In this sense, conflict resolution is a key element in understanding the evolution of morality because conflicts and conflict resolution provide the experiential basis by which moral judgments are formed in childhood.

Evolutionary Ethics

Debates about the evolution of morality began toward the end of the nineteenth century with strong disagreements between Huxley (1989 [1894]) and Kropotkin (1972 [1902]). Huxley believed that the harsh principle of natural selection could never have led to moral tendencies. He presented acts of altruism and sympathy as antithetical to nature: a human cultural innovation. Kropotkin, on the other hand, argued in *Mutual Aid* that Huxley had too narrow a view of nature. Where Huxley saw perpetual struggle and combat, Kropotkin argued that the term *struggle* in the phrase "struggle for existence" is not to be taken literally because survival is often achieved through cooperation rather than competition.

For a long time, Huxley's views were taken as the traditional evolutionist position despite the fact that they unequivocally contradicted Darwin's own opinions in *The Descent of Man.* Darwin made no secret that there was room in his theory for the origins of human morality. He did not even exclude the possibility of ethics in other animals, noting: "Any animal whatever, endowed with well-marked social instincts, the parental and filial affections being here included, would inevitably acquire a moral sense or conscience, as soon as its intellectual powers had become as well developed, or nearly as well developed, as in man" (Darwin 1981 [1871], pp. 71–72).

The first to integrate Darwin's ideas with those of philosophy, anthropology, and the social sciences was Westermarck (1912), who offered a compelling framework within which to discuss

what he termed "human moral emotions," ranging from sympathy to a tendency toward reciprocity. The term *moral emotion* is slightly misleading because it includes rational evaluations and judgments: these emotions clearly had a cognitive component. Darwin himself had been inspired by the important insights achieved a century before him by moral philosophers, such as Hume (1978 [1739]; see also Mercer 1972) and Smith (1937 [1759]), that humans have a natural, spontaneous caring capacity. Current research in human social psychology (e.g., Batson 1990; Wispé 1991) seems to support Smith's famous and succinct definition of sympathy: "How selfish soever man may be supposed, there are evidently some principles in his nature, which interest him in the fortune of others, and render their happiness necessary to him, though he derives nothing from it, except the pleasure of seeing it" (Smith 1937 [1759], p. 9).

What was much less known at the time of Darwin and Huxley, however, is how cooperative and intensely sociable some animals are, including our closest relatives. Parallels between human morality and animal behavior—implicitly present in Darwin's considerations—could be drawn explicitly only a century or more later.

The objective of such an approach is explicitly not the derivation of moral rules directly from nature. In the past this has been tried, resulting in a dubious genre of literature with usually biblical titles spelling out how moral principles contribute to survival (e.g., Wickler 1981 [1971]). Such attempts to derive ethical norms from nature are highly questionable. Known as the *naturalistic fallacy*, this problem was recognized centuries ago (Hume 1978 [1739]).

Our position is quite different. Although human morality does need to take human nature into account either by fortifying certain natural tendencies (such as sympathy, reciprocity, loyalty to the group and family, and so on) or by countering other tendencies (such as within-group violence and cheating), it is in the end the individual members of a society who decide, over a period of many generations, on the contents of a moral system. There is a parallel here with language ability: the capacity to develop and learn a very complex communication system such as language is naturally present in humans, but it is highly influenced by social experience. In the same way, we are born with a *moral ability* and a tendency to be influenced by the moral values of our social environment, but there is little evidence to suggest that we are born with a ready-made moral code in place. The content of moral systems is constructed out of social interactions and reflected in behavior patterns (in the case of nonhuman primates) or judgments (in the case of humans).

Comparative Approaches

Recently, there has been a convergence of perspectives on the social nature of humans and animals between developmental psychologists and primatologists (see de Waal 1989, 1993, 1996; Cords & Killen 1998; Verbeek et al., Chapter 3). In both lines of inquiry, research on conflict resolution has been particularly important in demonstrating what it means to be social and what it means to be moral. With human children, research has shown that children's social conflicts provide an important experiential basis for the development of moral concepts of fairness, justice, and equality. Research with nonhuman primates has shown that nonaggressive methods, such as reconciliation, are often used to resolve conflicts. This work has also provided a window into other forms of prosocial behavior in nonhuman primates, such as acts of sympathy and reciprocal exchange (de Waal 1996).

The obvious common ground between current evolutionary and developmental approaches is that,

instead of looking at human morality as coming from the outside—imposed by adults on the passive child, or imposed by culture on a fundamentally nasty human nature—it is generated from the inside. What we mean by "inside" is *not* that things happen in isolation from outside influences: evolution operates on the basis of ecological pressures, which come from the outside, and development takes place in constant interplay with the outside world. What we mean instead is that the decision making and emotions underlying moral judgments are generated within the individual rather than being simply imposed by society. They are a product of evolution, an integrated part of the human genetic makeup, that makes the child construct a moral perspective through interactions with other members of its species.

This is consistent with the early comparative approaches to developmental psychology in the early and mid-twentieth century by Baldwin (1915), Piaget (1970 [1950]), and Werner (1926) (see Langer & Killen 1998). Both Baldwin and Piaget studied how thought and experience construct reality, and they did this by comparing the stages of mental development ontogenetically and phylogenetically. Werner extended Baldwin's approach by asserting that comparative approaches must account for microgenesis (short-term local development) and pathogenesis (pathological development) as well as ontogenesis, phylogenesis, and ethnogenesis. Although these theorists did not explicitly argue for a comparative approach to the study of morality, their work laid the foundation for examining morality from a comparative perspective.

Stent (1978, p. 16), a neurobiologist, noted in the introduction to his edited book, *Morality as a Biological Phenomenon*, "It is the study of childhood development by cognitive psychologists that can presently offer one of the potentially most fruitful meeting grounds for biology and moral philosophy." Following this lead, our goal is to describe current research on morality in childhood and adolescence, compare it with research on social behavior in nonhuman primates, and speculate about the moral basis of behavior in nonhuman primates. Our approach is developmental, comparative, and evolutionary.

Morality from Outside or Inside?

In evolutionary biology it is fashionable to portray animals, including humans, as thoroughly selfish and competitive, a position derived from a narrow interpretation of the "struggle for existence" metaphor. But Darwin himself warned against such an interpretation, explaining that the term *struggle for existence* should not be taken literally, that it includes mutual dependencies between individuals (Darwin 1981 [1871]). Thus, whereas Darwin saw room within his theory for the origins of morality, some of his followers—starting with Huxley (1989 [1894]) but including contemporary biologists such as Williams (1988)—have tried to push morality outside the biological domain.

This is all the more surprising because the neo-Darwinian revolution emphasizes so-called altruistic behavior. Here the perspective on altruism is purely utilitarian, however, which seems to have led to a confusion between function and motive. If behavior in the long run serves the individuals who perform it, it is argued, the underlying motives must be selfish. Whereas the first part of this assumption is logical, the second is not. Self-serving behavior does not need to be selfishly motivated. We should retain a separation between how and why certain behavioral traits have been selected over millions of years and the actual motives and psychology activating the behavior. This classical separation, between ultimate and proximate causation, implies that not all

behavior is psychologically selfish. De Waal (1996) has reviewed this confusing debate in which the nonexistent emotions of genes (which, of course, can only be metaphorically "selfish") have been mistaken as reflecting actual human emotions.

If the intense sociality of some animals has been underestimated, so has that of children. In the early formulations of child development, children were routinely characterized as either motivated by selfish or aggressive impulses (Freud 1923 [1960], 1930 [1961]) or as asocial, a blank slate (Watson 1924; Skinner 1971). These approaches dominated popular societal outlooks of the moral abilities of young children for much of this century. The following quote from the American Academy of Pediatrics' guidebook, entitled *Caring for Your Baby and Young Child: Birth to the Age of Five* (Shelov 1991, p. 297), provides an illustration of this perspective: "By nature, children this age (2–3 years) are selfish and self-centered. They may refuse to share anything that interests them, and they do not easily interact with other children, even when playing side by side, unless it's to snatch a toy or to quarrel over one that someone else has grabbed from them. There may be times when your child's behavior makes you want to disown them, but if you take a closer look, you'll notice that all the toddlers in the play group are probably acting the same way."

Unlike his contemporaries, Piaget (1932) characterized children as predisposed to understand the social world through processes of reflection, abstraction, and judgment. Children reflected on and made judgments about their social exchanges with parents and peers. What made Piaget's writings so different from the predominant early- and mid-twentieth-century American views of child development was that he investigated how children develop morality as a result of reflecting on their interactions with others, particularly peers.

The child's social world, as he or she examined it, included authority-child relations, peer relations, and concepts of justice, fairness, and equality developed out of these social exchanges (see Damon 1977, 1983).

Despite Piaget's theoretical innovations, his formulations also constituted an underestimation of the child's moral abilities. Piaget characterized the child as initially oriented to authority's commands (i.e., acts are evaluated as right in terms of what authority states is right), and only during middle childhood would the child evaluate acts as right or wrong with respect to independent principles of justice. Kohlberg (1969, 1984) expanded Piaget's work by demonstrating how children evaluate acts in terms of the negative consequences to the self, such as punishment, and not solely in terms of authority expectations and rules (for reviews, see Turiel et al. 1987; Killen 1991; Smetana 1995; Turiel 1998). Again, while Kohlberg's theory reflected a dramatic departure from psychoanalytic or behaviorist formulations, Kohlberg's framework, too, underestimated young children's moral thinking. This was because, in part, he focused on the emergence of abstract, ethical (philosophical) reasoning and analyzed the ways in which this type of thinking manifested itself in adolescence.

Over the past 20 years, however, several lines of work have demonstrated that morality emerges in early development (see Kagan & Lamb 1987; Killen & Hart 1995). One line of research has examined moral judgment and moral reasoning following Piaget's work based on Kantian ethics (see Turiel et al. 1987; Killen 1991; Smetana 1995), and another line of research has analyzed prosocial development, including conceptions about empathy as well as emotional responses to others' distress, following Humean philosophical tenets and ethological methods (see Cummings et al. 1986; Eisenberg & Strayer 1986; Zahn-Waxler

& Radke-Yarrow 1990; Zahn-Waxler & Hastings 1999). The findings from these two approaches, albeit reflecting different theoretical traditions, have provided a wealth of evidence for the emergence of morality in early development. First, we will discuss the moral judgment studies, followed by a look at some of the prosocial research findings.

As mentioned above, Piaget distinguished himself from other psychologists in the early part of the century by defining morality as independent principles, not as cultural norms. The distinction between independent principles and cultural norms is important because when morality is defined by social norms, morality is relativistic; it is whatever the culture deems right or wrong. Moreover, when moral systems are confused with societal systems, then all rules and norms are potentially moral, and morality loses its status as a set of principles or maxims about how we ought to treat one another. Whereas some cultural rules constitute moral principles, in terms of their generalizability and unalterability (e.g., the cultural rule "Do not steal" is also a moral rule because it is not viewed as a matter of consensus), many cultural rules do not constitute moral principles (e.g., the cultural rule "No nude sunbathing" is viewed largely as a matter of consensus). Many philosophers have made the distinction between moral principles and societal rules (Searle 1969; Dworkin 1978), but only in the past 20 years have psychologists explicitly incorporated this distinction into their theories about moral reasoning and behavior.

Research over the past two decades has shown that morality constitutes one of several self-regulating systems that coexist in development with other areas of knowledge, such as societal and psychological knowledge (Nucci & Turiel 1978; Turiel 1978, 1983, 1998). The societal domain refers to regularities that are designed to promote the smooth functioning of social interactions within a group, and the psychological domain refers to self-concept, autonomy, and issues that are not regulated but are a matter of individual choice (see Nucci 1981, 1996; Turiel 1983, 1998; Turiel et al. 1987; Smetana 1995; Tisak 1995). Studies have shown that children, adolescents, and adults, in a wide range of cultures, evaluate moral issues using criteria that are consistent with moral philosophical definitions (Rawls 1971; Gewirth 1978; Nagels 1979, 1986; Williams 1981) that include obligatory, universalizable, unalterable, and impersonal (see Turiel 1983, 1998; Smetana 1995). This is in contrast to other types of social rules, such as conventions used to regulate social interactions within a group, which are determined by agreement, consensus, or institutional expectations. Thus, moral rules are not defined by cultural norms but are evaluated in terms of independent principles of justice, fairness, and rights.

It is particularly important to use criteria to assess how children evaluate social issues because, as Turiel (1983) has pointed out, methodological techniques used to assess morality are not always consistent with the definition of morality. Just as it would be erroneous to use a mathematical task to assess morality, it would also be inaccurate to use a conventional game (in which rules are determined by consensus and not by independent principles) to assess moral judgments. Children, adolescents, and adults differentiate moral rules from social-conventional rules, and this indicates that morality is not defined by the social system. Providing criteria for what constitutes a moral judgment is necessary in developmental and comparative work. Cross-species comparative studies of language ability (Savage-Rumbaugh 1998), self-concept (Parker & Gibson 1990; Parker 1998), theory of mind (Whiten & Byrne 1997; Whiten 1998), and conflict resolution (de Waal 1996; Cords & Killen 1998) have been particularly care-

ful about clarifying the criteria used to define social phenomena.

Further, although concepts of justice and others' welfare have provided much of the focus for the research on morality in children's thinking, prosocial concepts are also part of morality. Prosocial moral acts, which pertain to positive social actions such as helping, caring, and giving (see Eisenberg & Strayer 1986), have been shown to be part of morality because children and adults also view these acts as independent of rules, laws, and the dictates of authorities (criteria similar to justice and others' welfare). Moreover, prosocial behavior is broader than morality, and some of the earliest documentations of potentially relevant moral responses include sympathetic, empathetic, and caring responses by infants and toddlers toward others (Hoffman 1983; Eisenberg & Strayer 1986; Zahn-Waxler et al. 1986; Dunn 1988). Although not all prosocial acts are evaluated as obligatory or universalizable (e.g., acts of kindness may not be viewed as obligatory in that it is not wrong from a moral viewpoint to refrain from acting in a kind manner), prosocial acts are clearly differentiated from social-conventional rules by children and adolescents. This distinction is relevant to an examination of "morality" in nonhuman primates that have been shown to display a range of prosocial behaviors.

In general, studies of prosocial development have shown that very young children engage in acts of sharing, caring, helping, and altruism (see Zahn-Waxler et al. 1979; Radke-Yarrow et al. 1983; Eisenberg & Fabes 1998; Zahn-Waxler & Hastings 1999). Zahn-Waxler and Hastings (1999) argue that it is no longer helpful to conceptualize the origins of morality in extreme dichotomous categories, as either selfish and asocial (or antisocial), on the one hand, or altruistic, on the other hand. Rather, it is more accurate and more comprehensive to investigate how different orientations emerge and get coordinated in behavior and judgment. In her research program, Zahn-Waxler and her colleagues have shown that by one year of age, most children show comfort to another person in distress. By two years of age, comforting becomes more frequent, more differentiated, and more closely tied to the nature of the context. Zahn-Waxler has studied the affective, cognitive, and physiological dimensions of empathy in early development. In nonoptimal environments, children often experience delays that affect their empathic and prosocial responses. However, in optimal environments, prosocial responding occurs early in life and is manifested in many different ways. Eisenberg & Fabes (1998) differentiate empathy from sympathy by defining empathy as the display of an emotional reaction to the feelings of another that is similar to the other's feelings and sympathy as a concern for others based on a sense of the emotional state of the other; children's feelings of empathy have been shown to be related to their prosocial behavior (but not defined by it). Other forms of prosocial behavior include forgiveness (see Park & Enright, Box 17.1), helping, and caring.

There are different positions in the research literature on the relationship between moral emotions and moral judgment (see Turiel 1998, for a discussion of the controversies). Some theorists have argued for a primacy of emotions and state that moral emotions, such as empathy and sympathy, form the core of morality (not moral judgments, such as justice and fairness) (Hoffman 1991). Other theorists, however, assert that without information about intentionality it is difficult to interpret the moral status of emotions. A child may display empathy (e.g., understanding the emotions of another child) but still not make the judgment that it would be wrong to hurt the child (the moral judgment). There are a number of different facets to this view, many more than we

have room to discuss here. From our perspective, we assert that emotions such as empathy and sympathy provide an experiential basis by which children construct moral judgments. Emotional reactions from others, such as distress or crying, provide experiential information that children use to judge whether an act is right or wrong (see Arsenio & Lover 1995). For example, when a child hits another child, a crying response provides emotional information about the nature of the act, and this information enables the child, in part, to determine whether and why the transgression is wrong. Therefore, recognizing signs of distress in another person may be a basic requirement of the moral judgment process. The fact that responses to distress in another have been documented both in infancy and in the nonhuman primate literature provides initial support for the idea that these types of moral-like experiences are common to children and nonhuman primates.

As an illustration, de Waal (1996) documents reactions to distressed individuals and the tendency to share food with others. Generally, it seems that our closest relatives, the great apes, go further in this regard than more distantly related primates, such as the monkeys. For example, "consolation" has thus far been demonstrated only in chimpanzees despite systematic attempts to find it in monkeys. Consolation is defined as friendly or reassuring contact provided by a bystander to a recipient of aggression (de Waal & van Roosmalen 1979). In the chimpanzee, this kind of interaction typically consists of putting an arm around the victim or patting him or her gently on the back or shoulder. Because of the contrast between monkeys and apes in this regard, de Waal & Aureli (1996) have recently speculated that consolation may require empathy. Since higher forms of empathy and sympathy require the ability to take someone else's perspective, the difference may result from Hominoids (i.e., humans and apes) possessing this ability but not monkeys.

Theories of Acquisition of Morality

How is morality acquired? Is it taught by adults? Or is it constructed out of social interactions? Much research has shown that children acquire morality through a social-cognitive process; children make connections between acts and consequences. Through a gradual process, children develop concepts of justice, fairness, and equality, and they apply these concepts to concrete everyday situations (Killen & Hart 1995). From this view, morality is developed out of social interactions (with peers and with adults) and is not imposed on individuals from outside influences. This is consistent with an evolutionary view of morality in terms of morality being something that slowly emerges over time. In addition, there are multiple sources of influence on children's acquisition of moral judgment (Turiel et al. 1987; Grusec & Goodnow 1994; Killen & Nucci 1995; Smetana 1995; Tisak 1995). In this view, morality does not come from parents—nor does it "come from" peers. Rather than being transmitted, morality is constructed through social interactions and social judgments.

Yet, in general, children's moral thinking has long been regarded as external, adopted from the adult social environment. According to Freudian and Skinnerian viewpoints, left to their own devices, like the children in Golding's (1954) *Lord of the Flies*, children would never arrive at anything like morality. Freud's view of the acquisition of morality was that children incorporated parental values to develop a superego, the moral agency of the self. Skinner theorized that adults provide the environmental contingencies necessary to shape a moral being. In the classic theories, morality is a result of adult influence (incorporation of parental

---

BOX 17.1

Forgiveness across Cultures

Seung-Ryong Park & Robert D. Enright

Forgiveness is an element of all cultures, but its expression and effect vary. Comparisons between participants from three varied cultures, the United States, Korea, and Taiwan, reveal some interesting findings. Before looking at those comparisons, let us define forgiveness.

Ancient texts from varied cultures reflect common elements of forgiveness: (1) Forgiveness is a form of morality centered in mercy, not justice. Thus, the person who forgives is welcoming another despite the injustice suffered. (2) When one forgives, it is always in the context of another's (or others') unfairness toward the offended person. (3) Forgiveness usually involves anger or resentment toward the offending person, which eventually is replaced by more beneficent expressions, such as compassion, caring, and even moral love. Many ancient texts, including Hebrew and Jewish (Vine 1985), Christian (Vine 1985), and Islamic (Sch-he-rie 1984), presented various forms of forgiveness such as the deity-human relationship or one group "forgiving" another.

Some contemporary philosophers have presented forgiveness as a construct. North (1987) defined forgiveness: "If we are to forgive, our resentment is to be overcome not by denying ourselves the right to that resentment, but by endeavoring to view the wrongdoer with compassion, benevolence, and love while recognizing that he or she has willfully abandoned the right to these moral qualities" (p. 502). Richards (1988) noted that a forgiver would give up negative affect, such as anger, hatred, resentment, sadness, and/or contempt. Downie (1965) claimed that a forgiver would replace the negative with more positive affect such as compassion.

Ancient texts and philosophies on forgiveness led Enright et al. (1991) to define forgiveness as a psychological construct in which the offended party's

BOX 17.1 (continued)

Reconciliation

		Yes	No
Forgiveness	Yes	Forgiveness and reconciliation	Forgiveness without reconciliation
	No	Reconciliation without forgiveness (a truce or an interaction based on mutual interests)	No forgiveness and no reconciliation

The four possible combinations for the occurrence (Yes) or absence (No) of reconciliation and forgiveness.

sense of negative emotions, such as outrage and hostility, decreases and positive emotions, such as compassion or love, emerge. The offended experiences a cessation of or decrease in negative judgments and the increase in positive judgments. The offended also changes his or her behavior toward the offender. The offended may previously have acted out subtle revenge, but now he or she is ready to begin a friendly relationship with the offender.

Forgiveness is frequently confused with the following concepts believed to be synonyms of it: condonation, excusing, and reconciliation. People who condone realize that they were offended but deliberately refuse to retaliate, and they decide to put up with the injustice. Someone who forgives does not put up with injustice; instead, he or she actively resolves it by accepting the offender despite moral injury. When people excuse, they often minimize the wrong in question and hence, unlike those who forgive, do not fully acknowledge it as wrong.

Forgiveness is not reconciliation. Forgiveness is an internal release that a forgiver has achieved after much effort; reconciliation is a behavioral coming together that a forgiver and the forgiven may establish with trust. The offended presupposes that the offender has changed and that a more just relation will ensue. The key difference is that when the offended forgives, he removes all barriers that he may have had in blocking the relationship. Then the offender must remove her barriers for reconciliation to occur. Forgiveness is a prerequisite to

reconciliation. The figure accompanying this box illustrates quadrants that have the four combinations of forgiveness and reconciliation, forgiveness without reconciliation, reconciliation without forgiveness, and no forgiveness and no reconciliation. A caution is warranted here. Reconciliation without forgiveness is hardly entitled to constitute genuine reconciliation. It is rather a truce or an interaction based on mutual interests.

Forgiveness may vary among people and within an individual to the degree in which an offended offers that forgiveness. The Enright Forgiveness Inventory (EFI: Subkoviak et al. 1995) was developed as a reliable and valid measure of the degree of forgiveness toward an offending person.

The EFI, a 60-item scale, has 10 items for each positive and negative subscale of affect, behavior, and cognition. An example of an EFI item is "I feel hostile toward him/her," measuring negative affect. Participants are instructed to answer the EFI items on a six-point scale from Strongly Disagree (1) to Strongly Agree (6). The possible score ranges from 60 to 360. The higher the score, the more forgiving is the respondent.

In accordance with clinical observations (Fitzgibbons 1986), forgiveness is expected to correlate negatively with anxiety. Coyle & Enright (1997) found that male participants whose marital or romantic partners aborted their pregnancy, after intervention based on forgiveness, demonstrated a significant gain in forgiveness as measured by the EFI and significant decreases in anxiety,

BOX 17.1 (continued)

anger, and grief compared with a control group that had not yet received treatment.

The use of the EFI also highlights differences both *within* and *across* cultures. U.S. participants ($N = 400$) forgive as much by decreasing negative feelings as by increasing positive feelings toward the offender, whereas both Korean ($N = 400$) and Taiwanese participants ($N = 400$) forgive more by decreasing negative feelings rather than by increasing positive feelings toward the offender. Across cultures, there are significant differences in how people feel and behave toward and think of their offenders. U.S. participants have higher scores on most EFI subscales and the total score than those in Korea and Taiwan. The EFI score of 286, for example, for a Korean or Taiwan participant means that he or she is more than one standard deviation above the mean, while the same score, 286, for an American participant means that he or she is average on forgiving.

The inverse relationships between forgiveness and both anxiety and depression were examined in each culture. Subkoviak et al. (1995) found that those U.S. participants who forgive tend to have lower anxiety. However, this is marginally the case within the Korean sample (Park 1994) and not the case with the Taiwanese sample. Forgiveness was found to correlate negatively with depression for both U.S. parents and Korean parents reporting a hurt within a family context, but not in the Taiwanese sample. One unexpected finding in the Korean sample is that participants reporting a great deal of hurt showed statistically significant positive correlations between forgiveness and depression. This group appears to be more depressed as they forgive their offenders.

Weak or nonexistent relationships between forgiveness and mental health variables such as anxiety and depression found in the Taiwanese and Korean samples need to be explained. In Asian cultures, keeping one's composure even in adversity is regarded as virtuous. These Asian participants, reflecting this cultural norm, may have experienced anxiety or depression, or both, but may not have reported this. Huang's (1990) study suggests that this hypothesis may have merit. She observed anger in Taiwanese participants in two ways: through facial expression measurements and self-reports. There was no relationship between observed anger and self-reported anger. The observed anger was higher than that reported on a questionnaire. Degree of observed anger was inversely related to forgiveness. Future studies examining the relationship between forgiveness and emotion may require the direct observation of these variables in addition to paper-and-pencil questionnaires.

Reconciliation is important for effective conflict resolution. Forgiveness may be necessary for true reconciliation. Therefore, forgiveness may be essential in the process of conflict resolution, and further studies are warranted.

values or transmission, for Freud and Skinner, respectively). These views have remained widely influential, in both research and educational arenas. These views, however, are not consistent with an evolutionary approach to morality because they assume a total lack of a moral predisposition in our species.

Consistent with the view that there is such a biological predisposition, de Waal (1996), in his review of the literature, culled four dimensions of human morality for which evidence exists in nonhuman primates. In this chapter, we do not assert that any animal other than our own species is moral—there are no indications for moral reasoning in other animals or for explicit consensus building made possible by language—but we point out that each of these dimensions of human morality is to some degree present in chimpanzees and can perhaps also be found in other nonhuman animals. These moral dimensions are:

1. *Sympathy.* Caring responses stemming from attachment and emotional contagion. For example, there often is high tolerance toward

and special treatment of disabled and injured individuals.

2. *Social norms.* Prescriptive social rules can be found in many social animals. One can observe a sense of social regularity and expectation about how one ought to be treated. The degree of internalization of rules is variable.

3. *Reciprocity.* A concept of giving, trading, and revenge is evident in chimpanzees. For example, a recent study demonstrates that a few hours after a chimpanzee has been groomed by another, he or she tends to return the favor by sharing food specifically with this partner (de Waal 1997). Such a relation is hard to explain without memory of previous events and a psychological mechanism labeled *gratitude.* Reciprocity may also take a negative form, but retributive aggression against violators of reciprocity rules—an important part of human moral systems—has yet to be demonstrated.

4. *Getting along.* The presence of peacemaking and conflict avoidance in nonhuman primates and other animals is amply demonstrated in the present volume. These mechanisms of relationship repair permit the accommodation of conflicting interests through negotiation.

In each of these four areas, parallels can be drawn between the development of children and the behavior of various nonhuman primates. There can be little doubt that human moral development goes further in each of these areas than the levels reached by our closest relatives. Cognitive, social, and linguistic development adds to the complexity. Most important, we can assess intentional states in humans much easier than in nonhuman animals, enabling us to differentiate between prosocial behavior (moral emotions) and moral judgments. The precise differences are not well

understood, however, and despite controversies about the criteria for intentionality in nonhuman primates (see Tomasello 1998), there are clear similarities between species regarding social behaviors toward others.

Connection between Morality and Conflict Resolution

All the moral dimensions listed above by de Waal (1996)—sympathy, social norms, reciprocity, and getting along—relate to conflict and conflict resolution. The last dimension does so by definition, but the other three dimensions also function to mitigate conflict and allow for cooperation. Here we will investigate this connection with conflict resolution, first with nonhuman primates, followed by developmental studies with children.

Evolutionary biologists, such as Alexander (1987), conceptualize the relationship between morality and conflict resolution differently than the way developmental psychologists theorize about it. Although Alexander (1987) argues that morality comes into play in biology only if there are potential conflicts of interest among mutually dependent parties, he acknowledges that with humans moral principles often serve as mechanisms for resolving conflicts (by appeals to principles) and therefore are not defined solely as conflict-resolution techniques. According to Alexander (1987), potential conflicts of interest among mutually dependent parties require rules about acceptable versus unacceptable behavior and about the distribution of resources. Morality arises in the face of tensions between socially cohesive and potentially destructive tendencies: moral systems are designed to promote the first and control the second. Since the same tensions observable in humans are observable in nonhuman primate societies, he argues, these animals provide an excellent starting point to explore the possible origins of morality. The close taxonomic relationship between ourselves and other Old World primates implies that similarities with human moral tendencies most likely represent homologies rather than analogies. The first and foremost similarity is that primates actively seek to reduce social tensions and resolve conflicts that threaten social relationships. It is not surprising, therefore, that renewed interest in the evolution of morality has sprung from primate reconciliation studies.

The tension between conflict and cooperation is quite pervasive in the animal world. For animals to overcome conflict over food, through sharing, requires payoffs that serve not only the subordinate but also the dominant. What would otherwise prevent the dominant from claiming all food for him- or herself? Would not the dominant animal be better off competing instead of sharing? This would indeed be the case if subordinates had nothing to offer, but in societies based on cooperation subordinates have all sorts of leverage. For example, high-ranking males may owe their position to the support of lower-ranking individuals, and all males may need to remain united against neighboring males lest they and their territory be annihilated. This is the situation for wild chimpanzees. Because of the intricate exchange of support and services, ranging from alliances to sexual favors, chimpanzee society has been described as a web that ensnares the dominants as much as the subordinates (de Waal 1982). To a lesser degree the same applies to many other primates and probably also to cooperative non-primates.

These mutual dependencies have profound implications for conflict management: it means that it does not always pay to win a conflict, especially one with an individual with whom one commonly cooperates. Reciprocity thus becomes an alternative to competition, promoting selective tolerance and negotiation where other animals might fight.

Social norms are expectations about the behavior of others and about the way oneself ought to be treated. This is a very little studied area of animal behavior, but one with great potential, as animals who repeatedly interact are likely to develop conventions about how to settle issues between them. We notice the ability to follow rules in other animals most easily when the rules are of our own design, such as those that we apply to pets and work animals. Yet the remarkable trainability of certain species, such as sheepdogs and Indian elephants, hints already at the possibility of a rule-based order among these animals themselves. De Waal (1996, p. 89) gives the following example of rule enforcement:

> One balmy evening, when the keeper called the chimpanzees inside, two adolescent females refused to enter the building. The rule at Arnhem Zoo being that none of the apes will receive food until all of them have moved from the island into their sleeping quarters, the chimpanzees actively assist with the rule's enforcement: latecomers meet with a great deal of hostility from the hungry colony.
>
> When the obstinate teenagers finally entered, more than two hours late, they were given a separate bedroom so as to prevent reprisals. This protected them only temporarily, however. The next morning, out on the island, the entire colony vented its frustration about the delayed meal by a mass pursuit ending in a physical beating of the culprits. Needless to say, they were the first to come in that evening.

This particular group response concerned a regulation put into place by people. Chimpanzees also seem to develop rules of their own, however. For example, Nishida (1994) describes for wild chimpanzees how a subordinate may turn the tables on a dominant individual after an apparent rule violation, such as an attack from behind, without warning. Chimpanzees may also massively "woaow" bark in protest against certain actions in their midst—for example, an adult male punishing a juvenile too roughly—followed by physical intervention, sometimes even ostracism. One never hears this particular bark when a mother punishes her own offspring or when an adult male controls a tiff among juveniles, even if he uses force in the process. Therefore, not every fight triggers these calls. It is a reaction to a very particular kind of disturbance, one that seriously endangers lives or relationships.

The role of sympathy and empathy in conflict is also little understood, but if we assume these capacities, it is logical to expect their activation during and after hostile situations. Thus, whereas the predominant model to explain the occurrence of reconciliation is one based on the value of cooperative relationships (see, e.g., de Waal, Chapter 2; Cords & Aureli, Chapter 9; van Schaik & Aureli, Chapter 15), it is conceivable that the sight of a distressed or injured opponent also triggers sympathetic responses. In bonobos (*Pan paniscus*), for example, which are a particularly sensitive species, reconciliations are mostly initiated by the aggressor (de Waal 1987). Aggressors have been seen to return to their victim immediately after an attack, only to take the hand or foot they have just bitten in their hand to inspect and lick the spot. Such behavior suggests memory of previous actions and perhaps regret (de Waal 1996).

The emotions evoked by conflict are complex, far beyond the fear and aggression dimension usually assumed. Because conflict endangers peaceful, cooperative relationships and triggers expressions of emotion in the opponent that vicariously affect the other, conflicts entail an entire interaction chain of different emotions. For example, recent research has indicated that anxiety is not limited to the loser of an escalated conflict. Aureli (1997; Aureli & Smucny, Chapter 10) used behavioral indicators, such as self-scratching, to measure the level of anxiety following conflict. He and his co-worker found an elevation in both aggressors and victims. The elevation is especially high after conflict with a valuable partner, one with whom they often associated or cooperated. The observed anxiety therefore most likely concerned the state of the relationship.

If conflict induces anxieties in both parties, this means that conflict is not just about fight or flight, and winning or losing, but also about the future of the relationship. Conflict is thus, also emotionally, at the interface of competition and cooperation, which is the interface at which morality develops. Thus, moral systems, from an evolutionary viewpoint, can be looked at as an elaborate form of conflict resolution seeking fairness and social integration in a competitive world (Alexander 1987; de Waal 1996).

In humans, morality takes the form of judgments and expectations of how individuals ought to treat one another. It is theorized that this stems from social experiences that occur throughout development, beginning in infancy. Conflict and conflict resolution provide an important experiential source for the construction of social and moral categories (Killen 1991; Turiel 1998). Given that morality is assumed to be a product of inferences about social interactions, and not a result of direct adult-child transmission, it is necessary to understand the nature of children's social interactions and what they abstract from these experiences. Piaget (1932) postulated direct connections between adult-child relationships and peer relationships; the latter, but not the former, are necessary for the acquisition of moral judgment. However, as mentioned above, research on social interactions has moved away from making straightforward links between types of relationships (e.g., adult-child or peer) and types of social knowledge (rule orientations or moral orientations).

Instead, research has been conducted on the nature of social interactions (acts and consequences) and children's social interpretation of, and reflection on, these social experiences. *Both* adult-child and peer interactions have the potential to facilitate moral judgment. What matters is the nature of the interaction (unilateral, aggressive or mutual, cooperative) and how messages are communicated. Not surprisingly, research on social interactions, and particularly on conflict resolution, has also undergone a change in theoretical approach similar to the work on social judgments. Moving away from unitary stage descriptions of social interaction patterns (e.g., solitary, onlooker, parallel play), investigations of children's social interactions have examined the multiple considerations involved in children's play and social interactions as well as the contextual and cultural influences. This has been particularly evident in the area of children's conflicts and methods of conflict resolution (see Ross & Conant 1992; Shantz & Hartup 1992).

In the early studies on children's conflict, few distinctions were made between different types of conflicts and means of conflict resolutions. This was due, in large part, to the primary focus on children's aggressive conflicts (for a review, see Shantz 1987; Shantz & Hartup 1992), which were seen as potentially or actually maladaptive. However, as the definition of conflict broadened to include "a protest, resistance, or retaliation of one individual's reaction to another individual's behavior" (see Shantz 1987; Killen 1989, 1991; Ross & Conant 1992; Shantz & Hartup 1992), aggressive conflicts were viewed as just one form of interpersonal conflict. Whereas aggressive conflicts would not be predicted to be positively related to the acquisition of moral understanding, nonaggressive conflicts, such as object disputes or how to structure an activity, were theorized to play a positive role in development by enabling children to

develop constructive methods of conflict resolution involving negotiation, compromise, and perspective taking (see Killen 1991; Dunn & Slomkowski 1992; Ross & Conant 1992; Shantz & Hartup 1992).

For example, Hay & Ross (1982) provided evidence to show that young children's social conflicts were not "blind fights" but were largely about object sharing and embodied explicit communicative and social functions. Engaging in object disputes in which children use toys as bargaining tools for social interaction may provide experiential sources for moral development in ways that engaging in aggressive conflicts does not (see Fig. 17.1). Aggressive conflicts involve particular types of irreversible negative moral consequences (e.g., physical harm to another cannot be "undone"), whereas nonaggressive conflicts involve potentially reversible moral consequences (e.g., a toy grabbed away from someone can be returned) as well as social-conventional dimensions that may not involve negative intentions to another at all (e.g., how to structure the activity).

The natural course of conflict resolution in childhood seems to be one in which adults guide children by pointing to connections between acts and consequences and by fostering social interactional strategies among peers to promote peer methods of conflict resolution. The fact that young children are capable of resolving peer conflicts without the direct intervention from adults (Killen & Turiel 1991) indicates that they are capable of figuring out constructive methods on their own under optimal conditions. In fact, adult intervention may hinder the conflict outcome by not allowing children the opportunity to acquire negotiation skills (Bayer et al. 1995; Pitrowski 1995; Verbeek et al., Chapter 3). These findings suggest that morality is not "taught" by adults but, rather, that morality emerges slowly out of children's interactions with others and reflection on those experiences.

FIGURE 17.1. Object disputes are common in children's lives and can be studied in settings such as the one shown here. These conflicts typically do not disrupt the flow of interaction but serve as social bargaining tools. Photograph by Melanie Killen.

In fact, in the peer conflict context, techniques that are designed to assist children in working out conflicts by themselves are often most effective. Whereas direct methods include the use of punishment, commands, and rule statement, indirect methods include the use of dialogue, explanations of the relationships between acts and consequences, and the suggestion of alternative resolutions. Studies have indicated that it is more effective for adults to act as a facilitator or mediator than as an instructor when it comes to peer conflict resolution. When adults teach children how to negotiate and compromise, conflicts diminish. Teachers and parents recognize this when asked to evaluate direct and indirect methods of conflict resolution; teachers and parents state that, under ideal conditions, teachers should help children to work out conflicts themselves rather than to rely on the use of punishment or rule statements (Killen 1991). Thus, adult intervention techniques are more effective when a number of factors are taken into consideration, including the nature of the conflict, the child's interpretation of the adults' message, the appropriateness of the message, and the communicative style used to convey the message (Grusec & Goodnow 1994).

The types of conflict-resolution techniques used to respond to conflicts are also revealing about the ways in which conflicts provide an experiential source for moral development. Research on the outcomes and consequences of different types of reactions that children have to conflict situations has shown that children who deal with conflict using negotiation and compromise, rather than aggressive means, are liked by their peers whereas children who have difficulties resolving conflicts with friends are rejected by their peers (Corsaro 1985; Putallaz & Wasserman 1990). Do children continue playing together after engaging in conflicts? Research by Hartup et al. (1988) has shown that nonaggressive conflicts rarely disrupt the flow of interaction; most nonaggressive conflicts are frequently occurring events that do not alter the course of children's interactions and are generally low in affect (see also Verbeek et al., Chapter 3). These findings suggest that children are aware of the social sensitivity of other children and that this sensitivity helps toward creating and implementing positive methods of conflict resolution.

Further, recent work on children's conflict resolution has documented aspects of the social context that influence methods of conflict resolution. These social context variables have included social relationships (Hartup et al. 1988; Verbeek et al., Chapter 3), history of relationships (Hinde 1974, 1979; Dunn et al. 1995), gender composition (Leaper 1991; Killen & Naigles 1995), affective and emotional states (Dunn 1988; Arsenio & Killen 1996), and the social meaning of the behavior by the interactants (see Verbeek et al., Chapter 3). As a result of these findings, conflict resolution has come to be recognized as a potentially constructive event in children's lives with potential connections to the acquisition of moral knowledge (e.g., Shantz 1987; Killen 1991; Ross & Conant 1992; see Fig. 17.2).

Most of the research on conflict resolution described so far has concentrated on the role of conflict in early social development and its potential for facilitating moral development. In these

FIGURE 17.2. Conflicts about turn taking have the potential to provide children with the opportunity to develop negotiation skills. Here one girl wistfully looks on while she waits for her turn on the swing. Photograph by Melanie Killen.

studies the source of conflict has been shown to be primarily object disputes and the distribution of resources. As children get older, however, the bases for conflict shift from the concrete exchange of objects to the more abstract realm of psychological interactions (Shantz 1987).

In particular, intergroup relationships constitute a source of conflict as children become more aware of social group processes and social group dynamics (Putallaz & Wasserman 1990; Killen & Stangor in press). Studies have examined why some children are rejected from peer groups (focusing on individual characteristics; see Asher & Coie 1990) and how children form groups (Kindermann 1993). Recently, children's evaluations of exclusion based on social group membership have been investigated. Social groups can include gender, race, ethnicity, culture, and peer network. In a set of studies (Theimer et al. 1998; Horn et al. 1999; Killen & Stangor in press) it has been shown that children and adolescents believe that it is wrong, from a moral viewpoint, to exclude someone from a group on the basis of the characteristics of group membership (e.g., children judge that it would be wrong for an all-girl ballet club to exclude a boy just because boys don't usually do ballet). Yet, in complex situations, with age, children take both moral and group considerations into account when evaluating group inclusion and exclusion. At times, fairness (morality) is a priority, and at other times, group cohesiveness (social conventional considerations) is a priority. How children actually resolve conflicts about group inclusion and exclusion has not been the subject of much study. However, it is clear that these types of conflicts involve complexities that extend beyond the more concrete nature of object disputes and resources. Social group inclusion and exclusion involve moral considerations, such as fair and equal treatment, and social-conventional issues, such as group cohesiveness and group identity. With age, children's conflicts become complex, multifaceted, and increasingly psychological. Further, in-group/out-group considerations become very relevant in their interactions and in their moral decision making. (Although we do not have the space to elaborate on this issue, the topic of in-group/out-group distinctions is an area that is ripe for investigation from a comparative approach because in-group/out-group distinctions are also a very salient dimension of nonhuman primate interactions.)

Conflict and conflict resolution in early development promote negotiation skills in early social development, and we theorize that these types of experiences provide a foundation for resolving more complex social conflicts, such as the ones stemming from intergroup relationships, in later childhood, adolescence, and adulthood. In early social development, conflicts can often be characterized as based on moral (e.g., fairness) or social-conventional (e.g., structured activity) considerations. In middle childhood and adulthood, conflicts are more often multifaceted, involving a range of moral and social-conventional issues; these issues have to be weighed and coordinated by individuals in order to develop constructive methods of resolution.

In many but not all of these domains there are parallels with nonhuman primate conflict resolution and negotiation: to explore this common ground to the fullest, together with intensified research on the way children settle disputes, will provide us with the broadest possible perspective on the origins of human morality, one grounded in both evolution and development.

References

Alexander, R. D. 1987. *The Biology of Moral Systems.* New York: Aldine.

Arsenio, W., & Killen, M. 1996. Conflict-related emotions during peer disputes. *Early Education and Development,* 7: 43–57.

Arsenio, W., & Lover, A. 1995. Children's conceptions of sociomoral affect: Happy victimizers, mixed emotions, and other expectancies. In: *Morality in Everyday Life: Developmental Perspectives* (M. Killen & D. Hart, eds.), pp. 87–130. Cambridge: Cambridge University Press.

Asher, S., & Coie, J. 1990. *Peer Rejection in Childhood.* Cambridge: Cambridge University Press.

Aureli, F. 1997. Post-conflict anxiety in nonhuman primates: The mediating role of emotion in conflict resolution. *Aggressive Behavior,* 23: 315–328.

Baldwin, J. M. 1915. *Genetic Theory of Reality.* New York: Putnam.

Batson, C. D. 1990. How social an animal: The human capacity for caring. *American Psychologist,* 45: 336–346.

Bayer, C. L., Whaley, K. L., & May, S. E. 1995. Strategic assistance in toddler disputes, II. Sequences and patterns of teachers' message strategies. *Early Education and Development,* 6: 405–432.

Cords, M., & Killen, M. 1998. Conflict resolution in humans and non-human primates. In: *Piaget, Evolution, and Development* (J. Langer & M. Killen, eds.), pp. 193–218. Mahwah, N.J.: Lawrence Erlbaum Associates.

Corsaro, W. 1985. *Friendship and Peer Culture in the Early Years.* Norwood, N.J.: Ablex.

Coyle, C. T., & Enright, R. D. 1997. Forgiveness intervention with post-abortion men. *Journal of Consulting and Clinical Psychology,* 65: 1042–1046.

Cummings, E. M., Hollenbeck, B., Iannotti, R., Radke-Yarrow, M., & Zahn-Waxler, C. 1986. Early organization of altruism and aggression: Developmental patterns and individual differences. In: *Altruism and Aggression: Biological and Social Origins* (C. Zahn-Waxler, E. M. Cummings, & R. Iannotti, eds.), pp. 165–188. Cambridge: Cambridge University Press.

Damon, W. 1977. *Social World of the Child.* San Francisco: Jossey-Bass.

Damon, W. 1983. *Social and Personality Development.* New York: Norton.

Darwin, C. 1981 [1871]. *The Descent of Man, and Selection in Relation to Sex.* Princeton: Princeton University Press.

de Waal, F. B. M. 1982. *Chimpanzee Politics: Power and Sex among Apes.* London: Jonathan Cape.

de Waal, F. B. M. 1987. Tension regulation and non-reproductive functions of sex among captive bonobos (*Pan paniscus*). *National Geographic Research,* 3: 318–335.

de Waal, F. B. M. 1989. *Peacemaking among Primates.* Cambridge: Harvard University Press.

de Waal, F. B. M. 1993. Reconciliation among primates: A review of empirical evidence and unresolved issues. In: *Primate Social Conflict* (W. A. Mason & S. P. Mendoza, eds.), pp. 111–144. Albany: State University of New York Press.

de Waal, F. B. M. 1996. *Good Natured: The Origins of Right and Wrong in Humans and Other Animals.* Cambridge: Harvard University Press.

de Waal, F. B. M. 1997. The chimpanzee's service economy: Food for grooming. *Evolution and Human Behavior,* 18: 375–386.

de Waal, F. B. M., & Aureli, F. 1996. Consolation, reconciliation and a possible cognitive difference between macaques and chimpanzees. In: *Reaching into Thought: The Mind of the Great Apes* (A. E. Russon, K. A. Bard, & S. T. Parker, eds.), pp. 80–110. Cambridge: Cambridge University Press.

de Waal, F. B. M., & Johanowicz, D. L. 1993. Modification of reconciliation behavior through social experience: An experiment with two macaques species. *Child Development,* 64: 897–908.

de Waal, F. B. M., & van Roosmalen, A. 1979. Reconciliation and consolation among chimpanzees. *Behavioural Ecology and Sociobiology,* 5: 55–66.

Downie, R. S. 1965. Forgiveness. *Philosophical Quarterly,* 15: 128–134.

Dunn, J. 1988. *The Beginnings of Social Understanding.* Cambridge: Harvard University Press.

Dunn, J., & Slomkowski, C. 1992. Conflict and the development of social understanding. In: *Conflict in Child and Adolescent Development* (C. U. Shantz & W. W. Hartup, eds.), pp. 70–92. Cambridge: Cambridge University Press.

Dunn, J., Slomkowski, C., Donelan, N., & Herrera, C. 1995. Conflict, understanding and relationships: Developments and differences in the preschool years. *Early Education and Development,* 6: 303–316.

Dworkin, R. 1978. *Taking Rights Seriously.* Cambridge: Harvard University Press.

Eisenberg, N., & Fabes, R. A. 1998. Prosocial Development. In: *Handbook of Child Psychology,* Vol. 3: *Social, Emotional, and Personality Development* (N. Eisenberg, ed.), pp. 701–778. New York: Wiley.

Eisenberg, N., & Strayer, J. 1986. *Empathy and Its Development.* Cambridge: Cambridge University Press.

Enright, R. D., and the Human Development Study Group. 1991. The moral development of forgiveness. In: *Handbook of Moral Behavior and Development,* Vol. 1 (W. Kurtines & J. Gewirtz, eds.), Hillsdale, N.J.: Lawrence Erlbaum Associates.

Fitzgibbons, R. P. 1986. The cognitive and emotive uses of forgiveness in the treatment of anger. *Psychotherapy,* 23: 629–633.

Freud, S. 1960 [1923]. *The Ego and the Id.* New York: Norton.

Freud, S. 1961 [1930]. *Civilization and Its Discontents.* New York: Norton.

Gewirth, A. 1978. *Reason and Morality.* Chicago: University of Chicago Press.

Golding, W. 1954. *The Lord of the Flies.* New York: Capricorn.

Grusec, J., & Goodnow, J. 1994. Impact of parental discipline methods on the child's internalization of values: A reconceptualization of current points of view. *Developmental Psychology,* 30: 4–19.

Hartup, W. W., Laursen, B., Stewart, M. I., & Eastenson, A. 1988. Conflict and the friendship relations of young children. *Child Development,* 59: 1590–1600.

Hay, D. F., & Ross, H. S. 1982. The social nature of early conflict. *Child Development,* 53: 105–113.

Hinde, R. A. 1974. *Biological Bases of Human Social Behaviour.* New York: McGraw-Hill.

Hinde, R. A. 1979. *Towards Understanding Relationships.* London: Academic Press.

Hoffman, M. L. 1983. Empathy, guilt and social cognition. In: *The Relationship between Social and Cognitive*

Development (W. Overton, ed.), Hillsdale, N.J.: Lawrence Erlbaum Associates.

Hoffman, M. L. 1991. Empathy, social cognition, and moral action. In: *Handbook of Moral Behavior and Development*, Vol. 1: *Theory* (W. M. Kurtines & J. L. Gewirtz, eds.), pp. 275–301. Mahwah, N.J.: Lawrence Erlbaum Associates.

Horn, S., Killen, M., & Stangor, C. 1999. The influence of group stereotypes on adolescents' moral reasoning. *Journal of Early Adolescence*, 19: 98–113.

Huang, S. T. 1990. Cross-cultural and real-life validations of the theory of forgiveness in Taiwan, The Republic of China. Ph.D. diss., University of Wisconsin, Madison.

Hume, D. 1978 [1739]. *A Treatise of Human Nature.* Oxford: Oxford University Press.

Huxley, T. H. 1989 [1894]. *Evolution and Ethics.* Princeton: Princeton University Press

Kagan, J., & Lamb, S. 1987. *The Emergence of Morality in Young Children.* Chicago: University of Chicago Press.

Killen, M. 1989. Context, conflict, and coordination in early social development. In: *Social Interaction and the Development of Children's Understanding* (L. T. Winegar, ed.), pp. 119–146. Norwood, N.J.: Ablex.

Killen, M. 1991. Social and moral development in early childhood. In: *Handbook of Moral Behavior and Research Development*, Vol. 2: *Research* (W. M. Kurtines & J. L. Gewirtz, eds.), pp. 115–138. Mahwah, N.J.: Lawrence Erlbaum Associates.

Killen, M., & Hart, D. 1995. *Morality in Everyday Life: Developmental Perspectives.* Cambridge: Cambridge University Press.

Killen, M., & Naigles, L. 1995. Preschool children pay attention to their addressees: The effects of gender composition on peer disputes. *Discourse Processes*, 19: 329–346.

Killen, M., & Nucci, L. P. 1995. Morality, autonomy, and social conflict. In: *Morality in Everyday Life: Developmental Perspectives* (M. Killen & D. Hart, eds.), pp. 52–86. Cambridge: Cambridge University Press.

Killen, M., & Stangor, C. In press. Children's social reasoning about inclusion and exclusion in gender and race peer group contexts. *Child Development.*

Killen, M., & Turiel, E. 1991. Conflict resolution in preschool social interactions. *Early Education and Development*, 2: 240–255.

Kindermann, T. A. 1993. Natural peer groups as contexts for individual development: The case of children's motivations in school. *Developmental Psychology*, 29: 970–977.

Kohlberg, L. 1969. Stage and sequence: The cognitive-developmental approach to socialization. In: *Handbook of Socialization Theory and Research* (D. Goslin, ed.), pp. 347–480. Chicago: Rand McNally.

Kohlberg, L. 1984. *Essays on Moral Development: The Psychology of Moral Development.* San Francisco: Harper and Row.

Kropotkin, P. 1972 [1902]. *Mutual Aid: A Factor of Evolution.* New York: New York University Press.

Langer, J., & Killen, M. 1998. The comparative study of mental development. In: *Piaget, Evolution, and Development* (J. Langer & M. Killen, eds.), pp. 1–6. Mahwah, N.J.: Lawrence Erlbaum Associates.

Leaper, C. 1991. Influence and involvement in children's discourse: Age, gender, and partner effects. *Child Development*, 62: 797–811.

Mercer, P. 1972. *Sympathy and Ethics: A Study of the Relationship between Sympathy and Morality with Special Reference to Hume's Treatise.* Oxford: Oxford University Press.

Nagels, T. 1979. *Mortal Questions.* Cambridge: Cambridge University Press.

Nagels, T. 1986. *The View from Nowhere.* New York: Oxford University Press.

Nishida, T. 1994. Review of recent findings on Mahale chimpanzees. In: *Chimpanzee Cultures* (R. W. Wrangham, W. C. McGrew, F. B. M. de Waal, & P. Heltne, eds.), pp. 373–396. Cambridge: Harvard University Press.

North, J. 1987. Wrongdoing and forgiveness. *Philosophy*, 62: 499–508.

Nucci, L. P. 1981. Conceptions of personal issues: A domain distinct from moral or societal concepts. *Child Development*, 52: 114–121.

Nucci, L. P. 1996. Morality and the personal sphere of actions. In: *Values and Knowledge* (E. S. Reed,

E. Turiel, & T. Brown, eds.), pp. 41–60. Mahwah, N.J.: Lawrence Erlbaum Associates.

Nucci, L. P., & Turiel, E. 1978. Social interactions and the development of social concepts in preschool children. *Child Development*, 49: 400–407.

Park, S. R. 1994. Measuring interpersonal forgiveness in Korea. Master's thesis, University of Wisconsin, Madison.

Parker, S. T. 1998. The evolution and development of self-knowledge. In: *Piaget, Evolution, and Development* (J. Langer & M. Killen, eds.), pp. 171–192. Mahwah, N.J.: Lawrence Erlbaum Associates.

Parker, S. T., & Gibson, K. R. 1990. *"Language" and Intelligence in Monkeys and Apes*. Cambridge: Cambridge University Press.

Piaget, J. 1932. *The Moral Judgment of the Child*. New York: Free Press.

Piaget, J. 1970 [1950]. *Genetic Epistemology*. New York: Viking Press.

Pitrowski, C. 1995. Children's interventions into family conflict: Links with the quality of siblings relationships. *Early Education and Development*, 6: 377–404.

Putallaz, M., & Wasserman, A. 1990. Children's entry behavior. In: *Peer Rejection in Childhood* (S. Asher & J. Coie, eds.), pp. 189–216. Cambridge: Cambridge University Press.

Radke-Yarrow, M., Zahn-Waxler, C., & Chapman, M. 1983. Children's prosocial dispositions and behavior. In: *Carmichael's Manual of Child Psychology*, Vol. 1 (4th edn.): *Socialization, Personality, and Social Development* (P. H. Mussen, ed.). New York: Wiley.

Rawls, J. 1971. *A Theory of Justice*. Cambridge: Harvard University Press.

Richards, N. 1988. Forgiveness. *Ethics*, 99: 77–97

Ross, H., & Conant, C. 1992. The social structure of early conflict: Interactions, relationships, and alliances. In: *Conflict in Child and Adolescent Development* (C. U. Shantz & W. W. Hartup, eds.), pp. 153–185. Cambridge: Cambridge University Press.

Savage-Rumbaugh, E. S. 1998. The evolution of language development. In: *Piaget, Evolution, and Development* (J. Langer & M. Killen, eds.), pp. 145–170. Mahwah, N.J.: Lawrence Erlbaum Associates.

Sch-he-rie, M. 1984. *The Scale of Wisdom*, 6, Qum, Iran: Propagation Center (Arabic).

Searle, J. 1969. *Speech Acts*. London: Cambridge University Press.

Shantz, C. U. 1987. Conflicts between children. *Child Development*, 58: 283–305.

Shantz, C. U., & Hartup, W. W. 1992. *Conflict in Child and Adolescent Development*. Cambridge: Cambridge University Press.

Shelov, S., ed. 1991. *Caring for Your Baby and Young Child: Birth to the Age of Five*. New York: Bantam Books.

Skinner, B. F. 1971. *Beyond Freedom and Dignity*. New York: Knopf.

Smetana, J. G. 1995. Morality in context: Abstractions, ambiguities, and applications. In *Annals of Child Development*, Vol. 10 (R. Vasta, ed.), pp. 83–130. London: Jessica Kingsley.

Smith, A. 1937 [1759]. *A Theory of Moral Sentiments*. New York: Modern Library.

Stent, G. 1978. *Morality as a Biological Phenomenon*. Berkeley: University of California Press.

Subkoviak, M. J., Enright, R. D., Wu, C., Gassin, E. A., Freedman, S., Olson, L. M., & Sarinopoulos, I. 1995. Measuring interpersonal forgiveness in late adolescence and middle adulthood. *Journal of Adolescence*, 18: 641–655.

Theimer, C. E., Killen, M., & Stangor, C. 1998. Preschool children's evaluations of exclusion in gender-stereotypic contexts. Manuscript under review. University of Maryland, College Park.

Tisak, M. 1995. Domains of social reasoning and beyond. In: *Annals of Child Development*, Vol. 11 (R. Vasta, ed.), pp. 95–130. London: Jessica Kingsley.

Tomasello, M. 1998. Social cognition and the evolution of culture. In: *Piaget, Evolution, and Development* (J. Langer & M. Killen, eds.), pp. 221–246. Mahwah, N.J.: Lawrence Erlbaum Associates.

Turiel, E. 1978. The development of concepts of social structure: Social convention. In: *The Development of Social Understanding* (J. Glick & K. A. Clarke-Stewart, eds.), pp. 25–107. New York: Gardner Press.

Turiel, E. 1983. *The Development of Social Knowledge: Morality and Convention*. Cambridge: Cambridge University Press.

Turiel, E. 1998. The development of morality. In: *Handbook of Child Psychology*, Vol. 3: *Social, Emotional, and Personality Development* (N. Eisenberg, ed.), pp. 863–932. New York: Wiley.

Turiel, E., Killen, M., & Helwig, C. C. 1987. Morality: Its structure, functions, and vagaries. In: *The Emergence of Morality in Young Children* (J. Kagan & S. Lamb, eds.), pp. 155–244. Chicago: University of Chicago Press.

Vine, W. E. 1985. *An Expository Dictionary of Biblical Words*. Nashville, Tenn.: Thomas Nelson.

Watson, J. B. 1924. *Behaviorism*. Chicago: University of Chicago Press.

Werner, H. 1926. *Comparative Psychology of Mental Development*. New York: International Universities Press.

Westermarck, E. 1912. *The Origin and Development of the Moral Ideas*, Vol. 1. London: Macmillan.

Whiten, A. 1998. The mindreading system. In: *Piaget, Evolution, and Development* (J. Langer & M. Killen, eds.), pp. 73–99. Mahwah, N.J.: Lawrence Erlbaum Associates.

Whiten, A., & Byrne, R. W. 1997. *Machiavellian Intelligence II: Evaluations and Extensions*. Cambridge: Cambridge University Press.

Wickler, W. 1981 [1971]. *Die Biologie der Zehn Gebote: Warum die Natur uns kein Vorbild ist*. Munich: Piper.

Williams, B. 1981. *Moral Luck*. Cambridge: Cambridge University Press.

Williams, G. C. 1988. Reply to comments on "Huxley's evolution and ethics" in sociobiological perspective. *Zygon*, 23: 437–438.

Wispé, L. 1991. *The Psychology of Sympathy*. New York: Plenum.

Zahn-Waxler, C., & Hastings, P. 1999. Development of empathy: Adaptive and maladaptive patterns. In: *Moral Sensibilities and Education*, Vol. 1: *The Preschool Child* (W. van Haaften, T. Wren, & A. Tellings, eds.). Bemmel, the Netherlands: Concorde.

Zahn-Waxler, C., & Radke-Yarrow, M. 1990. The origins of empathic concern. *Motivation and Emotion*, 14: 107–130.

Zahn-Waxler, C., Radke-Yarrow, M., & King, R. 1979. Child rearing and children's initiations towards victims of distress. *Child Development*, 50: 319–330.

Zahn-Waxler, C., Cummings, E. M., & Iannotti, R. 1986. *Altruism and Aggression: Biological and Social Origins*. Cambridge: Cambridge University Press.

Conclusion

CHAPTER 18

Shared Principles and Unanswered Questions

Frans B. M. de Waal & Filippo Aureli

As the present volume attests, in a wide range of disciplines—from law to developmental psychology, and from anthropology to primatology—a shift in emphasis is under way, which makes social relationships a central aspect in our thinking about human and animal behavior. Instead of the traditional focus on behavioral output and how it is generated, the focus is now on behavioral interaction and how it unfolds within the relationship between two parties, how it modifies the relationship, and functions within it. In this perspective, conflict is not merely a matter of aggressive tendencies or of who wins or who loses. Conflict is part and parcel of relationships, generated by the colliding interests of individuals and constrained by their overlapping interests (de Waal, Chapter 2).

The disciplines represented in this volume do not always communicate with one another, and they use quite different vocabularies to address these issues, but a number of shared principles are easily recognizable. First, there is a need for peaceful coexistence among cooperative entities. Peace does not always have priority over conflict—if so, there never would be any open conflict—but the need to harmonize is urgent enough that it affects the course and intensity of conflict. Evolutionary pressures favoring cooperation have led to powerful management and reparative mechanisms. This has led to great complexity in the regulation and negotiation of social relationships. For example, the variation in dominance styles among nonhuman primates (Preuschoft & van Schaik, Chapter 5; Thierry, Chapter 6; Sapolsky, Box 6.1; Pereira &

Kappeler, Box 15.2), the evolution of human morality (Killen & de Waal, Chapter 17), and the way tantrums function within parent-offspring exchanges (Potegal, Box 12.1) are all best understood within the relational framework (de Waal, Chapter 2).

The second shared principle is the enormous flexibility of the mechanisms and behavior involved and their responsiveness to the environment (Aureli & Smucny, Chapter 10). This is visible in the cultural variation among human societies (Butovskaya et al., Chapter 12; Fry, Chapter 16; Park & Enright, Box 17.1), the modifiability of primate reconciliation behavior through social experience (Cords & Aureli, Chapter 9), the different coping strategies people and other primates employ under varying population densities (Judge, Chapter 7), and so on. Among the most plastic of social mechanisms, conflict and conflict resolution may be the primary ways in which we and other animals adjust to the pressures of the social and physical environment.

This represents a sharp departure from the old Lorenzian view of aggression as an "instinct." This term conjures up images of inevitability and uniformity, whereas what we see is an eminently responsive system that weighs the need for cooperation against the costs and benefits of competition. It results in an equilibrium unique for each society and probably each social relationship within it. It also breaks with a tradition in evolutionary biology that looks at conflict as a zero-sum game, ignoring the long-term investment both parties may have made in each other. If there is any common theme emerging from the contributions to this volume, it is that the cost of conflict increases with increasing value of the relationship between the contestants, a variable thus far ignored in evolutionary modeling (de Waal, Chapter 2; Matsumura & Okamoto, Box 5.1; Cords & Aureli,

Chapter 9; van Schaik & Aureli, Chapter 15; see Silk, Box 9.1, for a different emphasis).

Because of this, a whole psychology has evolved around the hidden cost of disharmonized relationships, which induces anxieties when future cooperation is in jeopardy and makes the parties seek each other out for repair (Aureli & Smucny, Chapter 10; Cheney & Seyfarth, Box 10.1) or at least show some spatial rapprochement after a relationship has been disturbed (Call, Box 9.2). This has been compared to processes of *homeostasis*, as if there is an equilibrium state for which all parties strive. Ultimately, in the case of our own species, it has led to judgments about how conflicts ought to be resolved in a fair manner and when aggression is justified or not. Thus the "ought" question, which has occupied moral philosophers over the ages, may reach all the way back to rather mundane mechanisms of how to get along and when to repair relationships (Killen & de Waal, Chapter 17).

In reviewing where we stand today, we can see the following areas with either a great need for additional data or a potential for theoretical development. We highlight each area below with a simple question without suggesting that answering these questions will be simple:

- *How similar is human conflict management to that of other animals?* As mentioned above, there is increasing evidence across disciplines that conciliatory mechanisms operate especially between cooperative entities (e.g., Yarn, Chapter 4; Cords & Aureli, Chapter 9; Fry, Chapter 16). There is a need for further research using similar methodology on subjects as disparate as cooperatively breeding birds, spider monkeys, human hunter-gatherers, business partners, and entire nations. These studies can be interdisciplinary collaborations resulting in a cross-fertilization of the techniques applied in a variety of fields. This comparative approach has been successfully applied in the study of conflict management among children by employing observational techniques first developed for the study of nonhuman primates (Verbeek et al., Chapter 3; Butovskaya et al., Chapter 12).

- *How general is conflict resolution in the animal kingdom?* Much empirical and theoretical progress has been made with regard to the conflict-resolution skills of apes and Old World monkeys (de Waal, Chapter 2; Veenema, Box 2.1; Cords & Aureli, Chapter 9; Appendix A). If theories about the functional significance of reconciliation developed on these primates are correct, there is no good reason not to expect similar mechanisms in other taxonomic groups. Perhaps reconciliation is as old as the combination of competition and cooperation, which would make it phylogenetically ancient indeed. Or perhaps these mechanisms evolved several times independently in cooperative species. This volume includes contributions on less-studied primate species (Schaffner & Caine, Chapter 8; Strier et al., Box 15.1; Pereira & Kappeler, Box 15.2) and on animals other than primates (Schino, Chapter 11; Rowell, Box 11.1; Samuels & Flaherty, Box 11.2; Hofer & East, Box 11.3). It is, however, clear from the preponderance of primate contributions that we still have a long way to go before we will know how widespread the phenomenon is and be able to explore its evolutionary path.

- *How does conflict management affect health and well-being?* Apart from the old fight-flight balance assumed to underlie all confrontation, there are indications that conflict among socially engaged parties involves far more complex emotions, such as anxiety about the relationship, social attraction and attachment, concern about hurting the

other, and forgiveness (Aureli & Smucny, Chapter 10; Weaver & de Waal, Box 10.2; Park & Enright, Box 17.1). Some of these variables have physiological correlates; hence the outcome of conflict may well have serious health consequences, and conflict management may prevent them (Sapolsky, Box 6.1; Aureli & Smucny, Chapter 10). Similarly, there is a budding field of research on human forgiveness that is keenly interested in its health impact.

- *How is conflict resolution acquired?* Despite flourishing conflict-resolution programs for school-age children, knowledge of natural conflict resolution in children and juvenile primates is absolutely minimal. We need adequate background data on both humans and other animals to help us understand how conflict-resolution skills are acquired (Weaver & de Waal, Box 10.2; Butovskaya et al., Chapter 12; Potegal, Box 12.1) and how they interact with other aspects of social development (Verbeek et al., Chapter 3; Killen & de Waal, Chapter 17). Only after we have developed a theoretical framework in this regard will we be able to evaluate the efficacy of conflict-resolution and -prevention programs. The gist of the thinking in this volume would be that there is little to be gained with an emphasis on peacemaking in the absence of attention to relationship value, and conflict-resolution programs should capitalize on this notion. In addition, more attention to the development of conflict management skills is needed in animal studies; much can be learned from recent studies in child psychology.

- *How is conflict prevented or avoided?* The most effective way to avoid disturbance of relationships is not reconciliation after a fight but either the successful management of a conflict by means of dominance or avoidance or, even better, the prevention of conflict altogether. Most of us who watch nonhuman primates have the feeling that potentially conflicting situations occur all the time but that only a small portion of them lead to actual confrontation. This is a very difficult area of research (as much of the behavior involved seems rather subtle, from some grooming to ease tempers to mere walking away from the scene) but potentially a most promising one. The present volume includes examples of this research: tolerance can be increased before conflict arises (Verbeek et al., Chapter 3; Kuester & Paul, Box 5.2; Colmenares et al., Box 5.3; Judge, Chapter 7; Koyama, Box 7.1; Fry, Chapter 16); specific lifestyles can minimize opportunities for overt conflict (Schaffner & Caine, Chapter 8; Strier et al., Box 15.1; Fry, Chapter 16); and dominance relationships can prevent conflict escalation (Preuschoft & van Schaik, Chapter 5).

- *How is conflict regulated in the wider social context?* Third parties intervene in ongoing conflicts; they pacify and mediate (Das, Chapter 13; Petit & Thierry, Box 13.1; Watts et al., Chapter 14; Castles, Box 14.1), culminating in the human cultural innovation of institutions that manage conflict (Yarn, Chapter 4; Fry, Chapter 16), even between entities such as ethnic groups or nations (Neu, Box 4.1). The involvement of society as a whole in conflict among its parts raises the issue of social integration at levels higher than the individual. There is a great need for studies of the higher integrative levels of societies.

- *How effective are the discussed mechanisms in managing conflict?* The term *reconciliation* is best regarded as a heuristic device: the implied function of the observed interaction lets itself be translated into a number of testable hypotheses (de Waal, Chapter 2). In the case of animals, but also that of our own species, we shouldn't take consequences

for granted. It is all too easy to assume that reconciliations or consolations function as such or that conflict disappears after it has been "managed." But we know it often doesn't, and we should keep a critical eye on presumed consequences of behavior. There is good evidence that friendly reunions between former opponents in some primate species serve as reconciliations because they have been shown to restore tolerance, reduce further aggression, and decrease tension and anxiety (Cords & Aureli, Chapter 9; Aureli & Smucny, Chapter 10). More research in this regard is urgently needed, however. With respect to human behavior, it is important to assess the effectiveness of conflict-resolution programs in schools, adult intervention in children's conflicts, marriage counseling, and the many institutions that manage conflict in different cultures and at different levels, from conflict between individuals in the same societies to international or ethnic conflict.

- *How does conflict resolution relate to morality and justice?* Two general trends in primate evolution, toward greater cognition and greater social complexity, have brought about the sort of societies we live in, which are unthinkable without elaborate systems of resource distribution and conflict management. Morality is the way we regulate inclusion in and exclusion from the benefits offered by the collective, and it is crucial how the values and judgments associated with it came into existence. There is still much work to be done on the evolutionary and cultural origins of systems of morality and justice (Yarn, Chapter 4; Fry, Chapter 16; Killen & de Waal, Chapter 17).

With Konrad Lorenz's *On Aggression* we entered several decades in which humans were routinely depicted as naturally aggressive. Even as we write, almost 40 years later, books are still coming out that one-sidedly emphasize the "dark side" of humanity or the evolutionary origins of violence. No biologist and few social scientists would deny that our species has natural aggressive potentials, but we possess many other natural tendencies as well, some of which serve to keep aggression in check. Thus, to call us naturally peaceful (which is, after all, a state observed far more often than war and strife) would be at least as justified as calling us naturally aggressive.

Why, then, don't most authors present a more balanced view? Are they too concerned about violence to see the "bright side"? We believe that a perspective that actually integrates conflict and cooperation and tries to understand how the one functions alongside the other is far more productive than one that looks at one or the other separately. To treat aggressive tendencies in isolation from all other behavior and from the wider social context is a bit like studying a fish on dry land. It is possible, and does provide useful infor-mation, but one will never understand what a fish is really about unless one places it in its nat-ural medium.

The present volume introduces the first studies approaching conflict and aggression from this broader perspective, coming from both evolutionary biology and the social sciences. We look at this as only the beginning: current research is only scratching the surface. Especially with regard to our own species, conflict resolution is often treated more as a goal (of therapy, of mediation, of education) than as a process that needs to be studied and analyzed in detail before it will be understood. As in aggression research, in which various approaches have traditionally supple-mented one another, we hope that the present volume will contribute to a balanced, multidisci-plinary analysis of processes without which complex societies would be unthinkable and without which social relationships would not last beyond their first conflict.

Appendixes

The Occurrence of Reconciliation in Nonhuman Primates

Species	Reconciliation	Selective Attractiveness	Reconciled Conflicts (percent)
PROSIMIANS			
Ringtailed lemur, *Lemur catta*[1]	no		—
Redfronted lemur, *Eulemur fulvus rufus*[1]	yes	yes	14–21
NEW WORLD MONKEYS			
Brown capuchin, *Cebus apella*[2]	yes	yes	21
Red-bellied tamarin, *Saguinus labiatus*[3]	no		—
OLD WORLD MONKEYS			
Sooty mangabey, *Cercocebus torquatus atys*[4]	yes		55
Vervet monkey, *Cercopithecus aethiops*[5]	yes		14
Patas monkey, *Erythrocebus patas*[6]	yes	yes	31
Golden monkey, *Rhinopithecus roxellanae*[7]	yes		43–54
Spectacled langur, *Trachypithecus obscura*[8]	yes		41–51
Gelada baboon, *Theropithecus gelada*[9]	yes		30–45
Olive baboon, *Papio anubis*[10]	yes	yes	16
Hamadryas baboon, *Papio hamadryas*[11]	yes		24
Guinea baboon, Papio *papio*[12]	yes		27
Chacma baboon, *Papio ursinus*[13]	yes		10–35
Stumptail macaque, *Macaca arctoides*[14,15]	yes	yes	26–53
Longtail macaque, *Macaca fascicularis*[16,17,18]	yes	yes	13–40
Japanese macaque, *Macaca fuscata*[17,19,20]	yes	yes	12–37
Moor macaque, *Macaca maurus*[21]	yes		40
Rhesus macaque, *Macaca mulatta*[22,23]	yes	yes	7–23
Pigtail macaque, *Macaca nemestrina*[24,25]	yes	yes	30–42
Black macaque, *Macaca nigra*[26]	yes	yes	40
Liontail macaque, *Macaca silenus*[27]	yes	yes	42–48
Barbary macaque, *Macaca sylvanus*[17,28]	yes	yes	28–33
Tonkean macaque, *Macaca tonkeana*[29,30]	yes		46
GREAT APES			
Mountain gorilla, *Gorilla gorilla beringei*[31]	yes, only male-female	yes	—
Bonobo, *Pan paniscus*[32]	yes		48
Chimpanzee, *Pan troglodytes*[33,34]	yes	yes	18–47

Sources: 1: Kappeler 1993; 2: Verbeek & de Waal 1997; 3: Schaffner & Caine 1992; 4: Gust & Gordon 1993; 5: Cheney & Seyfarth 1989; 6: York & Rowell 1988; 7: Ren et al. 1991; 8: Arnold & Barton 1997; 9: Swedell 1997; 10: Castles & Whiten 1998; 11: Zaragoza & Colmenares 1997; 12: Petit & Thierry 1994b; 13: Silk et al. 1996; 14: de Waal & Ren 1988; 15: Perez-Ruiz & Mondragon-Ceballos 1994; 16: Aureli et al. 1989; 17: Aureli et al. 1997; 18: Aureli 1992; 19: Aureli et al. 1993; 20: Petit et al. 1997; 21: Matsumura 1996; 22: de Waal & Yoshihara 1983; 23: Call et al. 1996; 24: Judge 1991; 25: Castles et al. 1996; 26: Petit & Thierry 1994a; 27: Abegg et al. 1996; 28: Aureli et al. 1994; 29: Demaria & Thierry 1992; 30: Thierry et al. 1994; 31: Watts 1995; 32: de Waal 1987; 33: de Waal & van Roosmalen 1979; 34: de Waal 1986.
Note: The table includes only studies comparing post-conflict samples with control samples for a given species (see Veenema, Box 2.1, for methodological details); if no controlled study was carried out for a given species, studies based only on post-conflict samples are reported. Experimental studies and studies on unstable groups are not considered.
Yes = it was found; no = it was not found. (See Schaffner & Caine, Chapter 8, and Pereira & Kappeler, Box 15.2, for the two species for which reconciliation was not found.)
Selective Attractiveness = post-conflict selective increase of interaction between the former opponents, not an increase indiscriminately involving all potential partners.
Reconciled Conflicts = percentage of reconciled conflicts out of total recorded conflicts. Percentages were calculated in different ways in the various studies, and they cannot be directly compared across studies. They are reported here to give a sense of the variability of the behavior.

References

Abegg, C., Thierry, B., & Kaumanns, W. 1996. Reconciliation in three groups of lion-tailed macaques. *International Journal of Primatology,* 17: 803–816.

Arnold, K., & Barton, R. A. 1997. Post-conflict behaviour in spectacled langurs (*Tachypithecus obscurus*). *Advances in Ethology,* 32: 153.

Aureli, F. 1992. Post-conflict behaviour among wild long-tailed macaques (*Macaca fascicularis*). *Behavioral Ecology and Sociobiology,* 31: 329–337.

Aureli, F., van Schaik, C. P., & van Hooff, J. A. R. A. M. 1989. Functional aspects of reconciliation among captive long-tailed macaques (*Macaca fascicularis*). *American Journal of Primatology,* 19: 39–51.

Aureli, F., Veenema, H. C., van Panthaleon van Eck, C. J., & van Hooff, J. A. R. A. M. 1993. Reconciliation, consolation, and redirection in Japanese macaques (*Macaca fuscata*). *Behaviour,* 124: 1–21.

Aureli, F., Das, M., Verleur, D., & van Hooff, J. A. R. A. M. 1994. Post-conflict social interactions among Barbary macaques (*Macaca sylvanus*). *International Journal of Primatology,* 15: 471–485.

Aureli, F., Das, M., & Veenema, H. C. 1997. Differential kinship effect on reconciliation in three species of macaques (*Macaca fascicularis, M. fuscata,* and *M. sylvanus*). *Journal of Comparative Psychology,* 111: 91–99.

Call, J., Judge, P. G., & de Waal, F. B. M. 1996. Influence of kinship and spatial density on reconciliation and grooming in rhesus monkeys. *American Journal of Primatology,* 39: 35–45.

Castles, D. L., & Whiten, A. 1998. Post-conflict behaviour of wild olive baboons. I. Reconciliation, redirection and consolation. *Ethology,* 104: 126–147.

Castles, D. L., Aureli, F., & de Waal, F. B. M. 1996. Variation in conciliatory tendency and relationship quality across groups of pigtail macaques. *Animal Behaviour,* 52: 389–403.

Cheney, D. L., & Seyfarth, R. M. 1989. Redirected aggression and reconciliation among vervet monkeys, *Cercopithecus aethiops. Behaviour,* 110: 258–275.

Demaria, C., & Thierry, B. 1992. The ability to reconcile in Tonkean and rhesus macaques. *Abstract Book, 14th Congress of the International Primatological Society,* p. 101. Strasbourg: SICOP.

de Waal, F. B. M. 1986. The integration of dominance and social bonding in primates. *Quarterly Review of Biology,* 61: 459–479.

de Waal, F. B. M. 1987. Tension regulation and nonreproductive functions of sex in captive bonobos (*Pan paniscus*). *National Geographic Research,* 3: 318–335.

de Waal, F. B. M., & Ren, R. 1988. Comparison of the reconciliation behavior of stumptail and rhesus macaques. *Ethology,* 78: 129–142.

de Waal, F. B. M., & van Roosmalen, A. 1979. Reconciliation and consolation among chimpanzees. *Behavioral Ecology and Sociobiology,* 5: 55–66.

de Waal, F. B. M., & Yoshihara, D. 1983. Reconciliation and redirected affection in rhesus monkeys. *Behaviour,* 85: 224–241.

Gust, D. A., & Gordon, T. P. 1993. Conflict resolution in sooty mangabeys. *Animal Behaviour,* 46: 685–694.

Judge, P. G. 1991. Dyadic and triadic reconciliation in pigtail macaques (*Macaca nemestrina*). *American Journal of Primatology,* 23: 225–237.

Kappeler, P. M. 1993. Reconciliation and post-conflict behaviour in ringtailed lemurs, *Lemur catta*, and redfronted lemurs, *Eulemur fulvus rufus*. *Animal Behaviour*, 45: 901–915.

Matsumura, S. 1996. Post-conflict contacts between former opponents among wild moor macaques (*Macaca maurus*). *American Journal of Primatology*, 38: 211–219.

Perez-Ruiz, A. L., & Mondragon-Ceballos, R. 1994. Rates of reconciliatory behaviors in stumptail macaques: Effects of age, sex, rank and kinship. In: *Current Primatology*, Vol. 2: *Social Development, Learning and Behaviour* (J. J. Roeder, B. Thierry, J. R. Anderson, & N. Herrenschmidt, eds.), pp. 147–155. Strasbourg: Université Louis Pasteur.

Petit, O., & Thierry, B. 1994a. Reconciliation in a group of black macaques. *Dodo, Journal of the Wildlife Preservation Trusts*, 30: 89–95.

Petit, O., & Thierry, B. 1994b. Reconciliation in a group of Guinea baboons. In: *Current Primatology*, Vol. 2: *Social Development, Learning and Behaviour* (J. J. Roeder, B. Thierry, J. R. Anderson, & N. Herrenschmidt, eds.), pp. 137–145. Strasbourg: Université Louis Pasteur.

Petit, O., Abegg, C., & Thierry, B. 1997. A comparative study of aggression and conciliation in three cercopithecine monkeys (*Macaca fuscata, Macaca nigra, Papio papio*). *Behaviour*, 134: 415–432.

Ren, R., Yan, K., Su, Y., Qi, H., Liang, B., Bao, W., & de Waal, F. B. M. 1991. The reconciliation behavior of golden monkeys (*Rhinopithecus roxellanae roxellanae*) in small breeding groups. *Primates*, 32: 321–327.

Schaffner, C. M., & Caine, N. G. 1992. Who needs to reconcile? Post-conflict behavior in red-bellied tamarins. *American Journal of Primatology*, 27: 56.

Silk, J. B., Cheney, D. L., & Seyfarth, R. M. 1996. The form and function of post-conflict interactions between female baboons. *Animal Behaviour*, 52: 259–268.

Swedell, L. 1997. Patterns of reconciliation among captive gelada baboons (*Theropithecus gelada*): A brief report. *Primates*, 38: 325–330.

Thierry, B., Anderson, J. R., Demaria, C., Desportes, C., & Petit, O. 1994. Tonkean macaque behaviour from the perspective of the evolution of Sulawesi macaques. In: *Current Primatology*, Vol. 2: *Social Development, Learning and Behaviour* (J. J. Roeder, B. Thierry, J. R. Anderson, & N. Herrenschmidt, eds.), pp. 103–117. Strasbourg: Université Louis Pasteur.

Verbeek, P., & de Waal, F. B. M. 1997. Post-conflict behavior of captive brown capuchins in the presence and absence of attractive food. *International Journal of Primatology*, 18: 703–725.

Watts, D. P. 1995. Post-conflict social events in wild mountain gorillas (*Mammalia, Hominoidea*) I. Social interactions between opponents. *Ethology*, 100: 139–157.

York, A. D., & Rowell, T. E. 1988. Reconciliation following aggression in patas monkeys, *Erythrocebus patas*. *Animal Behaviour*, 36: 502–509.

Zaragoza, F., & Colmenares, F. 1997. Reconciliation and consolation in Hamadryas baboons, *Papio hamadryas*. *Advances in Ethology*, 32: 158.

Key Terms Used in the Volume

The list is based on definitions from Hinde (1974), Mason (1993), and Cords & Killen (1998) and was finalized with the help of the following colleagues from various disciplines: Marina Cords, Douglas Fry, Peter Verbeek, and Douglas Yarn.

Aggression. Behavior directed at members of the same species in order to cause physical injury or to warn of impending actions of this nature by means of facial and vocal threat displays.

Appeasement. Actions directed at a potential aggressor serving to reduce the risk of being attacked; not necessarily a form of conflict resolution.

Conflict (i.e., interindividual conflict). Situation that arises when individuals act on incompatible goals, interests, or actions. Conflict need not be aggressive.

Conflict Management. Actions or adherence to conventions that settle conflict but do not necessarily end it. Conflict management serves to reduce the cost of conflict to one or both opponents. Some examples of conflict management that do not end conflict are the following. Ritualized dominance relationships, respect of possession, the development of social customs and routines, and forms of reassurance and appeasement before potential conflict are means to avoid the escalation of conflict to aggression and can be considered as peacekeeping mechanisms. Ritualized forms of aggression function in avoiding the escalation to physical aggression and potential injury. Redirection of aggression by the recipient to third parties and interference by powerful individuals often break up ongoing aggressive interactions. Third parties may console one or both opponents. Such consolations do not end the state of conflict but alleviate the distress associated with such a state. Forms of conflict management that end conflict are considered as conflict resolution.

Conflict Resolution. Actions that eliminate the incompatibility of attitudes and goals of the conflicting individuals. Resolution is obtained by direct communication between former opponents that restores interaction between them. Examples include compromise, sharing, and reconciliation or peacemaking. Opponents may also agree to disagree. In this case the differences at the basis of the conflict still persist, but the opponents have agreed on the significance of these differences. Agreeing to disagree is not the same as dropping the conflict. Only the former involves communication between the two parties.

Consolation. The calming of a distressed conflict participant by a third party through friendly interaction.

Post-Conflict Third-Party Affiliation. Friendly interaction between a bystander and an individual previously involved in a conflict. It may have various functions depending on the role of the individual in the conflict and the relationship between the bystander and either opponent. Potential functions are appeasement, consolation, reassurance, and recruitment of support.

Reassurance. Actions serving to reduce anxiety of (and restore confidence in) the recipient of the action; not necessarily a form of conflict resolution.

Reconciliation. Post-conflict friendly reunion of former opponents that restores their social relationship disturbed by the conflict. Reconciliation is one mechanism of conflict resolution. If this function is not implied or demonstrated, a more descriptive term (e.g., post-conflict friendly reunion) is used.

Redirection. An aggressive action by the recipient of aggression against a third individual. It may function to divert the attention of the former aggressor (and other group members) from the original recipient of aggression.

Social Dominance. An attribute of a dyadic social relationship characterized by a consistent outcome in favor of the same member of the dyad and a default yielding response of the other individual rather than escalation. The consistent winner is dominant over the consistent loser that is subordinate. Dominant individuals often have priority of access to resources.

Submission. Actions that combine elements of fear, appeasement, and attraction, usually performed by individuals who also grant the other precedence in competitive contexts. Submission indicates the absence of aggressive tendencies but is not resolution of conflict.

References

Cords, M., & Killen, M. 1998. Conflict resolution in human and nonhuman primates. In: *Piaget, Evolution, and Development* (J. Langer & M. Killen, eds.), pp. 193–219. Mahwah, N.J.: Lawrence Erlbaum Associates.

Hinde, R. A. 1974. *Biological Bases of Human Social Behaviour.* New York: McGraw-Hill.

Mason, W. A. 1993. The nature of social conflict: A psycho-ethological perspective. In: *Primate Social Conflict* (W. A. Mason & S. P. Mendoza), pp. 13–47. Albany: State University of New York Press.

Contributors

Kate Arnold, Scottish Primate Research Group, School of Psychology, University of St. Andrews, St. Andrews, United Kingdom

Filippo Aureli, Living Links, Yerkes Regional Primate Research Center, Emory University, Atlanta, Georgia, USA, and School of Biological and Earth Sciences, John Moores University, Liverpool, United Kingdom

Nilcemar O. Bejar, Estação Biologica de Caratinga, Ipanema, Minas Gerais, Brazil

Marina Butovskaya, Institute of Cultural Anthropology, Russian State University for Humanities, Moscow, Russia

Nancy G. Caine, Department of Psychology, California State University, San Marcos, California, USA

Josep Call, Max Planck Institute for Evolutionary Anthropology, Leipzig, Germany

Dennison S. Carvalho, Estação Biologica de Caratinga, Ipanema, Minas Gerais, Brazil

Duncan L. Castles, Hasegawa Laboratory, Department of Cognitive and Behavioural Science, University of Tokyo, Tokyo, Japan, and School of Life Sciences, Roehampton Institute, London, United Kingdom

Dorothy L. Cheney, Department of Biology, University of Pennsylvania, Philadelphia, Pennsylvania, USA

W. Andrew Collins, Institute of Child Development, University of Minnesota, Minneapolis, Minnesota, USA

Fernando Colmenares, Departamento de Psicobiología, Universidade Complutense de Madrid, Madrid, Spain

Marina Cords, Anthropology Department, Columbia University, New York, New York, USA

Marjolijn Das, Ethology and Socioecology, Department of Comparative Physiology, Utrecht University, The Netherlands

Frans B. M. de Waal, Living Links, Yerkes Regional Primate Research Center, and Psychology Department, Emory University, Atlanta, Georgia, USA

Marion L. East, Max Planck Institute for Behavioral Physiology, Seewiesen, Germany, and Institute for Zoo Biology and Wildlife Research, Berlin, Germany

Robert D. Enright, Department of Educational Psychology, University of Wisconsin–Madison, Madison, Wisconsin, USA

Cindy Flaherty, Biology Department, Woods Hole Oceanographic Institution, Woods Hole, Massachusetts, USA

Douglas P. Fry, Department of Social Sciences, Åbo Akademi University, Vasa, Finland, and Bureau of Applied Research in Anthropology, University of Arizona, Tucson, Arizona, USA

Willard W. Hartup, Institute of Child Development, University of Minnesota, Minneapolis, Minnesota, USA

Heribert Hofer, Max Planck Institute for Behavioral Physiology, Seewiesen, Germany, and Institute for Zoo Biology and Wildlife Research, Berlin, Germany

Peter G. Judge, Department of Psychology, Bloomsburg University, Bloomsburg, Pennsylvania, USA

Peter M. Kappeler, Department of Behavior/Ecology, German Primate Center, Göttingen, Germany

Melanie Killen, Department of Human Development, University of Maryland, College Park, Maryland, USA

Nicola F. Koyama, School of Biological and Earth Sciences, John Moores University, Liverpool, United Kingdom

Jutta Kuester, Department of Zoology and Neurobiology, Ruhr University Bochum, Bochum, Germany

Thomas Ljungberg, Division of Ethology, Department of Zoology, University of Stockholm, Stockholm, Sweden

Antonella Lunardini, Department of Ethology, Ecology, and Evolution, Division of Anthropology, University of Pisa, Pisa, Italy

Shuichi Matsumura, Primate Research Institute, Kyoto University, Kyoto, Japan

Joyce Neu, Conflict Resolution Program, The Carter Center, Atlanta, Georgia, USA

Kyoko Okamoto, Primate Research Institute, Kyoto University, Kyoto, Japan

Seung-Ryong Park, Department of Educational Psychology, University of Wisconsin–Madison, Madison, Wisconsin, USA

Andreas Paul, Institute of Zoology and Anthropology, Göttingen, Germany

Michael E. Pereira, Department of Biology, Program in Animal Behavior, Bucknell University, Lewisburg, Pennsylvania, USA

Odile Petit, Laboratoire d'Ethologie et Neurobiologie, Centre National de la Recherche Scientifique, Université Louis Pasteur, Strasbourg, France

Michael Potegal, Pediatric Neuropsychology Clinic, Fairview–University of Minnesota Medical Center, Minneapolis, Minnesota, USA

Signe Preuschoft, Living Links, Yerkes Regional Primate Research Center, Emory University, Atlanta, Georgia, USA

Thelma E. Rowell, Department of Integrative Biology, University of California at Berkeley, Berkeley, California, USA

Amy Samuels, Daniel F. and Ada L. Rice Conservation Biology and Research Center, Brookfield Zoo, Brookfield, Illinois, and Biology Department, Woods Hole Oceanographic Institution, Woods Hole, Massachusetts, USA

Robert Sapolsky, Department of Biological Sciences, Stanford University, Stanford, California, and Institute of Primate Research, National Museums of Kenya, Nairobi, Kenya

Colleen M. Schaffner, Department of Psychology, College of St. Benedict and St. John's University, Collegeville, Minnesota, USA

Gabriele Schino, Sezione Miglioramento delle Produzioni Animali, Ente per le Nuove Tecnologie l'Energia e l'Ambiente, Rome, Italy

Robert M. Seyfarth, Department of Psychology, University of Pennsylvania, Philadelphia, Pennsylvania, USA

Joan B. Silk, Department of Anthropology, University of California at Los Angeles, Los Angeles, California, USA

Darlene Smucny, Laboratory of Comparative Ethology, National Institute for Child Health and Human Development, National Institutes of Health Animal Center, Poolesville, Maryland, and Department of Anthropology, University of California at Los Angeles, Los Angeles, California, USA

Karen B. Strier, Department of Anthropology, University of Wisconsin–Madison, Madison, Wisconsin, USA

Bernard Thierry, Laboratoire d'Ethologie et Neurobiologie, Centre National de la Recherche Scientifique, Université Louis Pasteur, Strasbourg, France

Carel P. van Schaik, Department of Biological Anthropology and Anatomy, Duke University, Durham, North Carolina, USA

Hans C. Veenema, Department of Comparative Physiology, Utrecht University, The Netherlands

Peter Verbeek, Institute of Child Development, University of Minnesota, Minneapolis, Minnesota, and Department of Psychology, Illinois Wesleyan University, Bloomington, Illinois, USA

David P. Watts, Department of Anthropology, Yale University, New Haven, Connecticut, USA

Ann Ch. Weaver, Living Links, Yerkes Regional Primate Research Center, Emory University, Atlanta, Georgia, and Department of Psychiatry, University of Colorado, Denver, Colorado, USA

Douglas H. Yarn, Consortium on Negotiation and Conflict Resolution, College of Law, Georgia State University, Atlanta, Georgia, USA

Index

Page numbers followed by letters *b*, *f*, and *t* indicate material presented in boxes, figures, and tables, respectively.

post-conflict behavior in, 229, 235*t*
unusual traits of, 232*b*
Squirrel monkeys (*Saimiri sciureus*), subordinates in, 114*b*
Squirrels, reconciliation in, 225
Strangers
conflict management between, 79–84
peacemaking tendencies between, 252
Stress
aggression-induced, reduction of, 288
crowding and, 130, 144, 148
physiological measurement of, 189
subordination and, 114*b*
Stress management
male policing and, 295
redirected aggression and, 115*b*–116*b*
social affiliation and, 115*b*
Stress response, 202
effect of reconciliation on, 207
in post-escalation phase, 205
functions of, 206
Struggle for existence, 355
Stumptail macaques (*Macaca arctoides*)
aggression in, 27
conciliatory tendencies in, 27, 109*f*, 194
conflict management strategies in, 110
in crowded conditions, 134*t*
distance regulation in, 191*b*–193*b*
dominance indicators in, 93, 98*t*
geographic distribution of, 108*t*
infant handling in, 119
in prefeeding situations, 131*b*
third-party affiliation of aggressors in, 264*t*
Submission
definition of, 388
displays of, 85
in lemurs, 318*b*–319*b*
in macaque species, 110–11
in spotted hyenas, 232*b*
Subordinates
in despotic dominance relationships, 97
leverage of, 80*b*, 90, 363
opportunism of, 89
in tolerant dominance relationships, 97
Subordination
formal indicators of, 93–96
functions of, 97
as key to equality, 98
stress associated with, 114*b*
See also Dominance-subordination relationships
Sudan, Carter's mediation in, 66*b*–67*b*
Sunfish (*Lepomis gibbosus*), effects of crowding on, 143*t*

Swedish children, peacemaking in
initiative for, 250
strategies for, 249
tendency for, 247–48
Sympathy
as dimension of morality, 361–62
versus empathy, 358

Taiwan, forgiveness in, 361*b*
Taiwan macaques (*Macaca cyclopis*), geographic distribution of, 108*t*
Talean Zapotec culture
conflict management in, 343–45
conflict prevention in, 340
Tamarins (*Leontopithecus* and *Saguinus*)
distinguishing characteristics of, 156
See also Callitrichids; *specific species*
Tantrums
as behavioral adaptation, 253*b*
mediating role of, 213–14, 253*b*–255*b*
Targets of aggression
affiliation with. *See under* Post-conflict third-party interactions
peacemaking initiative by, in children, 250–51
Taurotragus euryceros. See Bongo antelopes
Teacher mediation, 45–46, 250
Teeth-chattering display, 110–11, 111*f*
Temperament
and conciliatory tendencies, 194
physiological correlates of, 114*b*–116*b*
and reactions to crowding, 145
variation in macaque species, 113–16
Temper tantrums. *See* Tantrums
Tension-reduction strategies
under crowding, 139–40
before feeding, 131*b*–132*b*
Termination of conflict, 35
bilateral engagement in, 37
strategies for, 35, 35*t*
Terminology, 8, 387–88
Territoriality, 85–86
and neighborliness, 86
Territory defense, and aggression, 16
Testosterone, and aggression, 204
Theory of mind, in primates, 290*b*
Theropithecus gelada. See Gelada baboons
Third-party interactions, post-conflict. *See* Post-conflict third-party interactions
Third-party interventions, 378
in ADR, 58*t*, 59
cognitive demands of, 289*b*–291*b*

Designer: Nicole Hayward
Compositor: Princeton Editorial Associates, Inc., Scottsdale, Arizona
Text: Weiss
Display: Weiss and Gill Sans
Printer and binder: Friesens, Altona, Manitoba, Canada